BUSINESS STATISTICS
TEXT ▪ CASES ▪ SOFTWARE

BUSINESS STATISTICS
TEXT ▪ CASES ▪ SOFTWARE

OWEN P. HALL, JR.
HARVEY E. ADELMAN
*Both of
Pepperdine University*

Homewood, IL 60430
Boston, MA 02116

IBM®PC and IBM®PS/2® are a registered trademarks of
International Business Machines Corporation.

MS™ DOS is a trademark of Microsoft Corporation.

Richard D. Irwin, Inc., makes no warranties, either expressed or implied, regarding the enclosed computer software package, its merchantability or its fitness for any particular purpose. The exclusion of implied warranties is not permitted by some states. The exclusion may not apply to you. This warranty provides you with specific legal rights. There may be other rights that you may have which may vary from state to state.

© RICHARD D. IRWIN, INC., 1991

All rights reserved. No part of this publication may be reproduced, stored in a retrieval system, or transmitted, in any form or by any means, electronic, mechanical, photocopying, recording, or otherwise, without the prior written permission of the publisher.

Senior sponsoring editor: Richard T. Hercher, Jr.
Project editor: Karen Smith
Production manager: Bette K. Ittersagen
Cover photographer: Russell Phillips
Text illustrators: Carlisle Communications, Ltd./Benoit Design
Compositor: Progressive Typographers, Inc.
Typeface: 10/12 Times Roman
Printer: R. R. Donnelley & Sons Company

Library of Congress Cataloging-in-Publication Data

Hall, Owen P.
    Business statistics: text, cases, and software / Owen P. Hall, Jr., Harvey E. Adelman.
      p.   cm.
    Includes bibliographical references and index.
    ISBN 0-256-06089-4 (PC version)   0-256-10056-X (PS/2 version)
    1. Commercial statistics—Data processing. 2. Statistics—Data processing.   I. Adelman, Harvey E. II. Title
HF1017.H23 1991
519.5′024658—dc20   90-38294

Printed in the United States of America
1 2 3 4 5 6 7 8 9 0 DOC 7 6 5 4 3 2 1 0

Preface

Statistical analysis has become an increasingly important management tool throughout all levels of business and government. The glowing success of the Japanese in effectively penetrating the international marketplace with products of high quality and high reliability can be directly attributed to their emphasis on statistical quality control. The need for a better understanding of basic statistical principles will continue to grow in concert with the advent of the information age. The primary purpose of this text is to provide the student with such an understanding using the latest developments in computer-based courseware (CBS).

A number of different approaches have been used in the preparation of statistical textbooks. They have ranged from the "theoretical" to the "intuitive." For the most part, however, the primary focus of these texts has been on solving simple problems using hand solution methods. The authors believe that this approach is inconsistent with the needs of modern management practice. Instead, we think the primary focus should be on problem formulation and results interpretation. The role of performing the computations should be left to the computer. The widespread proliferation of microcomputers on college campuses provides an effective vehicle for implementing this pedagogical strategy. Although some texts do acknowledge the usefulness of computers in solving statistical problems, few have modified their basic approach to incorporate the full potential of computer-based analysis. As an instructional aid, however, most of the technical chapters in these books contain an appendix that presents the basic analytical formulas along with appropriate examples.

The most important challenge in statistics, however, is not in performing the computations (either by hand or by machine) but in collecting the data and properly interpreting the computational results. Normally, the primary issue facing the business manager is not in developing an answer but in asking and formulating the proper question. Accordingly, a major focus of *this* text is to provide a contextual framework to aid the student in understanding the questioning process and to provide answers through computer-based analysis.

The modern manager uses statistics in three basic ways:

1. Collecting and processing data (i.e., descriptive statistics).
2. Estimating population characteristics from the collected data (i.e., statistical inference).
3. Developing relationships from the processed data (i.e., statistical forecasting).

In each application, the manager must relate the questions that have been asked to the data collected and the statistical analysis performed. This text is designed to assist the student in developing insights into the application of statistics in modern management practice through the application of real-world business cases.

TOPICS COVERED

The topics covered in the 15 chapters of this book are similar to those found in most first-year statistics texts. Within each chapter, however, the primary emphasis is on *problem formulation* and on *interpreting the results.* Simple graphical models are used to illustrate the basic principles before expanding to more realistic and interesting business management problems. This text features a wide array of statistical quality-control applications. Additionally, a number of internally based cases and examples are presented.

This text introduces the most frequently used statistical methods via specific business applications. Each subject area is introduced by describing examples of its use in a real organization. Formulation is emphasized, and each chapter contains examples of formulated problems and at least one formulated business case. The text contains over 800 problems and approximately 40 cases. The same basic outline, which is fully explained in chapter 1, is used in each chapter.

This text has been designed to provide the instructor with considerable flexibility in terms of selecting topics to meet specific course requirements. With several exceptions each chapter stands by itself, so the course can be taught with topics introduced in the order preferred by the instructor. The variety and extent of the problem sets provides the instructor with considerable flexibility in preparing a course ranging in length from one quarter to two semesters.

SUPPORTING MATERIALS

Accompanying this text is a complete package of support materials. These include:

- Computerized Business Statistics (CBS) Software Package.
- Instructor's Manual.

- Financial Data on Fortune 500 Firms and Key Economic Data.
- Solutions to Problems and Cases.

The financial and economic data and the solutions to problems and cases are available on data diskette.

ACKNOWLEDGMENTS

The authors would like to acknowledge the following individuals for their help and support in the creation of this text and the software *Computerized Business Statistics*.

Amir Aczel, *Bentley College*
Randy J. Anderson, *California State University, Fresno*
Charles Branyan, *Memphis State University*
Anthony A. Casey, *University of Dayton*
Gilbert Coleman, *University of Nevada, Reno*
Les Dlabay, *Lake Forest College*
Satyendra Dutt, *Delaware State College*
David L. Eldredge, *Murray State University*
Stewart Fliege, *Pepperdine University*
Edna Frye, *Governors State University*
Edward Y. George, *University of Texas at El Paso*
Stephen Grubaugh, *Bentley College*
Wendel Hewett, *University of Texas, Tyler*
Peter Hoefer, *Pace University*
Geoffrey B. Holmewood, *Hudson Valley Community College*
J. Marcus Jobe, *Miami University Ohio*
David D. Krueger, *St. Cloud State University*
Stan Malik, *Governors State University*
Clifton Miller, *University of Texas of the Permian Basin*
George Miller, *North Seattle Community College*
Kurt Moser, *Pepperdine University*
Lou Mottola, *University of Bridgeport*
Sumy Renjin, *Pepperdine University*
Peter Rob, *Tennessee State University*
Donald L. Schmidt, *American Graduate School of International Business*
John C. Shannon, *Suffolk University*
Susan A. Simmons, *Sam Houston State University*
Rex Snider, *Troy State University*
George Vlahos, *University of Dayton*
Edward J. Willies, *Tidewater Community College*
Robert S. Wu, *Longwood College*
Jack Yurkiewicz, *Pace University*

Owen P. Hall, Jr.
Harvey E. Adelman

Contents in Brief

1	INTRODUCTION	1
2	THE ROLE OF STATISTICS IN MANAGERIAL DECISION MAKING	9
3	DESCRIPTIVE STATISTICS	27
4	PROBABILITY THEORY	87
5	PROBABILITY DISTRIBUTIONS	133
6	SURVEY DESIGN AND DATA BASE MANAGEMENT	191
7	SAMPLING AND ESTIMATION	231
8	HYPOTHESIS TESTING	287
9	SIMPLE LINEAR CORRELATION AND REGRESSION	345
10	MULTIPLE REGRESSION ANALYSIS	399
11	TIME SERIES AND FORECASTING	467
12	CHI-SQUARE ANALYSIS	537
13	ANALYSIS OF VARIANCE	579
14	NONPARAMETRIC STATISTICS	629
15	MANAGERIAL DECISION ANALYSIS	677
	APPENDIX A: STATISTICAL TABLES	733
	APPENDIX B: COMPUTERIZED BUSINESS STATISTICS	757
	APPENDIX C: SOLUTIONS TO SELECTED ODD-NUMBERED PROBLEMS	773
	INDEX	791

Contents

1	**INTRODUCTION**		1
	1.1	What Is Business Statistics?	2
	1.2	New Developments in Business Statistics	4
	1.3	Trends in Teaching Business Statistics	4
	1.4	Organization of the Book	5
	1.5	Summary	7
	1.6	Teaching Supplements	7
2	**THE ROLE OF STATISTICS IN MANAGERIAL DECISION MAKING**		9
	2.1	Introduction	10
	2.2	Example Management Problem: National Cancer Institute	11
	2.3	Overview of the Scientific Method	12
	2.4	Managerial Decision-Making Process	18
	2.5	Total Quality Management	20
	2.6	Summary	22
	2.7	Glossary	22
	2.8	Bibliography	23
	2.9	Problems	23
3	**DESCRIPTIVE STATISTICS**		27
	3.1	Introduction	29
	3.2	Example Management Problem: Global Precious Metal Exchange	29
	3.3	How to Recognize a Descriptive Statistics Problem	30
	3.4	Descriptive Analysis	30
	3.5	Computer Analysis	56

xi

	3.6	Practical Applications	58
	3.7	Case Study: Goodfaith Emergency Clinic	62
	3.8	Summary	64
	3.9	Glossary	65
	3.10	Bibliography	65
	3.11	Problems	66
	3.12	Cases	83
4	**PROBABILITY THEORY**		**87**
	4.1	Introduction	88
	4.2	Example Management Problem: The California Lottery	90
	4.3	Basic Rules of Probability	90
	4.4	Fundamental Concepts	91
	4.5	Computer Analysis	110
	4.6	Practical Applications	111
	4.7	Case Study: Automobile Fuel Economy Standards	114
	4.8	Summary	117
	4.9	Glossary	118
	4.10	Bibliography	118
	4.11	Problems	119
	4.12	Cases	129
5	**PROBABILITY DISTRIBUTIONS**		**133**
	5.1	Introduction	134
	5.2	Example Management Problem: Seven-Day Tire Company	136
	5.3	Characteristics of a Probability Distribution	137
	5.4	Probability Distributions	137
	5.5	Computer Analysis	169
	5.6	Practical Applications	173
	5.7	Case Study: Drack Industries	176
	5.8	Summary	178
	5.9	Glossary	178
	5.10	Bibliography	179
	5.11	Problems	179
	5.12	Cases	187

6		**SURVEY DESIGN AND DATA BASE MANAGEMENT**	191
	6.1	Introduction	192
	6.2	Example Management Problem: Nautilus Health Spa, Inc.	196
	6.3	Questionnaire Design	197
	6.4	Computer Data Base Management Systems	211
	6.5	Practical Applications	214
	6.6	Case Study: Transpacific Airlines	218
	6.7	Summary	223
	6.8	Glossary	224
	6.9	Bibliography	225
	6.10	Problems	225
	6.11	Cases	227
7		**SAMPLING AND ESTIMATION**	231
	7.1	Introduction	232
	7.2	Example Management Problem: Wilcox Accounting Services	233
	7.3	Basic Characteristics of Sampling and Estimation	235
	7.4	Sampling and Estimation	235
	7.5	Computer Analysis	264
	7.6	Practical Applications	268
	7.7	Case Study: Bozart Investments Corporation	270
	7.8	Summary	271
	7.9	Glossary	272
	7.10	Bibliography	273
	7.11	Problems	273
	7.12	Cases	283
8		**HYPOTHESIS TESTING**	287
	8.1	Introduction	288
	8.2	Example Management Problem: Iowa Department of Motor Vehicles	289
	8.3	How to Recognize a Hypothesis Testing Problem	290
	8.4	Statistical Hypothesis Testing	290
	8.5	Computer Analysis	314
	8.6	Practical Applications	319
	8.7	Case Study: Heartwell Music Experiment	321
	8.8	Summary	323

	8.9	Glossary	325
	8.10	Bibliography	326
	8.11	Problems	326
	8.12	Cases	342

9 SIMPLE LINEAR CORRELATION AND REGRESSION 345

9.1	Introduction	346
9.2	Example Management Problem: Pacific Construction Company	347
9.3	How to Recognize a Simple Correlation and Regression Problem	347
9.4	Model Formulation	349
9.5	Computer Analysis	366
9.6	Practical Applications	371
9.7	Case Study: Pritikin Diet	374
9.8	Summary	376
9.9	Glossary	376
9.10	Bibliography	377
9.11	Problems	377
9.12	Cases	395

10 MULTIPLE REGRESSION ANALYSIS 399

10.1	Introduction	400
10.2	Example Management Problem: Far Filtration Company	401
10.3	How to Recognize a Multiple Regression Analysis Problem	402
10.4	Model Formulation	402
10.5	Computer Analysis	416
10.6	Practical Applications	422
10.7	Case Study: National Baseball League	430
10.8	Summary	431
10.9	Glossary	432
10.10	Bibliography	433
10.11	Problems	433
10.12	Cases	462

11 TIME SERIES AND FORECASTING 467

11.1	Introduction	468
11.2	Example Management Problem: Dialnet Telephone Exchange	469

11.3	How to Recognize a Time Series Problem	471
11.4	Classical Decomposition Model	471
11.5	Forecasting Models	476
11.6	Forecast Validation	494
11.7	Computer Analysis	496
11.8	Practical Applications	497
11.9	Case Study: Thermhouse Insulation Corporation	502
11.10	Summary	510
11.11	Glossary	511
11.12	Bibliography	512
11.13	Problems	512
11.14	Cases	533

12 CHI-SQUARE ANALYSIS 537

12.1	Introduction	538
12.2	Example Management Problem: Kwan Bottling Co.	539
12.3	How to Recognize a Chi-Square Problem	540
12.4	Model Formulation	540
12.5	Computer Analysis	551
12.6	Practical Applications	553
12.7	Case Study: Leaky Pen Company	559
12.8	Summary	561
12.9	Glossary	562
12.10	Bibliography	562
12.11	Problems	562
12.12	Cases	576

13 ANALYSIS OF VARIANCE 579

13.1	Introduction	580
13.2	Example Management Problem: Mitterand Cable Company	581
13.3	How to Recognize an Analysis of a Variance Problem	582
13.4	Model Formulation	583
13.5	Computer Analysis	593
13.6	Practical Applications	598
13.7	Case Study: Microchip Electronics, Inc.	600
13.8	Summary	603
13.9	Glossary	603

	13.10	Bibliography	604
	13.11	Problems	604
	13.12	Cases	626

14 NONPARAMETRIC STATISTICS — 629

	14.1	Introduction	630
	14.2	Example Management Problem: Perpetual Savings	631
	14.3	How to Recognize a Nonparametric Problem	632
	14.4	Basic Nonparametric Models	633
	14.5	Computer Analysis	644
	14.6	Practical Applications	647
	14.7	Case Study: Fujti Motors	651
	14.8	Summary	653
	14.9	Glossary	654
	14.10	Bibliography	655
	14.11	Problems	655
	14.12	Cases	674

15 MANAGERIAL DECISION ANALYSIS — 677

	15.1	Introduction	678
	15.2	Example Management Problem: Perpetual Investments Corporation	680
	15.3	How to Recognize a Decision Analysis Problem	682
	15.4	Formulating Decision Analysis Models	682
	15.5	Computer Analysis	703
	15.6	Practical Applications	705
	15.7	Case Study: Cleanall Corporation	709
	15.8	Summary	712
	15.9	Glossary	712
	15.10	Bibliography	713
	15.11	Problems	714
	15.12	Cases	729

APPENDIX A: Statistical Tables — 733

A.1	Areas under the Normal Curve	734
A.2	Student t Distribution	735
A.3	Critical Values of Chi-Squared	736
A.4	Critical Values of the F Distribution	737
A.5	Binomial Probability Distribution	740

	A.6	Poisson Distribution: Probability of Exactly x Occurrences	748
	A.7	Critical Values of the Durbin-Watson Test Statistic	750
	A.8	Critical Values of the Studentized Range Distribution (Tukey Test)	752
	A.9	Cumulative Distribution Function (Runs Test)	754

APPENDIX B: Computerized Business Statistics   757

APPENDIX C: Solutions to Selected Odd-Numbered Problems   773

INDEX   791

Chapter 1

Introduction

Statistical thinking will one day be as necessary for efficient citizenship as the ability to read and write.

H. G. Wells

CHAPTER OUTLINE

1.1 What Is Business Statistics?
1.2 New Developments in Business Statistics
1.3 Trends in Teaching Business Statistics
1.4 Organization of the Book
1.5 Summary
1.6 Teaching Supplements

"A formal planning system is the key to effective business management." This refrain is heard with increasing regularity throughout corporate America. An effective planning system incorporates both an internal and an external context, and the status of each context requires continuous updating. Here, statistical analysis plays an important role. Computer-based systems provide the firm with a steady stream of processed data (i.e., information) that can be used for improving corporate decision making.

A recent study investigated some potential ramifications of the use of computer-based planning systems.* This study clearly showed that companies that updated their planning system on a continual basis outperformed (as measured by differences in earnings and absenteeism) those firms that did not use systematic planning. Thus, these types of planning systems offer considerable promise in helping to improve the management decision-making process.

---

* B. S. Chakravarthy, "Tailoring a Strategic Planning System to Its Context," *Strategic Management Journal* 8, no. 6, pp. 517–31.

1.1 WHAT IS BUSINESS STATISTICS?

To most managers, *statistics* means "numerical descriptions" of specific business or technical data. For example:

- May Company announced that third-quarter net income rose 21% over the past year, to $104 million.
- The Bureau of Labor Statistics reported that the unemployment rate for March fell below 6%.
- GTE Corporation plans to eliminate 14,000 positions over the next five years.
- The Los Angeles Lakers basketball team had a winning percentage of nearly 65% for the 1987 season.

The primary objective of **business statistics** is to provide quantitative information for decision making. Statistics in general, and business statistics in particular, are often divided into two major categories: descriptive statistics and inferential statistics.

Descriptive statistics includes data collection, data classification, data display (i.e., graphics) and data processing (i.e., computations) such as:

- Product failure rates.
- Customer preference for a new fast-food product.
- Market share data.
- Average wage rates between industry groups.

Inferential statistics represents an important analytical tool for business decision making. The basic premise behind statistical inference is quite simple. Namely, descriptive statistics from a small sample are used to describe a larger, unseen group (i.e., a population). Statistical inference is a necessity in most business situations because data on a **population** of interest are unavailable or unattainable. Consequently, a **sample** is selected to represent the population. In this way the decision maker can infer population characteristics from what is usually a very small sample drawn from the population, as in:

- Identifying the winner of the presidential race after conducting an exit poll of 1% of the voters.
- Determining whether a batch of computer chips can be shipped or must be reprocessed based on a sample inspection of 2% of the lot size.
- Estimating the demand for a new detergent based on testing the product in 500 households.
- Forecasting revenues for next year based on sales data over the last three years.

One of the key business uses of statistical inference is in forecasting. The forecasting process has as its primary objective the prediction or estimation of future events. More specifically, forecasting is the attempt to estimate future changes based on a set of assumptions. There are a wide variety of forecasting methodologies, ranging from subjective approaches to very complex computer-based models. Forecasting can range from an exact science to a naive art. As such, the quality of the forecast depends heavily on the quality of the data and the accuracy of the facilitating assumptions.

HISTORICAL NOTE

The recording of data can be traced back to early man. The book of Numbers in the Bible contains several accounts of early census taking. One of the earliest applications of statistics to business was by the Englishman Thomas Watt (1705–1769) during the early part of the 18th century. Watt established a school with a particular emphasis on mathematics and mensuration. The interest in measurement was short-lived, however, and did not surface again for nearly two centuries.

1.2 NEW DEVELOPMENTS IN BUSINESS STATISTICS

Total quality management (TQM) is perhaps the single most important new development that has occurred in business over the past 10 years. One important aspect of TQM involves the use of statistical quality control. One of the major factors contributing to the overwhelming success of the Japanese in international commerce during 1980s has been their increased focus on product and service quality. Statistical analysis has played a crucial role in the Japanese success story. Through the use of statistics, the Japanese have been able to develop more accurate forecasts, reduce defectives and costs, and significantly improve quality. Today, many U.S.-based firms are attempting to adopt the Japanese approach to product and service quality.

A second development that has broad implications to the study of business statistics is the ever increasing internationalization of the marketplace. Economic and commercial decisions made overseas are having a growing impact on business operations within the United States. The economic integration of Western Europe by 1992 coupled with the dramatic changes occurring in Eastern Europe illustrates this new dynamic. U.S. firms will need to develop a better understanding of these trends if they are to remain competitive in the international marketplace. International market surveys and overseas financial analysis represent two of many areas where statistical analysis can be used to assist the decision maker.

In recognition of these new developments, the authors have attempted to incorporate throughout this text applications, problems, and cases involving both statistical quality control and international operations. More specifically most chapters contain examples of how statistical analysis can be used to support the TQM process. Furthermore, most chapters commence with an international vignette that depicts an actual business application involving the use of statistics.

1.3 TRENDS IN TEACHING BUSINESS STATISTICS

Many students take a course in statistics only because it is required. Usually, students do not have a very good understanding of the role of statistical analysis in modern business practice. In particular, they do not know:

- Why statistical analysis is important in business.
- What type of problems can be solved using statistical analysis.
- What benefits can result from the application of statistical analysis.

This lack of awareness is compounded by the normally heavy emphasis on computational mastery. The authors of this textbook believe the traditional focus on computational methods is counterproductive to the learning process and instead think the primary emphasis of teaching statistics should be oriented toward:

- Understanding how to apply statistical analysis to business applications.
- Asking the "right" questions.
- Identifying and collecting problem data.
- Formulating the appropriate model.
- Solving the problem using computer analysis.
- Interpreting the computer output in light of the specific application.

To meet these objectives, the primary focus in the classroom should be on problem formulation and results interpretation. This textbook has been written to support this process.

Computers have had a profound effect on the use and teaching of statistics. Most modern organizations make extensive use of computers for processing and interpreting data. The use of computers in this area is clearly on the rise. Accordingly, each student of statistics needs to become aware of the role computers play in the decision-making process. The authors believe the best way to accomplish this objective is by exposing the student to the use of computers and computer models throughout the course. This approach allows the student to formulate, solve, and interpret problems that are typically well beyond the scope of the traditional first-year statistics text. Furthermore, computer software service is a tireless tutor for reinforcing key principles and for examining a variety of alternative situations.

To support this goal, the authors strongly recommend the use of Computer Business Statistics (CBS). The CBS courseware consists of an instructional text and a software package. The text provides a general overview of the basic statistical models covered in this book as well as illustrative examples. The CBS software is available for the IBM® PC. A copy of the program disk can be found at the back of this book. The user-friendly menu-driven design ensures ease of use for students who are novices in computer operations.

1.4 ORGANIZATION OF THE BOOK

This book is organized to show why statistics is important to business, to provide examples of problems solved by statistics, and to describe the benefits that result from its use. The primary focus of most chapters is on the development of problem formulation skills. This chapter contains an introduction to business statistics. Chapter 2 introduces the scientific method and illustrates its relationship to statistics. The next 13 chapters constitute the main body of the book and describe a variety of statistical models. These chapters are organized in a similar fashion to enhance the learning process. The following paragraphs outline this organization.

The *introduction* contains a brief overview of the nature of a particular statistical process (e.g., hypothesis testing) and the type of problems it solves. An introductory business situation provides a nontechnical summary of an actual application of the process by a real organization. Each statistical process is then described in terms of an *example management problem.*

The next section of each chapter briefly describes *how to recognize* a problem of the type being discussed in the chapter. Each of these sections contains a list of the characteristics of such problems and the objectives of the statistical process. Also included is a list of the assumptions, where applicable, on which the process is based.

The *problem formulation* section shows how to formulate problems to be solved by the statistical process under investigation. The section on *computer analysis* describes how problems can be solved using CBS and how to interpret the results.

The *problem applications* section describes some important problems that can be solved with the model. Problem formulation examples are usually included for every type of application. A *case study* is presented at the end of every chapter to demonstrate how to formulate and solve problems taken from the business world. The *summary* section briefly highlights the key ideas that were introduced in the chapter. The *glossary* contains all the important terms introduced in the chapter.

A *bibliography* is included for those who wish to pursue a topic in more depth. Each chapter concludes with a set of *problems and cases.* The problems section begins with a set of simple problems designed to illustrate the basic principles covered in the text. Many of these problems can also be used as a bridge to the more complex issues contained in the cases.

The cases require moving to a level that is somewhere between the usual textbook problem and the actual problems faced by business decision makers. A few case problems require simplifying assumptions before the problem can be solved, and some problems must be formulated in slightly different ways from the standard problems that were solved previously. Few of the cases state the problem explicitly, so success in solving them often requires making assumptions about the data and relationships.

Whenever possible, specific business examples are used to enhance the learning process. Many of these are referenced so that more can be read about any that are of interest. Many unreferenced problems and cases use names of fictitious organizations. Some of these may inadvertently refer to a real company, but such references are purely coincidental. Some problems and cases use real examples as a starting point and then explore specific issues by extending beyond the original data. These extensions were developed for pedagogical purposes in this book and are not part of the original problem.

(Note that neither the original organization nor the original author(s) are responsible for this added material.)

In keeping with the growing interest in quality management and international operations, approximately 25% of the problems and cases found in this text deal with some aspect of these important subjects.

1.5 SUMMARY

The primary purpose of this book is to provide an introduction to the application of statistics to business. Business statistics can be defined as the science of collecting and processing data as an aid to management decision making. The subject of business statistics is often divided into two major categories: descriptive and inferential. Descriptive statistics includes data collection, data classification, data display, and data processing. Inferential statistics involves the use of data collected from a sample for the purpose of describing the entire population. Forecasting represents an important tool of inferential statistics. The teaching of statistics is becoming increasingly reliant on the use of computer analysis. A computer-based modelling approach to teaching statistics permits the student to focus more attention on problem formulation and results analysis while minimizing the traditional computational emphasis. Total quality management and international commerce represent two new areas where business statistics can be applied. Both areas represent important challenges to U.S. businesses throughout the remainder of this century and into the 21st century.

1.6 TEACHING SUPPLEMENTS

In addition to the text and the accompanying CBS computer disk (found in the rear jacket), this learning system contains the following teaching supplements:

- *Instructor's manual* with solutions to all of the problems and cases.
- *Data diskette* containing the following four large-scale data bases:
  1. Quarterly financial and operating performance for 20 Fortune 500 firms for the years 1980–1988.
  2. Financial and operating performance for the Fortune 500 for 1988.
  3. U.S. economic indicators and government expenditures, 1970–1988.
  4. International economic indicators and government expenditures by country, 1980–1985.

This data diskette also contains the input data files for the text problems and cases.

- *Conversion program* that can convert any ASCII-based data base into one that can be processed by CBS. Furthermore, the conversion program will convert a CBS data input file into an ASCII file per specifications given by the user.
- *Transparency masters* of the major figures and tables found throughout the text.

Chapter 2

The Role of Statistics in Managerial Decision Making

False facts are highly injurious to the progress of science.
Charles Darwin

CHAPTER OUTLINE

2.1 Introduction
2.2 Example Management Problem: National Cancer Institute
2.3 Overview of the Scientific Method
2.4 Managerial Decision-Making Process
2.5 Total Quality Management
2.6 Summary
2.7 Glossary
2.8 Bibliography
2.9 Problems

CHAPTER OBJECTIVES

The primary objectives of this chapter are to develop an understanding of

1. The relationship between the scientific method and statistics.
2. The role of measurement and numbers in business decision making.
3. How statistics supports the decision-making process.
4. The relationship between statistics and total quality management.

The basic idea behind the theory of biorhythms is that everyone's life is governed by three fundamental factors—physical, emotional, and intellectual—and these factors move in cycles of 23, 28, and 33 days, respectively. Although no scientific evidence has been made available to support these claims, many individuals take biorhythms very seriously (e.g., gamblers and health enthusiasts).

An investigation of this theory undertaken at Johns Hopkins University

focused on the incidence of driving accidents.* The study examined data on 205 automobile accidents where the driver was judged to be at fault. Accident rates were compared with the drivers' biorhythm cycles. The results demonstrated that the frequency of accidents during a driver's "low" period was no higher than at any other time in the cycle. This suggested that one's biorhythmic state is *not* a factor in automobile accidents.

2.1 INTRODUCTION

Key idea 1

The **scientific method** has emerged over the last 200 years as the principal engine for the general advancement of mankind. In a nutshell, the scientific method can be defined as a general strategy for gaining knowledge. The two basic characteristics that distinguish it from other forms of knowledge acqui-

---

* Shaffer, J. W., and Schmidt, C. W., "Biorhythms and Highway Crashes," *Archives of General Psychiatry,* vol. 35(1), pp. 41–46, 1978.

sition are *significance* and *precision*. The scientific method has had an impact on every phase of human development and will continue to do so in the foreseeable future. This trend is represented in American business by the increased reliance on science and technology. The slogan of Rockwell, Inc., "where science gets down to business," illustrates this point. The development and application of technology represents America's "trump card" in the growing international marketplace.

Research statistics is the primary mathematical discipline for implementing the scientific method. Research statistics can be defined as a systematic approach for investigating relationships between variables in a controlled environment. **Applied statistics,** on the other hand, involves the application of statistical theory and principles derived from the scientific method to business and related fields. Consequently, one needs to develop a basic understanding of the scientific method before proceeding to the study of statistical analysis.

Learning how to classify different statements into a basic logical structure is one of the most important aspects of scientific reasoning. Statements can be classified in terms of either an objective, a hypothesis, or a fact. For example, the statement, "the company wishes to increase sales by 20%" can be classified as an objective or goal. One of the primary applications of the scientific method is in analyzing causal hypotheses (e.g., an increase in advertising expenses will result in an increase in sales). As will be shown, the formulation and testing of hypotheses are key ingredients in business statistics. **Facts** are verified data used to support hypothesis-testing procedures (e.g., a 10% increase in advertising expenses resulted in a $2 million increase in sales).

The use of applied research in business decision making is undergoing a significant change due to the widespread availability of computer hardware and software. The growing sophistication of computer-based management information systems provides the decision maker with the capability to greatly expand the use of statistical analysis in the decision-making process. Additionally, there is a growing focus on the application of **total quality management** as a basic operating philosophy of which statistics plays a key role. These developments should, in turn, lead to significant improvements in corporate America's competitive performance.

2.2 EXAMPLE MANAGEMENT PROBLEM: NATIONAL CANCER INSTITUTE

The National Cancer Institute (NCI) has an ongoing interest in assessing the impact of secondary smoking on the general population. This issue emerged as an important health and public policy concern in the 1980s. Secondary

> **HISTORICAL NOTE**
>
> The word *statistics,* or *state arithmetic,* was first coined by the German academician Gottfried Achenwall (1719–1772) in 1749. The term, in its original context, meant the counting and calculating of activities required in the operation of a modern nation. The term *statistics* was popularized in the West by the Englishman John Sinclair (1754–1835) during the 1790s.

smoking is defined as the ingestion of tobacco smoke by a nonsmoker in a closed area (e.g., an airplane).

The institute believes that secondary smoke contributes to the incidence of cancer in the general population. To test this hypothesis, the institute has decided to undertake a comprehensive nationwide study. The study will consist of tracking two groups over an extended period of time. One group will serve as the control base, in which current association with smoking practices will continue. For the second group, every attempt will be made to eliminate the members' exposure to tobacco smoke. The institute plans to measure the incidence of cancer in both groups as an indicator of possible health effects of secondary smoke.

2.3 OVERVIEW OF THE SCIENTIFIC METHOD

The modern business manager—whether trained in engineering, science, or administration—constantly makes decisions based on a variety of factors. Typically, the decision-making process consists of both objective and subjective elements. Objective decision making is usually based on hard or factual data, whereas subjective decision making is based on personal experience. In objective decision making, the manager is interested in quantifying the identified alternatives. This quantification process requires data. These data are usually subject to a variety of arithmetic manipulations that provide the manager with a basis for making decisions.

Curiously enough, these procedures can be considered part of the scientific method, which is a practice generally ascribed to the scientist. It should be noted, however, that the general field of business is not a pure science in the sense of physics, chemistry, or biology. Rather, the basic data from which the manager operates may be taken from a combination of disciplines, such as economics, political science, sociology, psychology, or other more clearly identifiable sciences. Nevertheless, a manager employing this process may be termed an empirical scientist. In short, managers are constantly trying to improve their understanding of the business environment (e.g., the behavior of the marketplace) so that the quality of their decisions may improve.

FIGURE 2.1 Application of the Scientific Method to the National Cancer Institute Study
---

1. *Formulate hypothesis:* Those individuals that are exposed to secondary tobacco smoke will have a greater incidence of lung cancer than those who are not.
2. *Design experiment:* Establish two groups for measuring the impact of secondary smoking. Establish measures for determining the effects of secondary smoking on health. Establish the length of the experiment.
3. *Collect data:* Obtain mortality rates, morbidity rates, and health costs from the two groups.
4. *Analyze data:* Perform statistical hypothesis testing on the sample data from the two groups.
5. *Draw conclusions:* Establish the level of significance and decision criteria, and draw conclusions regarding whether differences exist between the two groups.
6. *Make decisions:* Decide, based on the preceding conclusions, whether or not secondary smoke is a health hazard.

Usually, managers can learn something about the behavior of the system if only they can know the proper questions to ask. If the question is not properly stated or is ambiguous, it is likely that the observations and data collection will also not be made properly. That is, if the questions are not suitably framed, no amount of interpretive genius is likely to lead to a proper answer.

Following are the basic steps normally assigned to the process of discovery via the scientific method:

1. Formulate and properly state the hypothesis.
2. Apply specific design principles to obtain data.
3. Assign numbers appropriate to the problem under study.
4. Interpret and analyze the data.
5. Draw conclusions.
6. Make decisions.

Figure 2.1 illustrates the use of the scientific method with respect to the National Cancer Institute problem.

A key aspect of the process shown in Figure 2.1 involves the inference from a *sample* to the defined *population*. For the National Cancer Institute experiment, the sample is the group of individuals in the study who were exposed to secondary smoking. Clearly, it is not possible to measure the impact of secondary smoking on everyone that might be exposed. The generalization of results obtained from a sample to the defined population is known as *inductive reasoning* and serves as the cornerstone to modern statistical analysis.

FORMULATING THE APPROPRIATE QUESTION

Managers frequently work backward in the sense that they use previously collected data as the basis for formulating questions. Such an approach could involve errors in logical design or in assignment of numbers to the phenomena in question, or could result in little reduction in uncertainty of interpretation. It is true that "the data really don't care what the question was." Nevertheless, the manager or researcher should specify the questions in advance to reduce the uncertainty in the final decision. One should note that in response to recurrent, open-ended questions such as, "What is my monthly profit?" or "What is my injury rate?" the manager generally receives reports on an ongoing basis. This state of affairs may account for why data files are retained in the desk drawer, in the files of the controller, or in the computer. The continuous collection of data that is not made in response to a specific question may also partially account for why managers are inundated with data of little value.

In many instances, "questions" are posed in the form of, "if X, then Y." For example, in the National Cancer Institute statement, "those individuals who are exposed to secondary tobacco smoke will have a greater incidence of lung cancer than those who are not," the secondary smoke is X, and the occurrence or absence of lung cancer is Y. The form of the statement "if X, then Y" is generally called a **hypothesis;** it contains two concepts, each operationally defined, and a statement of relationship. This model suggests the manner in which questions should be asked.

The major difference between a hypothesis and a question is that a hypothesis is deduced from theory. A question, on the other hand, may arise from the appearance of a specific problem (e.g., turnover or injuries). The manager's general reasoning may take the form of trying to account for that problem by relating it to a factor under his or her control. For example, the manager might claim that turnover is due to the autocratic management style exhibited by the firm's first-line supervisors. The question may then take the form of what relationship exists between turnover, as defined by monthly termination rates, and management style, as measured by the Reddin Test of Management Style.

The manager must phrase each question carefully to enhance the probability of obtaining a pertinent answer. For example, it is meaningless to ask what the effect of an equal opportunity program is. A better question asks what the relationship is between attendance of the EEO program and attitudes toward minorities as measured by a discrimination instrument. Another way of asking this question is, "What is the change in attitude as a function of attendance in the EEO program?" The latter form implies a measurement from which change might be inferred; the former does not.

The following guidelines will help in the formulation of questions.

- Concepts must be operationally defined, and exact measurements should be used (e.g., the number of hours exposed to secondary smoke).
- The question should comprise a minimum of two concepts and the relationships between them (e.g., a positive relationship exists between the amount of exposure to secondary smoke and mortality rates).

DESIGN PRINCIPLES

The "rules of evidence" used to resolve a given empirical question rests firmly on the logical design under which the data were collected. That is, the interpretation of any set of data largely depends on how thoroughly the effects of extraneous factors have been eliminated and how well the major effect, the relationship between the variables under study, has been isolated by the design. For example, in the classic Hawthorne study, the researchers sought to determine, among other things, if a relationship existed between increased ambient lighting (X) and production (Y). This study can be represented in the following manner:

- Old baseline: Production prior to the experiment
- Change: Increase in lighting (X)
- New baseline: Production after experiment

The study found that a significant increase in production did, in fact, occur. Can one then assert that the increase in lighting "caused" the increase in production? The answer is an unequivocal maybe. Before a definite assertion of relationship can be made, the following alternative research questions must be ruled out:

1. What factors other than X could have contributed to the results?
2. Was the increase noted due to practice, or taking measurement farther out on the learning curve?
3. Was the increase in production noted due to an increase in attention paid to the production workers?
4. Could the result be due to chance?

The general design rule for a study such as this one can be stated as: "The purpose of research design is to eliminate alternative interpretations to the extent that nature will allow." Had it been possible, the design for the Hawthorne study could have been augmented according to the scheme given in Figure 2.2. In this way, the various conditions for the three groups could be compared for their relationships to production. If groups II and III showed no increase in production, one could conclude that increased lighting levels did indeed lead to increased production, if all other factors were unchanged. As stated in the design rule, the researcher must rule out alternative interpre-

FIGURE 2.2 Proposed Experimental Design for Lighting Study

	Group		
Period	I (Experimental)	II (Control 1)	III (Control 2)
Pre-X	Production Increase in lighting	Production No lighting increase	Production Attention, no lighting increase
Post-	Production	Production	Production

tations. Thus, specific alternative questions may be accounted for in the design process by the judicious use of control groups.

A manager might have been content to make decisions based upon the first, inadequate, design illustrated. However, the chance for an error would be quite high. The principles of design clearly call for making comparisons, whenever possible, rather than collecting data from a single group. Thus, managers are strongly advised to split groups in seeking answers to questions even where complete control is not possible. For example, rather than sending all managers to a leadership course at one time, an executive should send only half the managers and use the other half as control. The executive will then have a better chance of determining the effect of the course. Even though all alternative interpretations of results may not have been systematically eliminated, the reduction in uncertainty is considerable.

MEASUREMENT AND NUMBERS

Science in general and business statistics in particular make extensive use of symbols. **Symbols** are an abstract representation of reality.* Symbols that can take on a variety of values are called **variables**. Variables can be classified as either **qualitative** (differing in kind) or **quantitative** (differing in degree). Some examples of both types of variables are listed below.

Qualitative	Quantitative
Sports preference	Calendar dates
Automobile models	IQ scores
Fortune 500 rank	Net worth
Corporate goals	Price/earnings (P/E) ratio

---

* For example, in the NCI problem, the symbol X represented those individuals exposed to secondary tobacco smoke.

The assignment of numbers to variables, which is called **measurement,** requires the use of a scaling system. The following discussion highlights, in order of increasing sophistication, the four basic scaling systems currently in use.

1. *Nominal scale.* The **nominal scale** is the most primitive. Typically, numbers are used as substitutes for symbols or names. Examples of variables measured on this type of scale include gender, football jersey numbers, and coffee brands. Although there is some argument about whether this type of system constitutes measurement at all, it is probably useful to classify it as scaling. Basically, the mathematical operation inherent in nominal scaling is substitution. Letters of the alphabet or words can be used for substitution instead of numbers.

2. *Ordinal scale.* The **ordinal scale** may be thought of as one that indicates an order, usually implying a *more than* or *less than* relationship. The mathematical operation associated with the ordinal scale is rank/order. The final standings of the National Basketball Association, the Fortune 500, and a company's objectives are examples of ordinal scaling. Combining the nominal and ordinal scales can yield very useful measurements. For example, a CEO may wish to relate production rate (ordinal) figures to management style (nominal) as a basis for hiring future managers. The intervals between numbers on the ordinal scale are either irregular or unknown. An example of the famous Likert scale is shown below.

The applicant should be granted credit.

Agree strongly	_____ (5)
Agree	_____ (4)
Undecided	_____ (3)
Disagree	_____ (2)
Disagree strongly	_____ (1)

In this case, each individual involved in the decision-making process is asked to render a judgment regarding the creditworthiness of the applicant.

3. *Interval scale.* The **interval scale** implies equal intervals. It has a convenient zero point but not an actual one. The lengths of service for several employees exemplifies the use of an interval scale. For example, one can state that Joe has six years more service than Fred, whereas Joe has three years less service than Joan.

4. *Ratio scale.* The **ratio scale** is the most advanced form of measurement and is commonly associated with the physical sciences. Revenues, costs, inventory, and market share are a few of the many ratio scales used in business. A real zero point is a requirement for the ratio scale. This characteristic permits the manager/analyst to describe differences between data points as ratios.

FIGURE 2.3 Overview of Data Measurement System

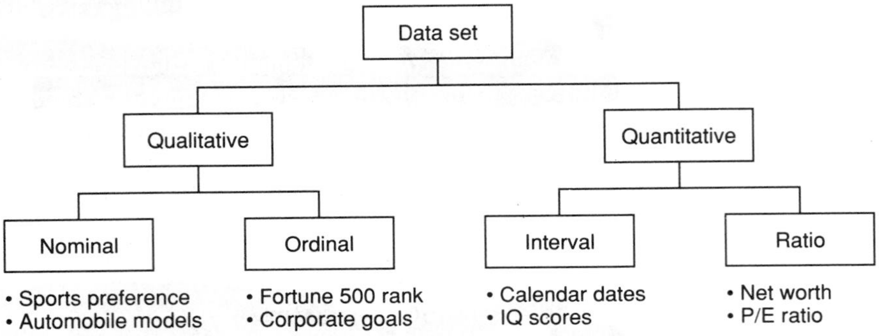

```
                          Data set
                  ┌──────────┴──────────┐
              Qualitative           Quantitative
            ┌──────┴──────┐       ┌──────┴──────┐
         Nominal        Ordinal  Interval       Ratio
```

- Sports preference
- Automobile models

- Fortune 500 rank
- Corporate goals

- Calendar dates
- IQ scores

- Net worth
- P/E ratio

All four types of measurement may be combined in a single application, as in the granting of a line of credit for a firm based on the following factors:

1. Ownership (nominal): Publicly or privately held
2. Credit rating (ordinal): On a 1-to-5 scale
3. Age (interval): Year of incorporation
4. Debt/equity (ratio): Comparison to industry norms

One should take note of the value provided by even the most primitive form of measurement. For example, comparing businesses by their credit rankings, (ordinal scale) is more fruitful than making decisions based on hard-to-define subjective criteria. Figure 2.3 provides an overview of the data-ordering system along with specific examples.

2.4 MANAGERIAL DECISION-MAKING PROCESS

Key idea 2

Statistics is often defined as the branch of science that is concerned with decision making in an uncertain environment. Consequently, there is a need to develop an understanding of the decision-making process and to identify the role of statistics in this process. The managerial decision-making process can be described in many different ways. One approach is outlined in Figure 2.4 and described below.

Define the Problem. Defining the problem is a critical and often-ignored part of the decision-making process. A problem does not arrive on a manager's desk in a package that says: "Here is a problem, and here is the data needed to solve it. Now solve it." Instead, it emerges—sometimes clearly, sometimes only vaguely—as a breakdown in a particular area of the organi-

FIGURE 2.4 Overview of Decision Analysis Process

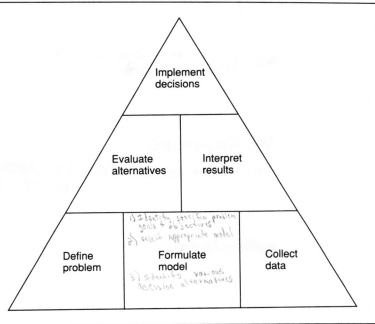

zation's processes or objectives. Managers normally do not see isolated problems. Rather, they see a continuous stream of events, data, complaints, needs, results, and symptoms. Based on this, they must make decisions, take actions, request more data, and generally do all the things that collectively define their jobs. It is in this environment that a manager must recognize the existence of a problem. For example, the human resources (HR) manager at a local technology company believes that the current high level of employee turnover represents a significant problem in terms of maintaining a high level of product quality.

Formulate Model. Formulating a model means translating the problem from an unstructured form into a conceptual (symbolic) representation. The first step in the formulation process is to identify the specific problem goals and objectives (e.g., minimize inventory cost). The second step is to select the most appropriate model; often more than one model is used to solve a particular problem. The third step is to identify the various decision alternatives. This is usually the most difficult task in the formulation process. For example, the HR manager may wish to develop a model for describing the relationship of employee turnover to many other factors (e.g., job satisfaction).

Collect Data. Data must be gathered according to the requirement of the model. Data collection can be expensive and time-consuming. In a market research problem, for example, the data gathering phase may include extensive test-marketing of a proposed new product. Each model has its own data requirements. Determining the level of accuracy is one of the key decisions in the data collection phase. Often, a problem will require a data base system to support the analysis process. For example, the HR manager would access the employee data base to obtain specific information regarding the conditions under which employees have been leaving the company.

Evaluate Alternatives. The evaluation of each decision alternative represents the basic technical step in the decision-making process. Here, the formulated model is used either to predict or to optimize the consequences (as measured by the problem goals) of each alternative. For example, the HR manager might wish to analyze the impact of an awareness program to acquaint the management with employee concerns.

Interpret Results. A problem is not necessarily solved even when all of the alternatives have been quantitatively analyzed. First, the results must be interpreted to make sure that they represent a solution to the original problem. Any factors that were not included in the formulated problem must be considered along with the model results. For example, the HR manager may conclude from an analysis of the results that increasing management awareness will reduce employee turnover.

Implement Decisions. Once the results are interpreted, a decision can be made and implemented. This is, in many respects, the most difficult phase of the problem-solving process. Only after implementing a decision does the manager find out if the "solution" has really solved the problem. Often, some aspects of the problem that were ignored in the formulation phase now become dominant features. In other situations, changes occur in the problem environment. In either case, the manager has a new problem to run through the problem-solving cycle. For example, on the basis of the analyzed alternatives, the HR manager could implement an ongoing management awareness program.

2.5 TOTAL QUALITY MANAGEMENT

The six-step analytical process outlined above presents a logical approach for analyzing business problems. This description, however, does not focus on any specific management philosophy regarding the operation of business. One management philosophy that is receiving increased attention throughout the industrialized world is based on the total quality management (TQM) concept. TQM focuses the responsibility for quality squarely on the producer of the product or service. The TQM approach seeks the highest level of excellence in all phases of the production process. The experiences of the 1980s have shown clearly that organizations that wish to remain competitive in the marketplace must provide high-quality products and services at competitive prices. As a result, quality has become an integral component of sound business strategy. Increasingly quality is having a strong impact on how the customer chooses among similar products and services. Poor quality usually has a negative impact on an organization's efficiency, not only through lower demand but also through higher costs. By using effective quality controls, suppliers ensure that their product or service meets specified standards of excellence.

An organization can maintain quality standards by implementing a TQM system that monitors reliability and controls production and service operations. W. Edward Deming, one of the key individuals responsible for transforming Japan into a world-class industrial power, developed the following 14-point TQM approach:

1. Adopt the TQM philosophy throughout the organization.
2. Develop a top management structure that supports the TQM philosophy.
3. Improve managerial supervision at all levels.
4. Promote the idea of total quality among departments, suppliers, and customers.
5. Incorporate quality in the basic design of the product or service.
6. Formulate work standards in which quality considerations play a key role.
7. Conduct on-the-job training for new employees.
8. Provide ongoing employee training.
9. Concentrate subcontracting to a few high-quality vendors.
10. Create a work environment where employees feel free to offer suggestions.
11. Provide employees with the capabilities to evaluate their own performances.
12. Empower employees to solve problems.
13. Utilize the most modern management decision-making systems.
14. Utilize statistical analysis to ensure that quality is designed into the product or service.

From this list, it is clear that statistics plays a key role in the TQM philosophy. Generally, the use of statistics in TQM is divided into process control and acceptance sampling.

Process control involves monitoring and maintaining quality during the manufacturing and service process. For example, a local telephone company is interested in determining the time a customer has to wait to contact an operator. A sample of the total number of daily calls to the operator could be taken and the average waiting time per call could be plotted per sample. This plot would show how the average waiting time varies over time. The telephone company could use this information to determine whether the response time on any given day was outside the company's service standards. A number of different statistical methods are used in monitoring product or service quality in this manner.

Acceptance sampling, as the term implies, involves the use of random sampling to determine whether to accept or reject an incoming or outgoing lot based on specific quality standards. For example, the current policy for a manufacturer of computers is that 95% of the chips purchased from vendors must meet certain reliability requirements. Instead of testing each chip, the manufacturer could test a sample from each batch in order to determine whether or not to accept the entire batch. One obvious advantage of this approach is the significant reduction in both time and costs.

2.6 SUMMARY

The scientific method is a generalized strategy for gaining insight into a wide range of natural and man-made phenomena. Research statistics is one of the primary tools for implementing the scientific method. Applied statistics extends the use of statistics to the mainstream of business and government decision making. The managerial decision-making process can be viewed as a stepwise approach that involves defining the problem, formulating an appropriate model, collecting data, evaluating the various alternatives, interpreting the results, and implementing the selected decisions. The total quality management (TQM) philosophy, which is one of the primary elements behind the success of Japanese industry throughout the world, relies heavily on statistical analysis. In this regard, the understanding and appropriate application of statistics will play an important part in helping the United States regain its competitive edge in a dynamic and growing international marketplace.

2.7 GLOSSARY

acceptance sampling A statistical procedure that uses random sampling to determine whether to accept or reject an incoming or outgoing lot based on specific quality standards.

applied statistics The application of statistics to business and related fields.

facts Data that have been observed and verified.

hypothesis A statement regarding the status of some characteristics of a population.

interval scale A measurement system that characterizes data in terms of a convenient zero point and equal intervals.

measurement The assignment of numbers to characteristics being observed.

nominal scale A measurement system that characterizes data in terms of some form of identification.

ordinal scale A measurement system that characterizes data in terms of rank order.

process control Monitoring and maintaining quality throughout the manufacturing and service process.

qualitative variable A variable that is primarily characterized by nonnumerical values.

quantitative variable A variable that is primarily characterized by numerical values.

ratio scale A measurement system that characterizes data in terms of a real zero point.

research statistics The primary tool for implementing the scientific method.

scientific method A generalized strategy for gaining knowledge on a systematic basis.

symbols An abstract representation of reality.

total quality management A systematic approach for improving product and service quality where statistics plays a key role.

variables Symbols that can take on a variety of values.

2.8 BIBLIOGRAPHY

Emory, C. W. *Business Research Methods.* 3rd ed. Homewood, Ill.: Irwin, 1985.

Giere, R. N. *Understanding Scientific Reasoning.* New York: Holt, Rinehart & Winston, 1979.

Heyel, C. "The Hawthorne Experiments." In *The Encyclopedia of Management.* 3rd ed. New York: Van Nostrand Reinhold, 1982, pp. 377–81.

Kerlinger, F. N. *Foundations of Behavior Research.* 3rd ed. New York: Holt, Rinehart & Winston, 1986.

Marquardt, D. W. "The Importance of Statistics." *Journal of the American Statistical Association,* March 1987.

Minton, P. D. "The Visibility of Statistics as a Discipline." *American Statistician* 37 (1983) pp. 284–89.

Murdick, R. G., and D. R. Cooper. *Business Research: Concepts and Guidelines.* Columbus, Ohio: Grid Publishing, 1982.

Peters, W. S. *Counting for Something.* New York: Springer-Verlag, 1987.

Simon, H. A. *The New Science of Management Decisions.* 2nd ed. Englewood Cliffs, N.J.: Prentice Hall, 1977.

2.9 PROBLEMS

1. Identify 10 current public policy issues, and state each one in the form of a hypothesis.
2. Define, in your own words, the scientific method.

3. Characterize the differences between research and applied statistics using business examples.
4. Discuss how statistics will play an ever-increasing role in American business.
5. How would an understanding of the history of statistical development help to improve management effectiveness or the teaching of statistics?
6. Discuss how the Japanese emphasis on statistical quality control has helped them penetrate the U.S. car market.
7. Comment on the following statement: "Statistical-based management systems are essential for competing effectively in the international marketplace."
8. Discuss the role that statistics plays in government as an instrument for formulating policy.
9. Could someone trained in management science and statistics ever become CEO of a major U.S. corporation?
10. Characterize the following statements as either a hypothesis (H), a fact (F), or an objective (O).
    a. The United States and the Soviet Union should embark on a joint mission to Mars. _____
    b. In 1988 Lee Iacocca was board chairman of Chrysler Corporation. _____
    c. Secondary smoking is hazardous to the public's health. _____
    d. Over 55% of the American public believes that the president is doing a satisfactory job. _____
    e. AIDS is the number 1 health problem in America. _____
    f. Information technology is becoming an increasingly important management tool. _____
    g. Increasing taxes will slow down economic growth. _____
11. Classify each of the following variables as qualitative or quantitative.
    a. Religious preference
    b. Letter grades
    c. After-tax profits
    d. The Dow Jones Industrial Average
    e. The winner of the 100-meter dash
12. Classify each of the following variables as qualitative or quantitative.
    a. Brands of coffee
    b. Country of origin
    c. Regional unemployment rate
    d. Results of a medical examination
    e. Inventory levels
13. Classify the following measurements as either nominal, ordinal, interval, or ratio.
    a. The finishing order of teams in the National Basketball Association
    b. Length of employment service
    c. Current assets relative to current liabilities
    d. Automobile models
    e. Refrigerator colors
14. Pollsters for one of the leading presidential contenders wish to determine their candidate's popularity in California. Identify *(a)* the appropriate target population, *(b)* the method of sampling, and *(c)* the type of statistical inference that could be made from the sample results.

15. The director of the Bureau of Labor Statistics wishes to assess the reliability of the current month's unemployment estimates. What data is the director required to collect to assist in this assessment?
16. The president of General Motors is preparing for a meeting of the board of directors. What types of descriptive statistics should he use if the company has performed well over the last quarter? If the company has performed poorly?
17. Classify the following measurements as either nominal, ordinal, interval, or ratio.
    a. Corporate organization chart
    b. Numbers assigned to football players
    c. Membership in the American Stock Exchange
    d. Inventory turnover
    e. Temperature scale
18. Identify which of the following involve nominal data.
    a. Corporate retained earnings
    b. Distance between Los Angeles and New York
    c. Hotel floor numbers
    d. Computer models
    e. Leading football passers
19. Discuss why IQ scores and calendar dates are examples of interval measurement.
20. Define in your own words the total quality management (TQM) philosophy.
21. Discuss the role statistical analysis plays in TQM.
22. Identify some of the key costs associated with poor quality.
23. In what ways can the six-step managerial decision-making process help promote the TQM philosophy?
24. Comment on the following definition of quality: "The degree of conformance with respect to the design requirements and customer needs."
25. Contrast the difference between process control and acceptance sampling.
26. Identify several U.S. industries where poor product or service quality has resulted in a reduced market share.
27. How can the TQM philosophy improve the performance of the organization and the contribution of its employees?

Chapter 3

Descriptive Statistics

There is nothing permanent except change.
Heraclitus

CHAPTER OUTLINE

3.1 Introduction
3.2 Example Management Problem: Global Precious Metal Exchange
3.3 How to Recognize a Descriptive Statistics Problem
3.4 Descriptive Analysis
3.5 Computer Analysis
3.6 Practical Applications
3.7 Case Study: Goodfaith Emergency Clinic
3.8 Summary
3.9 Glossary
3.10 Bibliography
3.11 Problems
3.12 Cases

CHAPTER OBJECTIVES

The primary objectives of this chapter are to develop an understanding of

1. How to develop frequency tables and draw frequency graphs.
2. The key measures of central tendency.
3. The key measures of dispersion and shape.
4. The role of descriptive statistics in quality control.
5. The role of computer analysis in descriptive statistics.

The U.S. federal government is a major source of funding for many community-level social programs (e.g., low-cost housing). Many federally funded grant programs require the recipient community to meet specific guidelines as a basis for continuing to receive federal funding. Many community officials believe that these mandate requirements are onerous, and they have communicated their concerns to federal officials. The Department of Housing and Urban Development (HUD), one of the largest granting agencies in

the federal government, decided to evaluate the overall performance of its mandated grant programs in response to these growing concerns.* A questionnaire was developed and sent to approximately 600 administrators in urban areas eligible for HUD funds. The response rate for the survey was nearly 70%.

The results from the survey indicated that 90% of those interviewed indicated that a HUD official had visited their community to review ongoing compliance. The data also showed that only 1 out of 440 respondents reported that HUD had reduced their funding because of lack of compliance, and only 8% reported a special condition on their grant. The use of these straightforward descriptive statistics revealed that most cities are in general compliance with the letter and intent of federal law.

---

* J. Massey and J. Straussman, "Another Look at the Mandate Issue: Are Conditions-of-Aid Really So Burdensome?" *Public Administration Review* 45 (March/April 1985), pp. 292–301.

> **HISTORICAL NOTE**
>
> The term *descriptive statistics* was first coined by G. T. Fechner (1801–1877) during the latter half of the 19th century. The German term he used, *kollektivmasslehre,* is very descriptive but unfortunately too long. The basis for Fechner's work was the design of a mathematical system to describe population characteristics using distributions.

3.1 INTRODUCTION

The role of numbers in business is obviously very pervasive. The use varies from the strict measurements of financial performance to the more subjective assessment of customer preference. In each of these cases, however, numbers are used for describing the situation. For example, suppose a manager computes an average production figure for one of his foremen. Note that the manager is merely attempting to "describe" a characteristic and is not trying to compare, generalize, or interpret. The manager might array the data, produce graphs, or use tables or charts, but is nevertheless merely describing a production tendency. This example illustrates the process of using description statistics.

The preparation of graphs and charts can help the manager develop a better understanding of a particular situation involving large quantities of data. Usually, graphs and charts are generated from frequency distributions. A frequency distribution is a method for classifying data into mutually exclusive categories. Additionally, even more sophisticated methods exist for describing data. The three most frequently used are measures of central tendency, measures of dispersion, and measures of shape. Measures of central tendency provide an indication of the central location of a data set. The mean is perhaps the best-known measure of central tendency. Measures of dispersion reveal the extent of the spread or scatter of the data set. Finally, summary measures of shape provide insight into the overall form of the frequency distribution.

3.2 EXAMPLE MANAGEMENT PROBLEM: GLOBAL PRECIOUS METAL EXCHANGE

The general manager (GM) at the Global Precious Metal Exchange wishes to develop a presentation for potential investors in precious metal bullion. The

GM believes that prices for gold, silver, and platinum will grow during the coming year, which should serve as an insurance policy against projected two-digit inflation. The GM has collected the following New York monthly price data for gold, silver, and platinum for the past year.

| | Price ($ per ounce) | | |
Month	Gold	Silver	Platinum
January	371.30	8.18	375.11
February	386.40	9.13	391.38
March	394.70	9.65	398.47
April	382.00	9.22	395.17
May	377.70	8.97	388.95
June	378.10	8.74	386.31
July	346.80	7.42	345.55
August	348.10	7.61	336.03
September	341.30	7.26	327.36
October	340.60	7.32	324.65
November	341.50	7.49	328.45
December	319.50	6.69	309.11

The GM wishes to assemble a number of different statistics and graphics to assist in the presentation process. Specifically, the GM wants to include both histograms and time line charts.

3.3 HOW TO RECOGNIZE A DESCRIPTIVE STATISTICS PROBLEM

The following highlights the general characteristics of most descriptive statistics problems:

- Data are collected in either frequency or measurement (metric) form.
- Graphics are used to depict relationships among the data.
- Measures of central tendency and variability are used to describe the data.
- Data can be presented in either raw or grouped form.

3.4 DESCRIPTIVE ANALYSIS

In descriptive statistics one is generally concerned with two types of data: frequencies and metric values. Frequencies represent the counting of things,

> **INTERNATIONAL VIGNETTE**
>
> Biral International, founded in 1919 by Bieri, manufactures and distributes a wide range of industrial and residential pumps throughout Europe and North America. At a recent meeting of the executive committee at corporate headquarters in Muensingen, Switzerland, the primary agenda item was a proposed cooperative agreement with a Hungarian manufacturer to produce Biral pumps for use in the Comecon countries. Biral had been approached by Magyarexport, one of Hungary's largest foreign trade cooperatives, to provide documentation and technical knowhow to produce a specific number of pumps. This data was to be turned over to Vilamos Manufacturing, a client of Magyarexport. Initially, Biral would supply the parts and Vilamos would assemble the pumps at its new factory located near Budapest. Over a period of two years, however, Vilamos would develop the capability to manufacture components.
>
> The proposed agreement had been negotiated by Mr. Vagano, coordinator for sales for Comecon, over a two-year period and had reached a critical stage. Biral's management realized that such an agreement could provide an efficient mechanism for significantly increasing sales to both Hungary and other Eastern block countries. Mr. Vacano had prepared an extensive set of descriptive statistics showing the current and projected growth in the Comecon market. Based on this data and the merits of the proposed contract, it was now up to the executive committee to make a decision.

and metric values represent the measurement of things. Frequencies are the number of objects in a category. The categories may be either quantitative or qualitative (e.g., sex is a qualitative classification). Knowledge-based systems proceed from categories and accompanying frequencies to the more tightly controlled measurements. Metrics imply the application of an arithmetic system by some rules. These systems have already been described in Chapter 2. Figure 3.1 summarizes the basic metrics used in statistics.

DATA PRESENTATION

Key idea 1

Many procedures exist for organizing, presenting, and graphing data. The definition of some of these techniques may aid in the understanding of their role in descriptive statistics. The growing availability of computer-based graphics makes the development of a variety of presentation methods extremely practical. CBS provides the user with a wide range of graphical presentation techniques, some of which are outlined below.

FIGURE 3.1 Summary of Basic Metrics

Measurement	Definition	Example
Nominal	The classification of things into categories.	Jersey numbers for the basketball team.
Ordinal	An ordering of data or things from lowest to highest.	The final team standings in the league.
Interval	A direct numerical measure with equal intervals.	The number of years a player has been in the league.
Ratio	A direct numerical measure with a real zero point.	The win/loss percentage of each team in the league.

Bar Charts. Simple bar charts are usually constructed with the class on the Y axis and the magnitude on the X axis (horizontal) or with the magnitude on the Y axis and the class on the X axis (vertical). For example, consider comparing the average price of each precious metal. A horizontal bar chart depicting these data along with the average price is given in Figure 3.2.

One admonition is that these charts should "stand alone." That is, the viewer should understand the intent of the chart without recourse to supplementary materials.

Multiple Bar Charts. Multiple bar charts are used to depict several variables at the same time. Usually, multiple bar charts are constructed with the classes on the horizontal axis and the magnitude on the vertical axis. Figure 3.3 presents a multiple bar chart showing the proportion of managers by sex

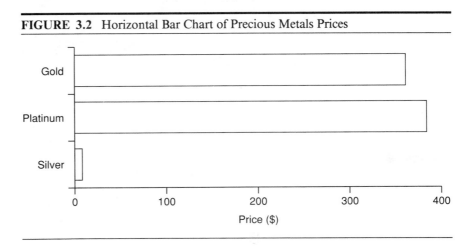

FIGURE 3.2 Horizontal Bar Chart of Precious Metals Prices

for selected industry groups—an example of a combined nominal scale and ratio scale measurement.

Pie Charts. Pie charts are commonly used to show the sizes of specific components relative to the total. A circle is drawn and is divided into slices corresponding to the percentages of the various parts. For example, the vice president at Global Metals may wish to portray the firm's annual budget as shown in Figure 3.4.

Frequency Tables. A frequency table is an organized method for classifying data into a set of mutually exclusive categories. Figure 3.5 presents the results of a final examination in freshman statistics for 100 students. The examination results are classified into six groups. The relative frequencies are obtained by dividing the number of counts for a given category by the total number of counts (e.g., $10/100 = 0.1$). The cumulative relative frequency column shows the proportion of observations that are less than or equal to the upper limit of a given class. For example, 70 out of the 100 students scored less than 80 on the statistics examination.

Histograms. The histogram is the most frequently used graphical presentation in business. A histogram is developed by drawing the class intervals on the X (horizontal) axis and the frequencies on the Y (vertical) axis. For

FIGURE 3.3 Multiple Bar Chart of Management Proportions by Sex for Selected Industry

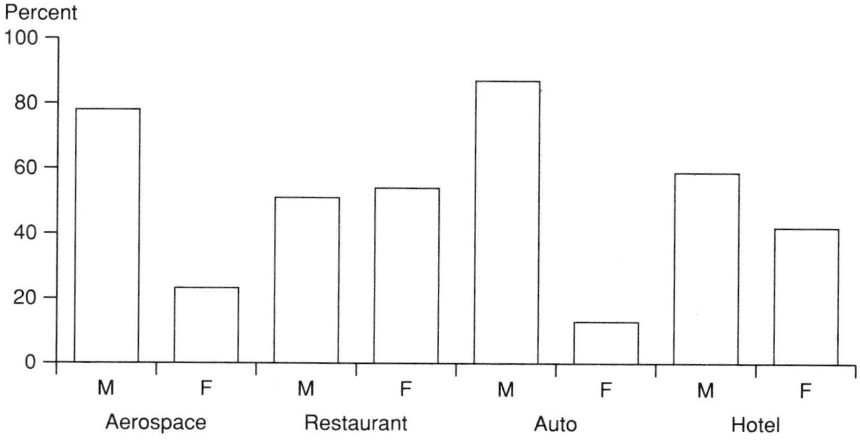

FIGURE 3.4 Pie Chart for Global Precious Metal Exchange Annual Budget

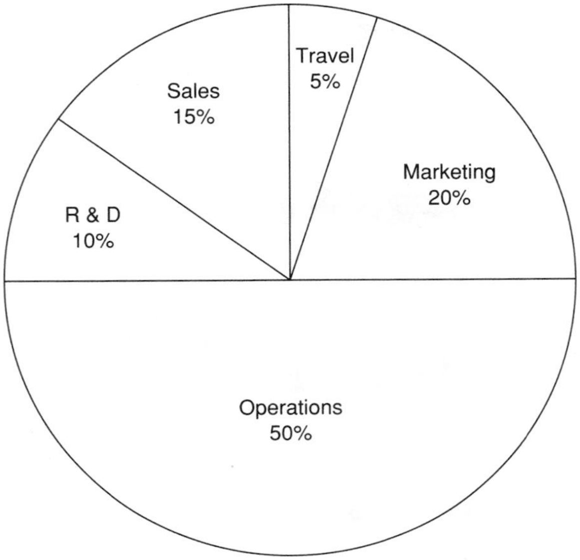

example, the GM at Global Metals may wish to develop a histogram of platinum prices as shown in Figure 3.6.

Frequency Curves. A frequency curve is a continuous curve for approximating a histogram based on connecting the midpoints of each class mark. These are used extensively in most aspects of statistical analysis. Figure 3.7 presents a frequency curve of the histogram given in Figure 3.6.

FIGURE 3.5 Frequency Table for Statistics Test

Test Score Range	Count	Relative Frequency	Cumulative Relative Frequency
≤50	0	0	0
50–60	10	0.10	0.10
60–70	15	0.15	0.25
70–80	45	0.45	0.70
80–90	25	0.25	0.95
90–100	5	0.05	1.00
Total	100		

FIGURE 3.6 Histogram of Platinum Prices for Global Precious Metal Exchange

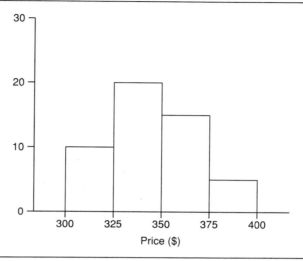

FIGURE 3.7 Frequency Curve of Platinum Prices

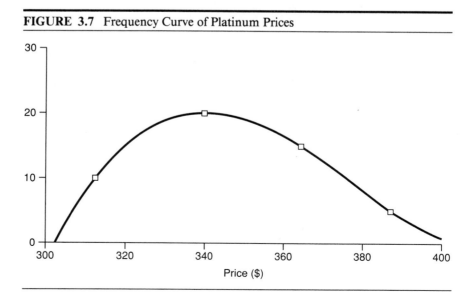

36 Chapter 3 Descriptive Statistics

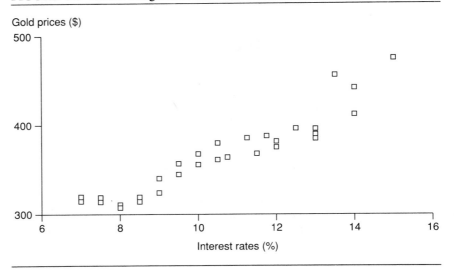

FIGURE 3.8 Scatter Diagram of Gold Prices versus Interest Rates

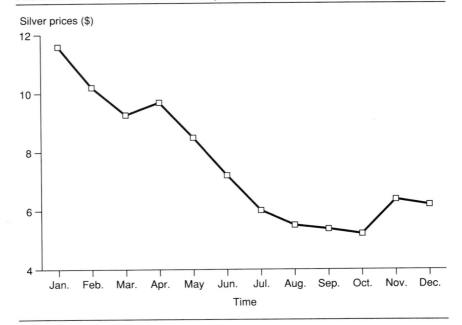

FIGURE 3.9 Time Chart of Monthly Silver Prices

Scatter Diagrams. A scatter diagram is a chart depicting the relationship between two variables of interest. Figure 3.8 shows a scatter diagram showing the relationship between gold prices and interest rates.

Trend or Time Series Charts. A specialized type of scatter diagram where time periods are placed on the X axis and magnitudes on the Y axis. The general purpose is to display the variable plotted over time. Figure 3.9 shows a time series chart of monthly silver prices recorded at the Global Precious Metals Exchange.

Pictographs. A pictograph presents data with the aid of a picture or symbols. Figure 3.10 shows a pictograph of the 1988 U. S. presidential election returns by state.

As in the case with the other methods, trend charts should stand alone. Additionally, there may be the need to correct the figures or numbers for any attenuation due to other economic factors, like the change in value of the dollar. Additional techniques are available for the eye-catching aspects of presentations (e.g., holographs). A detailed treatment of these techniques can be found in both statistics and communications texts that deal in displays for maximal understanding.

FIGURE 3.10 Pictograph of 1988 Presidential Election Returns

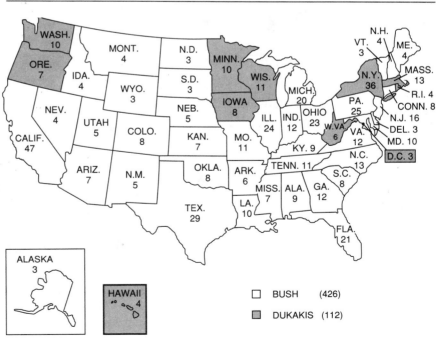

SUMMARY MEASURES

The preparation of tables and graphs provides the decision maker with a pictorial view of the assembled data. In many business situations, however, more exact measurements are needed. In these cases, summary measures are used to describe the data set. Unfortunately, the shift to summary measures involves a considerable amount of calculations. Fortunately, the widespread availability of user-friendly software renders even the most complex calculations relatively simple. As indicated above, the summary statistics used for describing data sets fall into three major categories:

1. Measures of central tendency.
2. Measures of dispersion or variability.
3. Measures of shape.

Measures of central tendency describe the center point of the data set. Measures of dispersion, on the other hand, characterize how the data are distributed. Measures of shape indicate the general form of the data.

Measures of Central Tendency

Key idea 2

Measures of central tendency are values that represent the center of the data set. The primary measures of central tendency used in business are the mean, median, and mode. These, along with several other measures, are illustrated below. It should be noted that the methods used for computing these measures are the same for population data and sample data.

Mean. The arithmetic mean is probably the most frequently used descriptive measure in statistics. Quite simply, it is the sum of all measures divided by the number of separate measures. It is mathematically computed as follows:

$$\overline{X} = \frac{\text{Sum of measurements}}{\text{Number of measurements}} = \frac{\Sigma X_i}{N}$$

This measure is also called the average. For example, a baseball player's batting average is the number of hits divided by the number of times at bat. The mean can be viewed as the point of balance in the data set.

The advantages of using the mean are:

1. Most people understand it.
2. It is unique and present in most data sets.
3. All observed values are used in its computation.
4. It is used extensively in other statistical procedures.

The disadvantages of using the mean are:

1. It is affected by extreme values because each number is used in computation.
2. It is time-consuming if you must process a large amount of ungrouped data without the aid of a computer.
3. It cannot be used for truncated distributions like categories in questionnaires where you may have a group called "45 years of age and older."

Figure 3.11 illustrates the method for computing the mean using the data for gold prices from the Global Precious Metal case.

Median. The median is the point in a distribution above and below which 50% of the measures lie. Typically, the median may not be one of the data points in the distribution. In general, the data set is organized in ascending or descending order, and the median is the value of which half of the measures are above and half are below. For an odd number of measures, the median is simply the middle figure. For an even number of measures, the median is the average of the middle two values. Figure 3.12 illustrates how to identify the median using the gold prices from the Global Precious Metal case for both 12 months (even) and 11 months (odd) of data. The general rule for locating the median is to add 1 to the number of data points and then divide by 2. In the case of 11 data points, this becomes $(11 + 1)/2$, or 6. Therefore, the sixth measure is the median (i.e., 348.10). The median is often represented by the symbol Md.

FIGURE 3.11 Computation of Mean Using Gold Price Data

Gold Price ($/oz)
$ 371.30
386.40
394.70
382.00
377.70
378.10
346.80
348.10
341.30
340.60
341.50
319.40
Sum $4,327.90

$\overline{X} = 4{,}327/12 = 360.66$

FIGURE 3.12 Computation of Median Using Gold Price Data

12 Months Gold Prices ($/oz)		11 Months Gold Prices ($/oz)
319.50		319.50
340.60		340.60
341.30		341.30
341.50		341.50
346.80		346.80
348.10	Median $= \dfrac{348.10 + 371.30}{2}$	348.10 = Median
371.30		371.30
377.70	$= 359.70$	377.70
378.10		378.10
382.00		382.00
386.40		386.40
394.70		

Advantages of using the median are:

1. It is not affected by extremely large or small measures.
2. It is arithmetically very simple to determine.
3. It can be used with truncated or open-ended categories.

Disadvantages of using the median are:

1. It is less comprehensible than the mean.
2. Arranging the arrays in ascending or descending order can require a great deal of time.
3. It has less use in statistical theory.

Mode. The mode is that class or measure having the highest frequency of occurrence. Thus, the mode is not usually calculated but simply observed. When a distribution has two modes, it is said to be *bimodal*. When there are more than two modes, the distribution is called *multimodal*. The mode has some of the same advantages as the median as a measure of central tendency. However, in many business situations, measures are not repeated and therefore no mode exists. The mode is often represented by the symbol Mo.

Additional Measures of Central Tendency. Several additional measures of central tendency frequently used in statistical analysis are:

- Weighted mean
- Geometric mean
- Harmonic mean

The *arithmetic mean,* as described above, assigns equal emphasis, or weight, to each data point. In some situations, data points do not have the same importance. For example, consider a consulting firm that bills its clients according to four standard billing categories of $60/hr, $75/hr, $100/hr, and $125/hr. Additionally, billable hours per category for this month are 400, 300, 200, and 100, respectively. Management wishes to know the average billable hourly rate. One approach would be to calculate the arithmetic as follows:

$$\overline{X} = \frac{60 + 75 + 100 + 125}{4} = \$90/\text{hr}$$

Unfortunately, this mean does not take into account the fact that billable hours for each category are different. To determine the correct mean in this case requires that each billing rate be weighted according to the number of hours per category. The general model for determining the weighted average is

$$\overline{X}_w = \frac{\Sigma w_i x_i}{\Sigma w_i}$$

where:

X_w = Weighted average
w_i = Weight assigned to ith category
Σw_i = Sum of all weights

Applying this model to the problem above yields the following revised figure for the average billable hourly rate:

$$\overline{X}_w = \frac{400 \times 60 + 300 \times 75 + 200 \times 100 + 100 \times 125}{400 + 300 + 200 + 100} = \$79/\text{hr}$$

Notice that this weighted average figure is somewhat smaller than the value computed by the arithmetic mean.

The estimation of average rates is another business application where the use of the arithmetic can yield overstated results. Consider, for example, an investor who wishes to determine the average annual interest rate received on an initial $10,000 investment over a four-year period. The annual interest rates during the four-year period were 14%, 10%, 8%, and 12%, respectively. The compound savings at the end of the fourth year were $15,168. The average interest rate based on the arithmetic mean is as follows:

$$\overline{X} = \frac{1.14 + 1.10 + 1.08 + 1.12}{4} = 1.11$$

Compounding this average rate over a four-year period results in an estimated savings of $15,181. Obviously, this result is incorrect. The actual average growth rate can be determined by multiplying the annual rates and then taking the fourth root of the product. This result is called the *geometric mean* and is the appropriate average for this class of problems.

The general model for determining the geometric mean is as follows:

$$\overline{X}_g = \sqrt[n]{X_1 X_2 X_3 \cdots X_n}$$

Applying this model to the problem above produces the desired results:

$$\overline{X}_g = \sqrt[4]{1.14 \times 1.10 \times 1.08 \times 1.12} = 1.1098$$

Thus the true average annual interest rate is slightly less than the figure computed by simply averaging the annual rates.

Suppose the dispatcher of a local trucking firm is interested in estimating the average speed for a delivery trip from Los Angeles to Bakersfield. The total distance is approximately 120 miles. The dispatcher estimates that the driver can maintain an average speed of 60 miles per hour when the road is clear. However, the average speed is reduced to 30 miles per hour during foggy conditions. The dispatcher estimates that one-half of the route will be foggy. Based on these estimates the average driving time is three hours. This translates into an average speed of 40 miles per hour. The average speed based on the arithmetic mean is

$$\overline{X} = \frac{60 + 30}{2} = 45 \text{ mph}$$

This result does not accurately describe the "true" average speed. That can only be determined using the *harmonic mean*. The general model for computing the harmonic mean is

$$X_h = \frac{N}{\Sigma(1/X_i)}$$

Using this model with the above data yields the correct result:

$$X_h = \frac{2}{(1/60) + (1/30)} = 40 \text{ mph}$$

Each of the above measures is used in calculating business index numbers (e.g., Consumer Price Index). Typically, they possess certain advantages over the more traditional measures (e.g., arithmetic mean). Index numbers are used frequently to compare conditions in one period with conditions in another. Consider, for example, computing the price index for a given period. The basic model is shown below:

$$I = \frac{\Sigma P_{in}}{\Sigma P_{i0}} \times 100$$

where:

I = Price index (simple average)
P_{in} = Unit price for ith item in period n

3.4 Descriptive Analysis

FIGURE 3.13 Computation of Price Index Using Simple Average

Commodity	Price (per bushel)		Quantity (billions of bushels)		Relative Price
	1979	1984	1979	1984	
Soybeans	$6.28	$7.75	3.09	2.53	1.23
Wheat	3.78	3.53	1.74	2.37	0.93
Corn	2.81	3.43	8.47	8.17	1.22

$$I(1984) = \frac{7.75 + 3.53 + 3.43}{6.28 + 3.78 + 2.81} \times 100 = 114$$

P_{i0} = Unit price for ith item in base period
N = Total number of price categories

Figure 3.13 presents selected farm prices for the periods 1979 and 1984 and illustrates the method for computing a price index using the simple average model. This result shows that the unweighted aggregate farm price index for the three commodities has increased by approximately 14% over the five-year period.

The models for computing the consumer price index using the weighted, geometric, and harmonic means are presented in Figure 3.14. The assumptions behind each of these models differ. For example, the weighted average model includes the actual amount produced or sold in calculating the index. This corresponds to the philosophy that each category should be weighted according to its relative importance. It is generally observed that quantity often provides the best indicator of relative importance. The commodity

FIGURE 3.14 Comparison of Weighted Average, Geometric, and Harmonic Price Indexes for 1984

Model	Relationship	1984 Estimate
Weighted average	$I = \frac{\Sigma(P_n \times Q_0)}{\Sigma(P_0 \times Q_0)} \times 100$	118
Geometric mean	$I = (\pi(P_n/P_0))^{1/N} \times 100$	112
Harmonic mean	$I = \frac{N}{\Sigma 1/(P_n/P_0)} \times 100$	114

where:

Q_0 = Weight for Nth commodity (based on quantity purchased)

44 Chapter 3 Descriptive Statistics

data given in Figure 3.13 were used in computing farm price indices for the three models for 1984.

Measures of Dispersion

Key idea 3

Measures of central tendency provide the decision maker with an estimate of the center location of the data set. However, these measures do not provide an indication of the variability of the data. For example, two data sets may have identical means, yet the data may vary considerably around the two means. Figure 3.15 shows a frequency plot for two investment alternatives with the same mean. Notice that the spread of the data for investment A is considerably larger than that for investment B. How the data are distributed can often be an important factor in the decision-making process. For example, most managers would tend to select alternative B due to the fact that the chances of a loss are quite low.

Obviously, additional statistics are needed to describe how the data are distributed. The literature refers to these as measures of dispersion or variation. Some of the measures of dispersion such as the standard deviation and the variance are fundamental to statistical inference. Other measures are

FIGURE 3.15 Frequency Distribution of After-Tax Return for Two Investment Opportunities

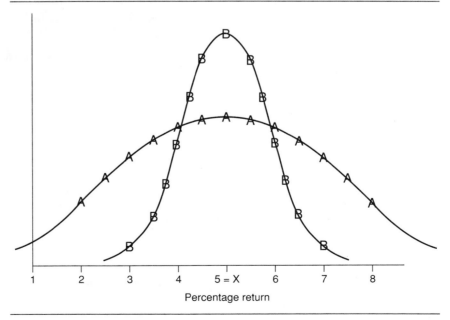

simply different descriptors currently in use. These will be treated definitionally. However treated, the concepts of dispersion need to be understood prior to getting into probability and the more advanced constructs of statistics (e.g., inference).

Range. The range is simply the highest observed value minus the lowest observed value. For example, the range for the gold price data is $72.20. Some statisticians divide that result by 2 to compute the average range. The range is obviously heavily influenced by extreme values and, as a measure of dispersion, is further limited when there are "open" classes such as "50 and over."

Quartiles. Although a measure of central tendency, the quartiles can also be viewed as measures of variability. The distribution is divided into four quartiles. The first quartile is that point above which 75% of the cases lie and below which 25% of the cases lie. The second quartile, or the median, is that point above *and* below which 50% of the cases lie. The third quartile is that point above which 25% of the cases lie and below which 75% of the cases lie. The fourth quartile is the upper 25%.

The above measures of dispersion appear frequently in the literature but are used less frequently in business. It should be noted that quartiles, fractiles, and percentiles are not influenced by the extreme measures and are therefore better descriptors of distribution than the range. To be truly useful, the measures of dispersion should be associated with some measure of central tendency such as the mean. This is the case with the next three measures of dispersion: the mean absolute deviation, the variance, and the standard deviation.

Mean Absolute Deviation. The mean absolute deviation (MAD) is the average of the sum of the absolute deviations. The procedure for computing the average deviation is given below:

$$\text{MAD} = \Sigma \frac{|X - \overline{X}|}{N}$$

The major advantage of the average deviation over the previously discussed measures is that all the measures contribute equally to the measures of dispersion. Thus, the "extreme" measures do not contribute disproportionately. However, the average deviation takes into account only the size of departures from the mean and not their direction. On that account it is less useful than the standard deviation and variance. The average deviation using the gold prices data is 21.

Variance. The variance represents a key measure of descriptive variability. As such it plays a significant role in all phases of statistical analysis. The variance is the average of the squared deviations from the mean. The computational relationships for determining the variance for a population or sample are given below:

$$\text{Var}(p) = \frac{\Sigma(X_i - \overline{X})^2}{N} \qquad \text{Var}(s) = \frac{\Sigma(X_i - \overline{X})^2}{N - 1}$$

This relationship, although somewhat imposing, is merely the average deviation of the data set. Note that the arithmetic operation of squaring removes the negative deviations and makes them positive. Thus the variance will never be negative. The relationship given above is for the variance of the population. In many business situations, however, the population mean is unknown. In these cases the population mean is estimated by the sample mean. When dealing with a sample (which is usually the case), the denominator in the relationship becomes $N - 1$ instead of N. The rationale for the $N - 1$ when using samples to estimate populations is as follows:

1. A sample makes it difficult to pick up the "extreme" cases. For example, a measure that occurs 1% of the time needs a sample of 100 to occur once.
2. Extreme cases make the variance larger.
3. The variance computation from a sample is therefore an underestimate of the population variance.
4. Subtracting 1 from the N in the denominator makes it smaller. A smaller denominator results in a larger value of the variance.

It is important to understand the variance from an intuitive point of view and not just the statistical perspective. The reader should think of the variance as a way of describing different people or events on the same measure. For example, people receive different grades in school, produce different amounts in business, and receive different salaries on their jobs. These differences are what a variance purports to describe. These differences in myriad forms underlie statistical testing, which is why the concept of variance is so important to statistics.

One should not get the impression from the foregoing that all variance is due to individual variation in ability, aptitude, or performance. Rather, variance may be a function of one or more of the following:

1. Real ability, behavior, performance, or some aspect of measurement.
2. Measurement error, or errors in the matching and assignment of numbers to phenomena.
3. Sampling error, or the distribution of the measurement in the sample and population.
4. Instrument error, or mistakes made by improper, unreliable, or invalid instruments.
5. Procedural error, or error occasioned by poor logic that leads from hypothesis to data not accurately reflecting intent.
6. Random error, or error that is unexplainable in any other way.

Thus, there are many potential sources of variance. The better the measure of the individual performances, the better the chance of having it "explained." Please note that the term *error* is not synonymous with the word *mistake*. Rather, it is a distributional variable produced by individual differences and other sources. Figure 3.16 shows the method for calculating the variance for gold prices.

Standard Deviation. The standard deviation is perhaps the most frequently mentioned descriptive measure of dispersion in the literature. One major reason for its widespread use is that it serves to describe the "famous" normal curve. It is simply the square root of the variance. The standard deviation can be for a population (p) or as a sample (S). The relationship for a population is

$$S(p) = \sqrt{\text{Var}(p)}$$

When the computed S is used for inferences to a population from a sample, the relationship is

$$S(s) = \sqrt{\text{Var}(s)}$$

The rationale of small sample correction is the same as used in computing the variance. Thus, the standard deviation for gold prices is

$$S = \sqrt{561.1553} = 23.68$$

FIGURE 3.16 Computation of Sample Variance for Gold Prices

Monthly Price ($/oz)	Mean ($/oz)	Difference ($/oz)	Squared Difference
371.30	360.70	10.60	112.36
386.40	360.70	25.70	660.49
394.70	360.70	34.00	1,156.00
382.00	360.70	21.30	453.69
377.70	360.70	17.00	289.00
378.10	360.70	17.40	302.76
346.80	360.70	−13.90	193.21
348.10	360.70	−12.60	158.76
341.30	360.70	−19.40	376.36
340.60	360.70	−20.10	404.01
341.50	360.70	−19.20	368.64
319.50	360.70	−41.20	1,697.44
	Sum	0	6,172.72

$$\text{Var} = \frac{6{,}172.72}{11} = 561.2$$

Measures of Shape

As outlined above, frequency distributions or curves are used to describe data. Typically, curves are characterized by the *degree of skewness* (*Sk*) and *degree of peakedness* (*Pk*). Figure 3.17 summarizes the basic ways to describe a curve in terms of its shape and also shows the relationship between the mode, median, and mean for the different distributions. Notice that when the distribution is symmetrical, the mode, median, and mean are the same. However, when the distribution is skewed, both the median and mean differ from the mode. For example, the distribution of automotive tire wear would tend to be symmetrical around the average. On the other hand, the distribution of salary incomes for a particular firm would tend to be skewed to the right. This is due to the fact that a few of the firm's employees (i.e., the management) make very large salaries while most make much less. Recall that only the mean calculation is influenced by the actual data values. Therefore, a few large salaries would tend to increase the mean while having little or no effect on the median or mode.

FIGURE 3.17 Curve Shape Characteristics

Symmetrical (normal)	Right and left sides are symmetrical (a round mode).	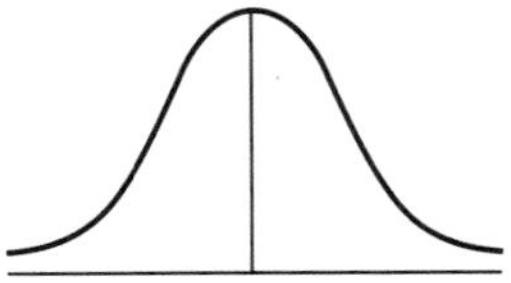 Mean / Median / Mode
Skewed left (negative)	More data on left side of mode than on right.	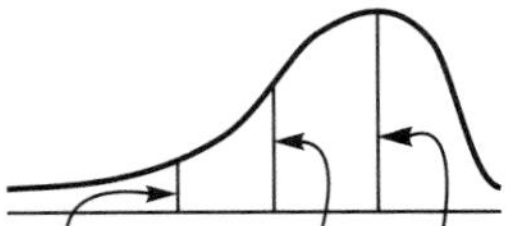 Mean — Median — Mode
Skewed right (positive)	More data on right side of mode than on left.	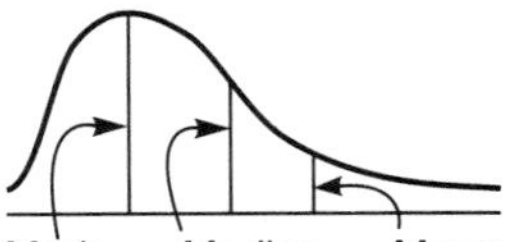 Mode — Median — Mean

Perhaps the most useful quantitative measure of skewness is the coefficient of skewness. However, the computational process for estimating this parameter is quite complex. Karl Pearson, an important contributor to modern statistical theory, developed the following formula for estimating the degree of skewness for moderately skewed distributions:

$$\text{Sk}(s) = \frac{3(\overline{X} - \text{Md})}{s}$$

When the difference between the mean and the median is small compared to the standard deviation, the curve tends to be symmetrical (it is exactly symmetrical when the mean equals the median). If, on the other hand, the difference between the mean and median is large compared to the standard deviation, then the curve is skewed: to the right if $\overline{X} > \text{Md}$ and to the left if $\overline{X} < \text{Md}$. The Pearson Coefficient of Skewness for the gold price data is 0.11. This relatively small value suggests that the data distribution is nearly symmetrical. CBS determines the coefficient of skewness based on the actual mathematical definition. Accordingly, the results may differ slightly from those developed from the above formula.

FIGURE 3.18 Curve Peak Characteristics

Curve type	Description	Graphic
Platykurtic	Broad peakness	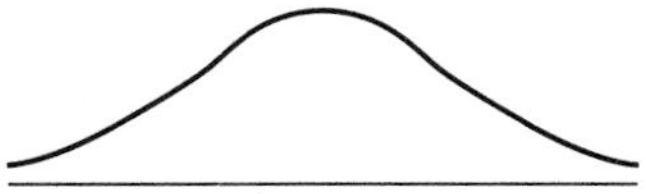
Mesokurtic	Moderate peakness	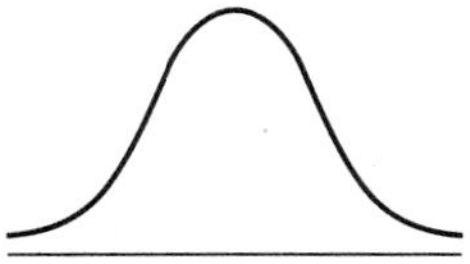
Leptokurtic	Slender peakness	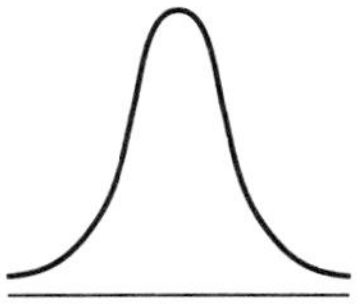

These descriptions of skewness, however, do not deal with the height or peakedness that represents the second dimension of description. The degree of peakedness (Pk) is called *kurtosis* and is highlighted in Figure 3.18.

The degree of kurtosis can be measured by the coefficient of kurtosis (k). The formula for determining this coefficient for sample data is:

$$Pk(s) = \frac{\Sigma(X - \overline{X})^4}{\dfrac{N}{s^4}}$$

This coefficient is included in CBS. Curves with a kurtosis coefficient of 2.5 to 3.5 are described as *mesokurtic*. Those with values above 3.5 are classified as *leptokurtic* (i.e., relatively sharp peak), and those with values below 2.5 are called *platykurtic* (i.e., relatively flat peak). The Kurtosis coefficient for the gold price data is 2.1, which suggests a relatively flat distribution.

SUMMARY MEASURES FOR GROUPED DATA

The statistical procedures discussed so far have been used for describing measures of central tendency and dispersion for raw or ungrouped data values. In many business situations, however, the data are in count or frequency form. For example, a study of executive salaries would typically report the results in terms of the frequency of occurrence of various salary categories. In these cases, special procedures are needed to develop descriptions of the classical measures of central tendency and variability. It is suggested that when both ungrouped and grouped data are available, the computation of the summary statistics should be based on ungrouped data. This is because the computation procedures will generally yield slightly different results. This is due to the fact that some information is lost when the data are reported in group form. Thus, the summary measures for group data can be viewed as approximating those developed based on ungrouped or raw data.

The data in Figure 3.19 present the results of a recent survey of executive

FIGURE 3.19 Group Presentation of Executive Salary Survey Results

Annual Salary ($000)	Frequency	Class Midpoint	Weighted Value
60–80	10	70	700
80–100	25	90	2,250
100–120	30	110	3,300
120–140	20	130	2,600
140–160	15	150	2,250
Total	100		11,100

salaries ($000) in the software industry. In order to retain anonymity, executives were asked to indicate their annual compensation in terms of one of the five salary categories. For example, 10 out of the 100 executives surveyed indicated that their salary was between $60,000 and $80,000. The class midpoint is assumed to represent the average of the values in a particular group, and all members are assumed to be equally distributed within the class. The weighted value figures are simply the product of the frequency times the class midpoint.

Group Measures of Central Tendency

The mean for group data is very similar to the weighted mean discussed previously. The group mean is simply the sum of the weighted values divided by the total number of counts. Mathematically, this relationship can be expressed as follows:

$$\overline{X} = \frac{\Sigma f_i \times M_i}{N}$$

where f_i is the frequency count for the ith group, M_i is the midpoint value for the ith group, and N is the total number of frequencies. The group mean for the salary survey is:

$$\overline{X} = \frac{11{,}100}{100} = \$111 \ (000)$$

The computation for the group median is somewhat more complicated. The basic model is given below:

$$\text{Md} = L + \frac{N/2 - F}{f} \times W$$

where:

L = lower limit of the median class
N = total number of frequencies
F = sum of the frequencies up to the median class
f = frequency of the median class
W = interval width

The group median for the salary survey problem can be computed as follows:

$$\text{Md} = 100 + \frac{50 - 35}{30} \times 20 = \$110 \ (000)$$

Thus the group median is very close to the group mean.

The modal class for group data is the category or categories that contain the largest number of frequencies. In the salary survey problem the modal

FIGURE 3.20 Computation of Group Variance and Standard Deviation for Executive Salary Survey

Annual Salary ($000)	Frequency	Class Midpoint	$f_i M_i$	$f_i M_i^2$
60–80	10	70	700	49,000
80–100	25	90	2,250	202,500
100–120	30	110	3,300	363,000
120–140	20	130	2,600	338,000
140–160	15	150	2,250	337,500
Totals	100		11,100	1,290,000

$$S^2 = \frac{1{,}290{,}000 - 100 \times (111)^2}{99} = 584.85 \text{ (variance)}$$

$$S = \sqrt{584.85} = 24.18 \text{ (standard deviation)}$$

class is the third category (i.e., 100–110), which contains 30 observations. The mode is defined as the midpoint of the modal class. For the survey problem this corresponds to a value of 105. In cases where there are two modal classes, the frequency distribution is called bimodal.

Group Measures of Variability

The two primary measures of variability for group data are the variance and the standard deviation. The model for determining the variance for group data is very similar to the one used for computing the variance for ungrouped data. The basic model for determining the sample variance is:

$$S^2 = \frac{\Sigma f_i M_i^2 - NX^2}{N - 1}$$

The standard deviation is simply the square root of the variance. Figure 3.20 illustrates the calculation of the variance for group data. The population variance is computed in a similar manner except that the denominator is N instead of $N - 1$.

THE USE OF DESCRIPTIVE STATISTICS IN QUALITY CONTROL

Key idea 4

As indicated above, variability is found throughout all phases of business operations. For example, it is hard to imagine that two parts made on the same machine will be identical or that two customers would receive the exact same level of service from a retail outlet. Accordingly, it is important to track or measure the level of variability associated with the production or service process. The standard approach for "tracking" the level of quality is through the use of process control charts. The basic idea behind the use of control charts is that they can be used to determine whether the process of interest is "in" or "out" of control. This section introduces two types of control charts frequently used in business; p-charts, and $\overline{X}$-charts. P-charts are used in the case of attribute measurements, e.g., the level of service is either acceptable or not acceptable. $\overline{X}$-charts are employed with processes where physical measurements are needed to determine product or service quality, e.g., the diameter of a ball bearing, or the length of time to complete a phone call. In either application, a control chart is prepared which shows the relationship between the actual measurements and the acceptable control limits (usually ± 3 standard deviations from the mean). Figure 3.21 shows control charts for several different applications. These graphics show that for panels A and B the process is out of control, that is, one or more of the data samples falls outside the control limits. The visual benefit of these descriptive statistics-based control charts is hopefully self evident.

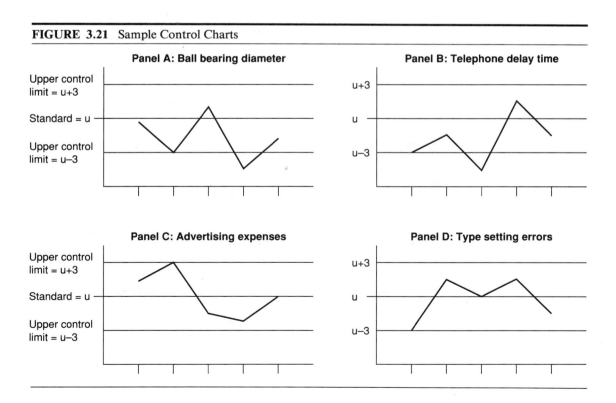

FIGURE 3.21 Sample Control Charts

P-charts. P-charts simply show a plot of the fraction of items in the sample that are rejected based on the specified attribute, the average fraction of defectives, and the control limits. The average fraction of defectives (p) of the samples is the total number of defectives from all samples divided by the product of the sample size and the number of samples. This relationship is given below:

$$\bar{p} = \frac{\text{Total number of defectives}}{\text{Sample size} \times \text{Number of samples}}$$

The corresponding standard deviation (Sp) can be computed as follows:

$$Sp = \sqrt{\frac{\bar{p} \times (1 - \bar{p})}{N}}$$

To illustrate the use of p-charts in a production setting, consider that the quality control manager at Telestar Pharmaceutical wishes to determine the level of contamination associated the manufacturing of a new vaccine. The manager has collected eight samples of 50 measurements each which indicate the number of packages found contaminated. This data is reported in Figure 3.22.

An approximate upper and lower limit for a p-chart can be developed as follows:

$$\text{Upper control limit (p)} = p + 3Sp$$
$$\text{Lower control limit (p)} = p - 3Sp$$

Figure 3.23 shows the resultant p-chart with the upper and lower limits. Notice that none of the sample proportions fall outside these limits and therefore the manager can conclude that the production process is currently in control.

$\overline{X}$-charts. Like the p-chart, the $\overline{X}$-chart presents a plot of the sample means drawn from the process under investigation, the average of the sample means, and the control limits. The average of the sample means ($\overline{X}$) is determined by summing the sample means and dividing the result by the number of samples.

FIGURE 3.22 Sample Data for Telestar Pharmaceutical

Sample Number	Number of Contaminated Packages	Sample Number	Number of Contaminated Packages
1	5	5	11
2	1	6	16
3	7	7	10
4	9	8	12

FIGURE 3.23 P-Chart for Telestar Pharmaceutical

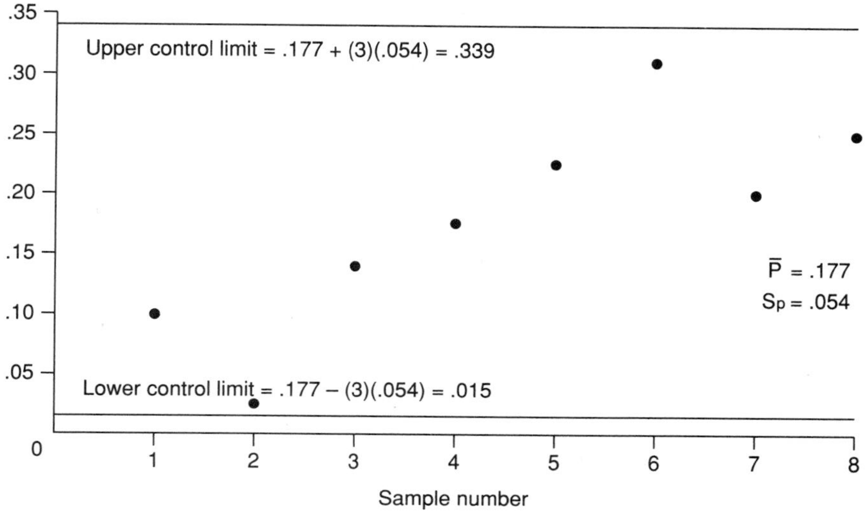

Upper control limit = .177 + (3)(.054) = .339

$\bar{P} = .177$
$S_P = .054$

Lower control limit = .177 − (3)(.054) = .015

Sample number

An approximate upper and lower limit for an $\overline{X}$-chart can be developed as follows:

$$\text{Upper control limit } (\overline{\overline{X}}) = \overline{\overline{X}} + 3S$$
$$\text{Lower control limit } (\overline{X}) = \overline{\overline{X}} - 3S$$

where S is the standard deviation of sample data and is computed using the formulas presented earlier in this chapter.

To illustrate the development of an $\overline{X}$-chart, consider that the operations manager at Havenhurst Manufacturing is interested in measuring the quality of the company's ball bearing production line. The manager has collected ten samples of 100 measurements of ball bearing diameters. This data is reported in Figure 3.24.

FIGURE 3.24 Sample Data on Average Ball Bearing Diameters

Sample Number	Average Diameter (inches)	Sample Number	Average Diameter (inches)
1	3.13	6	2.98
2	3.04	7	3.07
3	3.17	8	3.03
4	3.18	9	2.89
5	3.22	10	3.01

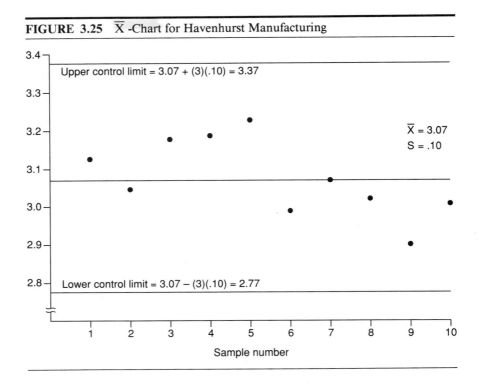

FIGURE 3.25 $\overline{X}$-Chart for Havenhurst Manufacturing

Figure 3.25 shows the resultant $\overline{X}$-chart with the upper and lower limits. Notice that each of the sample means falls within these limits and therefore the manager can conclude that the production process is currently under control.

3.5 COMPUTER ANALYSIS

Key idea 5

The widespread availability of the personal computer has brought about the opportunity to greatly expand the use of descriptive statistics throughout the decision-making process. Extensive graphic presentations can be generated with relative ease. Further, computer analysis greatly reduces the time required to process large quantities of data and improves the level of accuracy. Figure 3.26 illustrates a CBS analysis of gold prices for the Global Precious Metal Exchange problem. This analysis shows that the mean and median are nearly the same. The relatively small coefficient of skewness suggests that the data distribution is nearly symmetrical.

A CBS analysis of the same problem is given in Figure 3.27, where the data are grouped into five categories with an interval width of 20. Notice that the group mean is slightly less than the ungrouped mean, while the group median is nearly 10 points less than the ungrouped median. A CBS-generated

FIGURE 3.26 CBS Descriptive Statistics for Gold Price Data

```
CBS-Descriptive Statistics    I-GOLDA              07-05-1990 - 09:02:33
                              Results

   Mean:                                   360.6667
   Median:                                 359.7000
   Mode:                               none
   Range:                                   75.2000
   Variance (S):                           561.1553
   Standard Deviation (S):                  23.6887
   Coefficient of Skewness:                 -0.1203
   Coefficient of Kurtosis:                  1.4683


                              press ←┘
```

FIGURE 3.27 CBS Descriptive Statistics for Gold Price Data

```
CBS-Descriptive Statistics                         07-27-1990 - 15:49:14



   Mean:                                   360
   Median:                                 350
   Modal Class:             330    -    350
   Modal Class:             370    -    390
   Interval:                                20
   Variance (S):                           654.5455
   Standard Deviation (S):                  25.5841
   Coefficient of Skewness:                  0
   Coefficient of Kurtosis:                  1.3071


                              press ←┘
```

FIGURE 3.28 Gold Prices Histogram

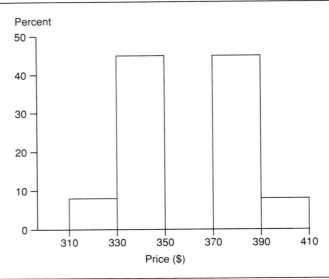

histogram showing the relative frequencies of the group data is given in Figure 3.28. This graphic shows the bimodal nature of the developed histogram. The largest frequencies occur in the second and fourth categories with a zero frequency for the middle category. This result points out the impact that the selection of the interval width can have on the calculations. For example, the selection of an interval width of 17 yields a more uniform distribution with a median nearly the same as the mean.

3.6 PRACTICAL APPLICATIONS

The following applications highlight several specific business examples involving the use of descriptive statistics.

Pedmont Electric Utility (Group Descriptive Statistics)

Gene Reynolds, the chief financial officer for Pedmont Electric, is interested in determining the distribution of average electric bills for customers within Pedmont's service area. Mr. Reynolds has selected at random the 20 customer invoices presented in Figure 3.29.

Reynolds would like to develop a frequency distribution consisting of five groups with an interval of $200. He would like to know the percentage of

FIGURE 3.29 Sample Invoice Data for Pedmont Electric Problem

Invoice	Amount	Invoice	Amount
560	$1020	4138	$1056
874	945	4653	1124
1233	987	5052	1795
1562	1003	5785	1412
1879	1645	6076	910
2061	1167	6231	1345
2456	1413	6978	1717
2877	894	7345	1312
3032	924	8147	1234
3461	812	8546	1189

customers with annual bills of $1,600 or more since they will be entitled to a 10% rebate. Further, he would like to estimate the average rebate per customer.

A CBS analysis of the raw data is presented in Figure 3.30. These results show that the mean of the invoices is $1,195.20. As requested by Reynolds, the raw data were converted to group data using the above specifications (i.e., five groups with a $200 interval width). A histogram of the group data is

FIGURE 3.30 CBS Analysis of Raw Data for Pedmont Electric

```
CBS-Descriptive Statistics                      07-27-1990 - 13:28:44
                          Results

    Mean:                                    1,195.2000
    Median:                                  1,145.5000
    Mode:                                       none
    Range:                                       983
    Variance (S):                            81,984.0620
    Standard Deviation (S):                   286.3286
    Coefficient of Skewness:                    0.6548
    Coefficient of Kurtosis:                    2.2315


                          press ←┘
```

FIGURE 3.31 Histogram for Pedmont Electric Problem

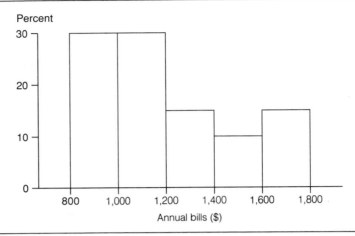

illustrated in Figure 3.31. The histogram indicates that the first two groups comprise 60% of the total number of invoices. A CBS analysis of this data is presented in Figure 3.32. These results show that the group mean is $1,200 or slightly more than the mean based on the raw data. An inspection of the relative frequency distribution shows that 15% of the sampled invoices are

FIGURE 3.32 CBS Analysis of Converted Group Data for Pedmont Electric

```
CBS-Descriptive Statistics                       07-27-1990 - 13:30:13



Mean:                                  1,200

Median:                                1,133.3334

Modal Class:               800    -    1000
Modal Class:              1000    -    1200

Interval:                              200

Variance (S):                         82,105.2660

Standard Deviation (S):                286.5402

Coefficient of Skewness:               0.5611

Coefficient of Kurtosis:               1.8780


                        press ←
```

FIGURE 3.33 Price and Quantity Data for Coffee, Tea, and Cocoa for 1980–1985

Year	Coffee Price	Coffee Quantity	Tea Price	Tea Quantity	Cocoa Price	Cocoa Quantity
1980	206.6	2,396,196	101.0	184,786	135.4	326,700
1981	159.4	2,185,260	91.7	190,254	108.5	539,000
1982	142.0	2,298,912	87.7	182,613	92.4	427,240
1983	140.0	2,171,268	105.5	170,451	92.1	398,640
1984	145.6	2,340,888	156.8	194,565	106.2	419,320
1985	142.4	2,468,136	90.0	174,617	98.7	585,420

$1,600 or more. The rebate can be determined by simply multiplying the average invoice value for the fifth group by the 10% discount factor. In this case, the average invoice value for the fifth group is $1,700. Therefore, each customer with a bill of $1,600 or more should expect to receive, on the average, a $170 discount.

International Coffee Exchange (Price Indexes)

The managing director of the International Coffee Exchange is concerned about the general decline in the prices and market share over the past few years. In particular, the director is interested in the relative decline in coffee prices and consumption with respect to its two main competitors: tea and cocoa. Figure 3.33 represents prices (in cents per pound) and imports (in thousands of pounds) for 1980 through 1985.

An inspection of Figure 3.33 shows that import quantities for coffee and tea have remained nearly constant over the reporting period, while import quantities for cocoa have fluctuated. Figure 3.34 presents a comparison of the relative price indexes for coffee and the three commodities combined for 1985. The aggregated price indexes were computed using simple, weighted, and geometric averages for the three commodities. The base year for these

FIGURE 3.34 Coffee and Aggregate Price Indexes, 1980–1985

Coffee	69
Commodity averages:	
• Simple	75
• Weighted	76
• Geometric	77

estimates was 1980. Interestingly, the aggregate price index is nearly the same for the three computational methods. Notice that the price index for coffee alone is 6 to 8 points lower (depending on computational method) than the aggregate commodity indexes. This suggests that coffee prices have been falling faster than the general market sector (i.e., coffee, tea, and cocoa).

3.7 CASE STUDY: GOODFAITH EMERGENCY CLINIC

Recent trends involving local hospitals closing or curtailing emergency care has created a growing concern among health care professionals and government officials. As usual, economics plays a key role in these developments. Typically, many individuals requiring emergency care do not possess health care insurance or other means to pay for the service. Further, many families use the emergency service in lieu of visiting a physician. To help counter these trends, the County Board of Supervisors has commissioned a feasibility study to evaluate the benefits and costs of establishing a group of privately owned and operated (though partially funded by the county) emergency clinics throughout the county. The primary role of the clinics would be to perform a preliminary diagnosis of the patient, leading to earlier treatment at the clinic or reassignment to a fully equipped hospital. Initial estimates indicate that approximately two thirds of the patients visiting a clinic could be treated at the facility. One important goal of this program would be to limit the cost of medical service to less than $100 per visit.

To assist in the evaluation, survey data was collected on projected hourly demand, cost per visit for a typical clinic, and average time waiting. These data are shown in Figure 3.35.

The county has authorized the medical staff to prepare a briefing on this project. The staff plans to make extensive use of descriptive statistics in preparing the plan. Figure 3.36 shows a histogram for patient demand. A CBS analysis of these data is highlighted in Figure 3.37. These results indicate that average demand is 11 patients per hour. Notice that the group mean,

FIGURE 3.35 Demand and Cost Frequency Data for Goodfaith Emergency Clinic

Demand	Frequency	Cost	Frequency
0–5	20	0–20	10
5–10	25	20–40	20
10–15	30	40–60	30
15–20	15	60–80	30
20–25	10	80–100	10

FIGURE 3.36 Patient Demand Histogram

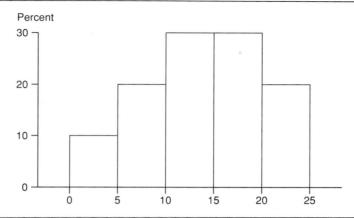

median, and mode are nearly identical. This suggests that the demand data tend to be nearly symetrical.*

Figure 3.38 shows a histogram for patient cost. A CBS analysis of these data is highlighted in Figure 3.39. These results indicate an average patient cost of $52.

FIGURE 3.37 CBS Descriptive Statistics for Patient Demand

```
CBS-Descriptive Statistics                    07-27-1990 - 13:41:32
                         Results

    Mean:                              11
    Median:                            10.8333
    Modal Class:               10   -   15
    Interval:                           5
    Variance (S):                      38.1313
    Standard Deviation (S):             6.1751
    Coefficient of Skewness:            0.2580
    Coefficient of Kurtosis:            2.1319


                         press ↵
```

---

* This observation is further supported by the relatively small coefficient of skewness (0.258).

FIGURE 3.38 Patient Cost Histogram

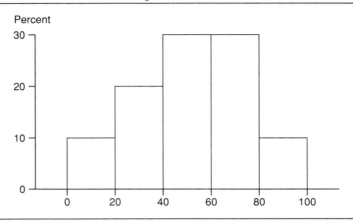

3.8 SUMMARY

This chapter has introduced a number of different ways of categorizing, displaying, and describing data. Specific techniques for categorizing and displaying data include histograms, pie charts, bar charts, time series graphs, scattergrams, and frequency polygons. The two basic ways of describing data are measures of central tendency and measures of dispersion. In modern

FIGURE 3.39 CBS Descriptive Statistics for Patient Cost

```
CBS-Descriptive Statistics                    07-27-1990 - 13:45:39
                              Results

   Mean:                                   52
   Median:                                 53.3333
   Modal Class:                  40    -   60
   Modal Class:                  60    -   80
   Interval:                               20
   Variance (S):                           521.2121
   Standard Deviation (S):                 22.8301
   Coefficient of Skewness:               -0.1936
   Coefficient of Kurtosis:                2.2014


                              press ←┘
```

statistics the mean and median are the most frequently used measures of central tendency, and the variance and standard deviation are the most frequently used measures of dispersion. Variance is a measure of the average squared difference between an individual value and the mean. The standard deviation is the square root of the variance. This measure is used in defining the normal curve and is fundamental to probability theory and analysis. Additionally, the shape of a curve can be described by its degree of skewness and peakedness. New techniques based on descriptive statistics are being used to monitor product and service quality throughout all phases of business operations. This chapter also introduced the use of computer analysis in displaying and describing data.

3.9 GLOSSARY

deviation The difference between an individual value and its mean.

group data Data depicted in frequency form using categories.

kurtosis The degree of peakedness of a data distribution.

mean A measure of central tendency that represents the arithmetic average of the data set.

median A measure of central tendency that represents the 50% point for an ordered data set.

median class The frequency class containing the median.

modal class The class containing the largest number of observations.

mode A measure of central tendency that represents most frequently occurring value of the data set.

parameter A numerical value that describes a specific characteristic of the population (e.g., population mean).

percentiles The division of a distribution into hundreds.

quartiles The division of a distribution into quarters.

range The difference between the highest and lowest values.

raw data Individual data values.

skewness The degree of the lack of symmetry of a data distribution.

standard deviation The square root of the variance.

statistic A numerical value that describes a specific characteristic of a sample (e.g., sample mean).

variance A measure of dispersion based on the average of the sum of the squares of the deviations.

3.10 BIBLIOGRAPHY

Becker, W. E., and D. L. Harnett. *Business and Economics Statistics.* Reading, Mass.: Addison-Wesley, 1987.

Chase, R. B., and N. J. Aquilano. *Production Operations Management.* 5th ed. Homewood, Ill.: Irwin, 1989.

Duncan, D. J. *Quality Controlled Industrial Statistics.* 5th ed. Homewood, Ill.: Irwin, 1986.

Gitlow, H. S.; S. Gitlow; A. Oppenheim; and R. Oppenheim. *Tools and Methods for the Improvement of Quality.* Homewood, Ill.: Irwin, 1989.

Gitlow, H. S., and R. Oppenheim. *Stat City.* 2nd ed. Homewood, Ill.: Irwin, 1986.

Hall, O. E., Jr., and H. E. Adelman. *Computerized Business Statistics.* 2nd ed. Homewood, Ill.: Irwin, 1990.

Hussain, D. *Information Processing Systems for Management.* Homewood, Ill.: Irwin, 1985.

Mason, R. D. *Statistical Techniques for Business Economics.* Homewood, Ill.: Irwin, 1985.

3.11 PROBLEMS

1. Identify several TV commercials that utilize histograms and pie charts.
2. Give business examples for the following terms:
   - mode
   - median
   - mean
   - range
   - variance
   - standard deviation
3. What is the difference between the population standard deviation and the sample standard deviation?
4. What is the difference between raw and group data? Provide specific business examples for both types of data.
5. For each of the following three distributions, describe the degree of kurtosis and skewness.

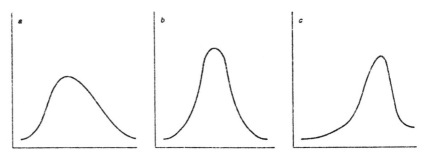

6. The statistics department at Pepperdine University recently conducted an experiment regarding the best method for teaching statistics. The experiment consisted of two sections of freshman statistics. One section (control group labeled C) received instruction according to the traditional method, which emphasizes hand calculations. The second section (experimental group labeled E) utilized computer courseware as a supplement to classroom lectures. The following graphic shows the result of the final examination for the two groups.

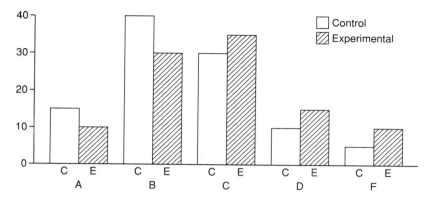

  a. Which method appears to achieve better results as measured by test performance?
  b. Which method yields the smallest variance in the test results?
  c. Which of the two distributions appears more symmetrical?
7. The following data were collected on length of employment for 40 selected employees of the Townhouse Corporation.

Tenure (Years)	Frequency Count
0–1	12
1–2	18
2–3	10
3–4	6
4+	4

  a. Develop a histogram and pie chart for this data.
  b. Describe the general shape of the histogram.
8. For the data given in problem 7:
  a. Determine the coefficient of skewness
  b. Determine the coefficient of kurtosis
  c. Describe the shape of the data using these parameters.
9. The following raw data is from a random sample of the number of outpatients treated per day at Getwell Hospital. Hospital management wishes to develop a histogram for the collected data.

```
45  50  36  59  28  42  67  33  35  40
32  18  23  41  31  21  43  47  39  51
45  35  29  44  53  61  49  53  34  46
```

10. For the data given in problem 9:
  a. Determine the coefficient of skewness.
  b. Determine the coefficient of kurtosis.
  c. Describe the shape of the data distribution using these parameters.

11. The incoming freshman class (55 students) at Pepperdine University recently took a statistics proficiency examination. A brief inspection of the results indicated that the scores range from 50 to 100. The proctor decided to describe these results by grouping them into five categories as indicated below:

Scores	Frequency
50–59	3
60–69	7
70–79	10
80–89	21
90–100	14

  a. Graph these data using a histogram.
  b. Graph a cumulative frequency distribution.
  c. What portion of the students scored 70 or above?

12. For the data given in problem 11:
  a. Determine the coefficient of skewness
  b. Determine the coefficient of kurtosis
  c. Describe the shape of the data distribution using these parameters.

13. The following information was generated from gas meter readings over the last month for a sample of 15 households:

Energy Use (therms)	Frequency Count
0–30	2
30–60	3
60–90	5
90–120	4
120–150	1

  a. Graph these data using a histogram.
  b. Graph a cumulative frequency distribution.
  c. What portion of the households consumed less than 90 therms?

14. The following annual salary information was obtained from a survey of 18 medical doctors through Los Angeles County:

$75,500	$57,500	$108,500
65,000	76,000	72,000
92,000	84,000	89,500
98,000	87,500	73,500
88,000	94,500	112,500
82,500	89,000	97,500

a. Develop a count and relative frequency distribution for this data.
  b. What proportion of doctors make more than $80,000?
15. The following data present a business traveler profile by characteristic and gender:

Characteristic	Men	Women
Average number of trips per year	5.1	3.5
First visit to place in past year	48	63
Other household member on trip	25	34
Also on vacation	21	28
Use own car	46	50
Use hotel	73	72
Spent $85+ a night for hotel	19	25

Discuss these statistics as they relate to differences observed by sex. For example, why would more women spend at least $85 per night for a hotel room?

16. Presented below are the results of a survey of 659 hotels showing the percentage of hotels providing the listed guestroom amenities:

Amenity	Percentage
Shampoo	86
Radio/alarm clock	76
Hair conditioner	57
Newspaper	55
Sewing kit	44
Shower massage	44
Hair drier	28
Bathrobe	23
Coffee maker	17

  a. Develop a bar chart showing these survey results.
  b. Do you think that these survey results reflect the hotel industry's increasing interest in attracting businesswomen?
17. Calculate a set of raw and group descriptive statistics for the data given in problem 9. Discuss any differences observed in the results.
18. Calculate a set of raw and group descriptive statistics for the data given in problem 14. Discuss any differences observed in the results.
19. In a recent telephone survey of 1,000 households, each participant was asked to state the five most important issues now facing the country. Presented are the

results of the survey that show the top five issues in order of first place ranking. For example, 32% of all those interviewed list crime/drugs as the most critical issue.

Issue	Percentage
Crime/drugs	32
Federal budget deficit	22
Affordable housing	14
Education	10
Nuclear war	8

   a. Why do the percentages sum to less than 100%
   b. Prepare a bar chart showing the results of the survey.
20. The following data were collected in a recent telephone survey on evaluating the President of the United States. The actual question asked was "Do you approve of the President's overall performance?"

Interviewer	Response	Interviewer	Response
1	Yes	7	Yes
2	Yes	8	No
3	No	9	No
4	Yes	10	Yes
5	Yes	11	No
6	No	12	Yes

   a. What type of descriptive statistics would best describe the results of the question?
   b. Calculate a mean and standard deviation for these data.
21. Presented below are data on the top 12 imported beer brands for 1987.

Brand	Sales (31-gallon barrels)	Market Share (%)
Heineken	2,254,880	24.1
Corona	1,705,645	18.2
Molson	998,900	10.6
Beck's	762,200	8.1
Moosehead	464,520	5.0
Labatt's	435,484	4.7
Amstel	281,322	3.0
Dos Equis	225,000	2.4
Foster's	225,000	2.4
St. Pauli Girl	222,967	2.4
Tecate	203,226	2.2
Guiness Stout	142,403	1.5

a. Prepare a pie chart showing the market share of the top three importers to the total.
b. Prepare a histogram showing sales for the top six importers.
c. Calculate the mean, median, and mode for both sales and market share.

22. Presented below are the total salaries paid out by each Major League baseball team in 1988 to players with guaranteed contracts who were terminated before the end of the season.

Team	Total Salaries ($000)	Team	Total Salaries ($000)
Atlanta Braves	$2,670	Los Angeles Dodgers	$950
Toronto Blue Jays	2,400	Milwaukee Brewers	700
New York Yankees	2,395	Montreal Expos	600
Cleveland Indians	2,350	Chicago Cubs	550
Kansas City Royals	2,500	Detroit Tigers	278
Chicago White Sox	1,730	Philadelphia Phillies	250
Cincinnati Reds	1,600	San Francisco Giants	225
Los Angeles Angels	1,400	Houston Astros	150
Baltimore Orioles	1,375	St. Louis Cardinals	28
San Diego Padres	1,100	Boston Red Sox	0
Texas Rangers	1,078	New York Mets	0
Minnesota Twins	1,005	Oakland Athletics	0
Pittsburgh Pirates	975	Seattle Mariners	0

a. Develop a set of descriptive statistics for this problem.
b. Which measure of central tendency best describes these data?
c. Construct a histogram with class widths of 400.

23. The following summarizes data for the top 10 offensive college football teams for the first half of the 1988 season.

Team	Total Plays	Total Yards	Total Yards/Game
Wyoming	698	4,856	540
Utah	652	4,160	520
Washington State	596	4,081	510
Nebraska	693	4,489	499
West Virginia	589	3,963	495
Oklahoma State	506	3,443	492
Air Force	547	3,868	484
Houston	528	3,360	480
UCLA	607	3,822	478
BYU	628	3,739	467

a. Develop a set of descriptive statistics for each variable.
b. Which measure of central tendency best describes each variable?
c. Rank each team by each variable.
d. Discuss the differences in ranking observed in part c.
e. How could the standard deviation be used in describing these data?

24. The following summarizes annual salary data for 20 NBA basketball players taken at random.

Player	Salary	Player	Salary
1	$1,200,000	11	$375,000
2	875,000	12	250,000
3	1,100,000	13	300,000
4	650,000	14	475,000
5	475,000	15	750,000
6	550,000	16	300,000
7	300,000	17	925,000
8	450,000	18	250,000
9	250,000	19	625,000
10	300,000	20	500,000

a. Which measure of central tendency best describes these data?
b. Which measure of central tendency would you use if you were representing the players' association in conducting negotiations with the NBA?
c. How would you describe the variance associated with this data base?

25. The following data present the annual yield on certificate of deposit (CD) and the fixed mortgage rate (FMR) for a sample of 12 California S&Ls as of November 11, 1988.

S&Ls	CD	FMR
Cal Federal	8.27	11.00
Coast Savings	8.11	10.90
Columbia Savings	8.00	10.76
Far West Savings	8.17	10.53
Glendale Federal	8.11	10.77
Great Western	8.09	10.87
Home Savings	8.15	10.87
Imperial Savings	8.19	10.63
Republic Federal	8.58	10.52
Valley Federal	8.14	10.71
Western Federal	8.46	10.53
World Savings	8.12	10.63

a. Rank the S&Ls by CD yield and FMR.
b. Discuss any differences observed in the ranking developed in part *a*.
c. Develop histograms for CD yield and FMR.
d. Develop a set of group descriptive statistics and discuss these results.

26. The following data represent 1986 annual net income ($000) for the world 20 largest firms (ranked in order of sales).

Firm	Income	Firm	Income
General Motors	$2,944,700	IRI	$ 197,118
Exxon	5,360,000	Toyota	1,717,733
Royal Dutch	3,725,779	Benz	831,600
Ford Motor	3,285,100	Du Pont	1,538,000
IBM	4,789,000	Matsushita	946,571
Mobile	1,407,000	Unilever	973,983
BPI	731,954	Chevron	715,000
General Electric	2,492,000	Volkswagen	286,133
AT&T	139,000	Hitachi	679,609
Texaco	725,000	ENI	342,275

a. Develop a frequency and relative frequency distribution to summarize these data.
b. Develop a histogram of this data set.
c. Prepare a set of descriptive statistics for this problem.

27. A survey of 200 shoppers yielded the following histogram for the average weekly prices paid for groceries. Determine the mean and standard deviation of the average grocery prices in this sample.

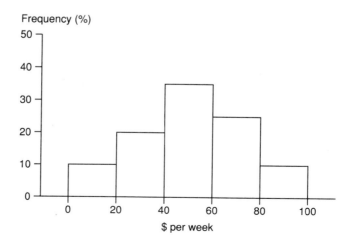

28. A survey of 150 service stations yielded the following histogram for the price of unleaded gasoline. Determine the mean and standard deviation for the price of unleaded fuel.

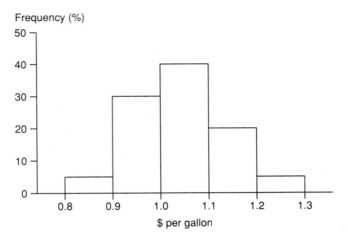

29. A survey of 20 taxable money market funds resulted in the following histograms for the 7-day yield. Determine the mean and standard deviation for the 7-day yield.

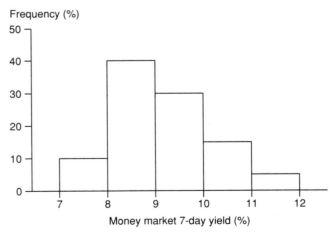

30. Develop a set of raw and group descriptive statistics for the silver and platinum prices data given in Section 3.2.
31. The management at Big Jack's Hamburger Stand is interested in determining the average weekly demand per customer for hamburgers and soyburgers (a new health food product). A customer survey was conducted over a two-week period at Big Jack's three retail outlets. The operations vice president, Ned Williams, has assembled the following survey results.

Hamburgers		Soyburgers	
Number	Frequency	Number	Frequency
0–1	2	0–1	1
2–3	4	2–3	3
4–5	3	4–5	4
6–8	1	6–8	3

Williams wishes to know which of the two products has the larger average demand and which has the smaller standard deviation.

32. Reported below are the pretax quarterly earnings for the Gilroy Corporation for the period 1984–1987.

Year	Earnings per Quarter ($000)			
	First	Second	Third	Fourth
1984	1,275	975	1,150	1,450
1985	1,325	1,025	1,275	1,425
1986	1,125	950	1,075	1,325
1987	1,125	925	1,100	1,300

  a. Calculate average earnings per quarter over the four-year period.
  b. Calculate average earnings per year.
  c. Demonstrate that the mean of average quarterly earnings is equal to the mean of average annual earnings.

33. The management at Snappy Lube, an automobile service firm specializing in fast oil changes, is interested in expanding service. Currently, the company has five service outlets and is considering building a new service facility near a major shopping center. Prior to undertaking this project, however, the management wishes to assess the demand for lube service in this area. The company has recently hired a consulting firm to undertake a survey of potential customers at the shopping center. The results of the survey are summarized below:

Frequency of Oil Change		Amount Willing to Pay	
Mileage (000)	Frequency	Charge	Frequency
1–4	21	$1–5	11
4–8	15	5–10	13
8–12	5	10–15	10
		15–20	8

Snappy Lube's management wishes to develop a set of descriptive statistics to help them decide on the new project. What would you recommend?

34. Presented below is the percentage of change in the real U.S. GNP from 1965 to 1984. Graph this data and discuss the general patterns.

Year	Growth (%)	Year	Growth (%)
1965	5.8	1975	−1.3
1966	5.8	1976	4.9
1967	2.9	1977	4.7
1968	4.2	1978	5.3
1969	2.4	1979	2.5
1970	−0.3	1980	−0.1
1971	2.8	1981	1.9
1972	5.0	1982	−2.5
1973	5.2	1983	3.5
1974	−0.5	1984	6.5

35. Presented below are wheat production and price statistics for the top 10 wheat-producing states for 1985.

State	Production (million bu)	Price ($/bu)
California	68.9	3.44
Colorado	139.3	2.91
Idaho	72.0	3.26
Kansas	433.2	3.05
Minnesota	142.4	3.23
Nebraska	89.7	3.00
North Dakota	323.3	3.32
Oklahoma	165.0	3.20
Texas	187.2	3.10
Washington	128.3	3.40

a. Rank the states in terms of production and price.
b. Does there appear to be a significant difference in the ranking?
c. Which of these two variables seems to have the larger variance?
d. Compute the standard deviation for both variables.
e. Compare the results obtained in parts c and d.

36. Presented below are R&D expenditures as a percentage of net sales and the number of R&D scientists/engineers employed by selected industry groups for 1984.

Industry Group	R&D Funds/ Net Sales (%)	R&D Scientists/ Engineers (000)
Aerospace	16.9	96.5
Electrical equipment	8.6	116.1
Chemicals	4.5	67.1
Motor vehicles	3.8	30.5
Scientific equipment	9.4	37.2
Primary metals	1.2	8.5
Fabricated metals	1.7	12.9
Paper products	1.2	7.9

   a. Rank the industry groups in terms of the two variables.
   b. Does there appear to be a significant difference in the ranking?
   c. Which of these two variables seems to have the larger variance?
   d. Compute the standard deviation for both variables.
   e. Compare the results obtained in parts c and d.

37. The marketing manager of Hallworth Corporation, a leading mail order firm, has prepared the following data on the current and projected sales volume by market territory:

Market Territory	1988 Sales (000)	1989 Growth (%)	1990 Growth (%)
North	$975	7.5	6.5
East	825	5.0	4.5
South	425	9.0	10.5
West	515	11.0	13.5
Overseas	375	14.5	22.5

   a. Calculate a weighted average for sales for 1989 and 1990.
   b. Rank the market territories using both sales and growth over the three-year period.

38. The law offices of Brown, Johnson, and Smith bill their clients according to the amount of participation of each partner. The current billing rates for Ms. Brown, Mr. Johnson, and Mr. Smith are $200, $150, and $100 per hour, respectively.

This year Ms. Brown billed 800 hours, and Mr. Johnson and Mr. Smith both billed 1,000 hours.

a. Determine the average billing rate for the firm.

b. The billable number of hours for next year is expected to increase by 10% for Ms. Brown and decrease by 5% for Messrs. Johnson and Smith. Estimate the average billing rate for next year.

39. A property management firm is interested in determining the average occupancy rate for five of their residential buildings. The estimated occupancy rates for the five properties are as follows: 78%, 92%, 100%, 84%, and 95%.

a. Estimate the mean occupancy rate using the arithmetic mean.

b. Estimate the mean occupancy rate using the geometric mean.

c. Compare the results obtained in parts a and b.

40. The management of the Los Angeles Raiders football team is currently negotiating contracts with its top two running backs. The following summarizes the rushing performance of each running back over his career with the Raiders.

Year	TCB	Total Yards	Average	TCB	Total Yards	Average
1983	195	895	4.59	NA	NA	NA
1984	315	1215	3.85	175	715	4.09
1985	335	1400	4.18	240	975	4.06
1986	275	950	3.45	270	1190	4.41
1987	225	925	4.11	305	1410	4.62

NA = Not available.

Which back should receive the higher salary based totally on

a. Average rushing yardage?

b. Consistency?

41. A race driver at the Indianapolis 500 times trials wishes to average 200 miles per hour for two laps around the 2.5-mile track. On his first lap he averaged 185 miles per hour. How fast must he go on the second lap to meet his goal?

42. The Cola Corporation markets a complete line of soft drinks. The following highlights product price and quantity (in cases) for 1982 and 1985:

Product	Prices per Case 1982	Prices per Case 1985	Quantity 1982	Quantity 1985
Cola	$3.25	$3.40	45,000	65,000
Sparkling water	1.75	2.25	10,000	18,000
O.J. drink	2.50	3.15	6,000	25,000

a. Compute the relative price for each product.
b. Compute a simple average index for 1985.
c. Compute a weighted average index.

43. The Consumer Price Index (CPI) consists of 400 items, including food, transportation, clothing, medical expenses, and housing. The data given below reflect several of the key items included in the CPI.

	Price per Gallon	
Item	1980	1986
Gasoline	$1.25	$0.90
Milk	1.10	1.70
Water utility	0.20	0.30

a. Compute the relative price for each product.
b. Compute a simple average index for 1986.
c. Compute a geometric-based index for 1986.

44. The management of the Automobile Club of Southern California is interested in determining the cost of operating an automobile in the Southern California area. The research department has collected the following operating cost data for 1980 (base year) and 1987.

		Price		Quantity	
Item	Units	1980	1987	1980	1987
Insurance	Annual policy	$250	$500	1	1
Gasoline	Gallons	1.20	0.95	1000	1200
Oil	Quarts	1.00	0.65	15	15
Maintenance	Annual repair	100	150	1	1
Tires	Tires	40	60	2	2

a. Compute the relative prices for each item separately.
b. Compute an unweighted aggregate price index for vehicle operating expense.
c. Compute a weighted aggregate price index for vehicle operating expense.
d. Compute an aggregate price index using the geometric mean.
e. Compare the results obtained in parts c and d.

45. The vice president of the Global Precious Metals Exchange is interested in computing a set of financial ratios for the firm. The following presents a balance sheet as of December 31, 1988.

Global Precious Metals Exchange
Balance Sheet

Assets		Liabilities	
Cash	$ 25,000	Accounts Payable	15,000
Accounts receivable	12,500	Notes Payable	25,000
Inventory	62,000	Accrued expense	5,000
Investments	4,500	Long-term debt	25,000
Fixed Assets	22,000	Stockholders' equity	55,000
Total assets	$125,000	Total Liabilities (including equity)	$125,000

The vice president plans to use the following formulas in preparing the required analysis:

$$\text{Current ratio} = \frac{\text{Current assets}}{\text{Current liabilities}}$$

$$\text{Quick ratio} = \frac{\text{Current assets} - \text{Inventories}}{\text{Current liabilities}}$$

   a. Determine the current ratio.
   b. Determine the quick ratio.
   c. Discuss why the measurements in parts a and b are examples of the ratio scale.
46. Use the balance sheet given in problem 45 to determine:
   a. Debt to total assets ratio.
   b. Debt to equity ratio.
47. The following data present the percentage share of total exported goods by the main manufacturing countries for 1980 and 1986.

Year	U.S.	Canada	France	Italy	Japan	U.K.	Germany	Other
1980	17.0	4.0	10.0	7.9	14.9	9.7	19.9	16.6
1986	14.0	5.4	8.8	8.2	19.4	7.6	10.7	25.9

   a. Develop pie charts showing the percentage share by country for the two years.
   b. Develop a pie chart showing the percentage change in share between 1980 and 1986.
48. The following data present automobile production and imports for selected countries for 1977 and 1986.
   a. Develop pie charts showing the percentage of world total by country for the two years.
   b. Develop a pie chart showing the percentage change in world total between 1977 and 1986.
   c. Develop pie charts showing the percentage of imports by country for the two years.

Country	Year	Actual (000)	Percentage of World Total	Imports (000)
United States	1977	9,214	29.8	2,791
	1986	7,829	23.4	5,400
France	1977	3,092	10.0	552
	1986	2,773	8.3	1,004
Japan	1977	5,431	17.6	41
	1986	7,810	23.4	73
United Kingdom	1977	1,328	4.3	698
	1986	1,019	3.1	1,072
West Germany	1977	3,791	12.3	949
	1986	4,311	12.9	1,295

49. The following data present trends in Gross Domestic Product (GDP) per employee for selected countries.

	Relative Level		Annual Growth
Country	1977	1986	1981–1985
United States	100	100	1.0
Canada	92	102	1.6
France	85	93	1.7
Japan	63	75	2.9
United Kingdom	56	77	2.6
West Germany	79	93	2.1

a. Develop bar charts showing the relative level of GDP per employee by country for the two years.
b. Develop a bar chart showing the annual growth rate by country.

50. The following data present the output and average percentage change in output per hour for selected countries.

	Relative Output per Hour		Average Annual Percent Change in Output, 1981–1984
Country	1973	1984	
United States	100	100	4.6
Canada	77	75	2.3
Japan	56	92	6.2
United Kingdom	48	42	5.5
West Germany	78	81	4.1

a. Develop bar charts showing the relative level of output per employee by country for the two years.
b. Develop a bar chart showing the average annual percentage change in output by country.

51. The QC manager at Intel Corporation is interested in determining the level of quality for the typesetting department. The manager has collected 10 samples with each sample containing 100 observations on the number of typesetting errors. The resultant sample data is reported below:

Sample Number	Number of Errors	Sample Number	Number of Errors
1	5	6	4
2	3	7	3
3	1	8	4
4	0	9	2
5	4	10	1

a. What type of control chart best describes this problem?
b. Develop the appropriate control chart.
c. Is the process in or out of control?

52. The administrator at South West Hospital is interested in determining the level of quality of the hospital's outpatient clinic. The administrator has collected eight sequential samples with each sample containing 100 observations of patient waiting times (in minutes). The resultant sample data is reported below:

Sample Number	Average Waiting Time (min)	Sample Number	Average Waiting Time (min)
1	12	5	21
2	18	6	12
3	7	7	8
4	9	8	13

a. What type of control chart best describes this problem?
b. Develop the appropriate control chart.
c. Is the process in or out of control?

53. A bank examiner for the Treasury Department is interested in determining the level of commercial loan defaults for a bank chain operating in Texas. The examiner has collected 10 sequential samples with each sample containing 1,000

observations on the number of commercial load defaults. The resultant sample data is reported below:

Sample Number	Number of Defaults	Sample Number	Number of Defaults
1	120	6	202
2	115	7	167
3	174	8	189
4	157	9	222
5	145	10	197

a. What type of control chart best describes this problem?
b. Develop the appropriate control chart.
c. Does the proportion of defaults appear stable or unstable?

3.12 CASES

3.12.1 Ross International

"For some reason people are always trying to second guess the stock market," stated Henry Clay, the managing partner of Ross International. "It must be basic to man's nature to try to steal money from the stock brokers. We have spent many years working out the perfect scheme for charging our clients . . . where we make money by the transaction no matter what the market does. Rises in the market only serve to get investors interested in providing us with new cash," he added.

"Let us circulate rumors about interest rates and their effects," said the company's market researcher. "Then when the rates rise or fall, we will know about what's happening in the market."

"Not so," said the firm's psychologist, "People respond to the *threat* of rising or falling interest rates and not to whether they actually rise or fall. My science realizes that the important variables in life are not different behaviors but attitudes toward them on a five-point scale."

The researcher said, "She means we should test the theory out and see what happens. That is, we need to determine the relationship between Federal Funds rates and the N.Y. Stock Exchange Composite. We can't do anything about the attitudes because we don't have existing data."

"Exactly what I said," said Henry. "We'll claim we were right anyway. No one can ever prove us wrong," laughed the psychologist.

The researcher was assigned the task of collecting data on the NYSE Composite and Federal Funds Rate for the period January 1984 to June 1985. These data are reported in Table 1 at the top of the next page.

TABLE 1 NYSE Composite and Funds Rate for 1984–1985

Date	NYSE Composite	Funds Rate
1984:		
January	96.16	9.56
February	90.60	9.59
March	90.66	9.91
April	90.67	10.29
May	90.07	10.32
June	88.28	11.06
July	87.08	11.23
August	94.49	11.64
September	95.68	11.30
October	95.90	9.99
November	95.85	9.43
December	94.77	8.51
1985:		
January	99.11	8.35
February	104.73	8.50
March	103.92	8.58
April	104.66	8.27
May	107.00	7.97
June	109.52	7.53
July	111.64	7.88

Suggested Questions

1. Find the mean, median, and mode for the data base.
2. Find the standard deviations for the two variables.
3. Develop a graphical representation for the two variables.
4. Which variable has the largest variance, and why?
5. Describe the degree of skewness and kurtosis for the two distributions.

3.12.2 Financial Analysis in the Construction Industry

Robert Morris Associates publishes an annual report that offers financial data from various segments of business. The *1985 Annual Statement Studies* is divided into five major categories: manufacturing, wholesaling, retailing services, contractors, and finance. Refer to Table 1 (on the next page) for a synopsis of those figures relating to overhead and revenue percentages in the commercial construction category. Note that as the size of the company increases, the overhead percentages decrease. It is important to note that the Morris group defines *revenue* as the "percentage of completion method of accounting," *gross profit* as net sales minus cost of sales, and *operating expenses* as all selling and general administrative expenses, including depreciation, but not interest expense.

TABLE 1 Contract Size and Income Data for Construction Industry

	Contract Size ($ millions)			
	0–1	1–10	10–50	50+
Number of firms	133	619	222	59
	Income Data (%)			
Contract revenues	100.0	100.0	100.0	100.0
Gross profit	27.2	14.7	9.4	7.3
Operating expenses	22.4	12.6	8.5	6.5
Operating profit	4.8	2.1	0.9	0.8
Other expenses	0.8	−0.3	−0.4	−0.6
Profit before taxes	4.1	2.4	1.3	1.3

The Associated General Contractors of America (AGCA) each year publishes a survey of over 10,000 participating general contractors throughout the country. Although this researcher could not get results of the latest study (due to its confidentiality), a 1977 AGC Operations Survey indicated that as revenues increased, overhead costs decreased. In the category of closely held corporations, firms doing $5 to $8 million in sales averaged overhead costs of 8% of revenue; firms doing $8 to $12 million, 6% of revenue; $12 to $20

TABLE 2 Overhead, Revenue, and Profit Data (1983–1985)

		Overhead as % of Revenue	Gross Profit as % of Revenue	Sales Revenue (millions)
1983	1st quarter	3.37	5.53	
	2nd quarter	2.81	3.75	
	3rd quarter	3.56	8.16	
	4th quarter	8.41	18.45	
	Yearly averages	4.20	8.12	$62.7
1984	1st quarter	5.74	7.02	
	2nd quarter	5.58	8.40	
	3rd quarter	3.10	9.33	
	4th quarter	6.36	17.03	
	Yearly averages	5.29	11.39	62.5
1985	1st quarter	6.43	8.64	
	2nd quarter	3.36	5.65	
	3rd quarter	3.21	4.61	
	4th quarter	3.63	5.45	
	Yearly averages	3.63	5.45	96.6

million, 5% of revenue; and over $20 million, 4% of revenue. Here again is strong evidence that as revenue increases, overhead costs decrease. Overhead, revenue, and profits data for the period 1983 to 1985 are presented in Table 2.

Suggested Questions

1. Develop a histogram that shows the relative size of the construction industry by level of business.
2. Develop a set of graphics to describe the trends in this industry.
3. Calculate the standard deviations for overhead and gross profit.
4. Develop a set of graphics that depict the seasonal variations in overhead and gross profits.

Chapter 4

Probability Theory

I think it would be useful to know that chance has rules which can be known, and that through not knowing these rules faults are made every day.

Montmort

CHAPTER OUTLINE

4.1 Introduction
4.2 Example Management Problem: The California Lottery
4.3 Basic Rules of Probability
4.4 Fundamental Concepts
4.5 Computer Analysis
4.6 Practical Applications
4.7 Case Study: Automobile Fuel Economy Standards
4.8 Summary
4.9 Glossary
4.10 Bibliography
4.11 Problems
4.12 Cases

CHAPTER OBJECTIVES

The primary objectives of this chapter are to develop an understanding of

1. The role of probability in business.
2. How to define probability.
3. The difference between objective and subjective probability.
4. How counting rules can be used to assign probability.
5. How to compute probability values for multiple events.
6. How to revise probability values using Bayes' theorem.

The use of probability analysis in judicial proceedings has a long history.* Probability analysis is important because most judgments are based on testimonial evidence. Consciously or subconsciously, probability is used in judging the credibility of witnesses, linking the testimony of several wit-

---

* S. L. Zabell, "The Probabilistic Analysis of Testimony," *Journal of Statistical Planning and Inference* 20, no. 3, November 1988.

> **HISTORICAL NOTE**
>
> The origins of probability theory can be traced, in large measure, to the work of the French mathematician Abraham de Moivre (1667–1754). He was responsible for the development of the general laws of probability multiplication and addition, the binomial distribution, and the normal curve. He also contributed significantly to the mathematics of life insurance. Perhaps one of his greatest contributions, however, was to free the science of probability from its exclusive domination by gambling and games of chance.

nesses, determining the chances of several events occurring simultaneously, and implementing the instructions of the judge. Typically, a judge instructs the jury that they must be convinced beyond any "reasonable doubt" before finding the defendant guilty in a criminal case. Clearly, this statement implies the assignment of probability. It is interesting to note, however, that the judge never indicates that the jury must be convinced beyond a certain percent (e.g., 99%) of the defendant's guilt to render a guilty verdict.

A common tactic used by both the prosecution and the defense is to indicate the unlikelihood or likelihood that another individual could have duplicated the same set of circumstances. For example, a defendant is charged with bank robbery. Eyewitnesses to the crime reported that the robber had red hair and a moustache, was at least six feet tall, and escaped in a red car. These characteristics exactly match those of the defendant. Suppose, for the sake of simplicity, that each of these characteristics has a 10% chance of occurrence in the general population. By multiplying these probability estimates together, the prosecutor could argue that there is only a .0001 chance that someone else exactly matching this description could have committed the crime. Although this line of reasoning has resulted in a number of both convictions and acquittals, it does suffer from several major potential flaws. First, there is the possibility that the probability estimates of individual characteristics are inaccurate, and second, each characteristic may not in fact be independent of every other characteristic.

4.1 INTRODUCTION

Key idea 1

Most managerial decisions are made in the face of some uncertainty regarding the possible consequences of these decisions. Examples include the introduction of a new consumer product, the expansion of an existing plant, and the substitution of the starting quarterback with the backup quarter-

> **INTERNATIONAL VIGNETTE**
>
> Sunwind AB, an operating division of Perstop Components AB, manufactures a line of interior trim products for Volvos and Saabs. However, sales to Volvo account for approximately 90% of total revenues. Sunwind currently maintains two factories—an older 20,000 square foot plant only 5 miles from Volvo's Torslanda facility and a second more modern 100,000 square foot plant located about 100 miles from the Volvo facility. Lars Larsen, the managing director of Sunwind, is under increasing pressure from corporate management to close the smaller plant and consolidate operations at the larger facility. Additionally, Larson is also under pressure from Volvo to improve product quality and delivery schedules.
>
> After analyzing the situation, Larson became convinced that Sunwind should initiate a just-in-time (JIT) delivery system for providing components from the local plant directly to the Torslanda facility. This approach would guarantee the continued operation of Sunwind's local plant and would increase overall profitability. Furthermore, the past problems involving quality and delivery could be solved. However, before embarking on this potentially costly program Larson needed to collect some data on the performance of similar JIT efforts and to develop more formal probability estimates on product demand from Volvo.

back. In each case, the decision maker is unsure of the outcome. These outcomes are usually referred to as **events.** Determining the chances of the various possible events occurring in advance should help in making the correct decision. In fact, most decision makers either consciously or subconsciously take into account the likelihood of various outcomes as part of the decision-making process.

The process of quantifying uncertainty or risk is accomplished through the assignment of probability. **Probability** is a measure of the chance that a certain event will occur and is measured on a scale from 0 to 1. A probability value of 0 indicates that the event cannot happen, and a probability value of 1 reveals that the event is sure to happen. Although probability theory is rooted in games of chance (i.e., gambling), it has emerged as a major tool in modern business practice. Specifically, probability estimates of future events along with other pertinent facts provide the decision maker with an analytical framework for evaluating the basic decision alternatives. For example, IBM's vice president of research and development estimates that there is a 60% chance that a new product will increase company profits. This probability estimate can be used for determining whether or not the new product should be introduced.

4.2 EXAMPLE MANAGEMENT PROBLEM: THE CALIFORNIA LOTTERY

The California lottery was initiated a few years ago as a means of raising funds for California's public school system. The lottery actually consists of a number of different games, including the "Big Spin." In this game, a contestant has an opportunity to spin a roulette wheel with 100 slots. The contestant wins the amount indicated by the slot containing the ball after the wheel has stopped spinning. Figure 4.1 shows the various monetary outcomes of the wheel, along with the number of slots containing those outcomes.

The lottery director is interested in determining the chances of various outcomes. More specifically, the director would like to know the chances of having to pay $1 million to at least one of the first three contestants and the chances of having to pay $1 million to each of the first three contestants.

4.3 BASIC RULES OF PROBABILITY

The basic rules governing probability theory are highlighted below:

- If S is the set of all possible outcomes for a given process, then $P(S) = 1$ (i.e., the probability of all outcomes is equal to 1).

FIGURE 4.1 Various Outcomes of the Big Spin Wheel

Payoff Amount	Number of Slots
$ 10,000	10
20,000	16
30,000	16
40,000	16
50,000	16
100,000	10
1,000,000	8
Double	8
	100

- If E is an event, then $0 \leq P(E) \leq 1$ (i.e., the probability of any event falls between 0 and 1).
- For an event E, $P(E) = 1 - P(\overline{E})$ where $\overline{E}$ is the complement of E (i.e., the probability of any event equals 1 minus its complement).

Additional probability rules involving two or more events are presented in the next section.

4.4 FUNDAMENTAL CONCEPTS

Generally speaking, probability is the chance that something will happen. Chance is a measure of the uncertainty associated with a given situation. Probability theory is the branch of mathematics that is used for quantifying uncertainty. Because most business situations involve uncertainty, it is essential to develop a basic understanding of the fundamental concepts associated with probability theory.

RANDOM EVENTS

A process that results in one of a number of well-defined outcomes that cannot be predicted in advance is called a *random experiment*. Examples include spinning the Big Wheel in the California lottery, tomorrow's weather, and the stock market's weekly closing average. The listing of all possible outcomes of a random experiment is referred to as the **sample space**. Any subset of a sample space is called a *random event*. Figure 4.2 presents several examples of experiments, sample spaces, and random events. For example, one can consider the activity on the New York Stock Exchange

FIGURE 4.2 Examples of Experiments, Sample Spaces, and Random Events

Experiment	Sample Space	Random Event
California Big Spin	All the slots on the wheel	The million-dollar slots
New York Stock Exchange	Fortune 500 stocks	Fortune 500 stocks with a P/E ratio above 10
Quality control check	Inspection lot size	Number of defective items
Presidential election	Voting electorate	Voting for the Republican candidate

(NYSE) as an experiment, with stock prices moving up and down in a somewhat unpredictable manner. One possible sample space of this experiment comprises the Fortune 500 stocks (the NYSE contains approximately 1,500 active stocks). One random event from this sample space would be those Fortune 500 stocks that have appreciated in price over the last six months. Obviously, there are many other possible combinations of sample spaces and random events that can be defined using the NYSE.

A special case, called a **collectively exhaustive event,** is represented by the collection of different random events that contain all of the basic outcomes of the sample space. In the case of the NYSE, a collectively exhaustive event would be the listing of all stocks on the exchange.

A single subset of a random experiment is called a *simple event*. For example, the million-dollar slots on the California Big Spin wheel represent a simple event. *Compound events* consist of two or more simple events. Obtaining a "double" on the California Big Spin followed by winning $100,000 illustrates a compound event. In this case, the first event is a "double," and the second event is $100,000. (In this situation, the contestant would actually win $200,000.) Typically, decision makers are interested in assessing the probabilities of both simple and compound events.

Compound events can be classified in terms of the level of interaction with each other. Events that have nothing in common are called **mutually exclusive.** For example, a quality control inspector will either accept or reject a specific production batch. In this case, "rejection" is the complement of "acceptance," and these two events define the entire sample space. This example illustrates the simplest form of mutually exclusive events. By way of contrast, sales at Ford Motor Company can be either higher or lower than sales for the previous year. The fact that sales are higher this year precludes the possibility that sales could also be lower at the same time. In this case, however, the events "higher" and "lower" do not define the entire sample space, since it is also possible for sales to remain unchanged. This situation represents a more complex case involving mutually exclusive events.

Two events that have something in common are often described in terms of the basic outcomes contained in either one event or the other or in both. In these cases, the interest is in defining the *union* of the two events. For example, an official in the U.S. Department of Commerce may be interested in knowing if foreign imports are going to be lower, or if the value of the dollar is going to be higher, or if both conditions will occur. Similarly, the same official may be interested in knowing if foreign imports are going to be lower *and* the value of the dollar is going to be higher. Here, the interest is in defining the *intersection* of the two events (i.e., the basic outcomes that are common to both events). Notice that both the union and the intersection are not confined to two events, but can be applied to any number of compatible events.

Each of these relationships can be depicted graphically using a *Venn diagram*, named after John Venn, a 19th-century English mathematician. These diagrams show the entire sample space in the form of a rectangle. Specific events are depicted as circles residing within the rectangle. If two events are mutually exclusive, then the circles representing these events will not overlap. On the other hand, if two events are compatible, the circles representing these events will overlap. Figure 4.3 presents a collection of Venn diagrams illustrating the various event relationships mentioned above.

FIGURE 4.3 Examples of Basic Event Interactions

Relationship	Venn diagram	Examples
1. Mutually exclusive events (E and F)	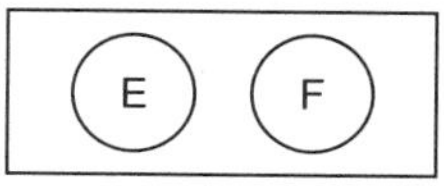 (E) (F) non-overlapping	E: Higher sales F: Lower sales
2. Complementary events (E vs. $\bar{E}$)	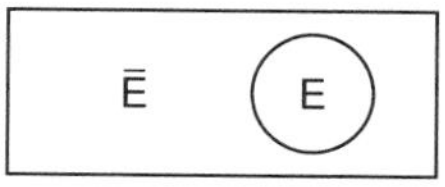 ($\bar{E}$) (E)	E: Accept batch $\bar{E}$: Reject batch
3. Union of two events (E or F)	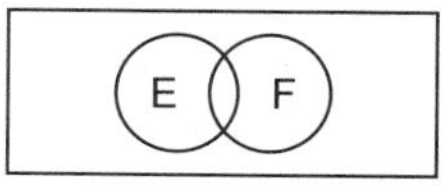 (E)(F) slight overlap	E: Reduced imports F: Higher dollar
4. Intersection of two events (E and F)	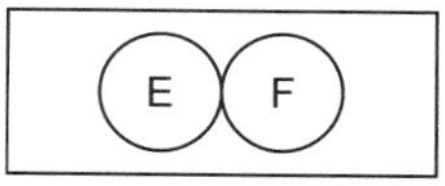 (E)(F) overlap	E: Reduced imports F: Higher dollar

DEFINING PROBABILITY

Key idea 2

Probability is defined on a measurement scale from 0 to 1. A probability value of 0 for a given event indicates that there is no chance of that event occurring. On the other hand, a probability value of 1 indicates that the event will occur with 100% certainty. For example, the probability that both a head and a tail will occur with a single flip of a coin is 0, and the probability that either a head or a tail will occur is 1. Basically, there are two ways of classifying probability: objective and subjective. **Objective probability** is based on a random experiment that can be repeated. The spinning of the lottery wheel is a random experiment that produces objective probability assignments. **Subjective probability,** on the other hand, is based on a random experiment that cannot be repeated because of changing conditions. Predicting the chances of rain tomorrow based on previous weather data is an example of a random experiment yielding subjective probability values. Each of these types of probabilities will be discussed since they both play an important role in the decision-making process.

Key idea 3

Objective Probability

Objective probability can be defined in one of two ways: *classical* or *relative frequency*. In the classical approach, the probability of an event E occurring is defined as follows:

$$P(E) = \frac{\text{Number of outcomes of the event}}{\text{Total number of possible outcomes}}$$

Notice that to use this definition, each outcome must have an equal likelihood of occurring. For example, a contestant may wish to know the probability of winning exactly $1 million (the event of interest) on a single spin in the California Big Spin. An inspection of the wheel indicates that there are eight slots containing $1 million (see Figure 4.1), and each of these slots has an equal likelihood of being selected. This becomes the numerator in the preceding model. The denominator is 100, which represents the total number of equally likely possible outcomes (i.e., the sample space). Mathematically, the chances of winning $1 million on a single spin are

$$P(\$1 \text{ million}) = \frac{8}{100} = .08$$

Observe that in using the classical approach, the probability of winning $1 million was obtained without ever spinning the wheel. Obviously, this is a major advantage. Unfortunately, the classical approach has a number of drawbacks that tend to limit its use in business applications. First, the chances of many business outcomes cannot be determined in advance. For example, an insurance company cannot determine in advance the chances of someone living to age 75 without obtaining actual mortality data on the population of interest. Second, the classical approach assumes equally likely outcomes. That is, if there are two possible outcomes (e.g., flipping a coin), then the probability of one or the other outcome occurring must be 50%. This condition is obviously not true in many situations. Consider, for example, the population of senior citizens. The classical approach would assume that there is an equally likely chance of selecting males and females from this population. This is clearly not true, since the proportion of females to males in the United States in the senior population is 2 to 1.

The alternative approach is to obtain data on the relative frequencies of past occurrences as a basis of assigning probabilities. Perhaps the simplest way to understand the basic differences between these two methods for assigning probability values is to consider an experiment involving the spinning of the California Big Wheel. Figure 4.4 compares the relative frequency of $1 million occurring, based on an increasing number of spins of the wheel, with the probability of obtaining $1 million based on the classical approach. These results show that as the number of spins increases, the relative frequency of occurrence of $1 million converges to the value obtained from the classical approach (i.e., .08). Clearly, the accuracy of the probability estimate increases as the number of observations increases. It should not be surprising that the frequency of occurrence of $1 million after only 25 spins is 20%. This is exactly the phenomenon that generates a few winners at the gambling tables in Las Vegas, even though the odds of winning, in the long run, are less than 50–50. It is interesting to note that another experiment involving spinning the Big Wheel would produce somewhat different results from the data reported in Figure 4.4. However, as the number of spins increases, the frequency of occurrence of obtaining $1 million again converges to .08. This

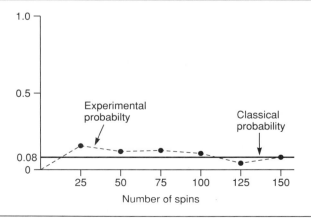

FIGURE 4.4 Relative Frequency of $1 Million Occurring

process of convergence is sometimes known as the *law of large numbers*. That is, the difference between the probability estimates determined empirically and those determined theoretically becomes smaller and smaller as the number of replications of the experiment increases.

Consider that the general manager (GM) at Allied Moving Van Lines is interested in determining the chances that a typical truck will average more than 10 miles per gallon. The GM had conducted an experiment where the fuel economy of the company's 100 trucks was measured over a 30-day period. The results of the experiment, presented in the form of a relative frequency distribution, are shown in Figure 4.5. The GM decided to classify the measurement data into three groups: less than 10 mpg, from 10 to less than 20 mpg, and from 20 to less than 30 mpg. The data indicate that there is a 70% chance that a truck selected at random will obtain at least 10 mpg. This probability estimate was derived simply by summing the chances for the second and third groups (i.e., 40% + 30% = 70%).

Subjective Probability

As indicated above, **subjective probabilities** are based on intuition or on previous facts, and not on the outcome of a repeatable random experiment. Nevertheless, subjective probabilities play an important role in business decision making. In fact, it may be true that subjective probabilities are used more frequently in assessing a situation than are objective probabilities. This is due to the fact that most business situations are so complex or individualized that they cannot be readily described or measured using a repeatable random experiment (e.g., market dynamics). Alternatively, the decision

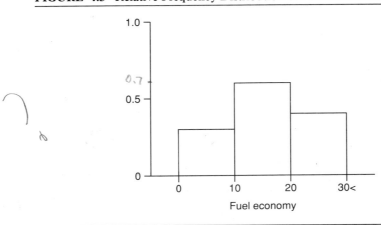

FIGURE 4.5 Relative Frequency Distribution of Fuel Measurement Experiment

maker may be unaware of the organization's capability to obtain objective probability estimates. For example, say the vice president for research and development at IBM is asked to estimate the chances of a new product "making it" in the marketplace. The vice president would no doubt draw on the company's experience in introducing similar products. Nevertheless, the probability estimate is clearly subjective since the conditions surrounding this experiment can never be fully duplicated.

The dividing line between objective and subjective probabilities, however, is somewhat unclear. Consider, for example, using historical weather data as a basis for forecasting tomorrow's weather. Under most circumstances, the historical data should provide insight into what may occur tomorrow. Yet the conditions that generate a particular weather pattern will, more than likely, not occur again. Even the probability estimates presented in Figure 4.5 are somewhat suspect in this regard; it is unlikely that the set of conditions under which the experiment was conducted could be exactly duplicated. However, for all practical purposes, the results can be classified as objective probabilities. The issue of objective versus subjective probabilities is discussed further in Chapter 15.

COUNTING RULES

Key idea 4

The random experiments presented so far have consisted of a relatively small number of outcomes. When the number of outcomes is small, the process of counting, as illustrated above, is efficient. However, in many experiments the number of outcomes is truly enormous, or the shape of the experiment is not well understood. In these cases, several alternative methods must be considered. The two basic approaches are:

1. Permutations and combinations
2. Sampling

Permutations and combinations are well-known counting techniques for experiments involving a large but well-defined set of outcomes. Sampling, on the other hand, is used when the outcomes of the experiment are not well defined (e.g., voting). The process of sampling and its implications for statistical inference are presented in Chapter 7.

Permutations and combinations indicate the different ways in which items can be arranged into specific subsets. The basic difference between permutations and combinations is that order is important in the former but not in the latter. For example, in designing a license plate, order is important. That is, the sequence ABC123 is quite different from the sequence CBA321. On the other hand, in the game of poker, order is not important. That is, the poker hand KKKAA is exactly the same as the poker hand AAKKK. The use of permutations and combinations, where applicable, significantly reduces the effort required in determining the total number of basic possible outcomes and the number of outcomes of interest for a given experiment. This, in turn, makes the computation of probability using the classical approach relatively straightforward.

Permutations

Permutations can be defined as the number of different ways a set (or subset) of objects can be arranged in a definite order. Figure 4.6 presents several different permutation models. The first model is used to determine the number of permutations when all n objects are used. For example, a pilot must make five different engine checks prior to takeoff. In how many different ways can the checks be carried out? The answer is 120 different sequences. The commonsense approach is to observe that any one of the five checks can be selected first, four ways are possible for the second check, three for the third check, two for the fourth check, and one way for the fifth and final check (i.e., $5 \times 4 \times 3 \times 2 \times 1 = 120$). The second model is applied when the interest is to determine how many different sequences can be generated with n objects taken r at a time. For example, a production manager wishes to know how many different ways six machines can be assigned to four available production teams. In this case, the second model given in Figure 4.6 can be used to determine the answer. That is, $6!/4! \times 2!$, or 15 different ways. Notice that the first model is a special case of the second model where $r = n$. The third model is applied when there are identical objects in the sample space. For example, a ship's signal officer has eight flags, of which three are blue, two are red, two are green, and one is black. How many different eight-flag sequences can the signal officer generate? The

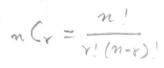

FIGURE 4.6 Alternative Methods for Determining the Number of Permutations

Model	Description
$P_n^n = n!$	Determines the number of sequences of n objects
$P_r^n = \dfrac{n!}{(n-r)!}$	Determines the number of sequences of n objects taken r at a time
$P_{x1,x2,\ldots,xk}^n = \dfrac{n!}{x_1! x_2! \cdots x_k!}$	Determines the number of sequences of n objects given k object types
$P_r^n = n^r$	Determines the number of sequences of n objects with replacement

answer is 1,680 (8!/3!2!2!1!). Finally, the fourth model is used to determine the number of sequences with replacement (unlike the previous three models). For example, in the game *Mastermind*, the objective is to guess the correct color sequence of four pegs. If there are six possible colors, how many different sequences are possible? The answer is 1,296 (i.e., $6 \times 6 \times 6 \times 6$).

Combinations

Combinations can be defined as the number of different ways a set of objects, or some subset, can be arranged independent of order. The basic model for determining the number of combinations that can be generated from a set of n objects taken r at a time is

$$C_r^n = \frac{n!}{r!(n-r)!}$$

The difference between this relationship and its counterpart given in Figure 4.6 (i.e., the second model) is the term $r!$ in the denominator. This suggests that for a given number of objects taken r at a time, there will be a factor of $r!$ fewer combinations than permutations. For example, a stockbroker has identified 15 promising stocks to his clients for future investment. If the broker recommends that a client invest in exactly 5 stocks, how many different combinations are possible? The answer is 3,003 [15!/(5! $\times$ 10!)].

In the California lottery, six numbered balls are selected at random without replacement. The numbers on the balls range from 1 to 49. The lottery's director wishes to know whether the game should be based on permutations or combinations. In the former case, the participant would need to select the correct numbers in the correct sequence. In the latter case, the participant would only need to select the correct numbers in any sequence. The number of sequences based on each approach is shown in Figure 4.7.

An inspection of Figure 4.7 indicates that the odds of winning a lottery based on permutations is slightly over 10 billion to 1. In contrast, the odds of

FIGURE 4.7 Computation of the Number of Permutations and Combinations for the California Lottery

Method	Model	Number of Sequences
Permutations	$P_6^{49} = \dfrac{49!}{43!}$	10,068,346,800
Combinations	$C_6^{49} = \dfrac{49!}{43!6!}$	13,983,815

winning based on the number of combinations is approximately 14 million to 1. Clearly, the odds of winning based on permutations is much too low relative to the target population of approximately 15 million adults (i.e., Californians and visitors). In fact, the actual lottery is based on combinations. Even here, the chances of winning the grand prize (i.e., selecting all six numbers) is still very low.

PROBABILITY LAWS

Key idea 5

The previous two sections showed how probability values can be assigned. The basic rules governing the assignment of probability for a single event were also introduced in Section 4.3. Many business situations, however, involve more than a single event (e.g., the level of imports and the value of the dollar). Several additional rules are required for computing probability values based on combining several events. The two basic rules are the *addition law* and the *multiplication law.* Both of these laws were first developed by Abraham de Moivre (see historical note). Figure 4.8 presents the basic form of the laws for two events. Each of these laws can be expanded to process an unlimited number of events.

Addition Law

The addition law is used to compute the probability of the *union* of two or more events. Recall that the union of two or more events is the event that occurs if one or more of the events occur. Traditionally, the symbol ∪ is used to indicate the union of two events and is equivalent to the term *or*. To illustrate the addition law, consider that the director of the California Big Spin contest is interested in determining the probability that a contestant would win either $1 million ($A$) or $10,000 ($B$). In this case, the director

FIGURE 4.8 Basic Addition and Multiplication Probability Laws for Two Events

Law	Model
Addition:	
General	$P(A \text{ or } B) = P(A) + P(B) - P(A \text{ and } B)$
Special*	$P(A \text{ or } B) = P(A) + P(B)$
Multiplication:	
General	$P(A \text{ and } B) = P(A) \times P(B|A)$
Special†	$P(A \text{ and } B) = P(A) \times P(B)$

* Mutually exclusive.
† Independent.

would use the *special law of addition* because these two events have nothing in common (i.e., they are mutually exclusive). Mathematically, the probability of either of these two events occurring can be expressed as follows:

$$P(A \text{ or } B) = P(A) + P(B) = .08 + .10 = .18$$

The probability of winning $1 million is .08, and the probability of winning $10,000 is .10 (see Figure 4.1). Therefore, the probability of the event (A or B) occurring is .18.

The same approach can be applied for determining the probability of (A or B or C). In this case, C represents winning $20,000. The probability of winning $20,000 is .10. The probability of the event (A or B or C) occurring is

$$P(A \text{ or } B \text{ or } C) = P(A) + P(B) + P(C) = .08 + .10 + .10 = .28$$

Again, the special law of addition is used because the events have nothing in common. That is, winning $1 million precludes the possibility of winning either $10,000 or $20,000.

In many situations, however, events do have something in common. In these cases, the *general law of addition* is used for computing the probability of multiple events occurring. For example, suppose the director of the California Big Spin wishes to determine the probability of one of any two contestants winning $1 million. Since winning $1 million on the first spin (A) does not preclude winning $1 million on the next spin (B), these two events are considered **independent** of each other. This means that the outcome of the first event has no impact on the outcome of the second event. When events are independent of each other, the special law of multiplication is used to evaluate the term $P(A \text{ and } B)$. The computation for determining the probability of A or B occurring is

$$P(A \text{ or } B) = P(A) + P(B) - P(A \text{ and } B)$$

$$P(A \text{ or } B) = P(A) + P(B) - P(A)P(B) = .08 + .08 - (.08)(.08)$$

$$P(A \text{ or } B) = .1536$$

Thus, the chance of either contestant winning $1 million is slightly over 15%. The same process could be applied in determining the chances of various other multiple outcomes.

Recall that the director of the California Big Spin was interested in determining the chances that one of any three contestants would win $1 million (i.e., A or B or C = $1 million). In this case, the appropriate addition formula is as follows:

$$P(A \text{ or } B \text{ or } C) = P(A) + P(B) + P(C) - P(A \text{ and } B) - P(A \text{ and } C) - P(B \text{ and } C) + P(A \text{ and } B \text{ and } C)$$

Notice the large number of intersection terms associated with this model and the fact that the intersection of the three outcomes (i.e., A or B or C) is added instead of subtracted. This is because the triple intersection has been subtracted out three times.

$$P(A \text{ or } B \text{ or } C) = .08 + .08 + .08 - (.08)(.08) - (.08)(.08) \\ - (.08)(.08) + (.08)(.08)(.08)$$

$$P(A \text{ or } B \text{ or } C) = .2213$$

Thus, the chance of at least one contestant winning $1 million is slightly over 22%. Typically, computer analysis is used in computing probability values for events with more than three outcomes. This will be demonstrated in Section 4.5.

Multiplication Law

The multiplication law is used to compute the probability for the *intersection* of two or more events. Recall that the intersection of two or more events is the event that occurs if all the events occur. Traditionally, the symbol ∩ is used to indicate the intersection of two events and is equivalent to the term *and*. The special multiplication law has already been demonstrated in computing the probability of two independent events. However, before proceeding further, it will be useful to define the following three types of probabilities often encountered in problems involving multiple events. These are **marginal, joint,** and **conditional probabilities.** Figure 4.9 summarizes the basic properties of these types of probabilities under conditions of independence and dependence for two events A and B.

The notion of independence has already been introduced in the preceding discussion. Basically, two events are independent if one event has no impact on the outcome of the other event. Consequently, two events are **dependent** if one event does impact the outcome of the other event. For example, the selection of an ace from a deck of 52 cards (without replacement) does influence the chances of drawing a second ace from the same deck. Dependent events are thus conditional on each other. The basic definition for computing conditional probabilities for two events is

$$P(B|A) = \frac{P(A \text{ and } B)}{P(A)}$$

Traditionally, the | in the term on the left stands for *given*. Thus $P(B|A)$ is read as the probability of event B given the event A. When two events are independent, the above expression can be simplified as illustrated in the following:

$$P(B|A) = \frac{P(A)P(B)}{P(A)} = P(B)$$

Thus, this relationship can be used to test the independence of two events. For example, the chances of winning $1 million on the second spin of the Big

FIGURE 4.9 Basic Properties of Marginal, Joint, and Conditional Probabilities for Two Events under Conditions of Independence and Dependence

Probability Type	Model Independence	Model Dependence	Definition
Marginal	$P(A)$	$\Sigma P(B\|A) \times P(A)$	The unconditional probability of a single event occurring
Joint	$P(A)P(B)$	$P(B\|A) \times P(A)$	The probability of two events occurring together
Conditional	$P(B)$	$P(A \text{ and } B)/P(A)$	The probability of one event occurring, given the occurrence of another event

Wheel, given the fact that $1 million was won on the first spin, are merely equal to the chances of winning on a single spin. Namely, the probability is .08. This is because the two events are independent of each other. The relationships presented in Figure 4.9, of course, can be expanded to incorporate as many events as defined by the problem.

As a second illustration of these different types of probabilities, consider the data presented in Figure 4.10. These data show the distribution, by sex and annual income, of 500 individuals surveyed at a shopping mall. For example, 25 of the individuals interviewed were males that made less than $15,000 per year.

The data in Figure 4.10 can be converted to a joint-marginal probability table as shown in Figure 4.11. This figure is called a *joint-marginal* table because it reports not only the probability of a single event occurring (e.g., annual income below $15,000), but also the chances of two events occurring simultaneously (e.g., male and annual income above $25,000). Marginal probabilities are denoted by an *M*, and joint probabilities are indicated by a *J*. The process of converting the survey data presented in Figure 4.10 is

FIGURE 4.10 Results of Shopping Mall Survey

Annual Income ($)	Male	Female	Total
<15,000	25	75	100
15,000–25,000	125	125	250
>25,000	100	50	150
Total	250	250	500

FIGURE 4.11 Joint-Marginal Probability Table for Shopping Mall Survey

Annual Income ($)	Male	Female	Total
<15,000	.05 (J)	.15 (J)	.20 (M)
15,000–25,000	.25 (J)	.25 (J)	.50 (M)
>25,000	.20 (J)	.10 (J)	.30 (J)
Total	.50 (M)	.50 (M)	1.00

accomplished by simply dividing each observed frequency by the total number of interviews (i.e., 500). For example, the marginal probability of the event "male" is merely 250/500, or .50. The results of this computational process are shown in Figure 4.11.

Suppose that the designers of the survey wish to know what is the probability that an individual earns over $25,000 annually, given that the individual is a male (i.e., $P(>\$25,000|Male)$). As indicated above, this is a conditional probability statement. The joint-marginal table presented in Figure 4.11 can be converted to a conditional probability table as shown in Figure 4.12. This is accomplished by merely dividing the marginal probability of male and female into each of the respected joint probabilities. Thus, the chances of someone earning over $25,000 annually given that he is a male is 40%. Notice that each column sums to 1.

The table in Figure 4.12 can be used to test whether sex and income are independent of each other. The fact that sex and income might be independent or dependent could impact the marketing approach taken by the mall operators. The procedure for testing for independence has already been discussed. Recall that two events are independent if $P(B|A) = P(B)$; otherwise, they are dependent. An examination of Figure 4.12 reveals the following evidence:

$$P(<\$15,000/Male) = .10 \neq P(\$15,000) = .20$$

whereas

$$P(\$15,000-\$25,000/Male) = .50 = P(\$15,000-\$25,000) = .50$$

FIGURE 4.12 Conditional Probability Table for Shopping Mall Survey

Annual Income ($)	Male	Female
<15,000	.10 (C)	.30 (C)
15,000–25,000	.50 (C)	.50 (C)
>25,000	.40 (C)	.20 (C)
Total	1.00	1.00

Notice that in the first case the conditional probability does not equal the marginal probability (implying dependence). However, in the second case, the conditional probability does equal the marginal probability (implying independence). The appropriate conclusion in these mixed cases is that the two events (i.e., sex and income) are dependent. A conclusion of independence for two or more events requires that all of the tests prove positive.

Probability Trees

Probability trees provide a graphical presentation of the various outcomes of an experiment along with their respective probabilities. More specifically, they are very useful for depicting both conditional and joint probabilities. A probability tree consists of nodes and branches. Recall that the director of the California Big Spin was interested, among other things, in determining the chances of three contestants in a row winning exactly $1 million. The probability tree presented in Figure 4.13 shows the various outcomes associated with spinning the wheel three times. The term *win* refers to winning $1 million, and the term *no win* refers to winning some other amount. Notice

FIGURE 4.13 Probability Tree Showing Chances of Three Contestants Winning $1 Million in California Big Spin

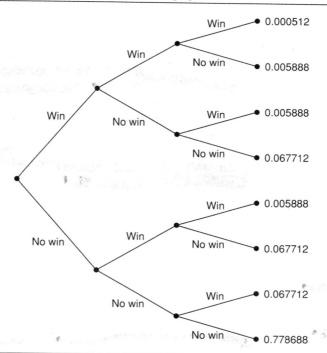

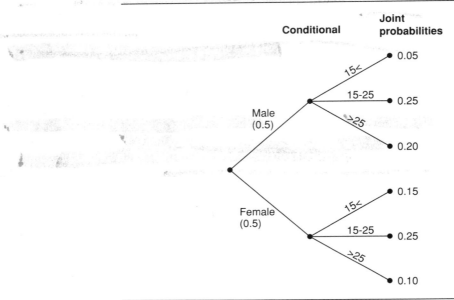

FIGURE 4.14 Probability Tree Showing Conditional and Joint Probabilities for Shopping Mall Survey

the ease of recognizing the set of outcomes that correspond to winning $1 million on three consecutive spins of the wheel. This set of outcomes is represented by the upper branch. The corresponding probability of this set of outcomes occurring (i.e., the event win-win-win) is .000512. This probability value was obtained using the special law of multiplication, since each spin of the wheel is independent of every other spin (i.e., $.08 \times .08 \times .08 =$ .000512). Obviously, the chance of three contestants winning $1 million in a row is quite remote. On the other hand, the chance that none of the three contestants will win $1 million is nearly .78 (i.e., $.92 \times .92 \times .92$).

In the preceding Big Spin problem, the conditional probabilities were merely the chances of a specific event occurring, since each event was independent of each preceding event. By way of contrast, Figure 4.14 shows a probability tree for the shopping mall survey problem. In this case, some of the events are not independent of each other (e.g., "female" and "income above $25,000"). Notice that the joint probabilities reported in Figure 4.14 match those given in Figure 4.12.

BAYESIAN ANALYSIS

Key idea 6

In many business situations, the decision maker is faced with the task of making a decision based on past or **prior probability** data. Often, the prior data are based on the decision maker's subjective estimate of the likelihood of various possible events. For example, a weatherman often uses past weather data in preparing a forecast for tomorrow. The historical data might indicate that there is a 30% chance of rain. However, is this estimate a good indicator that it will rain tomorrow? In some cases, the prior probability can be updated or revised with the help of additional information. Suppose, in addition to the prior data, the weatherman has access to a weather satellite that provides estimates of upcoming weather patterns. Armed with these additional data, the weatherman should be able to develop an improved estimate of tomorrow's weather.

Typically, the additional information is presented in the form of a conditional probability table. Such a conditional table presents the probability of a specific predictor or indicator, given the occurrence of a specific state. In effect, it provides a measure of the accuracy of a given prediction. Figure 4.15 presents several examples of conditional probability tables. Panel A shows the case when the prediction perfectly matches the actual state. For example, the probability that indicator I_1 accurately predicts S_1 is 1. Mathematically, this relationship can be expressed as follows: $P(I_1|S_1) = 1$. Similarly, $P(I_2|S_2) = 1$. Notice that the accuracy of a given indicator is always measured against the actual state.

The conditional tables given in Figure 4.15 are limit cases. They indicate the "best" and "worst" possible accuracy levels in terms of predicting the states of nature. On the surface, it would appear that Panel B represents the worst case since the indicator is always wrong. That is, the predictor indicated I_2 when S_1 occurs. This is equivalent to the situation where a barometer always predicts rain when the sun is shining. Therefore, in terms of predictability, one can conclude that Panel B is equivalent to Panel A. In fact, the worst case is represented by Panel C. Here, there is a 50-50 chance that the indicator will accurately predict the actual state.

The National Weather Service has recently placed in orbit, via the space shuttle, a satellite designed to improve weather prediction throughout the continental United States. Initially, the satellite will be limited to predicting rainfall in a given area. This limited capability will be upgraded as more sophisticated processing software is developed. Preliminary tests indicate that in 100 incidents of rain (S_1) in a given area, the satellite actually predicted rainfall (I_1) in 90 of these cases. Furthermore, in 200 cases when it was cloudy (S_2), the satellite predicted rain (I_1) 100 times, and in 100 cases when the sky was clear (S_3), the satellite predicted rain 20 times. These accuracy

FIGURE 4.15 Examples of Limit-Case Conditional Probability Tables

| | (A) | | (B) | | (C) | |
	I_1	I_2	I_1	I_2	I_1	I_2
S_1	1	0	0	1	.5	.5
S_2	0	1	1	0	.5	.5

FIGURE 4.16 Conditional Probability Table for Weather Satellite

	Indicator	
State	Rain (I_1)	No Rain (I_2)
Rain (S_1)	.9	.1
Overcast (S_2)	.5	.5
Clear (S_3)	.2	.8

estimates are reported in Figure 4.16. For example, there is a 50% chance that the satellite will predict rain, given that the weather is actually cloudy (i.e., $P(S_2|I_1) = .5$).

The manager of a local weather bureau would like to know the chances of various weather patterns occurring, given a specific prediction from the satellite. That is, the manager would like to revise the historical weather data with the information received from the satellite. The historical data reveal that, for this season of the year, it rains 30% of the time, it is cloudy 50% of the time, and it is clear 20% of the time. Unfortunately, the probability data presented in Figure 4.16 are not in a form that can be used directly by the manager.

The process of revising prior probability data with additional information is called *Bayesian analysis*. Figure 4.17 shows a schematic of the Bayesian model for revising probabilities. This diagram shows how the prior and the additional information are combined to generate the desired revised probability estimates.

The general model for developing revised or **posterior** probability estimates for the ith state is presented below:

$$P(S_i|I_i) = \frac{P(I_i|S_i) \times P(S_i)}{\Sigma P(I_j|S_j) \times P(S_j)} = \frac{P(I_i \cap S_i)}{P(I_i)}$$

FIGURE 4.17 Overview of Bayesian Analysis Process

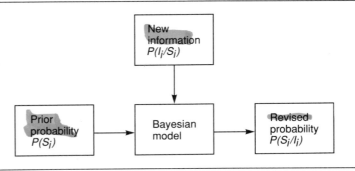

FIGURE 4.18 Probability Tree for Computing Revised Probabilities for the Weather Service Problem

```
                         States of nature
         ┌───────────────────┼───────────────────┐
         ↓                   ↓                   ↓
       Rain (R)           Overcast (O)         Clear (C)
         ↓                   ↓                   ↓
         .3                  .5                  .2
       ┌──┴──┐            ┌──┴──┐             ┌──┴──┐
       ↓     ↓            ↓     ↓             ↓     ↓
       .9    .1           .5    .5            .2    .8
       │     │            │     │             │     │
    Predict Predict    Predict Predict     Predict Predict
     rain   no rain     rain   no rain      rain   no rain
       ↓     ↓            ↓     ↓             ↓     ↓
      .27   .03          .25   .25           .04   .16
```

$P(r) = .27 + .25 + .04 = .56$
$P(nr) = .03 + .25 + .16 = .44$

where:

$P(S_i|I_i)$ = Probability of the ith state occurring, given a prediction of I_i
$P(I_i|S_i)$ = Probability estimate for the ith indicator, given the ith state
$P(S_i)$ = Prior probability for the ith state

Normally, the development of revised probabilities via the Bayesian model is performed using computer analysis. However, the following presentation is designed to illustrate the principles behind the computational process. Continuing the weather satellite example, Figure 4.18 shows a probability tree for computing the various joint probabilities. In this example, the states of nature are represented by an R for rain, an O for overcast, and a C for cloudy. Similarly, the indicators are r for a prediction of rain and nr for a prediction of no rain.

These results show that the probability of the satellite predicting rain (r) is 56%, and the probability of predicting no rain (nr) is 44%. These marginal probability values form the denominator in the Bayesian model [i.e., $P(I_i)$]. The numerator is represented by the joint probability of a given indicator and state of nature. For example, the probability of a prediction of rain and rain actually occurring is 27%. This joint probability can be seen as the leftmost branch in the probability tree. The revised probability of rain, given a prediction of rain, is

$$P(R|r) = \frac{.27}{.56} = .4821$$

110 Chapter 4 Probability Theory

Thus, there is an approximately 48% chance of rain, given that the satellite is indicating rain. This result is somewhat higher than the historical data on rainfall (i.e., 48% versus 30%). Nevertheless, these revised probability estimates are not very encouraging. Perhaps the satellite is in need of a tuneup! The remaining revised probability values are presented in the next section. In general, the total number of revised probability values is equal to the number of states of nature times the number of indicators. For the National Weather Service problem, the number of revisions is equal to 6 (i.e., 3 × 2). Clearly, the computational process becomes tedious when there are more than a few states of nature or indicators. Fortunately, the Bayesian model lends itself to computer analysis, as illustrated in the following section.

4.5 COMPUTER ANALYSIS

The analysis of certain types of probability problems can be greatly facilitated through the use of computer analysis. Specific examples include revising prior probabilities using Bayesian analysis and estimating the probabilities of joint events. This section presents several examples of the use of CBS in solving problems of this type.

National Weather Service

Figure 4.19 presents a CBS Bayesian analysis of the National Weather Service problem. An inspection of these results shows that there is a 56% chance that the satellite will indicate rain. Furthermore, there is a 36% chance that it will be clear, given a forecast of no rain. The latter prediction is taken from the revised conditional table shown in Figure 4.18.

The California Big Spin

Displayed in Figure 4.20 is a CBS probability analysis of the chances of one or more of the three contestants winning $1 million. The computer analysis shows that the chances are approximately 22% that one or more of the three contestants will win $1 million.

FIGURE 4.19 CBS Bayesian Analysis of National Weather Service Problem

Prediction	Marginal Probabilities	Revised Probabilities		
		R	O	C
r	.56	.4821	.4464	.0714
nr	.44	.0682	.5682	.3636

FIGURE 4.20 CBS Probability Analysis of Winning $1 Million in Three Spins of the California Big Wheel

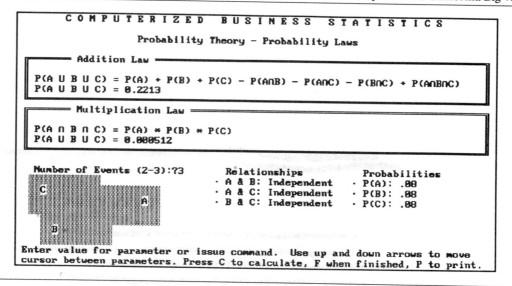

4.6 PRACTICAL APPLICATIONS

Presented in this section are several business situations involving the use of probability theory.

Western Computer Laboratories (Joint Probability)

The head of R&D at Western Computer Laboratories (WCL) has $3 million to spend this year on new product development. The director knows, however, that 60% to 90% of all new products fail by the end of the first year (McIntyre and Statman, 1982). One strategy for offsetting this high failure rate is to invest in several new products at the same time.

Accordingly, the director has identified three new R&D products for possible funding: a portable 30-megabyte hard drive, a portable optical scanner, and a 5 1/4" to 3 1/2" diskette converter. The director estimates that the chances of failure for any one project are 60%, based on an expenditure of $3 million, and that the success or failure of any one project is independent of any other project. Each project will require a minimum of $1 million and the probability of failure will increase by 5% for every $1 million decrease in the budget below $3 million. For example, if only $1 million is spent on a particular project, the chances of failure increase to 70%. Armed with these data, the director has prepared the estimates given in Figure 4.21.

FIGURE 4.21 Project Funding and Probability Estimates for WCL

Number of Projects	Budget per Project	Probability of a Single Failure	Probability that All Projects Will Fail
1	$3,000,000	.60	.60
2	1,500,000	.675	.46
3	1,000,000	.70	.34

These estimates were obtained through the use of the special multiplication law for independent events. For example, the probability that three projects out of three will fail is simply the probability of failure for the first project times the probability for failure of the second project times the probability of failure for the third project (i.e., $.7 \times .7 \times .7 = .34$). These results show that the risk of failure can be reduced considerably (i.e., from 60% to 34%) by funding more than one project, even though the probability of a single failure will increase as the funding level decreases. One of the key assumptions in this analysis is that the projects are independent. This assumption is supported by McIntyre and Statman, who state that "the correlation between new projects within the same firm is quite low."

Palmer Video (Bayesian Analysis)

The vice president of marketing at Palmer Videos is interested in forecasting the demand for the firm's video rewind unit. This product provides the user with a fast forward and rewind capability, thereby reducing the wear and tear on the head of the user's VCR. The vice president's preliminary estimate of product demand for the coming year is given in Figure 4.22. These estimates indicate that demand in the range of 30,000 to 50,000 units appears most likely. The vice president is somewhat unsure of these estimates, however, and has decided to undertake a market survey to improve these estimates.

FIGURE 4.22 Probability Demand Forecast for Palmer Video Rewind Unit

Market Size (000 units)	Demand Probability
<10	.2
10–30	.3
30–50	.4
>50	.1

FIGURE 4.23 Conditional Probability Table for Palmer Video Rewind Unit

Market Size (000 units)	Indicated Demand		
	Strong	Moderate	Weak
<10	.1	.2	.7
10–30	.1	.4	.5
30–50	.3	.4	.2
>50	.8	.2	0

More specifically, the vice president has decided to hire a consulting firm to conduct the survey. The historical performance of the consulting firm in estimating product demand is given in Figure 4.23. For example, 70% of the time when the consultants indicated that demand for the product would be weak, the actual demand was less than 10,000 units.

Based on the prior probability estimates and the conditional probability table supplied by the consulting group, the vice president can now develop a set of revised probabilities using Bayesian analysis. A CBS analysis of this problem is presented in Figure 4.24. These results show, for example, that there is a 29% chance that the consultants will estimate a strong demand for the product. Notice that these marginal probabilities sum to 1. The analysis also shows that there is a 55% chance that the demand for the rewind product is 30,000 to 50,000 units, given a forecast of a strong demand. These revised probabilities can be used by the vice president to improve the company's forecast. Unfortunately, these results are not very encouraging. That is, there is considerable uncertainty in the prediction of the various demand states.

FIGURE 4.24 CBS Analysis of Palmer Video

Prediction				Marginal Probabilities
Strong				.29
Moderate				.34
Weak				.37

Prediction	Revised Probabilities			
	<10	10–30	30–50	>50
Strong	.0690	.1034	.5517	.2759
Moderate	.1176	.3529	.4706	.0588
Weak	.3784	.4054	.2162	0.

4.7 CASE STUDY: AUTOMOBILE FUEL ECONOMY STANDARDS

The director of vehicle fuel economy standards at the U.S. Department of Transportation is interested in determining the probability of obtaining various fuel economy levels as a function of vehicle weight and driving environment. The director knows that vehicle weight and driving environment are not the only factors that contribute to variations in fuel economy (e.g., state of engine maintenance). However, these two variables are relatively easy to measure and should provide reasonable probability estimates. A test of 30 passenger cars was recently conducted, and the results are reported in Figure 4.25. These data show the weight of each vehicle tested, the driving environment (i.e., urban or rural), and the average fuel economy achieved over the test (measured in mpg).

As a first step in analyzing these results, the director used CBS to generate a set of histograms for the three variables. These results are shown in Figure 4.26. The computed average fuel economy is 22.8 mpg. Armed with these results, the director can now determine the probability of various test results. A selected set of probability estimates is given in Figure 4.27. These computations show the probability of obtaining various fuel economy results based on the sample. For example, the probability of achieving at least 19 mpg is 90%. This result was obtained by observing the frequency of occurrence of at least 19 mpg in the histogram (panel A) and dividing by the total sample size (i.e., 30). In contrast, the probability of obtaining at least the sample mean ($\sim$23 mpg) drops to approximately 53.3%.

FIGURE 4.25 Fuel Economy Test Data Results

Vehicle Weight (lbs.)	Urban or Rural	Fuel Economy (mpg)	Vehicle Weight (lbs.)	Urban or Rural	Fuel Economy (mpg)
3200	U	25	3800	U	24
4000	R	22	3500	U	19
4500	U	19	2800	U	25
4200	R	17	3500	U	21
3800	U	24	2800	R	27
3200	U	25	4000	R	21
3400	R	24	3350	U	22
2900	U	27	3100	U	25
3400	U	23	3500	R	22
4000	U	20	4100	U	18
3300	R	25	2900	R	28
3800	U	22	3400	U	24
3100	U	25	3300	U	22
3000	R	26	3100	R	24
3800	R	21	4300	U	17

FIGURE 4.26 Relative Frequency for Test Results

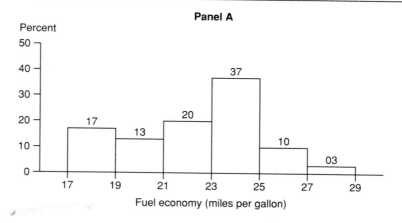
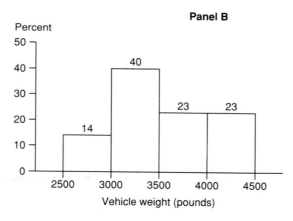
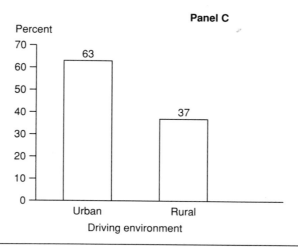

115

FIGURE 4.27 Probability Estimates for Various Test Results

$$P(\text{Mileage} \geq 21) = \frac{27}{30} = .900$$

$$P(\text{Mileage} \geq 23) = \frac{16}{30} = .533$$

$$P(21 \leq \text{Mileage} \leq 25) = \frac{20}{30} = .667$$

FIGURE 4.28 Marginal and Joint Frequency Table for Fuel Economy and Driving Environment

	Fuel Economy (mpg)		
Environment	<23	≥23	Total
Urban	9	10	19
Rural	5	6	11
Total	14	16	30

FIGURE 4.29 Marginal and Joint Probability Distribution for Fuel Economy and Driving Environment

	Fuel Economy (mpg)		
Environment	<23	≥23	Total
Urban	.30	.33	.63
Rural	.17	.20	.37
Total	.47	.53	1.00

FIGURE 4.30 Conditional Probability Table for Fuel Economy Given Driving Environment

	Fuel Economy (mpg)		
Environment	<23	≥23	Total
Urban	.47	.53	1.00
Rural	.45	.55	1.00

Determining the probability of occurrence for more than one variable is somewhat more complicated. In this case, the director would like to determine the probability of a certain fuel economy, given a specific value for one or both of the other variables. Problems of this type can be solved using joint-marginal probability analysis. Figure 4.28 presents a joint-marginal frequency distribution for fuel economy and driving environment. This table was developed by simply observing the frequency of occurrence of each fuel economy and driving environment combination reported in the test results. For example, 10 vehicles that experienced the urban driving cycle obtained at least 23 miles per gallon. The frequency table can be converted into a probability table by simply dividing each entry by the sample size. This result is shown in Figure 4.29. These data show, for example, that there is a 33% chance that a vehicle selected at random from the test fleet obtained at least 23 mpg while engaged in urban driving. Given this information, the director would like to know the probability of achieving at least 23 mpg, given that a vehicle was operated in an urban setting. Figure 4.30 features a conditional probability table derived from the joint-marginal table given in Figure 4.29. The conditional probability of obtaining at least 23 mpg for vehicles operating in an urban environment is approximately 53%. The closeness of this probability estimate to the one reported for all vehicles obtaining at least 23 mpg suggests that the type of driving environment (i.e., urban or rural) has very little impact on fuel economy. Although these results are interesting, the director should exercise some caution in drawing any final conclusions because of the relatively small sample size. Nevertheless, the test data should provide insight into the design of future experiments.

4.8 SUMMARY

This chapter has introduced some of the basic elements of modern probability theory. Probability theory is the branch of mathematics that is used to characterize uncertainty. Probability is an indicator of the chance that a certain event will occur and is measured on a scale from 0 to 1. A probability value of 0 indicates that the event cannot happen, and a probability value of 1 reveals that the event is sure to happen. Basically, there are two types of probability: objective and subjective. Objective probabilities are based on the outcome of a repeatable random experiment (e.g., flipping a coin). Subjective probabilities, on the other hand, are based on the outcome of a nonrepeatable random experiment (e.g., introducing a new consumer product). In other words, subjective probabilities are based on the beliefs of the individual making the probability assignments.

The sample space is defined as the set of all possible outcomes of an experiment and represents one of the basic building blocks of probability theory. Any subset of the sample space is called an event. Multiple events are defined as being either complementary, mutually exclusive, or compatible.

An important distinction of the latter category is whether the events are independent or dependent. This chapter introduced several methods of testing for independence. A number of basic rules are introduced for determining the probability of occurrence of multiple events. The two most important are the laws of addition and multiplication. Lastly, the process of revising prior probability data with additional information using Bayes' Theorem was presented. The capability of developing revised probability estimates represents an important tool in modern decision analysis.

4.9 GLOSSARY

collectively exhaustive events The list of all possible outcomes of an experiment.
conditional probability The probability of an event occurring given that another event has occurred.
dependent events The occurrence of one event does affect the occurrence of another event.
event One of several possible outcomes of a process or experiment.
independent events The occurrence of one event has no effect on the occurrence of another event.
joint probability The probability of two events occurring simultaneously.
marginal probability The probability of a single event.
mutually exclusive events Events that cannot occur simultaneously.
objective probability A probability estimate based on observations or facts.
posterior probability A probability estimate based on new information.
prior probability A probability estimate based on past information.
probability The chance that an event will occur.
probability tree A graphical method for depicting probability relationships.
sample space The set of all possible outcomes of a process or experiment.
subjective probability A probability estimate based on personal beliefs or intuition.

4.10 BIBLIOGRAPHY

Aczel, A.D. *Complete Business Statistics.* Homewood, Ill.: Irwin, 1989.

Ewart, P. J., et al. *Probability for Statistical Decision Making.* Englewood Cliffs, N.J.: Prentice Hall, 1974.

Feller, William. *An Introduction to Probability Theory and Its Applications.* New York: Wiley, 1971.

Heron House Editors. *The Odds on Virtually Everything.* New York: Putnam, 1980.

McIntyre, S. H., and M. Statman. "Managing the Risk of New Product Development." *Business Horizons* 25 (1982).

Ross, S. A. *A First Course in Probability.* 3rd ed. New York: Macmillan, 1984.

4.11 PROBLEMS

1. Classify each of the following as either an objective or subjective probability statement:
   a. The chance of Notre Dame winning the game is 50–50.
   b. Past experience indicates that 1 shopper out of 10 will buy something.
   c. The marketing manager reports that there is a 2 to 1 chance that sales will increase next year.
   d. The chances of having an IRS audit are 1 in 100.
   e. The National Weather Service reports a 30% chance of rain.

2. Classify each of the following as either an objective or subjective probability statement:
   a. The Los Angeles Dodgers are 3-to-2 favorites to win the World Series.
   b. A survey found that 60% of the voters support arms control.
   c. The probability of finding life on other planets is estimated at less than 10%.
   d. A participant has 1 chance out of 10 million of winning the lottery.
   e. An analyst estimates that there is a 60% chance that OPEC will raise oil prices.

3. Classify each of the following sets of events as either independent, dependent, or mutually exclusive.
   a. Vehicle weight and vehicle fuel economy.
   b. Purchasing a company and eating dinner.
   c. Attending a stockholders' meeting in New York and attending a New York Nicks basketball game at the same time.
   d. The size of the U.S. trade deficit and the value of the Japanese yen.
   e. Using a discount coupon to purchase a radio and using the same coupon to purchase a VCR.

4. What role do permutations and combinations play in probability theory?

5. Describe Bayes' Theorem in your own words, and discuss its importance in business.

6. Consider the sample space $W = (-2, -1, 0, 1, 2, 3, 4)$ as well as the following random events:

$$A = (-2, 4) \quad B = (-1, 0, 1) \quad C = (0, 2, 3, 4)$$

   Identify which of the following are mutually exclusive, totally exhaustive, both, or neither:
   a. A,B
   b. B,C
   c. A,C
   d. A,B,C

7. Consider the sample space W, which consists of the top 10 Fortune 500 industrial firms (1988 sales), as well as the following random events:

$$A = \text{P/E ratio above 10}$$
$$B = \text{P/E ratio less than or equal to 10}$$
$$C = \text{P/E ratio equal to 10}$$

   Identify which of the following are mutually exclusive, totally exhaustive, both, or neither:
   a. A,B
   b. B,C

c. A,C
d. A,B,C
8. Consider the sample space $W = (-4,-2,0,2,4,6,8)$ as well as the following random events:

$$A = (-2,0) \quad B = (-4,2,8) \quad C = (0,2,4,6)$$

Identify the union and intersection of these events:
a. A,B
b. B,C
c. A,B'
d. A,B,C
9. Bill Johnson supervises five production coordinators. How many different ways can Bill rank-order these five employees?
10. The mayor of Los Angeles wishes to form a blue-ribbon committee to investigate the rise in street crime in the city. The mayor has narrowed the list of candidates to five white males, five minority males, and five females.
    a. In how many different ways can a 10-person committee be formed?
    b. In how many different ways can a 10-person committee be formed if it must contain at least three minority males and three females?
11. The production manager at Webb Electronics wishes to establish a color-coding system for the 50 different major components used in manufacturing the company's compact disk unit. The company has eight distinct colors on hand. Furthermore, the manager wishes to formulate a distinct three-color code for each component.
    a. Does the manager have enough colors if different sequences of the same three colors can be used?
    b. Does the manager have enough colors if different sequences of the same three colors cannot be used?
    c. If the answer to part b is no, how many new colors are required to provide unique color coding to each of the 50 components?
    d. How many different arrangements could be made if the same color could be used more than once?
12. The management of Crane Manufacturing is interested in building four new plants in the sunbelt states over the next several years. More specifically, the management would like to construct two plants in 2 of the 13 southern states and two plants in 2 of the 6 southwestern states. How many different possibilities does the management have to consider?
13. The standard game of dice consists of rolling two dice and summing the amount indicated on the top face of each die.
    a. Identify the sample space for the game.
    b. What is the most frequently occurring sum?
    c. Develop a relative occurrence distribution for this game.
14. Monsieur Chevalier de Mere, a 17th-century nobleman, derived what he believed to be a guaranteed strategy for winning at dice. His strategy was based on the premise that there is a 67% chance that at least one 6 will appear on four rolls of a single die. De Mere became very rich using his strategy.
    a. What was de Mere's logic in believing that he had a 2/3 chance of winning?
    b. Is his logic correct?

c. Determine the actual probability of obtaining at least one 6 in four rolls.
d. Do any of the results derived above indicate why de Mere became a rich man?
15. The stud poker hand consists of 5 cards from a deck of 52 distinct cards.
    a. How many different five-card hands are possible?
    b. What is the probability of being dealt a royal flush (i.e., ace, king, queen, jack, and 10 of the same suit)?
    c. What is the probability of being dealt four aces?
    d. In addition to the monetary aspect of stud poker, what other factor makes the game so popular?
16. The distribution of the student body at Pepperdine University is as follows:

Freshmen	22%	Seniors	15%
Sophomores	20%	Graduates	25%
Juniors	18%		

   a. What is the probability of selecting two freshman from a sample of five students?
   b. What is the probability of selecting three graduates from a sample of six students?
   c. What is the probability of selecting four seniors from a sample of four students?
17. A local department store uses four billboards to advertise. Each billboard can be considered independent of all other billboards. A random sample of 100 shoppers were asked which, if any, of the billboards they had seen.

Billboard	Number Observed
#1	80
#2	75
#3	70
#4	90

What is the probability that:
a. All four were noticed?
b. None was noticed?
c. Billboards #1 and #4 were noticed?

18. The following data summarize the results of a recent survey of 100 shoppers at the Sears Store in South Bay. The survey shows the number of buyers and nonbuyers by sex:

Sex	Buyers	Nonbuyers
Male	27	25
Female	33	15

a. Are the events "male" and "buyer" independent or dependent?
b. Are the events "male" and "female" independent or mutually exclusive?
c. What is the probability of finding someone in the store who is a "nonbuyer"?

19. The following data summarize the results of a recent survey of the method of payment over the last month at the Lemon Tree Cafe by age group.

	Method of Payment	
Age Group	Credit Card	Cash
20–29	50	25
30–39	100	75
40–49	75	125
≥ 50	25	75

a. Are the events "paid by credit card" and "paid by cash" dependent, independent, or mutually exclusive?
b. Are the events "ages 20–29" and "paid by cash" independent or dependent?
c. What is the probability of a customer in the age group 40–49 paying by cash?
d. What is the probability of any customer paying by cash?

20. Consider an experiment of flipping a fair coin three times.
a. Develop a probability tree for this experiment.
b. What is the probability of obtaining two heads?
c. What is the probability of obtaining no heads?

21. The chances of the Los Angeles Lakers winning at home are 70%, whereas the chances of winning on the road are 50%. The Lakers are scheduled to play two home games followed by two road games in the coming week.
a. Develop a probability tree for this problem.
b. What is the probability of the Lakers winning all four games?
c. What is the probability of the Lakers winning three out of four games?
d. What are the chances of the Lakers losing all four games?

22. The director of Pro Tennis, Inc., is interested in developing a schedule for the upcoming "knockout" tournament. The tournament will match the world's top 16 tennis players in a one-set elimination bout.
a. Develop the play schedule for this tournament.
b. How many rounds will the tournament last?
c. How many total sets will be played?
d. If the probability of the top seed winning each game is 70%, what is the probability of the top seed winning the tournament?
e. If the probability of the top seed winning the first game is 90%, and the probability of winning decreases by 5% every round, what is the probability of the top seed winning the tournament?

23. The following table summarizes the promotion history at Weber Industries:

	Male	Female
Promoted	50	15
Not promoted	70	65

a. What is the probability of being promoted, given that you are a male? A female?
b. What is the probability of not being promoted?
c. Are there grounds for a discrimination lawsuit?

24. The following data summarize the results of an extensive clinical test of 30,000 participants, selected at random, for determining the impact of smoking on the incidence of lung cancer in the adult population.

	Lung Cancer	Free of Lung Cancer
Smokers	5,000	7,000
Nonsmokers	1,000	17,000

a. What is the probability that an individual selected at random does have cancer, given that the individual smokes?
b. What is the probability that an individual selected at random does not have cancer, given that the individual does not smoke?
c. What is the probability of selecting at random an individual who does not smoke?
d. What is the probability of selecting at random an individual who has lung cancer?

25. As part of the 1988 presidential campaign, the Republican party ran a profile of their candidate on both TV and radio. A recent survey of those who voted in the election indicated that 60% had seen at least one TV program featuring the Republican candidate. Furthermore, of that percentage, 80% also heard him on the radio. A total of 70% of those interviewed heard the candidate on the radio.
a. What is the probability that a voter saw the candidate on TV and heard him on radio?
b. What is the probability that a voter heard or saw the candidate?

26. Five out of 11 players on the Los Angeles Lakers basketball team are to be selected for the starting team.
a. In how many ways can the starting 5 be chosen, assuming all 11 can play any position?
b. In how many ways can the starting 5 be chosen, assuming that only 3 out of 11 can play center?

27. Four gentlemen have deposited their overcoats at a hat-check stand. Assuming that the coats are returned in a random fashion, what is the probability that
a. All four gentlemen receive their coats?
b. Exactly three gentlemen receive their coats?
c. Exactly two gentlemen receive their coats?
d. Only one of the gentlemen receives his coat?
e. None of the gentlemen receives his coat?

28. A box contains three cards. One card is white on both sides, another is black on both sides, and the third is black on one side and white on the other. What is the probability that a drawn card that is white on one side is also white on the other side?

29. Dr. Gallop has recently conducted a preference survey of 1,500 Americans with

regard to school prayer, gun control, and balancing the federal budget. The survey results show that 1,000 support school prayer, 300 support gun control, and 600 support balancing the budget. Additionally, 100 support all three issues, 200 favor school prayer and balancing the budget, 125 support school prayer and gun control, and 175 favor gun control and balancing the budget. What is the probability that an individual selected at random from the survey participants supports only

   a. School prayer?
   b. Gun control?
   c. Balancing the budget?

30. Four hundred employees at United Space Corporation were recently surveyed regarding a proposed stock ownership plan (ESOP). The responses were categorized according to management, technical, and staff personnel. Of the 60 managers, 45 were in favor of the plan. Of the 200 technical personnel, 100 were in favor. Finally, of the 140 staffers, 50 were in favor. Is job classification a determinant of employee attitude to the proposed plan?

31. A recent customer taste test of the top three soft drinks yielded the following results:

Category	Participant Preference (%)
Coke	25
Pepsi	25
7UP	8
Coke and Pepsi	12
Coke and 7UP	5
Pepsi and 7UP	4
Coke and Pepsi and 7UP	3

What is the probability of a participant selected at random preferring Coke or Pepsi or 7UP?

32. A recent driving test among the four leading intermediate automobiles (F = Ford, C = Chevy, P = Plymouth, T = Toyota) yielded the following results:

Vehicle	Participant Preference (%)	Vehicle	Participant Preference (%)
F	40	C&P	11
C	34	C&T	9
P	33	P&T	8
T	29	F&C&P	3
F&C	12	F&P&T	3
F&P	17	C&P&T	5
F&T	14	F&C&T	4
		F&C&P&T	2

What is the probability of a participant selected at random preferring Ford or Chevy or Plymouth or Toyota?
33. How many different grouping combinations can be developed from five classifications? Six classifications? (*Hint:* Derive a generalized model based on the patterns observed in problems 31 and 32.)
34. The California Lottery consists of selecting six numbered balls from a machine at random without replacement. The numbers range from 1 to 49. What is the probability of selecting at random
   a. All six numbers correctly?
   b. Five out of six correctly?
   c. Four out of six correctly?
   d. Three out of six correctly?
   e. Two out of six correctly?
   f. One out of six correctly?
   g. None correctly?
35. The California Big Spin consists of a wheel with various monetary outcomes (see Figure 4.1). What is the probability that
   a. Three out of the first five contestants will win $1 million?
   b. None out of the first five contestants will win $1 million?
   c. The first two of five contestants will win $1 million?
36. Referring to problem 35, what are the chances that a contestant will win at least $1 million? (Hint: Identify all of the different possible combinations using the "double" outcome.)
37. Astronomers have estimated that three conditions are required to produce life similar to that found on the earth. These are (1) the size of the star, (2) the existence of planets circling the central star, and (3) the distance of the planet from the central star. Recent estimates indicate that the earth's galaxy (Milky Way) contains 100 billion stars. Of that number, approximately 1 billion are about the same size as the sun. Furthermore, of that number, 10% are estimated to have planets. Finally, 2% of that number are located within the desired distance from the central star (i.e., not too close and not too far away).
   a. Classify the probability estimates for this problem as either objective or subjective.
   b. What is the probability of finding life on another planet in the galaxy, assuming that each of the above requirements is independent?
   c. What is the probability of finding life on another planet in the galaxy, assuming that the size of the star and the distance of the planet from the star are dependent?
   d. What are your impressions about finding life on another planet in the galaxy?
38. Determine the probability that two individuals out of a total of five will have the same birth date (assume a 365-day year).
39. Develop a general model for determining the probability that two individuals out of a total of N will have the same birthday.
40. The director of the California Department of Motor Vehicles wishes to design a new license plate consisting of a sequence of letters, numbers, or both. The director estimates that there are 11 million cars and light trucks currently operating in the state.
   a. What is the minimum number of positions required, based on the letters of the alphabet, to generate a sufficient number of unique license plates in California?

b. What is the minimum number of positions required, based on the first 10 digits (0–9), to generate a sufficient number of unique license plates?

c. What is the minimum number of positions required, based on a combination of letters and numbers, to generate a sufficient number of unique license plates?

d. What is the minimum number of positions required, based on a combination of three letters followed by three numbers, to generate a sufficient number of unique plates?

41. The number of cars in the state of California is predicted to increase by 5% per year.

  a. By what year would the proposed license plate design developed in option (c) of problem 40 become insufficient?

  b. What would be the simplest design change in the license plate to meet the new demand?

42. The producers of the movie *The Return of Flash Gordon* estimate that there is a 30% chance that the movie will be a success. The producers have recently received the reviews from a syndicated TV program that critiques new movies. The program's "track record" in predicting successes or failures at the box office is impressive: 80% of all movies that receive a positive rating end up being successful, and 90% of all movies that receive a negative rating are not successful. Unfortunately, the reviewers were not impressed with *The Return of Flash Gordon*.

  a. What is the probability that the new movie will receive a positive rating?

  b. What is the probability that the movie will be a success, given a positive rating from the reviewers?

  c. What is the probability that the movie will be a failure, given a negative rating from the reviewers?

43. The troubled S&L industry is initiating a new screening process for initiating commercial loans. Data supplied by the national S&L trade association indicate a 91% repayment rate. The screening criteria are based on the firm's debt-to-equity ratio (D/E). An examination of historical loan activity shows that 95% of all firms with a D/E ratio less than 1 completely repaid their obligations. On the other hand, only 85% of all firms with a D/E ratio of 1 or more completely repaid their debt. The association's records also reveal that 60% of all previous borrowers had a D/E ratio below 1.

  a. What proportion of repaid loans involve firms with a D/E ratio below 1?

  b. What proportion of nonrepaid loans involve firms with a D/E ratio of 1 or more?

44. This problem utilizes the fuel economy test data reported in Figure 4.25.

  a. Develop a joint-marginal probability table for fuel economy and vehicle weight.

  b. Develop a conditional probability table for fuel economy, given vehicle weight.

  c. Develop a joint-marginal probability matrix for fuel economy, vehicle weight, and driving environment.

  d. Develop a conditional probability matrix for fuel economy, given vehicle weight and driving environment.

45. The management at Blackthorn Pharmaceutical is considering a change in the

design of the company's logo. The firm's director of marketing estimates that there is a 60% chance this change would increase market share, a 20% chance there would be no change, and a 20% chance that market share would actually decrease. The marketing director is considering a pilot test of the new logo design. She estimates that there is an 80% chance that this test will indicate no change, given that market share would remain unchanged on a national basis. The director also estimates that there is a 10% chance that the test will indicate an increase. Furthermore, the director estimates that there is a 70% chance that the test result would indicate a decrease in market share, and a 5% chance that it would indicate an increase in market share, given that the new logo would actually result in a decline in market share. Finally, the director predicts that there is an 80% chance that the test result would indicate an increase in market share and a 20% chance that market share would remain unchanged, given that the new design would actually yield an increase in market share.

   a. What is the probability that the test marketing results will predict an increase in market share? No change? A decrease?
   b. What is the probability that market share will increase, given a positive test market result?
   c. What is the probability that market share will remain unchanged, given a negative test market result?
   d. What is the probability that market share will decline, given a negative test market result?

46. The National Weather Service provides daily weather forecasts to the major airports around the country. The basic report consists of an indication that weather at a specific locale will be clear, cloudy, or foggy. A forecast of foggy weather results in rerouting incoming aircraft to other airports. A cloudy forecast results in flight delays. The past performance of the National Weather Service in predicting weather conditions for Atlanta International is reported below.

Actual	Prediction		
	Clear	Cloudy	Foggy
Clear	.75	.20	.05
Cloudy	.15	.75	.10
Foggy	.05	.10	.85

The historical weather pattern for the Atlanta area for this time of the year is 60% clear, 30% cloudy, and 10% foggy.

   a. What is the probability of obtaining a prediction of cloudy weather?
   b. What is the probability of fog occurring, given a forecast of fog?
   c. What is the chance of flight delays, given a forecast of clear weather?

47. California has recently initiated a vehicle exhaust test program to help reduce smog. Preliminary estimates indicate that 75% of all vehicles tested should pass. Unfortunately, the test procedure is not 100% reliable. In fact, there is

a 10% chance that a vehicle with acceptable emission levels will fail the test, and a 20% chance that a vehicle with unacceptably high emission levels will pass the test.

  a. What is the probability that a randomly selected vehicle has acceptable emission levels, given that it passed the test?
  b. What is the probability that a randomly selected vehicle has acceptable emission levels, given that it failed the test?
  c. What is the probability of a randomly selected vehicle passing the test?

48. The Food and Drug Administration (FDA) has recently conducted a survey of a leading meat packing company. The firm classifies beef into three categories: choice, prime, and grade A. The current inventory at the plant consists of approximately 20% choice, 30% prime, and 50% grade A. An inspection of 100 graded sides of beef performed by the FDA inspectors revealed the following results:

Actual	Grade		
	Choice	Prime	Grade A
Choice	.9	.1	-0-
Prime	.1	.8	.1
Grade A	.1	.2	.7

What is the probability of finding a randomly selected side of beef of the following grades?
  a. Stamped choice, given that it is prime.
  b. Stamped grade A, given that it is grade A.
  c. Stamped choice.

49. The FDA has recently announced its approval of a new test procedure for indicating the presence of lung cancer at an early stage. The testing process has a 5% chance of indicating a "false negative" and a 10% chance of indicating a "false positive." Approximately 2% of the general adult population have incipient lung cancer.
  a. What is the probability that a randomly selected individual has lung cancer, given a positive test result?
  b. What is the probability that a randomly selected individual does not have lung cancer, given a negative test result?
  c. What is the probability that the test will indicate the presence of cancer?

50. Determine the probability that at least one project out of two will succeed at the Western Computer Laboratory (see Section 4.6) if $2 million is spent on the first project and $1 million is spent on the second. Is this a better strategy than the one analyzed in Section 4.6?

51. Reloux Corporation, located north of Paris, manufactures and markets a complete line of paper copiers. The following data were collected on the operating performance of the firm's current product line.

Model	Days Operational	Days Down
100	200	55
150	250	60
200	245	34
250	216	29
300	275	21

    a. Which is the "best" copier in terms of operating performance?
    b. What is the probability that all of the copiers will be down?
    c. What is the probability that any three of the copiers will be down?

52. The Sunworp SA Corporation, Stockholm, Sweden, manufactures and markets computer systems throughout northern Europe. The QC manager is interested in improving the failure resistance of the basic components. Historical data show that disk drive failures occur one-third as often as keyboard failures. Simultaneous failures of both have a probability of 10%. What is the probability of a keyboard failure?

53. Matia Pharmaceutical Limited of Bombay, India, utilizes four assembly lines to manufacture a line of prescription drugs. The QC manager has observed that line failures occur at a rate of 2% per month and are independent of each other.
    a. If the first line is operating out of specification, what is the probability that the second and third lines are also operating out of control?
    b. What is the probability that all four lines are operating within the design specifications?

4.12 CASES

4.12.1 Fluoride Test Program

Professor Blackwell of Pepperdine University's dental school has recently received a federal grant to study the effectiveness of various methods of applying fluoride to children's teeth. The benefits of fluoride in reducing tooth decay have been known for some time. However, the "best" method for using fluoride is still under debate. Accordingly, Dr. Blackwell has decided to conduct an experiment that will evaluate the three most frequently proposed application methods. Dr. Blackwell's experiment consists of four groups (one serves as a control group) of 200 participants (ages 8 to 12). Table 1 presents the basic program design along with the distribution of cavities prior to the program.

Each participant has agreed to follow the treatment program for a period of six months. The results of the experimental program are given in Table 2. These results show the number of new cavities observed during the final checkup.

TABLE 1 Experimental Program Design

	Number of Prior Cavities			
Group	0	1	>1	Treatment
1	30	80	90	No fluoride
2	20	70	110	Fluoride pills
3	25	50	125	Fluoride toothpaste
4	30	90	80	Fluoride applied directly to teeth

TABLE 2 Results of Experimental Fluoride Program

	Number of New Cavities		
Group	0	1	>1
1	50	100	50
2	70	90	40
3	80	80	40
4	100	80	20

Suggested Questions

1. What is the probability that a participant had no new cavities?
2. What is the probability of having no new cavities if you were a member of group 1? Group 2? Group 3? Group 4?
3. What is the probability that a participant had no new cavities, given that they started the program with no cavities?
4. What is the probability that a participant had no new cavities, given that they started the program with one cavity?
5. What is the probability that a participant from group 3 had more than one cavity, given that they started the program with no cavities?
6. Which of the three treatments do you think would be most effective in light of these results and the general convenience of each treatment?

4.12.2 Product Reliability in the Automotive Industry

U.S. car manufacturers, in order to remain competitive in the growing international automotive marketplace, have had to significantly improve product quality. A significant consequence of the steady improvement in reliability has been the introduction of a five-year—and in some cases a seven-year—unconditional guarantee. These guarantees, in turn, have provided an additional sales boost to the domestic car manufacturers. The overall reliability of any major automotive subsystem (e.g., brakes) can be achieved through improving the reliability of each system component, implementing parallel configurations, or both. Consider, for example, the schematic of a conventional automobile brake system shown in Table 1. The designers have estimated that the probability of failure of any one component is 5% for the first year, with a 1% increase in the failure rate for each subsequent year. The failure of any single component will cause the entire brake system to fail.

The probability that a brake system will operate successfully during the first year can be estimated by the special multiplication law for independent events. The assumption of independence among the four components is entirely reasonable for this application.

$$P(\text{System success}) = P(A) \times P(B) \times P(C) \times P(D) = (0.95)^4 = 0.81$$

Thus, the chances that a new car selected at random will not encounter brake problems during the first year of service is an unexciting 81%. An alternative brake system configuration to the one given above is shown in Table 2. With this configuration, the hydraulic line connecting the master cylinder with the wheel cylinder has been replaced with two hydraulic lines operating in parallel. The primary advantage of this approach is that a failure in either hydraulic line will not result in a system failure (i.e., the brakes will continue to function).

TABLE 1 Schematic of Conventional Brake System

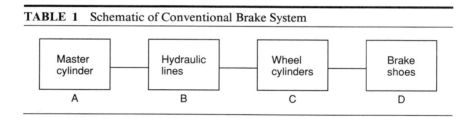

TABLE 2 Alternative Brake System Design

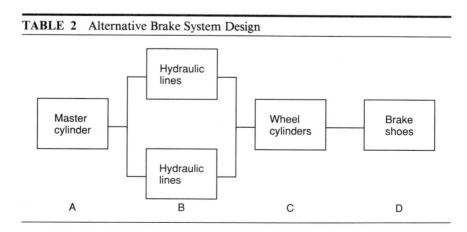

Suggested Questions

1. Develop an overall brake system probability of success for years 2 and 3.
2. What component reliability level is needed to ensure an overall success rate of 90%?
3. Discuss the assumption of independence of each component.
4. How many new cars will require brake service in the first year of operation, based on domestic sales of 10 million units?
5. What is the probability of success of the alternative configuration given in Table 2, assuming that the two components operating in parallel are independent?
6. Develop a new brake system where two of the components are operating in parallel. Compute the probability of operating success, assuming that the parallel components are independent and that the reliability of any one component is still 95% for the first year of operation.

Chapter 5

Probability Distributions

But to us, probability is the very guide to life.
B. J. Butler

CHAPTER OUTLINE

5.1 Introduction
5.2 Example Management Problem: Seven-Day Tire Company
5.3 Characteristics of a Probability Distribution
5.4 Probability Distributions
5.5 Computer Analysis
5.6 Practical Applications
5.7 Case Study: Drack Industries
5.8 Summary
5.9 Glossary
5.10 Bibliography
5.11 Problems
5.12 Cases

CHAPTER OBJECTIVES

The primary objectives of this chapter are to develop an understanding of

1. The general attributes of a probability distribution.
2. The differences between discrete and continuous probability distributions.
3. The basic characteristics of the normal, uniform, binomial, hypergeometric, and Poisson distributions.
4. How to make probability assignments using computer analysis.

The recent onslaught of Hurricane Gilbert (September 1988) has heightened interest in hurricane prediction and control. Alice was the largest hurricane observed in this century, with maximum winds measured at 180 miles per hour. In the early 1960s, the idea of seeding was introduced as a potentially effective strategy for reducing the impact of these monster storms. This strategy had its debut in 1969 with Hurricane Debbie. The results of two successive seedings of Debbie showed reductions in the peak wind speed of 31% and 15%, respectively.

> **HISTORICAL NOTE**
>
> Carl Friedrich Gauss (1777–1855) was one of the early pioneers in the development of probability theory. Gauss made important contributions in many fields of science and mathematics. He is credited with inventing the telegraph, the magnetometer, and the photometer. One of his most famous contributions involved the use of the so-called bell-shaped, or normal, curve. He was one of the first to describe the variations observed in the motions of planets with the normal distribution. To this day, in many parts of the world, the normal curve is often referred to as the Gaussian curve.

A study sponsored by the U.S. Department of Commerce in the early 1970s indicated that changes in the maximum surface wind speed 12 hours before landfall could be described by a normal probability distribution with a mean change of 0 miles per hour and a standard deviation of approximately 16 miles per hour.* This study was based on an average hurricane wind speed of 100 miles per hour. The study also related wind speed to the corresponding property damages, as shown in Figure 5.1. For example, a 100 mph hurricane will, on the average, yield approximately $100 million in damages (1972 dollars). On the other hand, a mere 16% increase (i.e., one standard deviation) in wind speed would nearly double damage costs. Armed with the information provided by the normal curve, the management of the U.S. Weather Service can better evaluate whether or not to employ the costly seeding strategy.

5.1 INTRODUCTION

The previous chapter introduced the notion of probability and demonstrated its role in managerial decision making. In many decision-making situations, however, the process of assigning probabilities using the classical theory outlined in Chapter 4 is inappropriate. Instead, many decision makers often turn to theoretically or empirically derived probability distributions for analyzing the impact of uncertainty on a given problem situation. Accordingly, probability distributions play an important role in busi-

---

* R. A. Howard, J. E. Matheson, and D. W. North, "The Decision to Seed Hurricanes," *Science* (June 16, 1972), pp. 1191–1202.

FIGURE 5.1 Distribution of Wind Speed and Property Damages

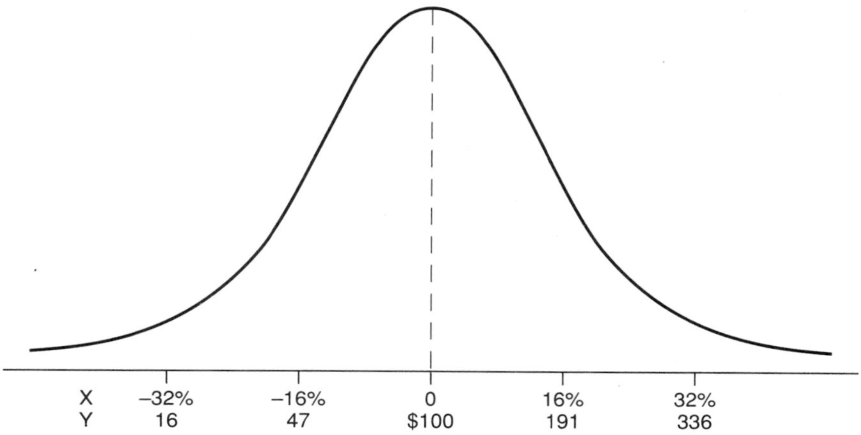

```
X    -32%   -16%    0      16%    32%
Y     16     47    $100    191    336
```

X = Percentage change in maximum wind speed.
Y = Resultant property damage ($ millions).

ness decision making. They provide the decision maker with the capability of assigning probability values to a wide range of decision-making situations. Examples include the probability of winning a government contract, the probability of completing a construction project on time, and the probability of getting more than 40,000 miles on a set of tires. Probability distributions are closely linked to frequency distributions. Recall from Chapter 3 that a frequency distribution is a tabular summary that reports the number of observations from the data set that fall into a specified number of mutually exclusive categories. In fact, one can view a probability distribution as a theoretical frequency distribution. That is, a frequency distribution describes the observed outcomes, whereas a probability distribution describes all possible outcomes. It is this characteristic that makes probability distributions so useful in the business decision-making process.

Unfortunately, no single probability distribution can describe all possible business situations or experiments. Consequently, there are a number of probability distributions, and the distribution used depends on the specifics of the situation at hand. In addition to providing an overview of the general characteristics of probability distributions, this chapter introduces the five most frequently used probability distributions: normal, uniform, binomial, hypergeometric, and Poisson.

5.2 EXAMPLE MANAGEMENT PROBLEM: SEVEN-DAY TIRE COMPANY

The Seven-Day Tire Company was founded nearly 25 years ago on the simple premise of offering convenience and value to the motoring public. The firm operates more than 30 outlets throughout the Southwest. Recently, as part of an extensive marketing effort, the company introduced a new 25,000-mile guarantee program. This program will replace, free of charge, any steel-belted radial tire that is deemed worn out after less than 25,000 miles of operation. The company's general manager, Jack Burton, is interested in determining the distribution of mileage for this class of tire. An inquiry into the company's management information system indicated an average of 30,000 miles with a standard deviation of 5,000 miles at the time of replacement. Specifically, Mr. Burton would like to know the percentage of customers who will

- obtain more than 40,000 miles.
- obtain between 28,000 and 35,000 miles.
- qualify for the company's 25,000-mile guarantee program.

INTERNATIONAL VIGNETTE

The management of the North Sea Machine Tool Factory (NSMT) was concerned about the level of orders for NSMT's special tool/transfer complex (TTC) product line. The TTC line, while accounting for less than 10% of the total systems produced at the factory, generated approximately 60% of the factory's revenue. Founded in 1948, NSMT was one of China's first state-owned enterprises to produce machine tools. With the implementation of "expansion of autonomy" throughout China in the early 1980s, NSMT's management and employees would bear the burden of any profit reductions due to lower sales. In this regard, Shi Pu Lan, the director of the Production and Planning Department, was evaluating the factory's operations and alternatives for the coming decade.

Director Shi was convinced that the lead time for manufacturing the TTC line was excessive. These production delays resulted in NSMT losing customers to other factories that were emphasizing customer service and faster deliveries. Director Shi was convinced, however, that NSMT's superior quality would permit the factory to regain market share if the production delays in the TTC line could be solved. A key aspect in reducing the delay time was to improve the forecasting for purchased parts. Typically, the time from order to delivery was about six months. The current policy was to place an order for approximately one-half of a year's parts requirements. Director Shi decided that an improved forecast of part requirements could be achieved through the use of a probability analysis of actual demand.

5.3 CHARACTERISTICS OF A PROBABILITY DISTRIBUTION

Key idea 1

The following list highlights the primary characteristics of a probability distribution:

- The probability of an individual event is between 0 and 1.
- The probability of all such events is equal to 1.
- The mean of the random variable describing the probability distribution (called the *expected value*) is determined by weighting each of the possible values by the corresponding probability value.

5.4 PROBABILITY DISTRIBUTIONS

Figure 5.2 summarizes the most commonly used probability distributions. Each of these probability distributions will be discussed in turn. Each presentation will include a description of the basic probability model along with

FIGURE 5.2 Summary of the Most Frequently Used Probability Distributions

Distribution	Type	Descriptive Parameters	Applications
Normal	Continuous	Mean; standard deviation	Tire wear; test scores
Uniform	Continuous	Upper and lower limits	Production output
Binomial	Discrete	Number of trials; probability of occurrence	Quality control; sports outcomes
Hypergeometric	Discrete	Number of trials; population size; number of successes	Lottery design; group organization
Poisson	Discrete	Mean	Waiting lines; car accidents

several business examples. First, however, the reader should become familiar with several important new concepts that play a key role in the use of probability distributions.

BASIC CONCEPTS

Key idea 2

As indicated in the introduction, a probability distribution describes the chances of occurrence of a given variable value. In general, any variable whose values are determined in a random manner is referred to as a **random variable.** A random variable may be viewed as a value that changes from occurrence to occurrence with no predictable pattern. Each possible value of a random variable, however, will occur with a certain probability. The probability of a specific value occurring is determined by its probability distribution. Typically, probability distributions are classified as either *continuous* or *discrete.* For a **continuous probability distribution,** the variable of interest can take on any value within a given range. For example, the fuel economy for a car, measured in miles per gallon, can take any value in a continuous range from a low of 10 mpg to a high of 40 mpg. A **discrete probability distribution,** as the name implies, can take on only specified values over a given range. For example, Figure 5.3 shows a discrete probability distribution where the random variable A represents the number of customer arrivals per minute at a fast-food drive-up window. Continuous probability distributions are usually described with a probability density function (PDF), whereas discrete distributions are described using a probability mass function (PMF). Both types of distributions will be featured in the following sections.

FIGURE 5.3 Probability Distribution for Drive-Up Window Example

Number of Arrivals (A)	Probability $P(A = x)$	Cumulative Probability
0	.37	.37
1	.37	.74
2	.18	.92
3	.06	.98
4	.02	1.00
5	.00	1.00
	1.00	

The information in Figure 5.3 can be presented in graphical form as illustrated in Figure 5.4. Panel A shows the probability distribution directly, and panel B depicts a **cumulative probability distribution** (CPD). As the name implies, a CPD shows the probabilities of random variables being less than or equal to a given value. A cumulative probability distribution for a random variable is developed by simply adding the probability value of the

FIGURE 5.4 Discrete (A) and Cumulative (B) Probability Distributions for Drive-Up Window Example

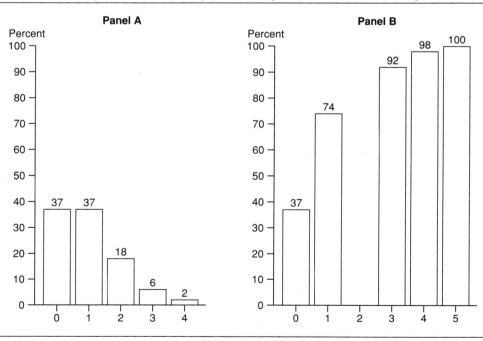

random variable of interest to all preceding probability values. For example, the probability that the number of arrivals is less than or equal to three per hour is 98% (i.e., .37 + .37 + .18 + .06).

Summary Measures for Random Variables

A random variable as described by a probability distribution can be summarized easily by using measures of central tendency and dispersion similar to those introduced in Chapter 3. The most important measure of location for a probability distribution is called the **expected value** of the random variable. The expected value of a discrete random variable, $E(X)$, is calculated in much the same way as the weighted mean. The calculation process is shown in Figure 5.5 for the drive-up window problem. Notice that the probability values correspond to the class frequencies used in computing the weighted average. In this case, the computed average arrival rate is 0.99 customers per minute.

$$E(X) = \Sigma X \times P(X)$$

The procedure for determining the variance of a random variable, $\text{Var}(X)$, is similar to the one used in calculating the expected value. Figure 5.6 shows the computational procedure for determining the variance and standard deviation, $SD(X)$, for the drive-up window problem. Here, the deviation between the random variable and the expected value is squared, and the result is multiplied by the probability of the corresponding random variable. The variance is simply the sum of the weighted values. The variance for the drive-up window problem is 0.60, and the standard deviation, the square root of the variance, is 0.77.

$$\text{Var}(X) = \Sigma[X - E(X)]^2 \times P(X)$$

FIGURE 5.5 Demonstration of Calculating Expected Value for Drive-Up Window Example

Number of Arrivals $(A = X)$	Probability $P(A = X)$	Weighted Average $P(A = X) \times X$
0	.37	0
1	.37	0.37
2	.18	0.36
3	.06	0.18
4	.02	0.08
5	.00	0
	1.00	$E(X) = 0.99$

FIGURE 5.6 Demonstration of Calculating Variance and Standard Deviation for Drive-Up Window Example

Arrivals (A)	Probability $P(A = X)$	Deviation $(X - E(A))$	Squared Deviation $(X - E(A))^2$	Weighted Value $P(A = X) \times [X - E(A)]^2$
0	.37	−0.99	0.98	0
1	.37	0.01	0	0
2	.18	1.01	1.02	0.18
3	.06	2.01	4.04	0.24
4	.02	3.01	9.06	0.18
5	.00	4.01	16.08	0
	1.00			$\text{Var}(X) = 0.60$
				$\text{SD}(x) = \sqrt{\text{Var}(X)} = 0.77$

NORMAL DISTRIBUTION

Key idea 3

The normal probability distribution is perhaps the most frequently used distribution in statistical analysis. As shown in Figure 5.2, the **normal distribution** is a continuous function described by a mean and a standard deviation. The basic characteristics of the normal distribution are listed below:

- The distribution is symmetrically shaped (i.e., bell-shaped) around a center line.
- The mean, median, and mode are all located at the center line.
- The two tails of the distribution extend infinitely in both directions (asymptotic to the baseline).

Many business situations can be described by the normal distribution (e.g., tire wear, automotive fuel economy, life of light bulbs, and diameter of machine parts). As outlined in Chapter 3, the normal curve is described by a mean and standard deviation. That is, the normal curve is a two-parameter

FIGURE 5.7 Examples of Population Distributions Described by the Normal Curve

Population Distribution	Mean	Standard Deviation
Annual income ($)	18,000	6,000
Tire mileage (miles)	30,000	5,000
Ball bearing diameter (in.)	0.5	0.1
Student GPA	2.5	1.0
Project construction time (mo.)	15	3
Cereal filling machine (oz.)	10	0.25

FIGURE 5.8 Normal Probability Distributions with Equal Means but Unequal Standard Deviations

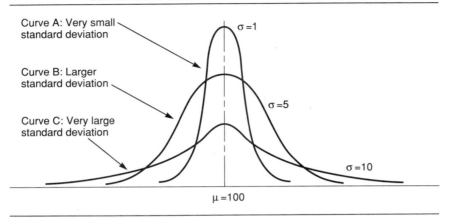

distribution. Figure 5.7 presents examples of population distributions that can be described by the normal curve.

Figure 5.8 shows several normal distributions with equal means but different standard deviations. Similarly, Figure 5.9 depicts several normal distributions with unequal means but with the same standard deviation. Finally, Figure 5.10 presents several normal curves with different means and different standard deviations.

A basic characteristic of the normal distribution, regardless of the value of the mean or standard deviation, is that the total area under the curve is 1. Furthermore, since the normal curve is symmetrical, half of the area falls to

FIGURE 5.9 Normal Probability Distributions with Different Means but Equal Standard Deviations

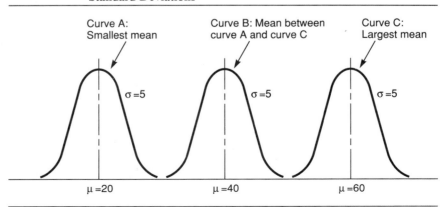

FIGURE 5.10 Normal Probability Distributions with Different Means and Different Standard Deviations

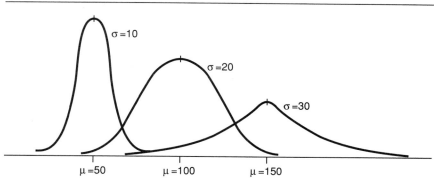

the left of the mean line, and half falls to the right. Because the area under the normal curve is equal to 1, specific areas can be viewed as probabilities. For example:

- Approximately 68% of the area under the curve is within ± one standard deviation of the mean.
- Approximately 95% of the area under the curve is within ± two standard deviations of the mean.
- Approximately 99% of the area under the curve lies within ± three standard deviations of the mean.

These three situations are shown graphically in Figure 5.11. Clearly, there are many other possible areas or probabilities of interest. Furthermore, there are an infinite number of mean and standard deviation combinations, each yielding a different normal curve. Consequently, the development of probability estimates for a problem involving a normal distribution requires the formulation and analysis of a specific curve. The process of formulating a normal curve for each new problem would obviously be very time-consuming and tedious. Fortunately, a simple formula is available to transform any normal curve to the so-called standard normal curve.

The Standard Normal Distribution

The **standard normal distribution** is a normal probability distribution with a mean of 0 and a standard deviation of 1. The formula for transforming any normal curve to standard normal is

$$Z = \frac{X - \mu}{\sigma}$$

FIGURE 5.11 Comparison of the Area under the Normal Curve for One, Two, and Three Standard Deviations from the Mean

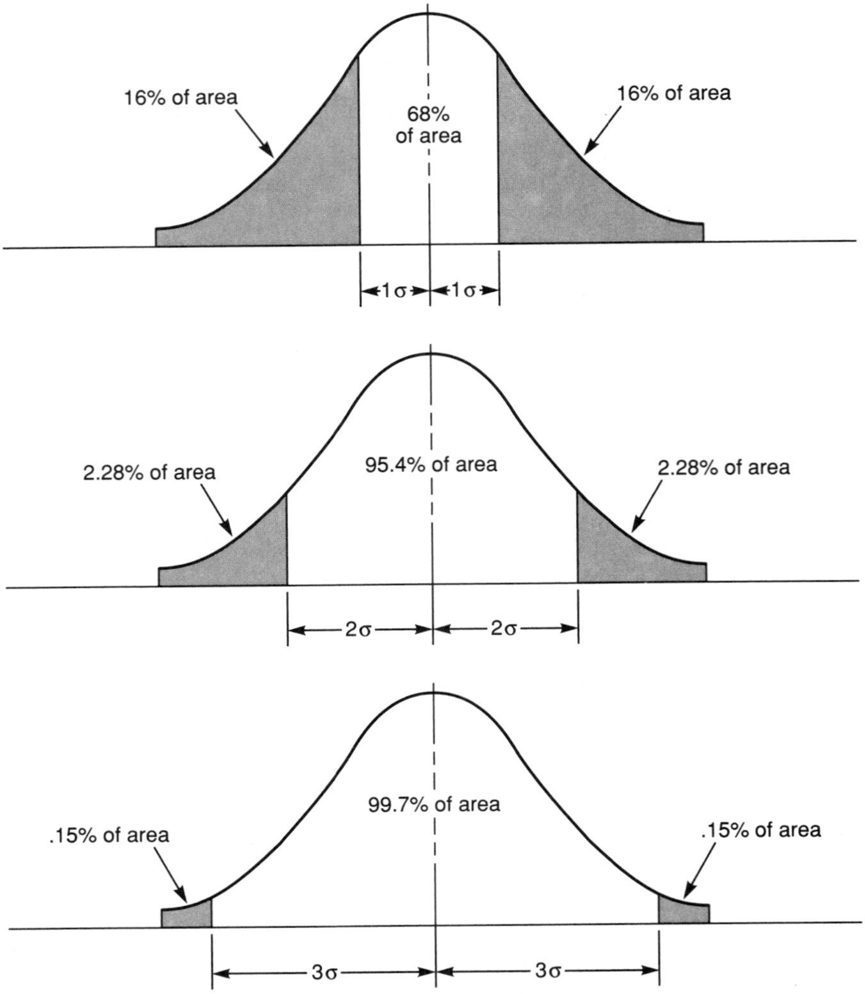

where X is the value of interest, μ is the population mean, and σ is the population standard deviation.

For example, consider a local bank that processes, on the average, 100,000 checks per week with a standard deviation of 10,000 checks. The bank manager wishes to know the probability of processing more than 110,000 checks in a week. The Z value for this problem, based on the above model, is 1 [i.e., (110,000 − 100,000)/10,000]. Therefore, the probability that the number of checks will exceed 110,000 is approximately .16. This estimate was derived from the earlier statement that approximately 68% of the area of the normal curve lies within ± one standard deviation of the mean. Consequently, about 34% of the area lies between the mean and the upper limit of 110,000. Therefore, about 16% is above the upper limit, since the area to the right of the mean is .5. Figure 5.12 shows the normal curve for this problem. The hatched area to the right of the curve represents the probability of interest for this problem.

In most situations, the computed Z value does not conform to either one, two, or three standard deviations from the mean. In these cases, the probability estimates are obtained from a standard normal distribution table. A standard normal table is given in Appendix A. An inspection of Table A.1 shows that the Z value, to one-tenth accuracy, is reported in the first column, and subsequent columns depict the second decimal. The entries reported inside the table are probabilities. For example, a Z value of 0.85 yields a probability or area of .3023. This probability value can be located via the intersection of the row containing 0.8 in the first column and the column headed by the value .05. The standard normal table can be used to derive probability values from any given Z value or a Z value from any given probability value. The use of the standard normal table has diminished somewhat in recent years as a result of the advent of the personal computer. Several exercises involving the standard normal distribution will be introduced in Section 5.5.

Most questions involving the normal distribution are cumulative in nature. That is, the interest is in determining the probability of a random variable being greater than or less than a given value, or between two given values. For example, the bank manager might wish to know the probability that fewer than 100,000 checks (or between 90,000 and 110,000 checks) would be processed in a given week. Occasionally there is interest in deter-

FIGURE 5.12 Normal Distribution for the Bank Check Processing Problem

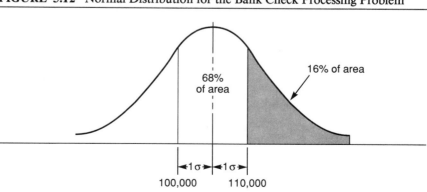

FIGURE 5.13 Normal Distribution for Cereal Filling Problem

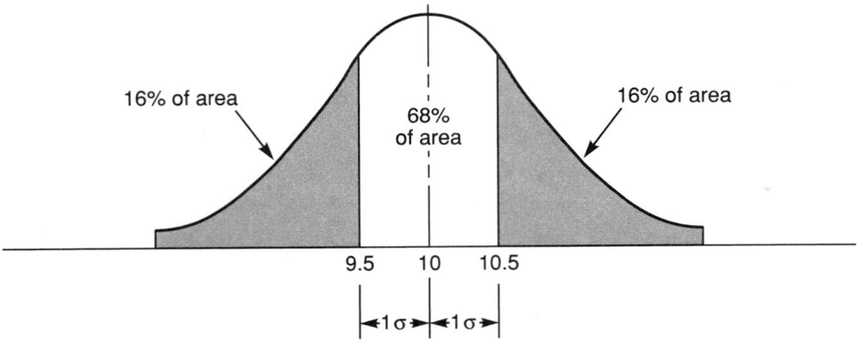

mining the probability of a single value occurring, even though, theoretically speaking, this probability is near 0 for a continuous distribution. To determine the probability in these cases, the value of interest is placed in a range of ±0.5. To illustrate, a cereal box filling machine has a mean filling volume of 10 oz and a standard deviation of 0.5 oz. The production manager wishes to know the chances of selecting a box at random that contains exactly 10 oz. The appropriate Z value for this problem is

$$Z = \frac{10.5 - 10}{0.5} = 1$$

Notice that the value of interest, 10 oz, was increased to 10.5. This suggests that any value between 10 and 10.5 would be considered 10 oz. Furthermore, a second Z value needs to be determined to cover the range between 9.5 and 10. This results in a Z value of -1. Thus, the chances of selecting a 10-oz box is represented by the range between $+1$ and -1 standard deviations from the mean, or approximately 68%. Figure 5.13 shows the normal curve for this problem. The white area represents the probability of obtaining a box containing 10 oz (i.e., 68%).

Probability Density Function

The normal curve can be described with the following density function:

$$f(x) = \frac{1}{\sqrt{2\pi} \cdot \sigma} e^{-0.5\left(\frac{x-\mu}{\sigma}\right)^2}$$

where:

$\pi = 3.14159$
$e =$ Base of natural log (2.182818)

μ = Population mean
σ = Population standard deviation

The density function $f(x)$ provides a measure of the height of the normal curve for different values of the random variable x. Notice that the relationship contains the two most important constants in mathematics: π and e. Although this model is somewhat imposing, particularly compared with the other probability functions presented in this chapter, it is basically described in terms of the mean and standard deviation (two very well known statistics). Perhaps of more importance is the fact that this function is not used directly. Instead, it is transformed using the Z model introduced above, which yields the following standard normal density function:

$$f(z) = \frac{1}{\sqrt{2\pi}} e^{-0.5z^2}$$

This density function also measures the height of the normal curve for various values of Z. For example, when $Z = 0$, the height of the curve becomes $1/\sqrt{2\pi}$, or approximately .3989. This function is used to generate the standard normal table presented in Appendix A. The actual probability values reported in Appendix A were obtained by integrating this function from $Z = 0$ to the Z value of interest (e.g., one standard deviation from the mean).

Seven-Day Tire Company

Recall that the manager of the Seven-Day Tire Company (see Section 5.2) is interested in determining

- The percentage of customers who will drive more than 40,000 miles.
- The percentage of customers who will drive between 28,000 and 35,000 miles.
- The percentage of customers who will qualify for the company's 25,000-mile guarantee program.

Each of these questions can be analyzed using the standard normal probability distribution. For example, the probability that a motorist will achieve more than 40,000 miles on a set of Seven-Day tires can be developed using the Z transformation shown below:

$$Z = \frac{40{,}000 - 30{,}000}{5{,}000} = 2$$

The probability value for a Z value of 2 (from Appendix A) is .4772. This corresponds to the chances that a motorist will drive between 30,000 and 40,000 miles. Therefore, the probability that a motorist will achieve more than 40,000 miles is .0228 (i.e., .5 − .4772). The normal distribution for this

FIGURE 5.14 Normal Distribution Showing the Probability of Obtaining More than 40,000 Miles

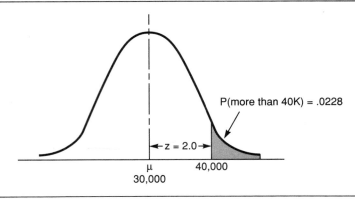

problem is shown in Figure 5.14. The hatched area to the right indicates the desired probability.

Determining the chances of a motorist driving between 28,000 and 35,000 miles requires the computation of two Z values, as shown below:

$$Z_1 = \frac{35,000 - 30,000}{5,000} = 1$$

$$Z_2 = \frac{28,000 - 30,000}{5,000} = -.4$$

A Z value of 1 corresponds to a probability of .3413 to the right of the mean.

FIGURE 5.15 Normal Distribution Showing the Probability of Obtaining between 28,000 and 35,000 Miles

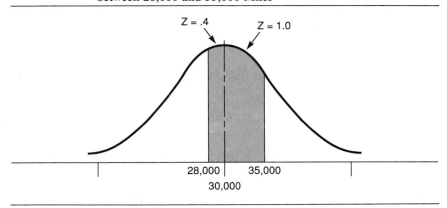

FIGURE 5.16 Normal Distribution Showing the Probability of Obtaining Less than 25,000 Miles

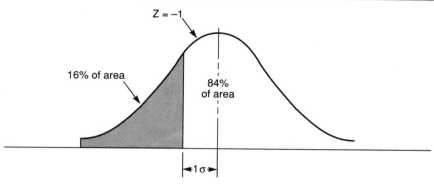

A Z value of $-.4$ corresponds to a probability of .1554 to the left of the mean. Summing these two estimates yields a combined probability value of .4972. This total suggests that there is approximately a 50% chance that a motorist selected at random will obtain between 28,000 and 35,000 miles on a set of tires purchased from Seven-Day. Figure 5.15 shows the normal distribution for this problem. Notice that the hatched area extends to both sides of the mean line.

Finally, the general manager wishes to know the percentage of motorists who will qualify for the company's 25,000-mile minimum guarantee program. The appropriate Z value for this problem is:

$$Z = \frac{25,000 - 30,000}{5,000} = -1$$

This result suggests that approximately 84% (i.e., .3416 + .5) of the customers who purchase Seven-Day's steel-belted radials will drive 25,000 miles or more. Consequently, approximately 16% will drive less than 25,000 miles and will therefore qualify for the replacement program. The normal distribution for this problem is shown in Figure 5.16. The area to the left of the -1 Z value represents the percentage of Seven-Day's clients who will receive replacement tires under the current policy.

UNIFORM DISTRIBUTION

Key idea 4

The **uniform distribution** is the simplest probability distribution used in statistical analysis. It is a continuous distribution, and its basic characteristic is that each value of the random variable within the given range has an equal chance of occurring. Examples include production output, annual unemployment rates, and arrival and departure schedules.

Probability Density Function

The probability density function for the uniform distribution is given below:

$$f(x) = \frac{1}{b-a} \quad \text{for } a \leq X \leq b; \text{ otherwise } f(x) = 0$$

This relationship defines the height of a horizontal line above the x axis. Figure 5.17 presents a uniform probability distribution representing national unemployment for limits $a = 4\%$ and $b = 8\%$. In this case, the height above the x axis is .25.

The area of probability under the entire distribution is, by definition, equal to 1. To determine probability estimates for specific values of the random variable X, the following cumulative probability function can be used:

$$P(X \leq L) = \frac{L-a}{b-a} \quad \text{if } a \leq X \leq b; \text{ otherwise } P(X \leq L) = 0$$

For example, if $L = 6$ then

$$P(X \leq 6) = \frac{6-4}{8-4} = .50$$

Thus the area between $X = 4$ and $X = 6$ is equal to 50% of the total area. This result is shown graphically in Figure 5.18 (i.e., the shaded area).

FIGURE 5.17 Uniform Probability Distribution for National Unemployment Rate

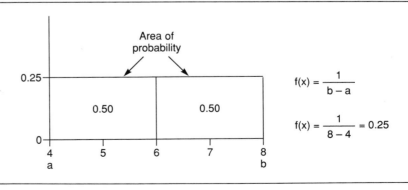

FIGURE 5.18 Cumulative Uniform Probability Distribution for National Unemployment Rate

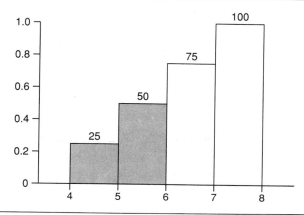

Summary Measures

The summary measures for the uniform distribution are relatively simple to determine and are reported in Figure 5.19. For example, the expected value for the preceding problem is calculated as follows:

$$\mu = \frac{a+b}{2} = \frac{4+8}{2} = 6\%$$

Thus, the average unemployment rate for the year is 6%. The variance and standard deviation for this problem are

$$\sigma^2 = \frac{(b-a)^2}{12} = \frac{(8-4)^2}{12} = 1.33$$

$$\sigma = \sqrt{1.33} = 1.15\%$$

FIGURE 5.19 Summary Measures for the Uniform Probability Distribution

Mean: $\mu = \dfrac{a+b}{2}$

Variance: $\sigma^2 = \dfrac{(b-a)^2}{12}$

Standard deviation: $\sigma = \sqrt{\sigma^2}$

FIGURE 5.20 Uniform Distribution for Ampex Manufacturing Problem

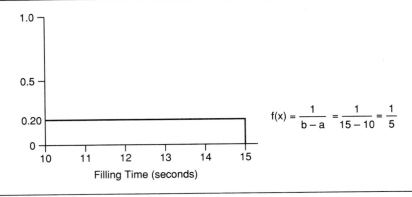

Ampex Manufacturing, Inc.

The production manager at Ampex Manufacturing is interested in determining summary statistics and the probability of various filling rates for a new packaging process. The new system takes between 10 and 15 seconds to fill a standard-size carton. The manager believes that the packaging process can be described by the uniform distribution presented in Figure 5.20.

The mean and standard deviation for this problem are as follows:

$$\mu = \frac{a+b}{2} = \frac{10+15}{2} = 12.5 \text{ seconds}$$

$$\sigma = \sqrt{\frac{(b-a)^2}{12}} = \sqrt{\frac{(15-10)^2}{12}} = 1.44 \text{ seconds}$$

Figure 5.21 presents cumulative probability estimates for various filling situations. These selected probability estimates were computed using the cumulative probability distribution reported above. For example, the probability that the time to fill a carton will take between 10 and 11 seconds is 0.20.

FIGURE 5.21 Probability of Occurrence of Various Filling Rates

Filling Time (seconds)	Probability of Occurrence
10–11	.20
12–14	.40
>14	.20
11.5–13	.30

BINOMIAL DISTRIBUTION

Key idea 5

Many business decisions result in one of two possible outcomes. Examples include winning or losing a contract, receiving or not receiving the promotion, and detecting or not detecting a defective part. Under some circumstances these types of problems can be described with the well-known **binomial probability distribution.** This discrete probability function was generated from an experiment called a **Bernoulli process** (named in honor of James Bernoulli, 1654–1705). The Bernoulli process possesses the following characteristics:

- A sequence of n identical trials of a random experiment.
- Only two possible outcomes for each trial.
- The probability of an outcome is fixed.

A sequence of coin tosses provides an ideal example. Each coin toss produces one of only two possible outcomes: a head or a tail. Furthermore, for a fair coin, the chances of obtaining a head or a tail remain constant at 50–50. Figure 5.22 presents the sample space and corresponding probability estimates for an experiment involving flipping a coin three times. The total number of permutations is 8 (i.e., 2^3), and since each event has an equal likelihood of occurring, the probability of a single event is 1/8. A number of different questions can be answered directly from this sample space. For example, what is the probability of obtaining exactly two heads? The answer is three (see Figure 5.22). Therefore, the probability of obtaining exactly two heads is 3/8. For sample spaces involving a small number of outcomes, this process of delineating the sample space and counting events of interest (e.g., two heads) is reasonably efficient. However, as the size of the number of trials increases, this process becomes less and less efficient.

A much more efficient alternative in most problem situations that can be described with the Bernoulli process is to use the **binomial probability distri-**

FIGURE 5.22 Sample Space and Probability of Occurrence for Coin-Flipping Experiment

Event	Probability
H,H,H	1/8
H,H,T	1/8
H,T,H	1/8
H,T,T	1/8
T,H,H	1/8
T,T,H	1/8
T,H,T	1/8
T,T,T	1/8

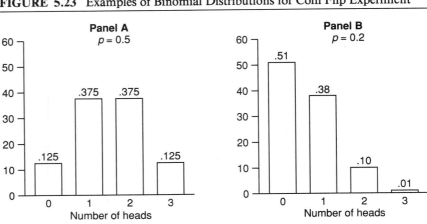

FIGURE 5.23 Examples of Binomial Distributions for Coin Flip Experiment

bution. Typically, this distribution can be accessed through either the probability mass function, a probability table, or computer software. All three alternatives yield exactly the same probability estimates for a given problem. To illustrate the use of the binomial distribution in the coin flip experiment, consider the graphics shown in Figure 5.23. These graphics depict the probability of occurrence of various outcomes for a fair coin (panel A, $p = 0.5$) and an unfair coin (panel B, $p = 0.2$).

These probability distributions were developed from the two sample spaces (one based on a fair coin and the other based on an unfair coin). Notice that the distribution in panel A is symmetrical, whereas the distribution in panel B is skewed to the right. This shift in the shape of the binomial distribution is based on the value of p. It should not be surprising that the probability of obtaining no heads should increase as the chances of obtaining a single head become smaller.

The continual struggle of the Los Angeles Dodgers provides a more interesting example of a Bernoulli process. Suppose the Los Angeles Dodgers are faced with a serious challenge if they are to advance to the league championship series with the New York Mets. They are tied with their arch rivals, the San Francisco Giants, with three games left in the regular season. To win their division, they must beat the Giants in two out of the last three games. This situation can be described by a Bernoulli process, assuming that the probability of winning a single game remains constant. This season the Dodgers have won 9 out of the 15 games played against the Giants. A probability tree for this problem is given in Figure 5.24. Notice that with this tree the win-win and lose-lose situations actually are terminated at the end of two nodes. That is, if the Dodgers win the first two games, they will win the division regardless of the outcome of the third game; similarly, if they lose the first two games, the Giants win the division. An inspection of the probability

5.4 Probability Distributions

FIGURE 5.24 Probability Tree for the Dodgers Series with the Giants

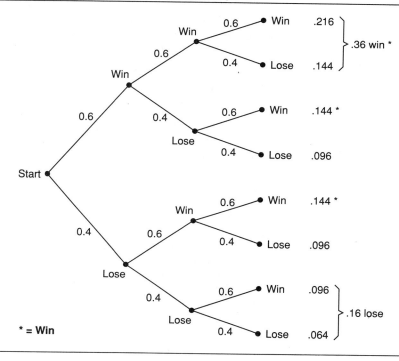

tree shows that the Dodgers have a 64.8% chance of winning the divisional championship (indicated by summing the probabilities marked *). Interestingly, the chances of the Dodgers winning the series are higher than the chances of them winning a single game, i.e., p (winning the championship) = 0.64 versus p (winning a single game) = 0.60.

Probability Mass Function

Fortunately, binomial probability estimates need not be developed from sample spaces or probability trees. Instead, the following probability mass function can be used:

$$P(x = S|n,p) = C_x^n \times p^x \times (1-p)^{n-x}$$

where:

S = Number of successes
n = Number of trials
p = Probability of a single success occurring
C_x^n = Number of combinations of n things taken x at a time
  = $n!/x!(n-x)!$

Notice that this relationship utilizes the combination model introduced in Chapter 4. This should not be surprising in view of the fact that this model simply indicates the number of combinations associated with a particular outcome of a given situation (e.g., for the Dodgers problem, with $n = 3$ and $x = 2$, $C = 3$; that is, there are three different ways that the Dodgers can win two games). To illustrate this relationship, consider the probability of the Dodgers winning three games in a row. The computations for this situation are

$$P(x = 3 | n = 3, p = .6) = C_3^3 \times (.6)^3 \times (.4)^0 = .216$$

This result agrees with the probability value shown in Figure 5.24 (i.e., top branch). The above probability mass function has been used to create a binomial probability table for selected values of n and p (see Table A.5 in Appendix A). This table is even more convenient to use than the mass function given above. Referring to Appendix A, when $n = 3$, $p = .6$, and $x = 3$, the corresponding probability is .216. Appendix A also includes a cumulative binomial probability distribution. This table can be used to develop probability estimates for x or fewer successes of a Bernoulli experiment. For example, when $n = 3$, $p = .6$, and $x \leq 2$, then the cumulative probability is .7840. That is, there is a 78.4% chance of having two or fewer successes with an experiment involving three trials where the probability of a single success is .6.

Summary Measures

The summary measures for the binomial distribution are illustrated in Figure 5.25. In summary, measures for the Dodgers baseball problem are shown below:

$$\mu = n \times p = 3(.6) = 1.8 \text{ games}$$
$$\sigma^2 = n \times p(1 - p) = 3(.6)(.4) = 0.72$$
$$\sigma = \sqrt{n \times p(1 - p)} = \sqrt{0.72} = 0.85 \text{ games}$$

These results show that the Dodgers, on the average, will win 1.8 games. This is slightly less than the required two out of three games required to make the playoffs. Will the Dodgers triumph? As a means of verifying these results, consider the computations depicted in Figure 5.26. These calculations show

---

FIGURE 5.25 Summary Measures for the Binomial Probability Distribution

Mean: $\mu = n \times p$
Variance: $\sigma^2 = n \times p(1 - p)$
Standard deviation: $\sigma = \sqrt{\sigma^2}$

Figure 5.26 Expected Value Computations for the Dodgers Baseball Problem

Number of Successes (X)	Probability P(X)	Weighted Value X × P(X)
0	.0640	0
1	.2880	.2880
2	.4320	.8640
3	.2160	.6480
		1.8000

the method for computing the expected value (i.e., mean) for the baseball problem. Notice that the expected value is identical to the mean value obtained using the summary measures method (i.e., 1.8 games).

Johnson Pharmaceutical

The quality assurance manager at Johnson Pharmaceutical is interested in assessing the quality of the firm's major bottle supplier. A review of the subcontractor's past performance indicates that approximately 5% of the supplied bottles have been defective. A recent sample of 10 bottles from a lot containing 200 bottles revealed 2 defective units. Figure 5.27 presents a binomial distribution for this problem ($n = 10$ and $p = .05$). The computed sample mean is 0.5 defectives. Furthermore, the chances of a sample of 10 bottles containing 2 or more defectives based on an inspection of Figure 5.27 is less than 10%. The quality assurance manager is clearly within company policy to reject the latest subcontractor shipment.

Approximating the Binomial Distribution with the Normal Curve

The normal curve can be used to approximate the binomial distribution in cases when the number of trials exceeds 20. The use of the normal curve in these cases eliminates the need for the large and cumbersome binomial tables. In fact, an inspection of the binomial table given in Appendix A reveals that the maximum size of the table is limited to 20 trials. Historically, binomial problems involving sample sizes above 20 were solved using the normal approximation. However, the advent of computer-based models of the binomial distribution has reduced the importance of this approximation. The process for transforming the binomial to the normal is given by the

158 Chapter 5 Probability Distributions

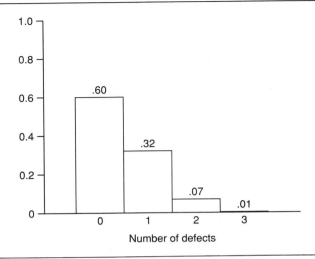

FIGURE 5.27 Binomial Distribution for Quality Acceptance Plan for Johnson Pharmaceutical

following equations:

$$\mu = n \times p$$
$$\sigma = \sqrt{n \times p \times (1 - p)}$$

For example, consider a procedure that tests a car for excessive exhaust emissions. Suppose that the historical data indicate that 40% of all vehicles undergoing the test fail. The manager of the facility would like to know the probability that at least half of the next 20 vehicles will not pass the test. An inspection of Table A.5 (the cumulative binomial probability distribution) with $n = 20$, $p = .4$, and $x \geq 10$ reveals that the chances of 10 or more vehicles failing the test is .1275 (1 − .8725). In contrast, the probability estimate based on the normal approximation is as follows:

$$\mu = .4 \times 20 = 8$$
$$\sigma = \sqrt{20 \times .4 \times .6} = 2.2$$

These estimates can now be transformed into a Z value using the Z transformation equation. One slight adjustment is made to reflect the fact that a continuous distribution is being used to estimate a discrete distribution. Specifically, a value of 10.5 is used for X instead of 10, since for a discrete distribution, the next value above 10 is 11. In general, this type of adjustment is required whenever the problem involves a cumulative-type analysis. The specific transformation is shown below:

$$Z = \frac{10.5 - 8}{2.2} = 1.14$$

The corresponding probability value for a Z value of 1.14 from the standard normal curve in Appendix A is .1271 (.5 − .3729). Notice how closely this probability estimate matches the one developed directly from the binomial distribution (.1275).

Some caution is suggested, however, in using the normal curve to approximate the binomial distribution. In many situations the binomial distribution can be highly skewed in either direction. This, of course, is in direct contrast to the normal distribution, which is always symmetrical around the mean. In these cases, the normal curve may not provide accurate estimates for the binomial distribution. As a general rule, the normal curve can be used whenever the estimated mean value (i.e., $n \times p$) exceeds 5.

The Bernoulli process is often characterized as sampling *with replacement* since the probability of an event occurring does not change from trial to trial. For example, a common practice in quality control (QC) is to inspect an item selected at random from a production batch to determine the incidence of defectives in the batch. Sometimes the item is returned to the batch after inspection. Consequently, the chances of selecting a defective item from the batch do not change from inspection to inspection. On the other hand, in some QC applications, the item is not returned to the batch (e.g., destructive testing). In these cases, the probability of selecting a defective item will change as sampled items are not returned. In these cases, a different probability distribution is used to estimate the chances of detecting defective items. This distribution is presented in the following section.

HYPERGEOMETRIC DISTRIBUTION

Key idea 6

The binomial distribution introduced in the previous section is used to find the probability of various outcomes in a Bernoulli process. One of the basic assumptions associated with the Bernoulli process is that the probability of a specific outcome remains constant throughout the experiment. For this to occur, the sampled item must be returned to the population being sampled. For example, the chances of drawing an ace from a standard deck of 52 cards is 4/52, regardless of the number of draws, if the selected card is returned to the deck (sampling with replacement). However, if the selected card is not returned to the deck (i.e., sampling without replacement), then the chance of obtaining a second ace is 3/51 if an ace was obtained on the first draw, or 4/51 if a card other than an ace was selected on the first draw. Although the difference in the probability values seems small in this case, the differences can be significant when the sample size increases.

A number of business situations involve sampling without replacement. A classic example is the quality control practice of destructive testing. Here, the product or component is destroyed during the testing process. For example, samples of sewer pipe from a production batch are routinely hydrostatically tested to determine whether or not the batch meets the stated pressure standards, and the process of testing the pipe could result in product failure.

Consequently, the probability of discovering a defective pipe changes as samples are withdrawn from the batch and not replaced. Using the binomial distribution to describe this situation (i.e., sampling without replacement) can yield erroneous results. In these cases, the **hypergeometric probability distribution** is used. This distribution is based on the following assumptions:

- The population contains N items consisting of S items with a certain characteristic (e.g., defective).
- A sample of size n is selected from the population *without* replacement.
- The probability of obtaining an item with a certain characteristic changes from trial to trial.

One of the simplest examples of the hypergeometric distribution involves the organization of groups. For example, consider that the district attorney is interested in forming a grand jury to investigate the growth of organized crime in southern California. Twenty voting-age adults, half of whom are from the minority community, have been prescreened by a district judge to ensure adequate competence. The district attorney wishes to form a jury consisting of four members selected entirely at random from the prescreened group. The district attorney would like to know the chances that minority citizens will make up at least 40% of the jury (i.e., two to four members). Figure 5.28 shows a probability tree for this problem, where "Min" represents a minority community member and "Maj" represents a nonminority community member. This tree is very similar to the one shown in Figure 5.24, except that the probability of selecting a minority member changes from selection to selection. For example, the probability of selecting four minority members for the group can be expressed as follows:

$$P(X = 4) = (10/20)(9/19)(8/18)(7/17) = .043$$

This result is depicted as the upper branch in the probability tree. The tree also shows the chances that the jury will contain at least two minority members. These situations are depicted by *. Thus, there is approximately a 70% chance that the jury will consist of two or more minority members.

Probability Mass Function

Fortunately, probability estimates for the hypergeometric distribution need not be developed from sample spaces or probability trees. Instead, the following probability mass function can be used:

$$P(x = L|n,N,S) = \frac{C_x^S \times C_{n-x}^{N-S}}{C_n^N}$$

$$= \frac{[S!/x!(S-x)!][(N-S)!/(n-x)!(N-S-n+x)!]}{N!/n!(N-n)!}$$

FIGURE 5.28 Probability Tree for Grand Jury Problem

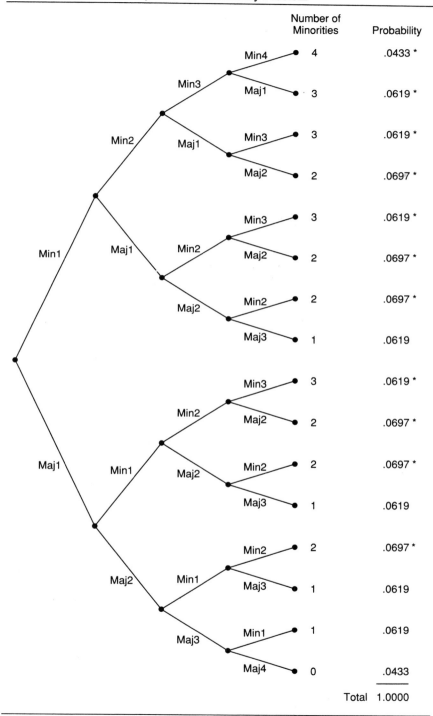

where x is the number of successes from a sample of n taken from a population of size N containing S successes.

For example, the probability that the entire jury will be composed of minority members (i.e., four) can be computed as follows:

$$P(x=4|4,20,10) = \frac{C_4^{10} \times C_0^{10}}{C_4^{20}} = \frac{210 \times 1}{4,845} = .0433$$

Similarly, the probability that the jury will consist of two or three members is

$$P(x=2|4,20,10) = \frac{45 \times 45}{4,845} = .4182$$

$$P(x=3|4,20,10) = \frac{120 \times 10}{4,845} = .2476$$

Notice that these results are consistent with those reported in the probability tree. In particular, the probability that the jury will consist of two or more minority members is approximately .7.

Summary Measures

Summary measures for the hypergeometric distribution are presented below.

Mean: $\mu = n \times p$

Variance: $\sigma^2 = n \times p(1-p) \times \frac{N-n}{N-1}$

Standard deviation: $\sqrt{\sigma^2}$

where $p = S/N$, the probability of success in the first trial.

The summary measures for the jury problem are as follows:

$$\mu = 4 \times .5 = 2 \text{ minority members}$$

$$\sigma^2 = 4 \times .5 \times (1-.5) \times \frac{20-4}{20-1} = 0.842$$

$$\sigma = \sqrt{0.842} = 0.918 \text{ minority members}$$

These results show that, on the average, the jury will contain two minority members. As a means of verifying these results, consider the computations depicted in Figure 5.29. These calculations show the method for computing the expected value (i.e., mean) for the jury problem. Notice that the expected value is identical to the mean value obtained using the summary measures method.

Finally, it should be noted that when the population is large relative to the sample size (i.e., $N \gg n$), the binomial distribution can be used to provide a

FIGURE 5.29 Expected Value Computations for the Jury Problem

Number of Minority Members (X)	Probability P(X)	Weighted Value $X \times P(X)$
0	.0433	0
1	.2476	.2476
2	.4182	.8364
3	.2476	.7428
4	.0433	.1732
	1.0000	2.0000

good approximation to the hypergeometric distribution. For example, Figure 5.30 compares the probability estimates for the jury problem based on the hypergeometric and binomial distributions. These results show that the binomial distribution does provide a fairly accurate estimate of the hypergeometric distribution, even though the population size (20) is not substantially larger than the size of the jury (4).

Danderberg International

The internal auditing team at Danderberg International plans to conduct an audit of the company's accounts receivable. The firm currently has 20 active accounts. The finance manager estimates that approximately half of these accounts are delinquent (i.e., 10 accounts). The standard practice at Dan-

FIGURE 5.30 Comparison of Hypergeometric and Binomial Probability Distributions of the Jury Problem

Number of Minority Members (X)	Hypergeometric Probability P(X)	Binomial Probability P(X)
0	.0433	.0625
1	.2476	.2500
2	.4182	.3750
3	.2476	.2500
4	.0433	.0625
	1.0000	1.0000

FIGURE 5.31 Hypergeometric Distribution of Sampled Accounts for Danderberg International Problem

Number of Delinquent Accounts in Sample (X)		Probability of Occurrence P(X)
0		.0163
1		.1354
2		.3483
3		.3483
4		.1354
5		.0163
	Total	1.0000

derberg is to perform a detailed audit of 25% of all active accounts. The manager of the auditing team wishes to know the chances that the sample will contain at least two delinquent accounts. This problem involves sampling without replacement, since all of the accounts to be audited are selected at once. Figure 5.31 presents a hypergeometric distribution that shows the chances of the sample containing various numbers of delinquent accounts. The probability estimates were developed using the hypergeometric probability mass function. These results show that the most likely number of delinquent accounts in the sample is either two or three. The data show, in fact, that there is nearly a 70% chance that the sample will contain either two or three delinquent accounts. The actual expected number is 2.5 (i.e., $n \times p$, or $5 \times .5$). Clearly, if the audit uncovers less than two accounts (about a 15% probability of occurrence), the manager of the audit team may wish to challenge the delinquent account estimate made by the firm's finance manager.

POISSON DISTRIBUTION

Key idea 7

Many business situations involve the occurrence of a series of single events (e.g., arrivals) in an unpredictable manner over time or space. This is in contrast to both the binomial and hypergeometric processes, which are based on the possibilities of two outcomes occurring over a sequence of trials or repetitions.

Under some circumstances these types of problems can be described with the well-known **Poisson probability distribution** (named in honor of Simeon Poisson, 1781–1840). The Poisson distribution is a discrete probability distribution with the following characteristics:

1. The probability of occurrence or nonoccurrence of an event is the same for each interval of a given length.
2. The occurrence or nonoccurrence of an event in any interval is independent of the occurrence or nonoccurrence in all other intervals.
3. The number of occurrences in any interval ranges between zero and infinity.
4. The simultaneous occurrence of two or more events in any period tends toward zero.

Customers arriving at a bank's drive-up window provide an ideal example of a Poisson process. Consider that customers arrive at the bank at an average rate of six per hour. The bank manager is interested in knowing the characteristics of customer arrivals in order to provide the optimal level of service (the bank can currently process up to eight customers per hour without significant delays). Figure 5.32 shows a Poisson probability distribution for an average arrival rate of six customers per hour (panel A). Panel A shows that the most likely number of arrivals is between five and six per hour. Furthermore, there is an 85% chance that eight or fewer customers will arrive in a one-hour period (this estimate was derived by summing the probabilities from $x = 0$ to $x = 8$. This suggests that approximately 15% of the time, customers arriving at the window will experience some delay.

Figure 5.32 also shows a Poisson probability distribution for an arrival rate of four customers per hour (panel B). Notice the general shift to the left relative to the distribution presented in panel A. In panel B, the most likely arrival rate is between three and four customers per hour. With an average

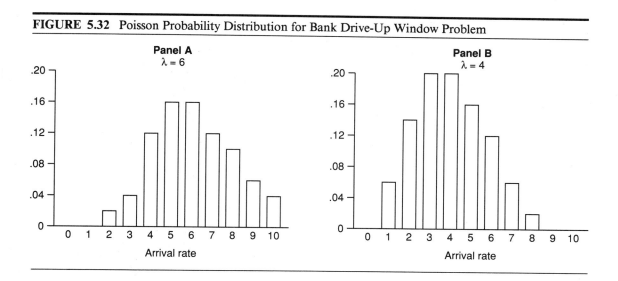

FIGURE 5.32 Poisson Probability Distribution for Bank Drive-Up Window Problem

arrival rate of four, there is a 98% probability that eight or fewer customers will arrive during the course of one hour. Obviously, one way of reducing customer delays would be to reduce the average arrival rate. (Another way would be to improve service efficiency.) Reducing the average arrival rate could be accomplished by extending the drive-up window hours.

Probability Mass Function

The probability estimates presented in Figure 5.32 were based on the following probability mass function:

$$P(x = A|\lambda) = \frac{e^{-\lambda} \times \lambda^x}{x!}$$

where x is the number of occurrences and λ is the mean arrival rate. For example, the probability of six customers arriving at the bank in one hour is

$$P(x = 6|6) = \frac{e^{-6} \times 6^6}{6!} = .1606$$

Thus, the chance of six customers arriving at the bank is approximately 16%. This result coincides with the probability estimate depicted in panel A of Figure 5.32 for $x = 6$. This formula is unwieldy for generating large numbers of probability estimates. Fortunately, a Poisson probability table (see Appendix A) has been developed. This table is similar to the one developed for the binomial table. Whereas the binomial distribution involves two parameters (n and x), the Poisson distribution involves a single parameter (λ). Figure 5.33 shows Poisson probability estimates (taken from Table A.6) for the bank drive-up window problem.

FIGURE 5.33 Poisson Probability Estimates for Bank Drive-Up Window Problem with $\lambda = 6$

Number of Arrivals (X)	Probability of Occurrence P(X)	Number of Arrivals (X)	Probability of Occurrence P(X)
0	.0025	8	.1033
1	.0149	9	.0688
2	.0446	10	.0413
3	.0892	11	.0225
4	.1339	12	.0113
5	.1606	13	.0052
6	.1606	14	.0022
7	.1377	15	.0009

FIGURE 5.34 Summary Measures for the Poisson Probability Distribution

Mean:	$\mu = \lambda \times t$
Variance:	$\sigma^2 = \lambda \times t$
Standard deviation:	$\sigma = \sqrt{\sigma^2}$

Summary Measures

The summary measures for the Poisson distribution are relatively straightforward to determine, as illustrated in Figure 5.34. Summary measures for the bank drive-up window problem are shown below:

$$\mu = \lambda = 6 \text{ arrivals/hour}$$
$$\sigma^2 = \lambda = 6$$
$$\sigma = \sqrt{\sigma^2} = \sqrt{6} = 2.449 \text{ arrivals/hour}$$

Not surprisingly, these results show that the mean is 6 arrivals per hour with a standard deviation of 2.449 arrivals per hour. As a means of verifying these results, consider the computations depicted in Figure 5.35. These

FIGURE 5.35 Computation of Expected Value for Bank Drive-Up Window Problem

Number of Occurrences (X)	Probability P(X)	Weighted Value $X \times P(X)$
0	.0025	0
1	.0149	.0149
2	.0446	.0892
3	.0892	.2676
4	.1339	.5356
5	.1606	.8030
6	.1606	.9636
7	.1377	.9639
8	.1033	.8264
9	.0688	.6192
10	.0413	.4130
11	.0225	.2475
12	.0113	.1356
13	.0052	.0676
14	.0022	.0308
		5.9779

calculations show the method for computing the expected value (i.e., mean) for the bank drive-up window problem. Notice that the expected value is nearly identical (slight difference is due to roundoff) to the mean value obtained using the summary measures method (i.e., 6 arrivals per hour).

Western Telephone Exchange Company

The general manager (GM) of the Western Telephone Exchange is concerned about the rate of line failures on the company's new fiber-optics network. The system is experiencing line failures at a rate of three per 100 miles of line per month. (This is an example of a problem involving length or distance intervals.) The GM would like to determine the chances of various line failures over a 50-mile segment connecting Dallas with Fort Worth, Texas. In this case, the average number of failures is 1.5 (i.e., 3 failures/100 miles × 50 miles). Figure 5.36 presents a Poisson probability distribution based on an average failure rate of 1.5. This distribution was derived from the Poisson table presented in Appendix A.

The company's current goal is a maximum of two line failures per 50 miles per month. Based on the results presented in Figure 5.36, the GM can conclude that there is an approximate 90% chance of meeting the company's goal.

Approximating the Poisson Distribution with the Normal Curve

Historically, the normal curve has been used to approximate the Poisson distribution in cases when the mean exceeds 20. The use of the normal curve in these cases eliminates the need for the large and cumbersome Poisson

FIGURE 5.36 Poisson Probability Estimates for Western Telephone Exchange Problem

Number of Failures (X)	Probability of Occurrence P(X)
0	.2231
1	.3347
2	.2510
3	.1255
4	.0471
5	.0141
6	.0035
7	.0008
8	.0001

tables. However, the advent of computer-based models for describing the Poisson distribution has reduced the importance of this approximation. The process for approximating the Poisson distribution by the normal distribution is similar to the approach outlined for approximating the binomial distribution with the normal distribution.

5.5 COMPUTER ANALYSIS

Key idea 8   CBS provides a convenient and efficient alternative to the traditional table look-up approach for developing probability estimates. This section illustrates the use of CBS in solving a number of the problems introduced in the preceding material.

Seven-Day Tire Company

Figure 5.37 shows the CBS input procedure for solving the Seven-Day Tire Company problem. Recall that this problem was analyzed using a normal probability distribution with an average of 30,000 miles and a standard deviation of 5,000 miles.

Figure 5.38 shows the corresponding CBS output for the tire problem. Notice that the developed probability distribution matches the one pre-

FIGURE 5.37   CBS Data Input for Seven-Day Tire Problem

```
        COMPUTERIZED      BUSINESS      STATISTICS
              Probability Distributions - Program Options Menu
         Options                    ----- Functions -----

      0.   Binomial          P(x ≥ a)      P(x = a)         P(a ≤ x ≤ b)
      1.   Poisson           P(x ≥ a)      P(x = a)         P(a ≤ x ≤ b)
      2.   Hypergeometric    P(x ≥ a)      P(x = a)         P(a ≤ x ≤ b)
      3.   Normal            P(x ≥ a)      P(mean ≤ x ≤ a)  P(a ≤ x ≤ b)
      4.   t                 P(x ≥ a)      P(mean ≤ x ≤ a)  P(a ≤ x ≤ b)
      5.   F                 P(x ≥ a)
      6.   Chi-Square        P(x ≥ a)
      7.   Quick Reviews
      8.   Exit to Main Menu
      9.   Exit to Operating System

                           -- Input Data --

   Enter mean, press ↵                                    30000
   Enter standard deviation (≥ 0), press ↵                5000
   Enter A, press ↵                                       40000_
```

170 Chapter 5 Probability Distributions

FIGURE 5.38 CBS Output for Seven-Day Tire Problem

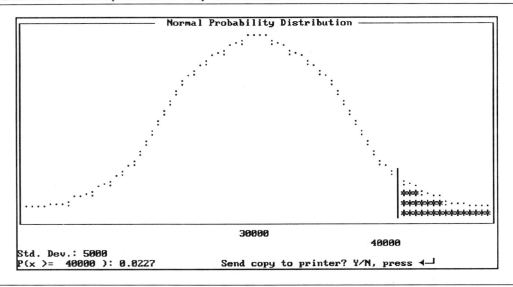

FIGURE 5.39 CBS Data Input for Johnson Pharmaceutical Problem

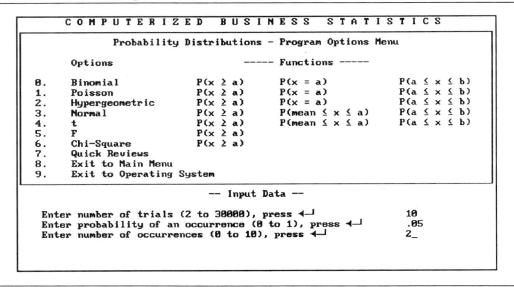

sented in Figure 5.14. These results show that the probability of obtaining more than 40,000 miles on a set of tires is only .0227 (i.e., the hatched area on the right).

Johnson Pharmaceutical

Figure 5.39 shows the CBS input procedure for solving the Johnson Pharmaceutical problem. Recall that this problem was analyzed using a binomial probability distribution with a sample size of 10 and a probability of .05.

Figure 5.40 shows the corresponding CBS output for the pharmaceutical problem. Notice that the developed probability distribution matches the one presented in Figure 5.27.

Danderberg International

Figure 5.41 presents the CBS input procedure for solving the Danderberg auditing problem. Recall that this problem was analyzed using a hypergeometric probability distribution consisting of a population of 20 accounts receivable (with an estimated 10 delinquent accounts) and a sample size of 5.

FIGURE 5.40 CBS Output for Johnson Pharmaceutical Problem

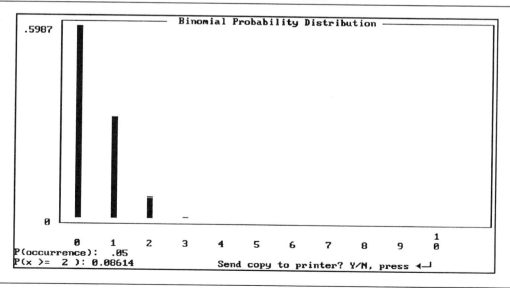

172 Chapter 5 Probability Distributions

FIGURE 5.41 CBS Data Input for Danderberg Auditing Problem

```
       C O M P U T E R I Z E D    B U S I N E S S    S T A T I S T I C S
              Probability Distributions - Program Options Menu

        Options                    ----- Functions -----
    0.   Binomial          P(x ≥ a)      P(x = a)          P(a ≤ x ≤ b)
    1.   Poisson           P(x ≥ a)      P(x = a)          P(a ≤ x ≤ b)
    2.   Hypergeometric    P(x ≥ a)      P(x = a)          P(a ≤ x ≤ b)
    3.   Normal            P(x ≥ a)      P(mean ≤ x ≤ a)   P(a ≤ x ≤ b)
    4.   t                 P(x ≥ a)      P(mean ≤ x ≤ a)   P(a ≤ x ≤ b)
    5.   F                 P(x ≥ a)
    6.   Chi-Square        P(x ≥ a)
    7.   Quick Reviews
    8.   Exit to Main Menu
    9.   Exit to Operating System

                          -- Input Data --

    Enter population size (1 to 20), press ↵              20
    Enter size of population subset (1 to 20), press ↵    10
    Enter sample size (1 to 10), press ↵                  5
    Enter number of occurrences (0 to 5), press ↵         2_
```

FIGURE 5.42 CBS Output for Danderberg Auditing Problem

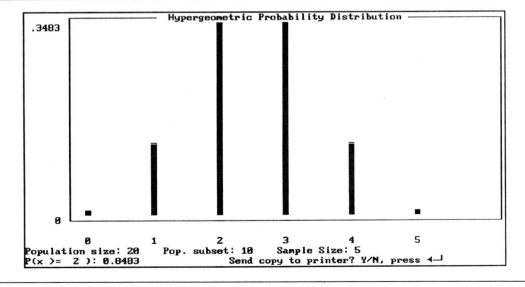

FIGURE 5.43 CBS Data Input for Western Telephone Exchange Problem

```
COMPUTERIZED  BUSINESS  STATISTICS
       Probability Distributions - Program Options Menu

    Options                       ----- Functions -----
 0. Binomial         P(x ≥ a)      P(x = a)         P(a ≤ x ≤ b)
 1. Poisson          P(x ≥ a)      P(x = a)         P(a ≤ x ≤ b)
 2. Hypergeometric   P(x ≥ a)      P(x = a)         P(a ≤ x ≤ b)
 3. Normal           P(x ≥ a)      P(mean ≤ x ≤ a)  P(a ≤ x ≤ b)
 4. t                P(x ≥ a)      P(mean ≤ x ≤ a)  P(a ≤ x ≤ b)
 5. F                P(x ≥ a)
 6. Chi-Square       P(x ≥ a)
 7. Quick Reviews
 8. Exit to Main Menu
 9. Exit to Operating System

                    -- Input Data --
 Enter average number of occurrences (.1 to 25), press ↵    1.5
 Enter number of occurrences (0 to 25), press ↵             2_
```

Figure 5.42 shows the corresponding CBS output for the auditing problem. Notice that the developed probability distribution matches the one presented in Figure 5.31.

Western Telephone Exchange Company

Figure 5.43 shows the CBS input process for solving the Western Telephone Exchange Company problem. Recall that this problem was analyzed using a Poisson probability distribution with an average failure rate (i.e., event rate) of 1.5 per 50 miles of cable.

Figure 5.44 shows the corresponding CBS output for the telephone problem. Notice that the developed probability distribution matches the one presented in Figure 5.36.

5.6 PRACTICAL APPLICATIONS

This section presents several business applications that illustrate the use of the binomial distribution.

FIGURE 5.44 CBS Output for Western Telephone Exchange Problem

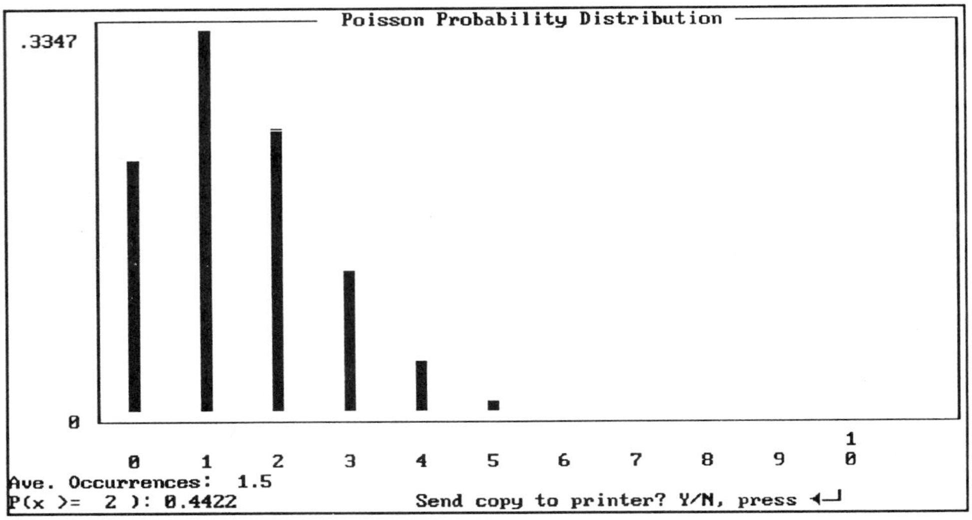

[Poisson Probability Distribution bar chart showing values at x = 0 through 10, with .3347 marked near the peak. Ave. Occurrences: 1.5. P(x >= 2): 0.4422. Send copy to printer? Y/N, press ↵]

South Eastern Charter Airlines (Ticket Forecasting)

The scheduling supervisor for South Eastern Charter Airlines is concerned about the growing number of "no shows" for the daily flight to the United States. The company operates a standard MD 80 with a seating capacity of 100 passengers. The supervisor estimates that there is a 10% chance that a ticketed passenger will be a "no show" on any given day. Furthermore, the decision of such a passenger not to show is independent of all other passen-

FIGURE 5.45 CBS Analysis of South Eastern Airline Problem

Number of Tickets Sold	Probability of Obtaining a Seat
100	1.0000
101	.9999
102	.9999
103	.9999
104	.9943
105	.9833
106	.9602
107	.9143

gers' decisions not to show. The supervisor would like to determine the number of tickets to sell (i.e., the level of overbooking) such that there is a 95% chance that everyone showing up with a ticket will have a seat.

This problem can be solved nicely using a **binomial** analysis. Symbolically, the desired relationship is as follows:

$$P(X \leq 100 | N = ?, P = 0.1) = .95$$

where N represents the number of tickets sold by the airline (i.e., trials).

A CBS analysis of this problem for various values of N is given in Figure 5.45. An inspection of these results indicates that the supervisor can overbook by six seats (i.e., 106) without violating the 95% criterion.

Mcquine Electric Limited (Acceptance Sampling)

The quality control manager at Mcquine Electric Limited is interested in establishing an acceptance plan for component shipments from the company's vendors. Naturally, it would be desirable if none of the components were defective; however, this is an unrealistic prospect. Furthermore, the only positive method for ensuring no defectives would involve a 100% inspection. This method is a very time-consuming and costly enterprise, and it is also subject to errors. Consequently, the manager would like to understand the implications of using sampling as an alternative to 100% inspection. The manager knows that in formulating a sampling plan, both the sample size and the acceptable number of rejects must be specified. For example, one inspection plan used throughout the industry consists of sampling 5% of the items, and if more than one item is found defective, then the lot will be rejected. However, is this the optimal plan for Mcquine?

Figure 5.46 presents the computations prepared by the manager with the help of the **binomial** distribution to illustrate the implications of various

FIGURE 5.46 Sampling Plan Analysis for Mcquine Electric

Sample Size (items)	Reject Level	Probability of Accepting Lot with Defective Rate	
		5%	10%
10	0	.60	.35
	1	.90	.72
	2	.97	.92
20	0	.36	.12
	1	.63	.39
	2	.82	.67

sampling plans for a lot consisting of 100 components. These results show that the probability of accepting a defective lot decreases as the sample size increases. For example, the chances of accepting a lot with 5% defectives based on a sample size of 10 is 60%, whereas the probability of acceptance drops to 36% when the sample size is doubled. Notice that the chances of accepting a lot containing defectives also increases as the reject level increases. Lastly, the chances of accepting a lot containing defectives decreases as the defective rate in the lot increases.

5.7 CASE STUDY: DRACK INDUSTRIES

Drack Industries distributes a wide range of pharmaceutical products to hospitals and medical centers. One of the products most in demand is the new computerized heart pacemaker. The manager of materials control, Alice Myers, is concerned about maintaining adequate inventory stocks in order to meet a probabilistic demand situation. A review of product demand for the previous two quarters reveals a mean of 150 pacemakers per week with a standard deviation of 20. The lead time for replenishing inventory from the manufacturer is two weeks. The current inventory policy is to order 500 pacemakers. This order quantity level provides Drack with a 2% discount on the $650 product.

Although the actual order quantity has been established (i.e., 500 units per order), the decision of when to order is still up in the air. The basic trade-off in determining when to order involves the cost of a stockout versus the cost of holding excess inventory. Although it is desirable to avoid stockouts, it is often unrealistic to think that they can be completely eliminated in a cost-effective manner. Ms. Myers estimates the cost of holding at $2 per pacemaker per month. She is less certain about the cost of a stockout. Therefore, she has decided to examine two order timing policies based on the following industrywide standards:

1. Maintaining a 95% service level.
2. Assigning a stockout cost of $10 per pacemaker.

The first alternative suggests that the company is willing to accept a stockout 5% of the time (i.e., 1 out of 20 replenishment periods). The second option provides a stockout cost estimate that can be used in determining the corresponding service level.

A plot of the lead time demand curve is shown in Figure 5.47. The shaded area on the right side of the curve represents the 5% stockout level. The general model for determining the reorder point (R), assuming demand can be described with a normal distribution, is

$$R = \text{Mean} + Z \times \text{Standard deviation}$$

For the Drack problem, the mean is 150 units, and the standard deviation is 20. The Z value for a 95% level of confidence is 1.64 (obtained from the

FIGURE 5.47 Product Demand Distribution over Lead Time

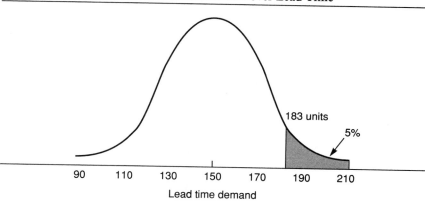

standard normal table or CBS). Substituting these values into the model yields the following reorder level:

$$R = 150 + 1.64 \times 20 = 182.8$$

By reordering when the inventory reaches 183 units, Ms. Myers can be 95% confident that she won't run out of pacemakers.

Usually, more precise estimates of the reorder level can be made when the stockout cost is known. For example, if the stockout cost is very large relative to the holding cost, then the service level for all practical purposes becomes 100%. On the other hand, if the stockout cost is very small, then the service level approaches 0%. The latter result suggests that no inventory should be maintained (i.e., rely entirely on back-ordering).

For this application, Ms. Myers has assigned a stockout cost of $10. This estimate, along with the holding cost, can be used to compute the stockout probability level. The basic model is

$$P_c = \frac{C_s}{C_h + C_s}$$

where:

P_c = Critical probability
C_s = Stockout cost
C_h = Holding cost

Substituting for C_h and C_s yields a critical probability, or confidence level, of 83.33%. Notice that this value is considerably lower than the 95% confidence value used with first option. The reorder point can now be determined using the same model, except that the Z value is .97 (corresponding to an 83.33% level of confidence).

$$R = 150 + .97 \times 20 = 169.4$$

Thus, the second option reduces the reorder point by approximately 13 units.

5.8 SUMMARY

This chapter has introduced a number of important probability distributions that are used extensively throughout business and government. Probability distributions provide an effective mechanism for describing the characteristics of a random variable. In other words, assigning probability values to various outcomes via a probability distribution is usually much more efficient than estimating by other means. The basic characteristics of a probability distribution are that the probability of an individual event ranges between 0 and 1 and that the probabilities of all events must sum to 1. The selection of a particular probability model requires a careful analysis of the problem situation. For example, applying the binomial distribution to a problem involving independent arrivals (i.e., Poisson process) can yield misleading results.

There are basically two types of probability distributions: continuous and discrete. A random variable described by a continuous distribution can take on any numerical value in a range, whereas a random variable described by a discrete distribution can take on only limited values. Examples of this type include the binomial, the Poisson, and the hypergeometric distributions. The normal and the uniform are examples of continuous distributions. Each probability distribution is described by a unique set of parameters. For example, the normal distribution is a two-parameter distribution requiring both a mean and a standard deviation.

This chapter has also illustrated how, under some conditions, the normal distribution can be used to approximate the binomial Poisson distribution. The primary advantage of using the normal approximation is that it greatly reduces the number of calculations. Finally, computer analysis provides an effective alternative to the traditional methods for assigning probabilities from a table for a given distribution.

5.9 GLOSSARY

Bernoulli process A random experiment consisting of n identical trials with only two mutually exclusive outcomes, where the probability of each outcome is constant and the outcome of one trial is independent of any other trial.

binomial probability distribution A discrete distribution used for describing the outcome of a Bernoulli process.

continuous probability distribution A distribution where the random variable can take on any numerical value.

cumulative probability distribution A distribution used to describe the probability that a random variable is larger or smaller than a given value.

discrete probability distribution A distribution where the random variable can take on only limited numerical values.

expected value The arithmetic mean of a random variable.

hypergeometric probability distribution A discrete distribution that provides probability estimates for a Bernoulli-type process where the outcome of one trial is dependent on preceding trials.

normal probability distribution A continuous, bell-shaped distribution that is described by a mean and a standard deviation.

Poisson probability distribution A discrete, one-parameter distribution used to describe the number of outcomes that occur over a specific time frame.

random variable A quantitative variable whose numerical value is determined by a random experiment.

standard normal probability distribution A normal distribution with a mean of 0 and a standard deviation of 1.

uniform probability distribution A continuous distribution where each value of the random variable has an equal chance of occurring.

5.10 BIBLIOGRAPHY

Aczel, A. D. *Complete Business Statistics.* Homewood, Ill.: Irwin, 1989.

Ingram, O., L. J. Gleser, and C. Derman. *Probability Models and Applications.* New York: Macmillan, 1980.

Mason, R. D. *Statistical Techniques in Business and Economics,* 6th ed. Homewood, Ill.: Irwin, 1986.

Mullet, G. M. "Simeon Poisson and the National Hockey League." *The American Statistician* (February 1977), pp. 8–12.

Springer, C. H., et al. *Probabilistic Models.* Homewood, Ill.: Irwin, 1977.

5.11 PROBLEMS

1. Describe how probability distributions can be used in the business decision-making process. Give examples.
2. Identify whether each of the following probability distributions is discrete or continuous.

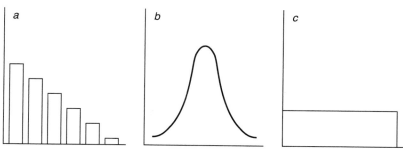

3. Under what conditions can the normal distribution be used to approximate the
   a. Binomial distribution?
   b. Hypergeometric distribution?
   c. Poisson distribution?
4. Describe the differences between the normal and standard normal probability distributions. Why is the standard normal distribution so useful in textbook problems?
5. Match the most appropriate probability distribution with the following applications:

Distribution	Application
a. Continuous	1. Number of cars arriving at a car wash.
b. Hypergeometric	2. Waiting time for an elevator.
c. Uniform	3. Failure rate for a computer chip.
d. Poisson	4. Composition of two independent groups.
e. Binomial	5. Sales revenues for the coming quarter.

6. Given a normal distribution with $\bar{X} = 100$ and $s = 10$, determine the following probabilities:
   a. $P(X > 100)$ b. $P(90 < X < 105)$ c. $P(X < 90)$
7. Given a normal distribution with $\mu = 0$ and $\sigma = 1$, determine the following probabilities:
   a. $P(X > 0.2)$ b. $P(0.8 < X < 0.9)$ c. $P(X < 0.5)$
8. The claims department for the Bridgeline Insurance Company estimates that the company should experience, on the average, 10,000 claims for 1989, with a standard deviation of 3,000 claims. Determine the probability that the company will experience:
   a. More than 12,000 claims.
   b. Less than 6,000 claims.
   c. Between 7,000 and 11,000 claims.
9. The average price-to-earnings (P/E) ratio for the Fortune 500 firms is 10 with a standard deviation of 1. How many firms have a P/E less than 8.5? How many firms have a P/E greater than 11? How many firms have a P/E between 10 and 12?
10. In a recent shipment of 10,000 microchips, 1,000 were found to be defective. Based on a random sample of 100 microchips, what is the probability that:
    a. Less than 10 will be found defective?
    b. Between 15 and 25 will be found defective?
    c. More than 30 will be found defective?
11. The average daily cost for Medicare patients at South Bay Hospital is $275 with a standard deviation of $100. The hospital receives a flat $350 from the government for each patient visit.
    a. What is the probability that South Bay will lose money on a per-patient basis?
    b. What is the probability that South Bay will net at least $5,000 after processing 100 Medicare patients?
    c. What level of government reimbursement is necessary to ensure a 95% probability that South Bay will make money on each visit?

12. The general manager for the City of Los Angeles Department of Water and Power (DWP) estimates that average monthly residential electrical power usage is 1,000 kWh with a standard deviation of 100 kWh.
    a. What is the probability that a randomly selected account uses more than 1,000 kWh per month?
    b. What is the probability that a randomly selected account uses between 900 and 1,050 kWh per month?
    c. The general manager estimates that there are approximately 1 million residential users within the DWP service area. What is the total revenue from residential accounts if energy use is billed at $0.06 per kWh?
13. The manager at Don's Hobby Shop estimates that the company monthly income is normally distributed with a mean of $2,000 and a standard deviation of $500. Assuming that the cost of running the shop is $1,500 per month, how much cash should the manager keep on hand at the beginning of each month to ensure that there is only a 10% chance of running out by the end of the month? A 5% chance?
14. The inventory manager at the All Night Auto Parts Store has collected the following information on products sold in the store:

Item	Weekly Demand		Beginning Weekly Inventory
	Mean	SD	
Jumper cables	50	5	60
Gas cans	40	3	50
Batteries	100	8	150

   a. What is the probability of running out of gas cans?
   b. How many sets of jumper cables should be stocked to ensure that the probability of running out is less than 2%?
   c. Assuming that the demand for gas cans and batteries is independent, what is the probability of running out of both products in a given week?
15. The amount of ice cream in a typical one-gallon carton can vary as a result of anomalies in the filling process. A recent industry study showed that the standard deviation for an ice cream filling machine is 0.1 gallons. Government regulations require that each carton be at least 95% full of ice cream by volume.
    a. What fill volume setting should be used to ensure that 98% of all cartons meet the government regulation?
    b. What is the probability that a carton will contain at least one gallon based on the setting developed in part a?
16. A recent investigation by the U.S. Treasury Department of Federal Savings (a bankrupt S&L) found that 3% of the outstanding loans were made to directors and stockholders. The guideline established by the Treasury Department for limiting the proportion of S&L loans made to directors and stockholders is 1.5% with a standard deviation of 0.5%. Should the government auditors be concerned with their findings?
17. Given a uniform distribution with $a = 10$ and $b = 20$, determine the following probabilities:
    a. $P(X > 12)$
    b. $P(12 < X < 16)$
    c. $P(X < 18)$

18. American Airlines flight 790 from Chicago to Los Angeles is scheduled to arrive at 5:00 P.M. However, due to abnormal headwinds, the flight could be delayed up to 30 minutes. Determine the probability that the flight will be:
    a. Only 10 minutes late.
    b. More than 20 minutes late.
    c. Between 5 and 20 minutes late.
19. Given a binomial distribution with $n = 15$ and $p = .5$, determine the following probabilities:
    a. $P(X = 5)$ b. $P(10 < X < 20)$ c. $P(X \leq 12)$
20. Given a binomial distribution with $n = 10$ and $p = .2$, determine the following probability:
    a. $P(X > 2)$ b. $P(3 < X \leq 6)$ c. $P(X = 4)$
21. Given a binomial distribution with $n = 100$ and $p = 0.45$, determine the following probabilities:
    a. $P(X > 20)$ b. $P(50 < X < 75)$ c. $P(X < 35)$
22. The quality control inspector at Miller Electric has obtained a batch of 15 computer chips. Historical data indicate a 5% failure rate. Determine the following (where X represents the number of failures):
    a. $P(X = 0)$ b. $P(X = 2)$ c. $P(X > 4)$
23. The Break the Habit Institute claims a 65% success rate in "curing" their patients of a smoking habit.
    a. Define the random variable.
    b. Is the random variable discrete or continuous?
    c. What probability distribution best describes this problem situation?
    d. What is the probability that 8 or more patients out of a total of 12 will be cured?
    e. What is the probability that 2 or fewer patients out of a total of 12 will be cured?
24. In the recent presidential election, the winner received 52% of the vote. From a sample of 10 individuals who cast ballots, what is the probability that:
    a. None voted for the winner?
    b. Three or more voted for the winner?
    c. No more than seven voted for the winner?
25. Many airlines engage in overbooking seats as a way of offsetting potential "no-shows." Current industry evidence indicates that 8% of those individuals making a reservation do not show up for the flight. The present level of overbooking is approximately 5%. Assuming seats for 100 passengers:
    a. What is the probability that no more than two ticketholders will be dropped from an overbooked flight?
    b. What is the probability of five or fewer no-shows on a completely booked flight?
    c. What is the net effect of overbooking and no-shows on a completely booked flight?
26. Develop summary measures for problem 25.
27. An unprepared student is given a spot quiz in statistics involving 15 true/false questions. Assuming that the student guesses at each question, what is the probability of the student:
    a. Obtaining a perfect score on the test?
    b. Answering 10 or more questions correctly?

c. Answering fewer than 7 questions correctly?
d. Answering exactly 8 questions correctly?
28. Develop summary measures for problem 27.
29. A new drug for treating asthma victims has resulted in a 50% cure rate nationwide. From a random sample of 100 asthmatic patients using the new drug, what is the probability that:
a. At least 25 will be cured?
b. Between 50 and 75 will be cured?
c. More than 20 will not be cured?
d. Less than 15 will be cured?
30. Compute the variance using the data given in Figure 5.26, and compare the result with the summary measures reported for the Dodgers problem.
31. Solve problem 21 using the normal approximation to the binomial. Compare the results with those developed in problem 21.
32. Results obtained from the California smog inspection program revealed that approximately 10% of all cars inspected fail to pass the smog test. Using the normal approximation to the binomial distribution, determine the probability that between 100 and 200 vehicles out of the next 1,000 inspected will fail the test.
33. Given a hypergeometric distribution with a population size (N) of 20, a sample size (n) of 10, and 10 successful items (S), determine the following probabilities:
a. $P(X = 5)$ b. $P(X > 7)$ c. $P(3 \leq X < 8)$
34. The Armax Corporation owns 12 stamping presses. Currently, 10 of the 12 presses are in operating order.
a. Define the random variable.
b. Is the random variable discrete or continuous?
c. What probability distribution best describes this problem situation?
d. What is the probability that three presses from a sample of six will be operating?
e. What is the probability that two or fewer presses out of a sample of eight will be operating?
35. The president of the United States is planning to establish a special "blue ribbon" task force to study organized crime. The task force will consist of the chief justice and four members of Congress. The president has narrowed his list to four Democrats and four Republicans.
a. What is the probability that the committee will be made up of all Democrats?
b. What is the probability that the committee will consist of at least two Republicans?
c. What is the probability that the committee will consist of an equal number of Democrats and Republicans?
36. Solve problem 33 using the normal approximation to the hypergeometric distribution. Compare the results with those developed in problem 33.
37. Compute the variance using the data given in Figure 5.26, and compare the result with the summary measures reported for the jury problem.
38. Given a Poisson distribution with $\lambda = 12$, determine the following probabilities:
a. $P(X > 3)$ b. $P(7 < X < 14)$ c. $P(X = 6)$
39. Given a Poisson distribution with $\lambda = 0.5$, determine the following probability:
a. $P(X > 1)$ b. $P(0.5 < X < 0.75)$ c. $P(X = 0.5)$
40. The manager at ToyMark is considering stocking a new computer game for the

Christmas season. A review of the historical data indicates that a similar product, when introduced last year, sold at a rate of five per week. The manager would like to know the probability of selling:

a. No games.
b. At least three games.
c. Between two and six games.

41. Accidents at a local oil refinery occur at a rate of two per month.
    a. What is the expected number of accidents per year?
    b. What is the probability that no accidents will occur next month?
    c. What is the probability that at least 10 accidents will occur next year?

42. The maintenance staff at American Airlines has found that two tires, on the average, must be replaced on a 747 during each quarterly inspection.
    a. What is the probability that no tires will need to be replaced?
    b. What is the probability that at least three tires will need to be replaced?
    c. What is the probability that the airline will spend more than $800 on tire replacement if each tire costs $200?

43. The phone system at Devesto Electric is designed to accommodate, on the average, 15 incoming calls per minute. The maximum number of lines is 20.
    a. What is the probability of a caller receiving a busy signal?
    b. What is the probability that the system is idle?
    c. If the company wishes to maintain a 95% service level, how many additional lines should be installed?

44. The City of Los Angeles has implemented a computer-based communication system for improving the dispatching of police officers. On a typical weekday evening, the system receives, on the average, 20 requests per hour for police assistance. What is the probability that less than 5 calls are received in a one-hour period? What is the probability that at least 10 calls are received?

45. Absenteeism at Kelly Investments averages two employees per day out of a total work force of 175 employees. What is the probability that five or more employees will be absent on a given day? What is the probability that exactly two employees will be absent?

46. The arrival rate at P.V. National Bank during the Friday afternoon rush period is 60 customers per hour.
    a. What is the probability that 20 or more customers will arrive within a 15-minute period during the rush hour?
    b. What is the probability that three or fewer customers will arrive at the bank during this period?
    c. What is the most likely number of arrivals during this period?
    d. Determine the expected value and standard deviation of the number of customers arriving during a 10-minute period.

47. Solve problem 43 using the normal approximation to the Poisson distribution. Compare the results with those developed in problem 43.

48. The Telex answering machine fails to properly record incoming calls 1% of the time. Based on 20 incoming telephone calls, what is the probability that:
    a. All the calls will be received?
    b. Between 16 and 18 will be received?
    c. Fewer than 18 will be received?

49. Solve problem 44 using the normal approximation.

50. Solve problem 46 using the normal approximation.
51. Solve problem 48 using the normal approximation.
52. Compute the variance using the data given in Figure 5.35, and compare the result with the summary measures reported for the bank drive-up window problem.
53. The production of corn per acre in the United States between 1975 and 1986 can be characterized by the following frequency distribution:

Yield per Acre (bu)	Frequency
80–89	2
90–99	3
100–109	4
110–119	2
120–129	1

   a. Construct a probability distribution from the frequency distribution.
   b. Compute the expected value for corn yield per acre.
   c. If a production yield of 100 bu per acre is required to cover costs, what is the probability that a farmer will at least break even?

54. The Packard Company manufactures and markets a home video training film entitled "How to Prepare Your Own Income Tax Return." The market department has prepared the following estimates on sales value (in units) for the coming year:

Sales	Probability
10,000	.2
20,000	.3
30,000	.2
40,000	.15
50,000	.10
60,000	.05

   a. What is the most likely sales level?
   b. What is the expected sales level?
   c. What is the probability that sales will be above 40,000 units?
   d. What is the probability that sales will be between 20,000 and 50,000 units?

55. The chief engineer at General Electric's lighting division is interested in determining the life for the standard 100-watt bulb. The following results were obtained from a recent experiment that tested the operating life of the 100-watt bulb.

Operating Life (hours)	Frequency of Occurrence
0–19	22
20–39	47
40–59	33
60–79	19
80–99	7

    a. Develop a probability distribution using these results.
    b. What is the expected value and standard deviation?
    c. What is the probability that a bulb will average 60 hours or more?
    d. What is the probability that a bulb will average less than 40 hours?

56. State Farm Insurance offers an earthquake insurance policy for residents living in the West. The current policy has a $10,000 deductible; that is, the policyholder must pay the first $10,000, and State Farm pays for the remaining repair cost, to a maximum of $200,000. The company has prepared the following table, which shows the damage amounts and estimated cumulative probability of occurrence on an annual basis for an individual policyholder.

Damage Amount ($)	Occurrence Probability
0–9,999	.990
10,000–49,999	.994
50,000–99,999	.997
100,000–149,999	.999
150,000–200,000	1.000

    a. What is the expected amount that State Farm would pay out per policy on an annual basis?
    b. How much should State Farm charge for this policy?

57. Determine whether each of the following statements is true or false. Explain your answer.
    a. The average arrival rate at a drive-up bank is 20 cars per hour. The probability of more than five cars arriving within 10 minutes is greater than 40%.
    b. The rejection rate for typesetting is three errors for every 10 pages. The probability of more than one error for every five pages is less than 50%.
    c. The monthly demand for TV sets at the local appliance store is normally distributed with a mean of 20 and a variance of 9. The probability that demand will exceed 35 sets per month is less than 5%.
    d. The mayor is organizing a committee of five members from the local community. There is a 50–50 chance that the membership will consist of all males.
    e. A single missile has a 40% chance of hitting the target. At least six missiles are required to ensure a 95% probability of at least one hit.

58. The Los Angeles Dodgers are preparing to face the Oakland Athletics in the 1988 World Series.

a. What is the probability of the Dodgers winning the series if each team has a 50–50 chance of winning a single game?
b. What is the probability of the Athletics winning the series if they have a 60% chance of winning a single game?
c. What is the probability of the Dodgers winning according to the actual outcome of the 1988 series?

59. Assume that the scheduling supervisor for South Eastern Airlines (see Section 5.6) has sold 106 tickets. What is the probability that the aircraft will take off with:
a. Exactly 95 passengers?
b. Between 90 and 100 passengers?
c. More than 100 passengers?

60. Referring to the South Eastern Airlines problem in Section 4.6, how many overbooked seats could the supervisor sell if the no-show rate is 20%?

61. The U.S. Department of Defense is currently redesigning the Space Defense Initiative (i.e., "Star Wars"). The new design proposal calls for the capability of the system to destroy up to 50 incoming ballistic missiles at a time. The secretary of defense wishes to have a 95% confidence that the system will meet these specifications. If the probability of a defensive missile destroying an enemy missile is 90%, what is the minimum number of defensive missiles required to achieve the secretary's goal?

5.12 CASES

5.12.1 State of California Lottery

In 1986 California initiated a lottery as a means of generating additional resources for public education. The lottery actually consists of several different games, including the "Big Spin." This game involves a roulette wheel with 100 slots. The minimum payoff is $10,000, and the maximum payoff is $1 million. Table 1 presents the various payoffs and their frequency of occurrence.

TABLE 1 Relative Frequency Distribution for California Big Spin Contest

Payoff	Number of Slots Containing Payoff Amount	Relative Frequency
$ 10,000	10	10%
20,000	16	16
30,000	16	16
40,000	16	16
50,000	16	16
100,000	10	10
1,000,000	8	8
Double	8	8
	100	100%

In order to play the Big Spin, one must first purchase a $1 lottery ticket. If the ticket contains the word *Big Spin,* the purchaser is entitled to participate in a random drawing. The chances of obtaining such a ticket are 1 in 1,000. The random drawing determines the actual contestants for the Big Spin. The chances of being selected from the random drawing are 1 in 500.

Suggested Questions

1. What is the chance that a purchaser of a $1 lottery ticket will participate in the Big Spin?
2. What is the expected value of the game for an individual who buys a $1 lottery ticket?
3. What is the expected value of the game for an individual selected from the random drawing?
4. What is the probability that a contestant will spin three doubles in a row?
5. What is the probability that a contestant will win $4 million during a single game?

5.12.2 Jensen Musical Instrument Company

The Jensen Musical Instrument Company, founded in 1895, produces a complete line of pianos (i.e., uprights to grands). The current manufacturing technology relies almost entirely on the handicraft of 20 senior workers, many of whom are reaching retirement age. The company's management is faced with two serious long-term challenges: the aging work force and increased competition from the Japanese. A recent survey performed by the company revealed that the work force in the local area is more interested in "high-tech" jobs that offer more promise for growth. Many of Jensen's senior craftsmen have performed basically the same set of tasks for over 30 years.

The production manager is also faced with a growing problem of absenteeism among the senior craftsmen. These employees know that they are unlikely to be dismissed because of the difficulty of finding qualified replacements. The manager estimates that the mean absentee rate is one employee per day. The impact on

TABLE 1 Absentee Rates and Production Impacts for Jensen Company

Absentee Rate (employees)	Production Impact (% reduction)	
	Manual	Semiautomatic
0	0	0
1	0	0
2	0	0
3	5	0
4	15	0
5	25	0
6	40	5
7	55	10

production becomes significant whenever three or more employees are absent. The manager has prepared the information shown in Table 1, which reports the impact on production of various absentee rates. For example, production is reduced by 15% whenever four employees are absent.

To counter this situation, the production manager is considering the installation of a semiautomated production line that will significantly reduce the impact of employee absenteeism. The proposed system will cost approximately $750,000 plus $50,000 in nonlabor operating expenses. The impact of absenteeism on the semiautomatic line is also shown in the table. The manager estimates that the company loses $100 per day for every 1% reduction in production.

Suggested Questions

1. What is the probability that more than three workers will be absent on a given workday? more than seven?
2. What is the expected loss per day for the two manufacturing alternatives?
3. Should the company purchase the proposed system? Assume a 10-year life and a 200-day working year.
4. In light of the current absenteeism rate, how many senior craftsmen should the company retain?
5. Identify several alternatives for helping the company solve its absenteeism problem.

Chapter 6

Survey Design and Data Base Management

Research is an organized method for keeping you reasonably dissatisfied with what you have.

C. F. Kettering

CHAPTER OUTLINE

6.1 Introduction
6.2 Example Management Problem: Nautilus Health Spa, Inc.
6.3 Questionnaire Design
6.4 Computer Data Base Management Systems
6.5 Practical Applications
6.6 Case Study: Transpacific Airlines
6.7 Summary
6.8 Glossary
6.9 Bibliography
6.10 Problems
6.11 Cases

CHAPTER OBJECTIVES

The primary objectives of this chapter are to develop an understanding of

1. The differences between a census and a sample.
2. How surveys are used to collect data.
3. The basic differences between hard and soft measurements.
4. How to design a survey instrument.
5. The role of database management in survey design.

Today, most citizens take for granted the accuracy of presidential polls. In times past, however, things were not always as rosy. In the 1936 presidential campaign, the *Literary Digest* predicted an easy win for Alfred M. Landon, the Republican challenger, over the incumbent president, F. D. Roosevelt.*

---

* M. C. Bryson, "The *Literary Digest* Poll: Making a Statistical Myth," *American Statistician,* November 1976.

> **HISTORICAL NOTE**
>
> L. H. C. Tippet (1900–1965) is generally given credit for publishing the first table of random numbers. Surprisingly, Tippet's random number table of 41,600 digits did not appear until 1927. These numbers were taken from British census data on the area of local parishes (the first and last digits were omitted). Dr. Tippet began his statistical career in 1925 working for the British Cotton Industry Research Association. He had the opportunity to study under the two giants of statistical development in the 20th century: Karl Pearson and R. A. Fisher. In addition to his work in random numbers, Dr. Tippet made a number of important contributions to the newly developing field of statistical quality control.

The predicted margin of victory was 57% to 43%. The actual outcome was quite different. President Roosevelt buried Mr. Landon in a landslide victory. The final tally gave President Roosevelt 63% of the votes cast. Clearly, something had gone wrong with *Literary Digest*'s survey process.

Two fundamental biases plagued the *Digest*'s nationwide poll: selection and nonresponse. The *Digest* had mailed questionnaires to 10 million Americans (the largest mailing in the history of political polling). Unfortunately, the names included in the mailing were taken from lists of individuals owning telephones and automobiles. This selection process tended to overrepresent higher-income individuals, who historically voted Republican, and to underrepresent lower-income voters, who would be expected to support the Democratic nominee (remember that the country was in the midst of the Great Depression). Additionally, only 25% of the 10 million questionnaires were returned. In this regard, those who responded tended to have a higher level of education than those who did not respond to the survey. These individuals, in turn, tended to vote Republican. Thus, a much larger proportion of nonrespondents were actually supporters of President Roosevelt. The combination of these two errors (selection and nonresponse) proved fatal to the *Digest*. The company folded within a few months of President Roosevelt's inauguration. Interestingly, a survey based on only 50,000 voters was taken at the same time by the unknown George Gallop. Mr. Gallop's poll accurately predicted President Roosevelt's victory.

6.1 INTRODUCTION

Managerial decision making is becoming increasingly dependent on data. Thus, the collection and processing of data represents a major step in the decision-making process. (*Data* represents a compilation of facts; data that has been processed is known as *Information*.) Chapter 2 introduced the four types of data measurements (i.e., nominal, ordinal, interval, and ratio). Basically, there are two methods used for collecting data: surveys and experiments. A **survey** represents the collection of data without the impact of controls. For example, the manager of a local health spa wishes to survey members regarding their reaction to several newly proposed programs and services. On the other hand, an **experiment** involves the collection of data where some or all of the factors influencing the outcome of the experiment are controlled. For example, the spa manager wishes to measure the impact of a new high-protein diet on the men who belong to the spa. In this situation, the manager has incorporated controls in the form of limiting the experiment to males and ensuring that those participating in the experiment are exposed to the same diet. Clearly, in many situations, experiments are preferred to surveys. However, the complexities of business often limit the use of experiments for generating new data. This is why the collection of data through surveys plays such an important role in modern business practice.

Key idea 1

Basically, there are two types of surveys: a **census** (a complete survey) and a sample (a partial survey). A **census** represents an attempt to measure certain characteristics of interest in the total population. When the population is relatively small (e.g., the number of aircraft operated by United Airlines), a complete census is quite practical and, in many cases, most desirable. For example, the FAA may require United Airlines to inspect all of its aircraft for cracks. However, when the size of the population is large (e.g., the population of the United States), measuring all the members of the population becomes infeasible. For example, it is becoming increasingly likely that many individuals living in the United States will not be counted in future population censuses. This is due in part to the growing population, but even more importantly to the fact that many groups within the country do not wish to be included in the census.

In many business situations, although a complete census would not be physically or technically impossible, it may be extremely expensive. In these cases, a **sample** survey represents a viable alternative to a complete census. In fact, sometimes a sample survey is preferred to a census independent of cost or physical considerations. For example, consider that a manufacturer of flash bulbs wishes to test the performance of the product. Clearly, if the manufacturer tests all of the bulbs (i.e., a census) no units of the product would be available for sale. On the other hand, a good indication of product quality could be obtained by testing a random sample of bulbs produced. Also, a sample survey can often provide more information on a population than a complete census can, given time or cost limits. For example, suppose a market research firm wishes to determine the buying habits of the residents of a small town (estimated population 6,000) within a period of one week. The manager estimates that the survey team can spend at most 5 minutes interviewing each of the town's residents, compared with a total of 50 minutes surveying a random sample of the population.

Key idea 2

Unfortunately, the process of collecting data via a survey is subject to errors. These errors can arise from a variety of sources. Basically, two fundamental types of errors can be encountered: systematic and random. **Systematic errors** arise as a result of bias in the selection, measurement, or reporting process. A systematic error is defined as the difference between the true population value and the average obtained based on a census of the population. For example, suppose the health director at a local university wishes to measure the height of all male students using a "yardstick" that is only 95% of a standard yardstick. Clearly, the reported results will overestimate the true height of the male population. A **random error,** on the other hand, occurs as a result of sampling a proportion of the population. It is defined as the difference between the value obtained by taking a single random sample and the value generated from a complete census. Random errors occur as a result of chance. For example, suppose the health director decided to survey only a sample of the male population using a standard yardstick. In this case, it is possible that the sample contains a larger proportion of "shorter" students than found in the general population. Consequently, the reported average height could understate the true population mean.

The following model illustrates the relationship between an individual measurement and both types of measurement errors.

$$\text{Individual measurement} = \text{True value} + \text{Systematic error} + \text{Random error}$$

An inspection of this relationship reveals that it is possible for an individual measurement to contain both systematic and random errors. This relationship is shown graphically in Figure 6.1, using the results from the height

FIGURE 6.1 Comparison of Random and Systematic Errors for Height Measurement Survey

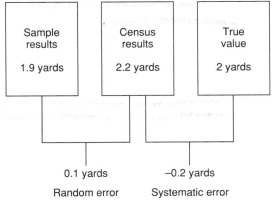

measurement study. The results from the sample revealed an average height of 1.9 yards. In contrast, the results obtained from the census yielded an average height of 2.2 yards. Suppose the true height of the male population is 2 yards. Therefore, the **systematic error is 0.2 yards** (i.e., the difference between the census results and the true value). Finally, it is possible, under some circumstances, for the systematic and random errors to cancel each other out. This possibility, however, should not be a consideration in designing or implementing surveys.

As indicated in the discussion of the 1936 presidential election, bias played a fundamental role in undermining the validity of the survey. In general, bias can occur in either the planning stage, the collection stage, or the processing stage. Bias can be found in either census data or sample data. Basically, there are three types of bias that need to be understood and controlled: selection, response, and nonresponse. Figure 6.2 summarizes these types of bias and provides several examples.

The primary approach for minimizing **selection bias** is to ensure random sampling. **Response bias,** on the other hand, can be reduced or eliminated through the careful construction of the measurement instrument (i.e., questionnaire). Similarly, **nonresponse bias** can be controlled through an effective questionnaire design and, in some instances, by providing incentives for participating in the survey.

As indicated above, the same types of biases can also occur in the data collection and data processing stages. Amazingly, even with the advent of computer data base systems, a large number of errors or biases can occur during the computational stage. In general, a strong quality control program is essential to minimize the impact of biases.

FIGURE 6.2 Summary of the Types of Survey Bias

Bias	Definition	Examples
Selection	The systematic propensity to include selected elements of a population with a particular characteristic while excluding other members with other characteristics	1936 *Literary Digest* survey A survey of company workers regarding their reaction to a new product
Response	The systematic propensity for answers to survey questions to be incorrect	IRS survey of the public's response to new tax code An exit interview of workers leaving the company
Nonresponse	The systematic propensity for certain elements selected for a survey not to respond	1936 *Literary Digest* survey A survey on crime in a minority community

Key idea 3

Most surveys involve the collection of so-called **soft data,** which include such diverse factors as attitudes, preferences, decisions, and intentions, in contrast to the **hard data** associated with business performance (e.g., employee turnover, corporate revenues, and price/earnings ratios). Most managers prefer hard data because of the higher level of confidence associated with such measurements. However, even hard data are prone to biases and errors. These remain, however, issues that can only be answered by soft measurements. The remainder of this chapter is devoted to the process of planning, collecting, and processing problems involving soft data.

6.2 EXAMPLE MANAGEMENT PROBLEM: NAUTILUS HEALTH SPA, INC.

The Nautilus Health Spa, founded in the early 1980s, specializes in fitness and weight control programs for middle-aged professionals. Presently, the company has over 30 spas operating throughout California and Arizona. This rapid growth has brought about a number of concerns regarding the future direction of the business. Specific issues include marketing strategy, expansion opportunities, and program quality. The spa's general manager

> **INTERNATIONAL VIGNETTE**
>
> The International Red Cross is known worldwide as a humanitarian organization that offers aid and assistance during times of international conflicts, internal strife, and natural disasters. Mr. Francois Perez had been recently appointed the head of the Red Cross's central tracking agency (CTA) after returning from a nineteen-week executive training program. The primary mission of the CTA is information management. One of the key sources of information is the registration of prisoners, displaced persons, and refugees. Historically, data cards were used in the registering and tracking process. However, prior to the arrival of Mr. Perez this process had been computerized. The computerized system had significantly reduced the number of registration errors and had increased the department's response time.
>
> Nevertheless, Mr. Perez was very concerned about his department's inability to accurately estimate the level of registering and tracking activities on an annual basis. These estimates were used in preparing the budget requests to support the department's operations. In this regard, Mr. Perez believed that his main task was to improve the administrative efficiency of the department through increased training. In addition, he decided to prepare and distribute a questionnaire to the various administrative units within the Red Cross for the purpose of better identifying the future requirements of the CTA.

has decided to retain a marketing consulting firm to help analyze some of these issues. The consulting firm has proposed to undertake a survey of both past and present members of the spa. The primary purpose of the survey will be to measure attitudes and perceptions concerning current program quality and future program design. More specifically, the survey will attempt to determine:

- Program strengths and weaknesses.
- Program selection.
- Opportunities for new programs.

The results from the survey will be used by the general manager in revising, as appropriate, the present marketing strategy and in improving program quality.

6.3 QUESTIONNAIRE DESIGN

For purposes of simplicity, the design of soft instruments will be presented under the title of *questionnaires*. This title will be used even to refer to research instruments or market surveys, regardless of whether or not paper-and-pencil or personal interview methods are used to obtain the data. Even

198 Chapter 6 Survey Design and Data Base Management

Key idea 4

though the data may be put to different uses, and even though the procedures may vary, the logic for the development of an instrument is generally the same. The primary admonition in developing soft instruments is that they are to be used only where other suitable methods are not available. For example, if a researcher wishes to relate personal variables to automobile ownership, he is perhaps better off analyzing the files of the Department of Motor Vehicles than by conducting a survey.

Figure 6.3 summarizes the basic design principles used in developing a questionnaire.

PRIMARY PURPOSE

The first step involves specifying the purposes to be served by the instrument. It is necessary to keep the final objective in mind in order to avoid errors in methods and questions. For example, if the objective is to measure the change in attitude resulting from a training course, one would design a different sort of instrument from one that examines attitudes of voters. This step essentially limits the population of likely items to those that bear on the primary purpose. Specific business examples include:

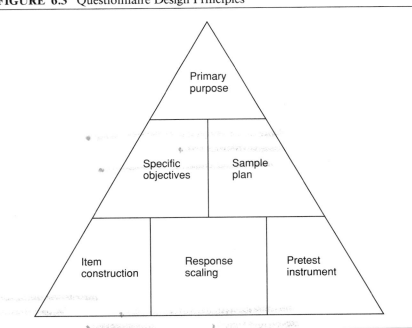

FIGURE 6.3 Questionnaire Design Principles

- To assess the economic damage caused by corporate collusion.
- To assess the impact of community standards on minority hiring.
- To determine buying preferences for soft drinks.
- To evaluate the effect of a corporate wellness program.
- To determine the market demand for a new disk drive unit.

Although some of these examples unquestionably require further definition, they nevertheless help focus the direction of the research. Furthermore, the articulation of basic study goals helps define the type of instruments and analytical methods to be used. Recall that the three primary purposes for the Nautilus Health Spa case are:

- Define program strengths and weaknesses.
- Assess program selection.
- Identify opportunities for new programs.

SPECIFIC OBJECTIVES OR DATA CATEGORIES

The next step in developing a questionnaire is to define those goals listed as the major purposes of the study. These can be specified in hypothesis form or simply as categories of data that, in aggregate, fulfill the intent of the study or the needs of the decision maker. For example, the following categories of data could fulfill this requirement for the Nautilus Spa study:

- Personal data:
    Sex
    Age
    Education level
    Income
- Attitudes toward spa:
    Staff
    Services
    Cleanliness
    Operating hours
    Membership fees

Please note that these categories are not necessarily items to be used in the completed instrument. Rather, they are merely definitions of the major intent of the study, that is, to determine the general attitude of the members regarding the operation of the spa. Additionally, they can lead to the definition of specific items. Thus, each category may lead to one or more actual items that, when aggregated, supply the data required by the decision maker. The category of services, for example, could be further defined into the following specific items:

Attitudes toward:

- Effectiveness of instruction
- Adequacy of instruction schedule
- Availability of equipment
- Adequacy of equipment
- Availability of staff for personal counseling

It should be clear that the first two steps follow the basic logic of questionnaire development. That is, the researcher or decision maker proceeds from the general need to the more specific application. This should help in drawing bounds around the study as well as providing criteria for deciding what items have to be included in the study.

SAMPLE PLAN

Prior to the specification of the actual items to be included in the questionnaire, the character and size of the sample should be considered. The first consideration involves the various characteristics of the population of interest. For example, if the study is a general population survey, the developer may be forced to consider multilingual or multicultural items (e.g., the manager of the spa may wish to develop questionnaires in both English and Spanish). Furthermore, the items must be understandable to all segments of the population. Thus, the designer needs to know a great deal about the educational level, socioeconomic status, and other related factors associated with the population of interest. The second consideration is size and representation, as they affect the items to be considered. For example, a sample of all supervisors of a plant would require a range and choice of questions different from those required from a survey of the workers.

The procedure for determining the size of the sample and the manner in which the sample is chosen is discussed in the next chapter. As a general principle, however, samples should be randomly selected. The statistics to be used in analysis and the generalities to be made all assume a random process. In general, the method used is some form of random **stratified** proportionate sampling. In that process, strata are developed, and proper proportions of those strata are used as sample parameters. Some strata generally used are

- Sex
- Age
- Occupation
- Education
- Salary range

If the population to be studied is made up of 51% female and 49% males, these are the correct percentages to select for the sample. The numbers required by stratified proportionate sampling are not necessarily so exact. Rather, the sample numbers should not be significantly different from the expected numbers. If the actual numbers differ significantly from those expected, then the sample results should not be used for inference to the larger population. However, the study may still be useful in terms of describing the actual sample results.

One might reasonably ask how the researcher decides on the strata to be used for sampling. Generally, that decision is aided by two interrelated sources: (1) relevant literature in the field; and (2) identification of those personal or demographic items that relate to the measurements being taken.

Suppose, for example, the spa manager wants to describe the attitudes of members toward the helpfulness of the staff. The literature clearly shows that these attitudes vary by age, sex, education, and income. These, then, become the sampling strata. For each of the strata used, every population element should have an equal opportunity of appearing in the survey. This statement represents a basic definition of a **simple random sample**.

ITEM CONSTRUCTION

Item construction is generally viewed as one of the most difficult tasks in the survey process. The simple reason is that words do not have the same meaning to each respondent. Respondents therefore react to their own understanding of the meanings of the words. Where the interview procedure is used with a set of items, clarifications can be made on the spot. In this case, it is hoped that these clarifications do not bias the results. On the other hand, mail procedures do not offer the same opportunities. Therefore, for these procedures a great deal of care must be exercised in selecting the wording most likely to ensure unambiguous communications.

It seems almost superfluous to discuss the manner in which unilateral meaning may be enhanced. However, prior to the discussion of measurement scales, it may be useful to review some of the things that have been discovered after many years of questionnaire use. The following descriptive guidelines should be used particularly in those questionnaires where face-to-face interaction is not possible:

1. Use clear, concise, and simple language.
2. Avoid complex, interrelated questions.
3. Provide an attractive and well-organized layout.
4. Make the questions easy to figure out.
5. Provide adequate spacing for the alternative answers.
6. Avoid compound subjects or objects.
7. Avoid leading questions.
8. Avoid phrasing questions in only one direction (either positive or negative).

If these simple guidelines are not observed, the manager/researcher may be confronted by bias in the form of measurement error or low response rate.

At a minimum, the researcher must decide the purpose of the data. For example, a researcher who wishes to determine the relationship between two variables will be forced to obtain responses in cardinal numbers. Thus, the respondent will not be asked to check an age range but will simply be asked his or her age. Similarly, the researcher seeking to establish relationships may wish to categorize potential responses to correspond to the literature as well as to the purpose at hand. The questionnaire may therefore use scales such as "more than – same – less than" rather than a numerical scale.

The categorization of potential responses should be given careful consideration vis-à-vis analysis. These few hints should help the designer make categorization decisions in item development:

1. Develop categories similar to those used in the relevant literature.
2. Determine, if possible, the actual population distribution prior to defining the categories.
3. If no categorization data are available, ask the question in an open-ended fashion.

SCALING FOR RESPONSE

The scales to be chosen obviously are interrelated with the wording of the items and cannot be isolated. Nevertheless, there may be a variety of different types of scales that are likely to elicit the desired information. Figure 6.4 lists some of the more popular scales used in questionnaire design.

Dichotomous Scales

The dichotomous scale is perhaps the simplest scale used in survey questionnaires. Basically, this scale provides the respondent with two alternatives. An example of such a scale is shown on the next page.

FIGURE 6.4 Typical Scales Used in Questionnaire Design

Type of Sale	Description
Dichotomous	Offers the respondent one of two alternatives
Open-ended	Permits the respondent to answer the question in his/her own words
Multiple choice	Respondent chooses from more than two options
Fill-in-the blanks	Respondent fills in an answer (e.g., Age: _____)
Rank/order	Respondent ranks a preset number of items, usually for importance

Do spa services meet your expectations?   Yes_____ No_____

Observe that this question is attitudinal in nature and forces the response into one of two dichotomous choices. Although this type of question has been shown to provide adequate scales for factual material, it does not enjoy the same success with attitudinal measures. Thus, unless additional scaling is done beforehand, the yes/no scale may not be adequate for respondents to express intensity of response.

Open-Ended

Many researchers use open-ended questions. Basically, they use such stimuli so that the respondents will not feel constrained to fit their perceptions into narrow, artificially created scales. Examples of two types of open-ended stimuli are provided below:

Please comment on the helpfulness of the spa staff in the spaces provided below.

_____
_____
_____

This is an open-ended statement and may elicit a wide variety of responses ranging from rates of production to assessments of morale. Consequently, it may be very difficult to analyze this type of result or to summarize the findings in meaningful information categories.

Please list the three most important objectives you have set for yourself at the spa for 1990.

1. _____
2. _____
3. _____

This type of statement is more structured than the former example. It is designed to elicit the same wide-ranging responses but in a more focused manner. Presumably, it should be simpler to analyze such responses because the variability of the subject matter has been reduced.

Open-ended questions are used predominantly as a starting point for the development of closed-end statements. It should be noted that, in spite of its advantages, the open-ended item is not recommended for any except the most experienced researcher. The typical methods of processing these item forms may be termed *content analysis* and may require some special training and talent. At a very minimum, content analysis should employ several different analysts to determine the reliability of the analytic categories. Should this procedure be required by the study objective, the more focused approach is recommended.

Multiple Choice

Multiple choice is the most often used rating scale. The literature concerning the use of this type of scale is very rich. Presented below are several of the most popular multiple-choice scales used in questionnaire design.

Likert Symmetrical 5-Point Scale. The Likert scale is perhaps the most frequently used scale in the measurement of attitudes. It is a fundamentally sound scale and is used both with and without the "undecided" category. Usually, each question or item is analyzed separately. When a given result involves adding five or six Likert scales together, the results may be weakened appreciably. No matter what numbers are assigned to each response category, the overall figure may be derived in a large number of different ways. This, in effect, would tend to lump unlike individuals in the same numerical categories and produce an attendant loss of measurement precision. A second criticism is that the wording of the item itself may influence the response since people normally tend to agree with statements. Consequently, it may be possible to influence results by virtue of strategic wording and placement of items. In spite of these criticisms, this type of scale is highly recommended for ordinary attitudinal studies. An example of the Likert scale is as follows:

I think the spa facilities are adequate for my needs.

Strongly agree	Agree	Undecided	Disagree	Strongly disagree

Three-Point Symmetrical Scale. The 3-point symmetrical scale is often used for collecting comparative data. Its advantage over the Likert scale is the smaller number of categories a respondent must consider. However, in some instances the smaller number of response alternatives may be viewed as undesirable because of the lack of sufficient detail. An example of a 3-point scale is presented below:

Please rate the quality of the spa compared to last year.

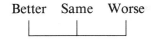

Asymmetrical Scales. In general, asymmetrical scales may be characterized as having more categories in one direction than in the other. The most noticeable weakness of these scales is the greater possibility that the final result will be in the positive direction than in the negative. Through the use of this kind of scale, it is possible to "engineer" the results. Therefore, many institutions, in providing self-ratings, may use this scale to show that they are operating effectively. An example of an asymmetrical scale follows.

Please rate the overall performance of your aerobics instructor.

    Excellent   Good   Fair   Poor
    |_____|_____|_____|

Nondescriptive Scales. A nondescriptive scale is often used when interval or ratio data are required. An example of a nondescriptive scale is given below:

Please rate the level of training at the spa.

    High   Medium   Low
    |____|____|____|____|____|____|

The respondent is asked to choose with no definition of the specific alternatives other than the end and middle points. Usually, significant variability is generated. Nevertheless, this form of scaling is not recommended for most questionnaires. First of all, the results are difficult to describe since the means and medians may fall in scale areas where no descriptors are available (e.g., between "high" and "medium"). Thus, a mean midway between "medium" and "high" has no predetermined description. Second, the variance one usually gets from this form of measurement is typically error variance.

Fill-in-the-Blanks

A fill-in-the-blanks scale is typically used where more convenient scales are not available. An example is given below:

Please state the number of years you have attended the spa. _____

The major problem with this type of scale is that respondents tend not to bother filling in statements that require them to do independent assessments. Even questions regarding sex or age, as easy as they are to answer, are omitted more often than those that have been scaled by the researcher. Thus, each item should be prescaled whenever distributions are known. For example, age categories from the population census could be used to design the item or question. The preceding question could be scaled as shown below.

Please check the number of years you have attended the spa.

    _____ <1 year
    _____ 1–2 years
    _____ 3–4 years
    _____ 4–5 years
    _____ >5 years

The spa manager found out from the responses, however, that most of the members surveyed had been with the spa for less than three years. The admonition here is to use meaningful categories that reflect an actual distribution. If you do not know the actual distribution, you should use the fill-in type of response, even though it may reduce the response rate for that item.

Rank/Order

A widely used response scale is the rank/order type, in which the respondent is asked to rank, in order of importance, specific items listed in the questionnaire. This form of scale is generally effective so long as the number of responses is small (i.e., five or fewer). An example of this type of scale is given below:

Please rank the following potential changes in spa services in order of importance to you (1 = most important, 4 = least important).

Rank	Objective
_____	Increase operating hours
_____	Increase locker room area
_____	Expand running track
_____	Add weight training class

When the number of items becomes large (> 5), respondents tend to skip over the entire question. In these situations the following approach can be used:

Please indicate the relative importance of each of the following items in your decision to join the spa:

Item	Very Important	Important	Unimportant	Very Unimportant
1. Price				
2. Operating hours				
3. Friendliness				
4. Parking				
5. Equipment				
6. Training classes				

Note that each of the items is rated independently. Nevertheless, by assigning numbers to the importance scale, it is possible to obtain natural ranks. This approach may not provide results identical to those obtained through the rank/order approach. However, it is possible to obtain ranks for longer lists of alternatives without suffering the traditional shrinkage in response rate.

In summary, the following is recommended for users of survey and questionnaire technology:

1. Employ a symmetrical scale.
2. Provide descriptors for all points of the scale.
3. Make each item easy to complete.
4. Avoid open-ended questions.
5. Avoid long lists of items.
6. Study distribution of items prior to the development of response categories.

INSTRUMENT PRETEST

The pilot test approach has received considerable attention in the literature. Few professionals would proceed with the administration of a survey without first engaging in a thorough pilot test and analysis. The manager, however, may be under time pressure and may not wish to engage in these sometimes very lengthy procedures. Nevertheless, it is recommended that some form of pilot test be undertaken no matter how unsystematic the procedures. There are some major insights an instrument designer might gain from a pilot test. These relate to:

- Ambiguities in wording
- Scaling adequacy
- Measurement adequacy
- Likely answer distributions
- Respondent comprehension

Figure 6.5 presents a CBS generated questionnaire for the Nautilus Spa Survey made up of items from the previous sections.

ADMINISTRATION AND ANALYSIS

The administration of the survey and the subsequent analysis of the data obviously play keys roles in ensuring a successful study. A variety of techniques are available for analyzing the data developed from a survey, and these are covered in detail in the following chapters of this text. Basically, there are four primary methods for collecting survey data: mail surveys, face-to-face interviews, telephone procedures, and computer-based interrogation. Figure 6.6 provides a comparison of these data collection methods.

FIGURE 6.5 CBS Generated Questionnaire for Nautilus Spa Member Survey

Nautilus Spa Member Survey

1. Your age:
   - _____ <20
   - _____ 20–29
   - _____ 30–39
   - _____ 40–49
   - _____ 50–59
   - _____ >59

2. Your sex: _____ Male _____ Female

3. Your annual income:
   - _____ <$15,000
   - _____ $15,000–$24,999
   - _____ $25,000–$44,999
   - _____ $45,000–$59,999
   - _____ >$60,000

4. Please check the number of years you have attended the spa.
   - _____ <1
   - _____ 1–2
   - _____ 3–4
   - _____ 4–5
   - _____ >5

5. I think the spa facilities are adequate for my needs.

Strongly agree	Agree	Undecided	Disagree	Strongly disagree

6. Please rate the level of training at the spa.

   High ———————— Medium ———————— Low

7. Please rate the overall performance of your aerobics instructor.

   Excellent ——— Good ——— Fair ——— Poor

8. Please rank the following spa services in terms of your interest.

Rank	Objective
_____	Increase operating hours
_____	Increase locker room area
_____	Expand running track
_____	Add weight training class

9. Please indicate the importance of each of the following items in your decision to join the spa:

Item	Very Important	Important	Unimportant	Very Unimportant
1. Price				
2. Operating hours				

FIGURE 6.5 *(concluded)*

3. Friendliness
4. Parking
5. Equipment
6. Training classes

10. Please comment on the helpfulness of the spa staff in the space provided below.

_____
_____
_____

Mail Surveys

A mail survey can be defined as a method of data collection using a written questionnaire administered through the mail. The following suggestions are provided to help reduce potential bias and increase the response rate.

1. Provide clean, professionally done instruments using as few pages as possible.
2. Always include a stamped envelope with the return address or a business reply envelope.
3. Include a cover letter that provides the background for the study.
4. Offer to share a summary of the results with those who might be interested (where appropriate).
5. Provide incentives for returning questionnaires (where appropriate).

Even using all these hints, the researcher should expect a response rate between 15% and 35% for the general unsolicited questionnaire. This can be

FIGURE 6.6 Comparison of Basic Data Collection Methods

	Data Collection Method			
		Interview		
Factor	Mail Survey	Face-to-Face	Telephone	Computer Interrogation
1. Response rate	Poor	Good	Fair	Fair
2. Time required	Medium	Large	Small	Small
3. Respondent anonymity	Good	Poor	Fair	Good
4. Flexibility	Poor	Good	Fair	Fair
5. Scaling	Good	Fair	Poor	Good
6. Cost	Low	High	Medium	Medium

enhanced by incentives, reminders, and appeals to a community of common interests. The low rates typically experienced in mail-back methods suggest that the researcher should prepare a defense against charges of bias or self-selection in the results.

Face-to-Face Procedures

A face-to-face interview involves a question-and-answer interaction between the interviewer and respondent(s). Again, a great deal of literature exists about the interview process. The following are a few simple recommendations that may aid in the interview process.

1. Solicit cooperation prior to the interview.
2. Allow the respondent to fill in the appropriate scales.
3. Explain words only when asked; then only clarify, and do not add anything unique.
4. Give the appearance of being a member of the group being interviewed.
5. Train all interviewers in how to conduct interviews.

The response rates based on the use of this procedure are typically very high (over 75%). However, this procedure tends to be very costly. The trade-off therefore is generally cost against allegations of bias as a function of nonresponse. If the sampling is adequate and procedures are carefully designed, little bias will presumably be introduced. Bias can occur, however, when the interviewer subtly affects the direction of the response. There are few safeguards offered in the literature to obviate this kind of effect. Knowing that it exists, however, should aid the interviewer's performance.

Telephone Procedures

A telephone interview is a two-way communication via the telephone between the interviewer and the respondent(s). The typical telephone inquiry leads to a response level somewhere between that of the mail survey and that of the face-to-face interview. There are two primary advantages of the telephone interview: speed of data collection and cost-effectiveness. Generally, a telephone interview is very ineffective for scaled items but adequate for informational questions. It is for the latter that telephone procedures are recommended.

Computer Interrogation

The use of computers for data collection is relatively new, but it is projected to increase significantly over the next several years. One technique involves a respondent filling out a computer-readable form. Examples include in-flight airline surveys, hotel ratings, and course evaluations. In another approach, the respondent is prompted through a series of questions via a computer terminal (e.g., at a shopping mall). This approach has the advantage of storing and processing the data in near-real time. However, many potential respondents are fearful of computers and may avoid participating in the survey.

6.4 COMPUTER DATA BASE MANAGEMENT SYSTEMS

Typically, data obtained from a survey is organized into a data base. A computer data base management system (CDBMS) is a software package designed for inputting, storing, and retrieving data. CDBMSs are playing an ever increasing role in survey design and analysis. Figure 6.7 presents a schematic of an interactive CDBMS showing the relationship between the overall survey process and the data base system. The CDBMS serves as a buffer between the survey design and data collection phases and the analysis

FIGURE 6.7 Schematic of Interactive CBDMS

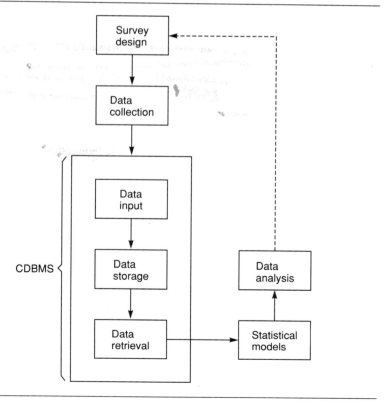

FIGURE 6.8 Example of Data Schema for Nautilus Health Spa

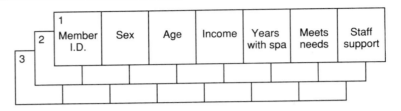

of the collected data. The "dotted" feedback line indicates that the results from the analysis can be used to revise the survey design.

Typically, a logical structure or schema is used in designing the data base. A **schema** describes the overall organization of the data in the data base. Figure 6.8 presents an example of a schema for the Nautilus Spa survey. A schema can consist of one of several logical data structures. The three basic structures for organizing data are hierarchical, network, and relational.

A **hierarchical** structure relates data through an "upside down" tree diagram. A tree structure indicates the relationship between one level of the data base to the next through the use of multilevel nodes. Figure 6.9 presents an example of a hierarchical data base structure for multiple spa surveys. Here the data is partitioned by spa (level 2) and membership status (level 3). Two conditions must be met to organize data in a hierarchical structure: (1) the tree must have a single beginning node; and (2) all nodes other than the beginning node must be related to exactly one higher level node.

In a **network** structure, data can be related to any other data within the data base. A network does not contain the rigorous structure found in a tree.

FIGURE 6.9 Example of Hierarchical Structure Data Base for Spa Surveys

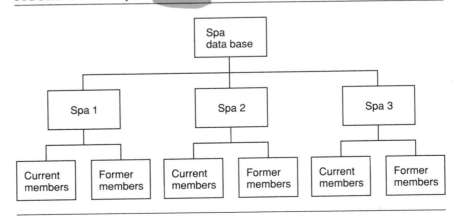

Key idea 5

FIGURE 6.10 Example of Network Structure Data Base for Spa Surveys

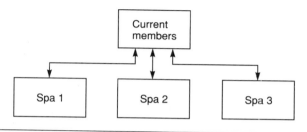

Figure 6.10 shows an example of a network data structure that relates the current members' responses for the three spas.

A **relational** structure presents the data in a row and column format, that is, a table structure. In a survey, the rows represent the individual respondents, and the columns represent the attributes of interest (e.g., age). Each row, which is analogous to a record, must be distinct; the same record should not appear more than once. Similarly, each column must also be unique. Figure 6.11 presents an example of a relational structure for the Nautilus Spa survey. The results from most individual surveys are organized using a relational structure. Organizing the data in this way facilitates the data analysis process. In this regard, one important attribute of a CDBMS is the ability to convert hierarchical and network data structures to relational ones.

A variety of CDBMSs are commercially available. Two of the most popular on university campuses are LOTUS 1-2-3® and dBase IV®. The CBS software package also contains a relational data base management system. The CBS system is designed to accommodate up to 200 records (rows) and 10 attributes (columns) per data file. For larger problems, a hierarchical structure consisting of multiple data files can be used.

FIGURE 6.11 Example of Relational Data Base Structure for Spa Survey

Member I.D.	Years with Spa	Meets Expectations (1 = Yes, 0 = No)	Aerobics Instructor (1 = Excellent, 4 = Poor)
1120	1	1	2
1175	1	0	3
1237	2	0	1
1415	3	1	4
⋮			
1812	1	1	1

6.5 PRACTICAL APPLICATIONS

This section presents several business examples involving the use of questionnaires.

Viking Aircraft Employee Termination Interview

The vice president of Human Resources at Viking Aircraft wishes to develop a better understanding of the factors contributing to employee turnover. The company has experienced a growing rate of turnover over the last several years, particularly among skilled workers, and it has had increasing difficulty in replacing skilled employees who have left the company. Even when a skilled worker is replaced, however, the net effect is costly. The vice president estimates that Viking invests approximately $100,000 in training costs per skilled employee over the first two years of employment. To help address this important issue, the vice president has decided to conduct an in-depth interview of 100 previous employees selected at random who have left over the past year. This figure constitutes approximately 10% of the number of employees that have left the company over this period. Figure 6.12 presents a CBS generated questionnaire that will be used in conducting the interviews.

In addition to the interview data, the vice president also wishes to record the personal data that can be obtained from the company's data base management system such as time with the company, salary at time of termination, job classification, and security classification. A summary of selected results from the interview process is presented in Figure 6.13. A preliminary inspection of these results suggests that management needs to do a better job in communicating with employees.

FIGURE 6.12 CBS Generated Employee Termination Interview

Note to interviewer: Please ask the individual being interviewed to respond to the first six questions using the following 5-point Likert scale.

Strongly agree	Agree	Indifferent	Disagree	Strongly disagree
1	2	3	4	5

1. I felt that company goals were effectively communicated to me.
2. I felt that any suggestions made by me were given careful consideration by my management.
3. I felt that my immediate supervisor provided me with the information needed to perform my job.

FIGURE 6.12 *(concluded)*

4. I felt that there was good rapport between the management and workers in the organization.
5. I felt part of a winning team.
6. I felt that the reward system was fair.
7. Please rank order the factors involved in your decision to leave the company from among the ones given below (1 = most important, 7 = least important):

   a. _____ Higher pay
   b. _____ Promotion
   c. _____ Better working conditions
   d. _____ More responsibility
   e. _____ Move to another area
   f. _____ Management
   g. _____ Other

8. Did you discuss your decision to leave with your management?

   a. _____ Yes   b. _____ No

9. Did the company make an attempt to change your decision to leave?

   a. _____ Yes   b. _____ No

10. If the answer to question 9 is yes, in what ways?

    _____
    _____
    _____

11. Please list any other considerations you feel might be helpful in understanding why someone might decide to leave the company.

    _____
    _____
    _____
    _____

FIGURE 6.13 Selected Results from Termination Interviews

Question	Response
1. Company goals and objectives	3.5
4. Good rapport	3.0
5. Winning team	3.8
7. Higher pay	55%
8. Discussed decision	23%
9. Attempt to change decision	24%

Palos Verdes Community Bank Customer Survey

The president of Palos Verdes Community Bank is considering several new products and services. These include a bank debit card, a home banking service, and expanded hours of service for the drive-up window. Before choosing one or more of these alternatives, the president wishes to survey the bank's current 400 customers. He has therefore asked the bank's market research department to develop and implement a customer survey. The manager of the research department used CBS to prepare the survey instrument reported in Figure 6.14.

FIGURE 6.14 CBS Generated Bank Customer Survey

Dear preferred bank customer: Palos Verdes Community Bank is currently considering several new products and services. We would very much like your help in this evaluation process. Accordingly, we would appreciate it if you would take a few minutes to complete this questionnaire. The results from this survey will be used in helping shape the future direction of the bank.

1. Rank on a scale from 1 to 5 the following proposed new products and services (1 = most important, 5 = least important):
    a. _____ Bank debit card   d. _____ Bank credit card
    b. _____ Home banking service   e. _____ Saturday banking
    c. _____ Expanded drive-up
               window hours

2. What type of new product(s) or service(s) would you like the bank to offer?
   _____
   _____

3. How often do you use the drive-up window?
    a. _____ Never   d. _____ Often
    b. _____ Rarely   e. _____ Usually
    c. _____ Sometimes

4. Which of the following time options would be most convenient?
    a. Saturday mornings   c. Early weekday mornings (7:00 A.M.)
    b. Saturday afternoons   d. Early weekday evenings (7:00 P.M.)

5. Do you currently have a computer at your home?
    a. _____ Yes   b. _____ No

6. Your age _____

7. Your sex _____

8. Your income:
    a. _____ Under $15,000   d. _____ $45,000–$59,999
    b. _____ $15,000–$29,999   e. _____ $60,000–$74,999
    c. _____ $30,000–$44,999   f. _____ Over $74,999

FIGURE 6.14 *(concluded)*

9. Your education (list highest level attained):
   a. _____ Attended high school   d. _____ Undergraduate degree
   b. _____ High school degree   e. _____ Graduate studies
   c. _____ Undergraduate studies   f. _____ Graduate degree

A summary of selected survey results is presented in Figure 6.15. These results are based on a total of 250 responses.

The results from the survey have been reported to the president along with the statement that the demographic data (e.g., age and income) closely match the profit of the bank's customer base. The president, after a careful

FIGURE 6.15 Selected Results from Bank Customer Survey

1. Relative importance of proposed new products and services:
   a. Bank debit card (3.5)*   d. Bank credit card (5.0)
   b. Home banking service (2.5)   e. Saturday banking (8.5)
   c. Expanded drive-up window
      hours (7.0)

2. (Not listed.)

3. Drive-up window usage:
   a. Never   10%   d. Often   30%
   b. Seldom   15   e. Usually 20
   c. Sometimes 25

4. Convenient times:
   a. Sat. morning   45%   c. Early mornings 10%
   b. Sat. afternoon 5   d. Early evenings 40

5. Home computer availability:
   a. Yes   25%   b. No   75%

6. Average age = 32.6 years

7. Male = 45%   Female = 55%

8. Customer income:
   a. 5%   d. 24%
   b. 13   e. 17
   c. 33   f. 8

9. Customer education:
   a. 5%   d. 20%
   b. 35   e. 10
   c. 20   f. 10

* The numbers in parentheses represent the average ranking.

review of these results, has decided to initiate Saturday banking and to increase the hours for the drive-up window. The president has also directed the manager of the market research department to undertake a more in-depth study of the future possibilities of home banking.

6.6 CASE STUDY: TRANSPACIFIC AIRLINES

Transpacific Airlines, founded in 1985, provides air service throughout the Pacific. The company operates five widebody DC-10s, using Honolulu as a regional hub. In 1987 the company experienced 500,000 passenger boardings. This represents a 79.4% loading factor, which is considered very good by airline standards. Currently, Transpacific offers single-class nonstop service to Los Angeles (twice daily), Tokyo, Sydney, and Anchorage. Figure 6.16 shows the present flight schedule.

At last quarter's board meeting, one of the key topics of discussion involved the possibility of expanding service. The company's chairman, Mr. Al Smith, observed that Transpacific's primary marketing focus had been on the business passenger. This, he stated, was in keeping with the general forecast for increased international trade throughout the remainder of this century. Honolulu, he believed, would become an important trade center for the Pacific basin. In support of this claim, the MIS director, Ms. Sue Brinks, reported that the proportion of business fliers had nearly doubled over the last three years (20% to 35%). Although everyone at the meeting seemed to support the concept of expanding service, no specific proposals were presented. Seizing the moment, the vice president of marketing, Ms. Jane Burton, suggested that her department conduct a market survey of on-board

FIGURE 6.16 Current Flight Schedule for Transpacific Airlines

Depart	Time*	Arrive	Time*
Honolulu	9:00 A.M.	Los Angeles	4:00 P.M.
Los Angeles	7:00 P.M.	Honolulu	10:30 P.M.
Los Angeles	9:00 A.M.	Honolulu	12:30 P.M.
Honolulu	4:00 P.M.	Los Angeles	11:00 P.M.
Honolulu	10:00 A.M.	Tokyo	2:45 P.M. (ND)†
Tokyo	7:00 P.M.	Honolulu	6:00 A.M.
Honolulu	12:30 A.M.	Sydney	7:00 A.M. (ND)
Sydney	9:00 A.M.	Honolulu	10:00 P.M.
Honolulu	8:00 A.M.	Anchorage	2:00 P.M.
Anchorage	5:00 P.M.	Honolulu	11:00 P.M.

* Local time.   † ND = Next day.

passengers to determine which services might be expanded. The chairman accepted Ms. Burton's proposal and requested that she provide an in-depth report of her findings at next quarter's board meeting.

Ms. Burton immediately called a meeting of her staff to outline the process for collecting the desired data. Everyone agreed that an in-flight questionnaire that took less than 10 minutes to fill out was the best approach. However, one of the issues raised at the meeting involved the question of language. Early company surveys had revealed that upwards of 25% of Transpacific's passengers did not speak or write English (of that amount, nearly 80% wrote and spoke Japanese). Accordingly, Ms. Burton decided to produce both an English and a Japanese version of the questionnaire. The basic form of the questionnaire was worked out at the meeting and was pretested on one of Transpacific's Honolulu-to-Tokyo flights. The final questionnaire, developed with the aid of CBS, is presented in Figure 6.17. Ms. Burton has decided to utilize the questionnaire over the next 30 days on all Transpacific's routes. At this time of the year, this should provide approximately 20,000 responses (Jane assumed that approximately 50% of the passengers would participate in the survey).

FIGURE 6.17 CBS Generated Customer Questionnaire

This questionnaire is designed to help Transpacific improve its service. We would appreciate it if you would take a few minutes to fill out the questionnaire. If you have any questions please feel free to ask one of our in-flight representatives.

Thank you.

_____
Jane Burton
Vice President of Marketing

1. Travel purpose:

   _____ Business   _____ Vacation
   _____ Other (please specify)

2. Age:

   _____ Under 20   _____ 50–59
   _____ 20–29   _____ 60–69
   _____ 30–39   _____ Over 69
   _____ 40–49

3. Sex:

   _____ Male   _____ Female

4. Income level:

   _____ Under $15,000   _____ $60,000–$74,999
   _____ $15,000–$29,999   _____ $75,000–$89,999
   _____ $30,000–$44,999   _____ Over $89,999
   _____ $45,000–$59,999

FIGURE 6.17 *(concluded)*

5. Citizenship:

    _____ United States
    _____ Australia
    _____ Japan
    _____ Other (please specify)

6. How did you hear about Transpacific?

TV	_____	Newspaper	_____
Radio	_____	Billboard	_____
Magazine	_____	Other	_____

7. How often do you fly?

    Less than once per month   _____
    1–2 times per month   _____
    More than twice per month _____

8. Why did you select Transpacific (check all that apply)?

    Food service   _____
    Frequency of scheduled departures _____
    Convenient departure times   _____
    Low fares   _____
    On-time departures   _____

9. Transpacific is considering incorporating one or more of the following services. Please rank them in order of importance (1 = most important, 5 = least important):

Telephone/FAX service	_____	Market report service	_____
Sleeper service	_____	Computer terminal service	_____

    Other (please specify) _____

10. What other cities would you like to see added to Transpacific's flight schedule. Please rank them in order of importance (1 = most important, 6 = least important)?

Singapore	_____	San Francisco	_____
Manila	_____	Wellington	_____
Hong Kong	_____		

    Other (please specify) _____

The survey data were collected at the end of the 30-day period and processed by the company's data base management system. The demographic data collected from the survey were found to be in close agreement with historical passenger profile data. A set of summary results was developed in preparation for the next quarterly board meeting. This task required approximately two weeks.

At the next board meeting, Mr. Smith asked Ms. Burton to report the

FIGURE 6.18 Income Histogram for Transpacific Passenger Survey

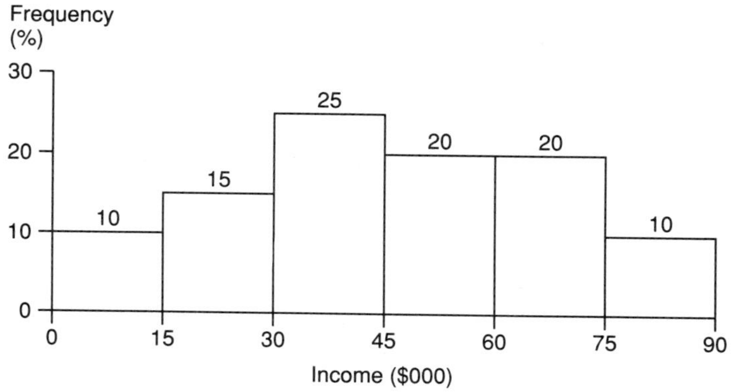

```
Frequency
(%)
30 ┐
                    25
              ┌─────┐
20 ┤                │     20    20
              │     ├─────┬─────┐
         15   │     │     │     │
         ┌────┤     │     │     │
    10   │    │     │     │     │   10
10 ┌─────┤    │     │     │     ├─────┐
   │     │    │     │     │     │     │
   │     │    │     │     │     │     │
 0 ┼─────┴────┴─────┴─────┴─────┴─────┤
   0    15   30    45    60    75    90
              Income ($000)
```

results of her survey. Jane started by presenting the survey instrument shown in Figure 6.17. Next, she showed a histogram of passenger income as seen in Figure 6.18. She reported that this histogram compared very favorably with the company's historical passenger income profile. The chairman asked Jane to explain the relevance of her statement. She responded that since the demographic data taken from the survey compared with the historical profile, the reported findings could be considered an accurate representation of the passengers' views.

Jane then turned to a series of charts highlighting the survey findings. The first focused on how passengers heard about Transpacific. These data, broken out by language of response, are shown in Figure 6.19. Jane concluded, not surprisingly, that TV advertising represented the most effective vehicle for acquainting the public with Transpacific, regardless of language. Mr. Smith observed, however, that airline reservations for most business travelers are made by the company. Furthermore, he noted that TV advertising is very expensive compared to the other media forms listed. Jane responded

FIGURE 6.19 Primary Media for Hearing about Transpacific

Media	English	Japanese
TV	44%	40%
Newspaper	19	17
Radio	14	23
Magazine	8	7
Billboard	4	3
Other	11	10

FIGURE 6.20 Ranking of Basic Airline Features

Feature	First-Place Ranking
Convenient departure times	39%
Low fares	34
On-time departures	15
Frequency of schedule departures	9
Food service	3

that Transpacific had an ongoing program to acquaint the business community with the company's flight program. Additionally, she stated that 65% of the passengers were traveling for nonbusiness reasons. Finally, Jane agreed with the chairman (a smart move!) that TV advertising was expensive. However, she argued that the amount of TV viewing would continue to increase at the expense of several of the other media forms (e.g., newspapers).

Jane then presented a graphic, shown in Figure 6.20, showing the first-place ranking of the five key airline features. Jane reminded the board that the objective of this question was to identify the reasons passengers had selected Transpacific. These results indicated that convenient departure times and low fares received nearly 75% of the first-place votes. The chairman mused that perhaps a reduction in food service might allow for a lowering of ticket prices.

The next chart Jane presented showed the relative ranking of several new in-flight services currently under consideration by the company for the business traveler (see Figure 6.21). The two most popular features were telephone and sleeper service. However, the idea of providing one or more on-board conference rooms also seemed to be of some interest, especially to the Japanese.

Jane's last chart focused on expanding Transpacific's current route structure. These data are shown in Figure 6.22. Singapore and San Francisco

FIGURE 6.21 Ranking of In-Flight Service by Language

Feature	First-Place Ranking	
	English	Japanese
Telephone/FAX service	35%	24%
Sleeper service	26	33
Market service	15	18
Computer service	11	7
Other (conference room)	13	18

FIGURE 6.22 Ranking of New Routes

Location	First-Place Ranking
Singapore	32
San Francisco	30
Hong Kong	18
Wellington	5
Manila	4
Other (Tahiti)	11

emerged as the two most attractive candidates for possible expansion. The chairman jumped into the discussion and stated that perhaps Singapore could be added to the Honolulu-to-Tokyo run and San Francisco to the Honolulu-to-Los Angeles route. Jane responded that the latter option would be quite feasible with the current fleet; however, an additional plane would be needed if Singapore was to be incorporated into the route structure. At the end of the presentation, Mr. Smith congratulated Jane on a fine effort and suggested that she work up a proposal for implementing one or more of the study findings.

6.7 SUMMARY

This chapter introduced the use of survey design and data base management as important tools for the modern business manager. The survey design process provides the manager with the capability of obtaining problem-specific data (e.g., customer preferences) that cannot be obtained by other means. Typically, two types of errors are found in survey data: systematic and random. Systematic errors are due to the bias in survey design. Random errors, on the other hand, are the result of measurement inaccuracies due to chance. Both of these types of errors can be controlled through an effective survey design. A key step in the design of an effective survey involves the development of a collection instrument or questionnaire. This chapter presented a six-step approach for developing and using questionnaires. Item construction represents perhaps the most difficult step in this process.

This chapter recognized the growing importance of computer data base management systems (CDBMS). A CDBMS is a software package that is designed to input, store, and retrieve data. More specifically, CDBMS provides an interface between the users and the data base. Basically there are three types of data base structures: hierarchical, network, and relational. Within each of these structures, the schema describes the logical relationship between data elements in the data base. It specifies the names, characteristics and relationships that exist between data members. A CDBMS can be designed to interact directly with the survey process. In the future, computer-based systems will play an ever more important role in the design and implementation of surveys.

6.8 GLOSSARY

census A method for measuring certain characteristics of interest from the total population.

computer data base management system (CDBMS) A software package designed to store, control, and retrieve data.

data A compilation of facts.

data base A collection of data.

experiment The collection of data with controls.

hard data Data obtained from an objective measurement system (e.g., corporate earnings).

hierarchical structure Data records are dependent and are arranged in multilevel structures.

information Data that have been processed.

network structure Multiple relationships are permitted between data records.

nonresponse bias The tendency for specific elements in the population with certain characteristics not to respond to the survey.

random error Measurement inaccuracies due to chance.

relational structure Data records are stored in title form of simple row and column tables.

response bias The tendency to answer survey questions incorrectly due to a fault in the questionnaire design.

sample A method for measuring certain characteristics of interest from a subset of the population.

schema A method for describing the overall logical organization of the data.

selection bias The tendency to include in a survey elements of a population with a certain characteristic while excluding elements with other characteristics.

simple random sample A method of sampling in which each member of the population of interest has an equal chance of being selected.

soft data Data obtained from a subjective measurement system (e.g., consumer attitudes and preferences).

stratified sample A method of sampling where the population of interest is divided into strata and from each strata a simple random sample is selected.

survey The collection of data without controls.

systematic error Measurement inaccuracies due to bias in the survey design.

6.9 BIBLIOGRAPHY

Bellenger, D. N., and B. A. Greenberg. *Marketing Research: A Management Information Approach.* Homewood, Ill.: Irwin, 1978.

Davis D., and R. M. Cosenza. *Business Research for Decision Making,* 2nd ed. Boston: PWS-Kent, 1988.

Emory, E. W. *Business Research Methods,* 3rd ed. Homewood, Ill.: Irwin, 1985.

McLeod, R. *Management Information Systems.* Chicago: SRA, 1987.

O'Brien, J. *Computer Concepts and Applications,* 3rd ed. Homewood, Ill.: Irwin, 1988.

———. *Computers in Business Management,* 3rd ed. Homewood, Ill.: Irwin, 1982.

———. *Information Systems in Business Management.* Homewood, Ill.: Irwin, 1988.

6.10 PROBLEMS

1. Discuss the advantages and disadvantages of each of the following methods for conducting a presidential survey:
   a. Mail
   b. Telephone
   c. Personal interview
   d. Computer interrogation
2. Discuss the advantages and disadvantages of each of the following methods for conducting a population census:
   a. Mail
   b. Telephone
   c. Personal interview
   d. Computer interrogation
3. Under what conditions might a survey be more accurate than a complete census?
4. Identify some of the major problems associated with the 1990 national census.
5. Identify whether a survey or a census would be more appropriate in each of the following situations:
   a. A politician wishes to know the general public reaction to a smoking ban in restaurants.
   b. The owner of a retail electronics store wishes to know the current inventory level.
   c. A market research firm wishes to know the potential demand for a new razor.
   d. A quality control manager wishes to know the level of defects in a production batch.
6. Identify whether a survey or census would be more appropriate in the following situations:
   a. A maintenance supervisor wishes to know the status of the braking system on the company's fleet of 747 jet airliners.
   b. The U.S. FDA wishes to know the potential side effects of a new drug to combat AIDS.

c. An auditing firm wishes to know the number of delinquent accounts receivable for one of its largest clients.
d. A professor wishes to know how much the students learned at the end of the course.
7. Distinguish between systematic and random errors. Provide examples of both types of errors. Which type of error is influenced by sample size?
8. Develop several measurement examples that contain systematic, random, or both types of errors.
9. Define the term *instrument design*. Discuss the relationship between measurement and instrument.
10. Identify several of the concerns associated with question phrasing and give examples.
11. Specify the flaw in each of the following questions:
    a. Do you exercise regularly?
    b. How large a sales increase are you expecting?
    c. What do you think of our latest computer system?
12. Specify the flaw in each of the following questions:
    a. Do you provide scholarships for minorities or women?
    b. Do you prefer small-size cars or some other size?
    c. Do you study more or less than eight hours per week?
13. What role do demographic data play in survey design?
14. Provide examples of the following types of measurement scales:
    a. Dichotomous
    b. Open-ended
    c. Multiple choice
    d. Fill-in-the-blank
    e. Rank order
15. What is a data base management system? Give examples.
16. What conflicts or trade-offs might arise in designing a common data base? How might these conflicts be reduced or eliminated?
17. Describe the differences between hierarchical, network, and relational data structures.
18. What are the four basic elements of a DBMS?
19. Construct a 5-item Likert scale (5-point) to measure public attitudes on controlling assault weapons.
20. Construct a 5-item Likert scale (5-point) to measure employee attitudes regarding a 10-hour, 4-day work week.
21. Construct a rank/order scale to measure the relative importance of the top five social issues facing the United States.
22. Construct a rank/order scale to measure the relative importance of the top five foreign policy issues facing the United States.
23. Develop a 10-item personal interview questionnaire for determining the student demand for computing services at your local computer center.
24. Develop a 20-item telephone questionnaire for measuring the attitudes of your community regarding the quality of the local school system.
25. Develop a 10-item telephone questionnaire for measuring the interest in real estate brokers for a computer-based residential pricing model.
26. Develop a 15-item mail questionnaire for measuring employee interest in a new flex-benefits package.
27. Why is pretesting so important in the design of effective surveys?

28. The manager of Proton Micro Computers is interested in determining the level of customer satisfaction with the products and services offered by the firm. More specifically, the manager would like to obtain customer attitude data on the following areas:
    1. Computer characteristics:
        a. Speed
        b. Down time
        c. Design growth
    2. Software characteristics:
        a. Ease of use
        b. Compatibility
        c. Expansion capability
    3. Service support:
        a. Cost
        b. Response time
        c. Reliability
    4. Customer training:
        a. Cost
        b. Effectiveness
        c. Availability

    a. Use CBS to develop a questionnaire for measuring the items listed above.
    b. What demographic items would you use and why?
    c. What kind of sampling plan would you use and why?
    d. What administrative procedures would you use and why?

29. The manager of High Ridge Car Wash is interested in determining the level of customer satisfaction with the present service. The manager would like to obtain customer data on the following:
    1. Staff courtesy
    2. Staff helpfulness
    3. Pricing
    4. Cleanliness (waiting area and restroom)
    5. Service time
    6. Accessories for sale
    7. Service currently provided
    8. Potential new services

    a. Use CBS to develop a questionnaire for measuring the items listed above.
    b. What demographic items would you use and why?
    c. What kind of sampling plan would you use and why?
    d. What administrative procedures would you use and why?

30. A local newspaper wishes to conduct a public opinion survey on these issues:
    1. Balancing the federal budget
    2. Capital punishment
    3. Gun control

    a. Use CBS to develop a questionnaire for measuring the items listed above.
    b. What demographic items would you use and why?
    c. What kind of sampling plan would you use and why?
    d. What administrative procedures would you use and why?

6.11 CASES

6.11.1 Sports Shoe, Inc.

Sports Shoe, Inc., founded in 1984, manufactures and markets a complete line of footware for the serious amateur as well as for the professional athlete. The company currently owns and operates 30 retail outlets in the greater New York area. Each outlet has a store manager and two assistant managers. Revenues for 1988 are estimated at $25 million. Table 1 summarizes corporate growth over the past five years. The company's general manager is concerned about this rapid growth particularly

TABLE 1 Sales Revenues and Number of Sports Shoe Stores

Year	Revenues ($000)	Number of Stores
1984	3,500	5
1985	7,000	11
1986	12,000	16
1987	18,500	25
1988	25,000 (est)	30

with respect to communication and level of control. Historically, corporate operations were tightly controlled by the general manager's office. However, the general manager is considering adding a new layer of management between his office and the store managers. The GM believes that the creation of five regional managers will enhance corporate flexibility and set the stage for future expansion.

However, before embarking on this reorganization, the GM would like to obtain the opinion of Sports Shoe's management team. The GM has decided to use a mailed questionnaire to solicit suggestions since there are over 100 managers and staffers. More specifically, the general manager would like to know what the managers and staffers feel about:

- The direction in which Sports Shoe is moving.
- The need to reorganize.
- How the company should be organized.
- The level of autonomy for store managers.
- The general policy for promotion.

In addition to these organizational issues, the GM would also like to know:

- What new products the company should consider introducing.
- How well the company is responding to the growing competition.
- Whether the company is doing an effective job with its customers.

The general manager hopes that the survey can be completed in approximately 30 days.

Suggested Questions

1. Identify the appropriate scale for each of the general manager's basic questions.
2. What additional questions might be added to the questionnaire?
3. What incentives might the GM use to ensure a large response?
4. What demographic questions should be included in the questionnaire?
5. How might the general manager use the results of the survey?

6.11.2 War Games International

War Games International, founded in the mid-1980s, designs and markets microcomputer-based war simulations. The company has concentrated primarily on simulations involving World Wars I and II. However, there appears to be a growing

TABLE 1 Historical and Projected Revenues for War Games International

Year	Annual Revenues ($000)	Number of Simulations
1986	10	1
1987	22	2
1988	48	4
1989	155 (estimated)	6
1990	310 (projected)	8
1991	675 (projected)	10

interest in simulations involving more modern situations (e.g., the 1982 Falkland Island conflict). The product is geared toward historians, military personnel, and educators. Preliminary marketing estimates indicate that there are about 500,000 such individuals in the United States and a like amount around the rest of the world. The game package includes a game diskette, battle manual, and operating instructions. A typical game is priced at $49.95. The company ultimately plans to offer approximately 20 military simulations, including "The Battle of Midway," "D-Day," and "Action in the North Atlantic." Table 1 presents historical and projected revenues through 1991. Also shown is the number of games offered. The company's chief game designer estimates that all 20 games will be completed by early 1995.

One of the major problems facing the company (in addition to rapid growth) is the need for a computerized data base management system. Currently, all orders and inquiries are processed by hand. This system was barely adequate when the number of annual orders was less than 1,000. However, the projected increase to nearly 10,000 orders per year by 1990 would completely overwhelm the present system. The company's president would like a data base system with the ability to:

- Process orders, including the printing of invoices.
- Track inventory levels.
- Accept credit card orders.
- Provide descriptive statistics of sales activity by region and game.
- Offer product information to customers via modem.
- Poll customers on their response to current and future products.

The president would like a proposal to implement a data base system with these characteristics within 30 days.

Suggested Questions

1. What type of data base system would you recommend?
2. What would be the basic steps in developing the proposed system?
3. What are some of the major problems likely to be encountered in developing the system?
4. What type of questionnaire design would best meet the company's requirements?
5. How long should the data base system take to implement?

Chapter 7

Sampling and Estimation

I have a great subject to write upon, but feel keenly my literary incapacity to make it easily intelligible without sacrificing accuracy and thoroughness.

Sir Francis Galton

CHAPTER OUTLINE

7.1 Introduction
7.2 Example Management Problem: Wilcox Accounting Services
7.3 Basic Characteristics of Sampling and Estimation
7.4 Sampling and Estimation
7.5 Computer Analysis
7.6 Practical Applications
7.7 Case Study: Bozart Investments Corporation
7.8 Summary
7.9 Glossary
7.10 Bibliography
7.11 Problems
7.12 Cases

CHAPTER OBJECTIVES

The primary objectives of this chapter are to develop an understanding of

1. The logic behind statistical inference.
2. Alternative sampling methods.
3. The nature of sampling distributions.
4. The central limit theorem.
5. How to develop point and interval estimates.
6. How to determine the size of the sample based on error considerations.

Every 10 years the U.S. government undertakes a national census to determine demographic, economic, and social trends. The results are used to help formulate public policy at the local, state, and national levels. The planned census for 1990 faces a large number of challenges, including the reporting of minority groups, which are widely dispersed throughout the population. Many members of these groups do not wish to be included in the census from fear of the government. One approach used to overcome some of the problems associated with minority groups involves sampling by referral*. This technique was used in a study of a group of Hispanic-Americans that constitutes approximately 3% of the total population of the city of Omaha. Probability sampling would have required 10,000 households to be interviewed in order to find 300 Hispanic-American households. The use of referral sampling reduced the number to fewer than 3,000 households. A comparison of the demographic data obtained from the sample with the census tract for Omaha revealed very few differences.

7.1 INTRODUCTION

Key idea 1

Chance and uncertainty play an important role in the business decision-making process. The usual method for quantifying chance and uncertainty in decision making is through the assignment of probability. The previous two chapters have laid the foundation for the application of probability theory to business decision making. In most business situations decisions are usually based on limited information rather than complete knowledge of a problem. The available information is often obtained by sampling the business environment instead of taking a complete census. In these cases, the decision maker must estimate the consequences of various alternatives from the information provided by the sample. Therefore, sampling and estimation are fundamental to modern business practice. The process of estimating is called **statistical inference**. Its premise is that information gained from a relatively small sample can be used to estimate the general environment.

Sampling and estimation are used extensively in business. For example, a marketing manager might use it to learn the potential demand for a new product, a political analyst might use it to learn the popularity of a specific candidate, or a quality control inspector might use it to learn the proportion of defective parts being produced. These decision makers are typically interested in learning something about a particular population without having to measure it in its entirety. The process of inferring the value of an unknown population **parameter** (e.g., mean or standard deviation) from a sample is known as *estimation*.

There are two basic types of statistical estimators: (1) point and (2) interval. An **estimator** is a sample statistic used to estimate a population parameter. A **point estimate** is a single numerical value that represents the "best" indicator of an unknown population parameter (e.g., mean). An **interval estimate** is a range of values that has a high probability of containing the population parameter. This chapter introduces the process of developing point and interval estimates for both means and proportions. The study of

---

* S. Welch, "Sampling by Referral in a Dispersed Population," *Public Opinion Quarterly* (Summer 1975), pp. 237–45.

> **HISTORICAL NOTE**
>
> William S. Gossett (1876–1937) pioneered the development of small-sample statistics. In 1899 he went to work for the Guiness Brewing Company. One of the problems he encountered was the influence of both temperature fluctuations and material variability on production quality. Unfortunately, the state of the art in statistics at that time was based on rather large samples. In brewing, however, Mr. Gossett found that the sample sizes would be quite small. He set to work to develop a sampling distribution based on small samples, which was published in his famous paper in 1909 under the pseudonym "Student."

sampling and sampling distributions offers an important insight into the process of statistical estimation. Accordingly, this chapter begins with a treatment of these two important concepts.

7.2 EXAMPLE MANAGEMENT PROBLEM: WILCOX ACCOUNTING SERVICES

Wilcox Corporation provides temporary accounting professionals to a variety of small-to-medium businesses operating throughout California. The job-scheduling manager at Wilcox Corporation, Mr. Fred Barnes, has received a number of complaints from several of the company's best clients regarding the seemingly high rate of absenteeism among the professionals sent by Wilcox. Yet Wilcox has traditionally had an excellent reputation in the industry for providing reliable workers. Recently Mr. Barnes has also received complaints from several community groups regarding the low percentage of women employed by Wilcox. He has therefore decided to examine the company's records over the past year to obtain a quantitative feel for these problems. An in-depth study of absenteeism among the company's 4,500 employees undertaken two years ago revealed an overall average absenteeism rate of 42.5 hours per year with a standard deviation of 19.4 hours. These results fall within the company's general guidelines of 30 to 60 hours per year. The analysis also showed that 40% of the company's employees were female. Because of the expense of undertaking another major study, Mr. Barnes has decided to conduct a survey of 30 employee files selected at random. From each file, Mr. Barnes has noted both the absenteeism rate (measured in hours per year) and the sex of the employee. These data are reported in Figure 7.1. Mr. Barnes would like to know (1) how the absenteeism rate generated from the sample compares with the data obtained from the previous study and (2) the proportion of women in the work force.

FIGURE 7.1 Absenteeism and Gender for 30 Randomly Sampled Employees at Wilcox Accounting Services

Absentee Rate (hours per year)	Sex	Absentee Rate (hours per year)	Sex
24	M	32	F
20	F	42	M
45	F	102	M
60	M	54	F
12	M	62	M
82	F	70	F
50	M	45	M
38	M	34	M
40	F	18	M
72	F	38	F
60	M	48	M
54	M	36	F
32	F	50	F
27	M	7	M
33	F	31	M

> **INTERNATIONAL VIGNETTE**
>
> Henry Pitt, construction project manager for the Royal Clyde Biscuit Co., was in a reflective mood as the last major section of the new cookie factory neared completion. The new Royal Clyde Biscuit factory was specifically designed to manufacture the Sloop sandwich cookie. Sloop, one of the world's "great" cookies since its introduction in 1928, was the United Kingdom's leading brand, with sales approaching 40,000 product tons per year. The construction project had not gone smoothly. Late equipment deliveries had accounted for some of the delays in completing the new plant. However, the major reason for the significant delay in starting up the new plant rested with getting the equipment to operate as planned.
>
> Some of the problems associated with the equipment could be traced directly to the lack of an adequate management information system (MIS). Pitt knew that the current MIS system was perhaps the most neglected aspect of the construction effort. For example, the current system had inconsistencies in controlling the production process. Pitt concluded that a fully integrated computerized information system was needed for improving product quality at the plant, especially during the start-up phase. Suddenly, he thought of a proposal from Hewlett-Packard, which had been collecting dust on his desk for the past several months. This proposal, ironically rejected for its high costs, would have already paid for itself by reducing the project delays.

7.3 BASIC CHARACTERISTICS OF SAMPLING AND ESTIMATION

The following list highlights some of the primary characteristics of sampling and estimation problems:

1. Sampling is used to develop an understanding of an unknown population.
2. The appropriate sample size is determined, based on specific error and economic considerations.
3. The data collection process involves some form of random sampling.
4. Population parameters are estimated from the sample.

7.4 SAMPLING AND ESTIMATION

As mentioned previously, *sampling* is the process of collecting data from a population without measuring it in its entirety; *estimation* is the process of inferring the values of unknown population parameters from a sample. In practice it is seldom possible to analyze an entire population before making a decision, nor is it often economically feasible to take large samples from that population. As a result, decision makers base their estimates of population mean (μ) and population standard deviation (σ) on a single sample: The sample mean $(\overline{X})$ provides an estimate of μ, and the sample standard deviation (s) provides an estimate of σ.

RANDOM SAMPLING

A basic requirement in sampling is to select a representative sample of the entire population. **Simple random sampling,** although not the only approach, is one of the most common methods in statistical analysis. It is the process of selecting items from a population in such a way that each item has an equal chance of occurring. The main reason for selecting random samples is to avoid unintentional bias. For example, suppose that Mr. Barnes of Wilcox Accounting Service were simply to review the personnel files of 30 of his friends at the company rather than draw a random sample. Would this sample be representative of the entire company? The answer is, probably not. Mr. Barnes's friends at the company might behave differently from the general employee population; they might exhibit different levels of absenteeism. Therefore, any conclusions or decisions based on this nonrandom sample would be incorrect.

One way to select a simple random sample of Wilcox employees would be to write the names of each employee on individual slips of paper, place them into an urn, and thoroughly mix them. Thirty slips of paper could then be selected from the urn, one at a time. This approach represents the ideal for selecting a random sample and works well for small groups. In practice, however, this complicated process is not necessary. Instead, a sample can be provided much more efficiently by a random number table, which operates on the following two principles:

- The probability of occurrence of a given number is the same as that of any other number (i.e., the distribution is uniform).
- A given number cannot be predicted from any other number (i.e., no pattern exists).

Figure 7.2 presents a table of CBS-generated five-digit random numbers. The strictly arbitrary five-digit grouping is designed primarily for ease of use. In cases where more than five digits are needed, two blocks of five-digit numbers can be merged. For example, one might need six digits to indicate the position of an individual's name in a large telephone directory. The first three digits could represent the page number, the next two the row number, and the final one the column number. To illustrate, consider the first six digits from the first two blocks of numbers (167761). This sequence translates to page 167, row 76, column 1.

Figure 7.3 shows how the random number table in Figure 7.2 can be used to identify the first six employees to include in the sample. The figure also

FIGURE 7.2 CBS-Generated Table of Five-Digit Random Numbers

```
COMPUTERIZED    BUSINESS    STATISTICS
                    Random Numbers
         1.       2.       3.       4.
  1.    16776    17808    54380    44412
  2.    18816    30852    82751    81785
  3.    64242    46943    02374    44228
  4.    37972    67575    40014    57700
  5.    43381    38873    88548    49820

  6.    78863    21902    96028    82677
  7.    39397    87982    79746    10649
  8.    43103    38999    65293    63210
  9.    82809    52718    50627    52743
 10.    02613    53094    95126    07355




                         Send copy to printer? Y/N, press ⏎
```

includes data on each employee's sex and rate of absenteeism. For this problem, only the first four digits of the five-digit number are used. For example, the first random number, 16776, is truncated to form employee number 1677. Thus, the personnel file for employee 1677 is included in the sample. Notice that the third random number, 54380, when truncated, is greater than the maximum number of employees (4,500) and therefore does not correspond to an element in the population. In such a case, the selection

FIGURE 7.3 A Sample of Wilcox Employees Selected by the Random Number Table

Random Number	Employee ID	Last Name	Absenteeism Rate	Sex
16776	1677	Johnson	24	M
18816	1881	Marks	20	F
64242	Ignore	—	—	—
37972	3797	Devesto	45	F
43381	4338	Banks	60	M
78863	Ignore	—	—	—
39397	3939	Murphy	12	M
43103	4310	Edeson	82	F

is discarded and another number is selected. This process continues until the desired sample size has been obtained. It is also possible for a number already chosen to reappear in the random number table. In this situation, the number is discarded and another selected.

SAMPLING METHODS

Key idea 2

Although simple random sampling is one of the most common sampling procedures, several other methods are often employed to approximate simple random sampling when economic, physical, or time-related considerations demand. Three of the most popular are systematic, stratified, and cluster sampling. The computational procedures for processing data collected using either stratified or cluster sampling are somewhat complex and will not be presented here. However, both sampling methods are included in the sampling and estimation model of CBS. Figure 7.4 summarizes these alternative methods.

Systematic Sampling

In some sampling situations, particularly ones involving a large population, it is often inconvenient to apply simple random sampling directly. Instead, items of a population are selected on the basis of a random initial starting point and a specified interval. This approach is called **systematic sampling** or convenience sampling. Suppose Mr. Barnes randomly selects a starting number of 300. This would result in an interval of 140 [i.e., (4,500 − 300/30]. Thus, Mr. Barnes would select 30 employees, starting with the 300th employee and continuing with every 140th employee (i.e., 300, 440, 580, . . .). Clearly this method is easier to use than simple random sampling.

FIGURE 7.4 Schematic of Systematic, Stratified, and Cluster Sampling

Method	*Procedure*	*Example*
Systematic	Select a starting point based on a random number and systematically select subsequent elements	Telephone surveys
Stratified	Select a simple random sample of size n from each group	Demographic studies
Cluster	Select a simple random sample of size n from each of the specified clusters	Auditing

Systematic sampling is often considered a good approximation of simple random sampling because its starting point is selected at random. However, systematic sampling can introduce error into the sampling process. This is particularly true when a consistent pattern exists. For example, suppose that a milling machine, due to an adjustment malfunction, generates a defective part on every 20th item produced (e.g., items 1, 21, 41, . . .). A sampling plan based on the selection of every 20th item for inspection would uncover either all or none of the defective parts depending on whether it started with the first item or another item. In contrast, a sampling plan based on simple random sampling would ordinarily show that 5% of the parts in the sample were defective.

Stratified Sampling

Stratified sampling is based on the understanding that the population can be divided into similar groups or strata that are as much alike (homogeneous) as possible in one or more characteristics. For example, the IRS often conducts random audits of taxpayers' returns by dividing the population into strata based on income-reporting level and randomly selecting elements from each stratum. The sample data obtained from each stratum are then weighted based on the percentage of taxpayers in each group. The result is a weighted average of taxpayer behavior for the entire population.

The following table illustrates a stratified sample design for an IRS audit of a community of 100,000 taxpayers based on five income levels or strata:

Stratum	Number of Tax Returns (N_i)	Sample Size (n_i)	Sample Mean ($\overline{X}_i$)	Sample Standard Deviation (s_i)
Under $20,000	25,000	25	16,500	3,500
20,000–39,999	35,000	35	29,500	4,000
40,000–59,999	20,000	20	41,000	3,800
60,000–79,999	15,000	15	68,500	4,600
80,000 and over	5,000	5	94,500	8,700
Total	100,000	100		

Stratified sampling is used when the variation within each group is small compared to the variation between groups. When the within-group variance is small, stratified sampling will provide results nearly identical to those of simple random sampling. This method can significantly reduce the cost of sampling, therefore, because it can achieve accurate results from smaller samples.

Cluster Sampling

In **cluster sampling** the population is divided into groups that are representative of the total population (i.e., heterogeneous). For example, a market research firm might use cluster sampling to determine the average number of VCRs per household in a large metropolitan area. In this case, the area would be divided into a series of similar representative clusters, such as all apartments in a block. Each apartment in a randomly selected subset of the total number of clusters would be surveyed. The results obtained in this manner would be used to estimate the number of VCRs in the total population.

The following table illustrates a cluster sample design for the market research firm's VCR survey of 5 apartment complexes out of a total of 100.

	Apartment Complex				
	1	2	3	4	5
Number of units	150	200	175	125	150
Number of VCRs	38	45	51	32	41

Cluster sampling is used when the variation within each group is large compared to the variation between groups. Although cluster sampling often requires a larger sample than either simple random sampling or stratified sampling, it can yield more precise results with significantly less time and cost. However, when the variation within groups is small (e.g., similar preferences on public policy), cluster sampling should be avoided as a basis for inference.

SAMPLING DISTRIBUTIONS

Key idea 3

Recall that Mr. Barnes, the job-scheduling manager at Wilcox Accounting, selected a simple random sample of 30 employees to determine the level of absenteeism. From this sample, summary measures were determined and reported (i.e., $\bar{X} = 44.20$ and $s = 21.08$). Suppose Mr. Barnes now decides to select another random sample from the employee population. Would the summary statistics computed from the two samples be the same? Intuition suggests that it is highly unlikely. In fact, taking 100 samples of 30 employees each would produce a distribution of sample means. Figure 7.5 shows a portion of the summary results developed from 100 simple random samples of 30 Wilcox employees. Not surprisingly, each of these sample results shows different summary measures.

FIGURE 7.5 Sample Summary Statistics from 100 Samples of 30 Employees

Sample Number	Sample Mean (hours per year)	Sample Standard Deviation (hours per year)	Sample Proportion (women)
1	44.20	21.08	.43
2	31.45	18.76	.37
3	51.21	23.12	.50
4	67.43	25.09	.63
5	24.31	19.45	.47
.			
.			
.			
100	47.56	20.52	.30

Figure 7.6 presents a CBS-generated frequency distribution of sample means based on the 100 samples. Additionally, Figure 7.7 shows a CBS-developed relative-frequency histogram of the sample means.

Sampling Distribution for Means

Figure 7.8 compares the theoretical sampling distribution of $\overline{X}$ and the relative-frequency histogram of sample means from 100 simple random samples of 30 Wilcox employees developed in Figure 7.7. Observe how closely the normal curve matches the frequency distribution. To determine the theoretical sampling distribution of $\overline{X}$, one must compute the mean and the standard deviation. This so-called "mean of means" can be obtained

FIGURE 7.6 Frequency Distribution of the Sample Means from 100 Simple Random Samples of 30 Wilcox Employees

Absentee Rate (hours per year)	Frequency	Relative Frequency
Less than 20	4	.04
20–39	22	.25
40–59	42	.42
60–79	29	.27
80–99	2	.02
Total	100	1.00

FIGURE 7.7 Relative-Frequency Histogram of Sample Means from 100 Simple Random Samples of 30 Wilcox Employees

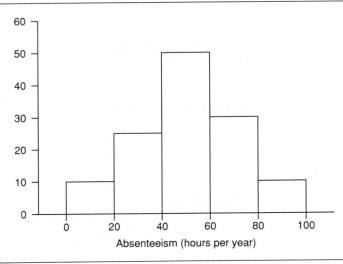

FIGURE 7.8 Comparison of the Theoretical Sampling Distribution of $\overline{X}$ and the Relative-Frequency Histogram of Sample Means from 100 Simple Random Samples of 30 Wilcox Employees

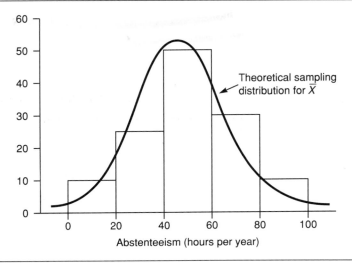

as follows:

$$\text{Mean of } \bar{X} = \frac{\Sigma \bar{X}_i}{n} = \frac{4{,}265}{100} = 42.65$$

The value 4,265 merely represents the sum of the 100 sample means. Notice that the mean of $\bar{X}$ compares very closely with the actual population mean of 42.5 days. This result should not be surprising since 3,000 (30 × 100) employees were sampled from a population of 4,500. More important, it can be shown that the expected value of $\bar{X}$ is equal to the actual population mean when simple random sampling is used. That is:

$$E(\bar{X}) = \mu$$

Similarly, the standard deviation of the sample mean $\bar{X}$ can be determined in this way:

$$\text{Standard deviation of } \bar{X} = \sqrt{\frac{\Sigma(X_i - \bar{X})^2}{n-1}} = \sqrt{\frac{1{,}131.02}{99}} = 3.38$$

Observe that this value is not even close to the population standard deviation of 19.4 hours. This is because it is an estimate of the standard deviation of the sampling distribution of the sample mean $\bar{X}$ and *not* the population standard deviation of the random variable X. In fact, the standard deviation of the sample mean $\bar{X}$, $\sigma_{\bar{x}}$, is related to the population standard deviation σ in the following manner:

$$\sigma_{\bar{x}} = \frac{\sigma}{\sqrt{n}}$$

The term $\sigma_{\bar{x}}$ is often referred to as the **standard error of the mean**. For the Wilcox problem, the theoretical standard error becomes:

$$\sigma_{\bar{x}} = \frac{19.4}{\sqrt{30}} = 3.54$$

Notice that the standard error value computed using the 100 samples compares very closely to the theoretical value (i.e., 3.38 versus 3.54). Again, this should not be surprising since the sample estimate was based on 3,000 observations. When the population standard deviation is unknown, the standard error can be computed using the sample standard deviation as an estimate for the population standard deviation as follows:

$$s_{\bar{x}} = \frac{s}{\sqrt{n}}$$

$$s_{\bar{x}} = \frac{21.08}{\sqrt{30}} = 3.85$$

The relationship between the population standard deviation and the standard error of the mean suggests that the standard deviation becomes smaller and smaller as the sample size increases. In summary, the curve presented in Figure 7.8 represents the sampling distribution of $\overline{X}$ with the following properties:

- The mean of the sampling distribution is equal to the population mean ($E(\overline{X}) = \mu$)
- The standard deviation of the sampling distribution (standard error) is equal to the population standard deviation divided by the square root of the sample size (i.e., $\sigma_{\overline{x}} = \sigma/\sqrt{n}$).
- The sampling distribution is normally distributed.

The preceding discussion has been based on the assumption of an infinite population, or sampling with replacement. In many business situations, however, the decision maker is dealing with a population of finite size for example, a class of 30 students. Such cases require an adjustment to the basic standard error model. This adjustment is called the *finite-population multiplier* and has the following form:

$$\text{Finite multiplier} = \sqrt{\frac{N-n}{N-1}}$$

where N is the size of the population and n is the size of the sample. This factor is multiplied by the standard error to obtain the corrected value. In general, the use of the finite multiplier will reduce the size of the standard error. Figure 7.9 presents a comparison of finite multipliers based on various population values using a constant sample size of 30. Not surprisingly, the multiplier approaches unity as the population size increases relative to the sample size. For example, the finite multiplier for the Wilcox problem is 0.9967 which is very cose to unity (i.e., $\sqrt{(4,500 - 30)/(4,500 - 1)}$. CBS includes the finite multiplier into the computational process in developing a value for the standard error.

Sampling Error of Proportions

It is often necessary to estimate the proportion of occurrence of a specific event in the population. For example, the U.S. Department of Labor provides monthly estimates of the unemployment rate through extensive nationwide sampling. The sample proportion $\overline{p}$ is a random variable that describes the outcome of a binomial process. In the case of Wilcox Accounting, Mr. Barnes needs to estimate the proportion (p) of women employed at the company. This is clearly a binomial process since there are only two possible outcomes (i.e., men and women). To develop such an estimate requires a sampling distribution. Fortunately, the process for developing a sampling distribution for a proportion is similar to the one used for developing a sampling distribution for the mean. A histogram of the proportion of

FIGURE 7.9 Comparison of Selected Finite Multipliers Based on Various Population Sizes with a Sample Size of 30

Population Size	Sample Size	Finite Multiplier
100	30	.841
500	30	.971
1,000	30	.985
4,500	30	.997

women employees at Wilcox is given in Figure 7.10. This distribution is based on 100 samples of size 30.

The sample taken by Mr. Barnes yields a proportion point estimate $\bar{p}$ of .43. This represents the proportion of women in the sample and is Mr. Barnes's best estimate of the proportion of women in the work force. However, different samples produce different sample proportions. Therefore, as in the case of sample means, the objective is to develop a sampling distribution to account for this variability. Recall from Chapter 5 that the binomial distribution can be approximated by a normal distribution whenever the sample size is relatively large. More specifically, the two basic criteria for

FIGURE 7.10 Relative-Frequency Histogram of Sample Means from 100 Simple Random Samples of 30 Wilcox Employees

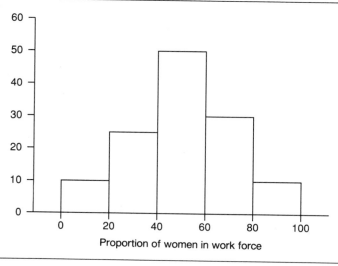

measuring the adequacy of the sample size are as follows:

$$n \times p \geq 5$$
$$n(1 - p) \geq 5$$

In these cases, the normal distribution can be used to approximate the sampling distribution of $\bar{p}$.

The standard deviation of the sampling distribution is determined as follows:

$$\sigma_{\bar{p}} = \sqrt{\frac{p(1-p)}{n}}$$

This expression is often referred to as the **standard error of the proportion.** Typically, the population proportion (p) is unknown. In these cases, the sample proportion ($\bar{p}$) is used as a substitute for the population proportion. For example, Mr. Barnes obtained a sample proportion of .43 (i.e., $p = .43$) from his survey of 30 employees. This means that 43% of the employees surveyed were women (i.e., 13 out of 30). Thus, the best estimate of the population proportion is .43. The standard error of the proportion becomes:

$$\sigma_{\bar{p}} = \sqrt{\frac{p \times q}{n}} = \sqrt{\frac{.43 \times .57}{30}} = .09$$

Figure 7.11 presents a comparison of the theoretical sampling distribution

FIGURE 7.11 Comparison of the Theoretical Sampling Distribution of $\bar{p}$ and the Relative-Frequency Histogram of Sample Proportions from 100 Simple Random Samples of 30 Wilcox Employees

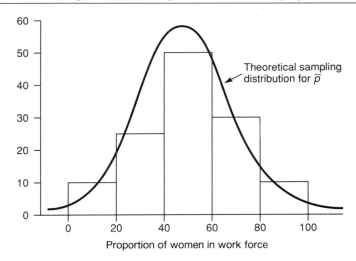

based on a sample mean of .43 and a standard deviation of .09 with the sampling histogram based on 100 samples of 30 Wilcox employees. In this case, the sampling distribution for the sample proportion closely approximates the relative-frequency histogram shown in Figure 7.11.

In summary, the curve presented in Figure 7.11 represents the sampling distribution of $\bar{p}$ with the following properties:

- The expected value of the sample proportion $\bar{p}$ is equal to the population proportion (i.e., $E(\bar{p}) = p$).
- The standard deviation of the sampling distribution (standard error of the proportion) is equal to the population standard deviation divided by the square root of the sample size (i.e., $\sigma_{\bar{p}} = \sigma/\sqrt{n}$)
- The sampling distribution is normally distributed.

For cases involving a finite population, the finite multiplier, introduced earlier, is used to adjust the standard error of the proportion in the same way it was used to adjust the standard error of the mean.

THE CENTRAL LIMIT THEOREM

Key idea 4

The **central limit theorem** (CLT) is the most important theoretical construct in statistical inference and perhaps in all of statistics. The CLT can be stated as follows:

The sampling distribution of the mean of a random variable taken from a population approaches a normal distribution as the sample size increases, regardless of the shape of the population distribution.

Clearly this is a very powerful statement. It suggests that the well-known normal curve can be used to solve a whole new class of statistical problems. With the CLT it becomes possible to make inferences about population parameters without knowing the shape or characteristics of the frequency distribution of the population.

Figure 7.12 graphically depicts the relationship between a sampling distribution and a normally distributed population. Not surprisingly, the sampling distribution takes on the shape of the normal curve. However, very surprising indeed is the graphic presented in Figure 7.13. In this case, the population distribution is not normal, yet the corresponding sampling distribution is normal. It is this property that makes the CLT so powerful and important.

As a final demonstration of the CLT, consider the uniform distribution with a range of 0 to 99 given in Panel A of Figure 7.14. Clearly this distribution is not normal. Suppose that a sample of 100 numbers were selected from this distribution. What would the expected value be? One might estimate a value near 50 since all the numbers have an equal likelihood of being selected (this is a characteristic of the uniform distribution). Suppose additional

FIGURE 7.12 Comparison of Population and Sample Distribution When the Population Distribution Is Normal

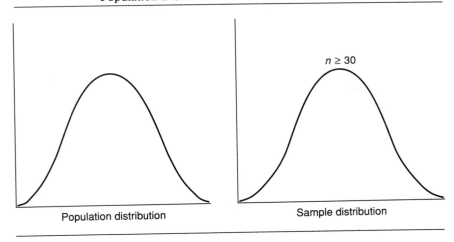

samples were selected. There would be a few cases in which the sample mean is very small (i.e., most of the numbers sampled are in the range of 0 to 20) and a few cases in which the sample mean would be quite large (i.e., most of the numbers sampled are in the range of 80 to 99). However, most of the sample means would tend toward the expected value (i.e., 50). This is what occurs as the size of the sample increases. For example, it would appear

FIGURE 7.13 Comparison of Population and Sample Distribution When the Population Distribution Is Not Normal

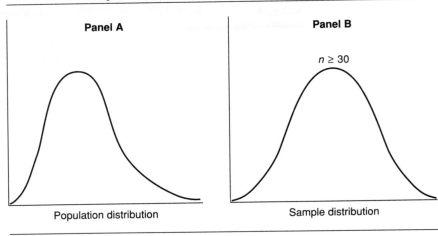

FIGURE 7.14 Comparison of Uniform Population Distribution and Sample Distribution

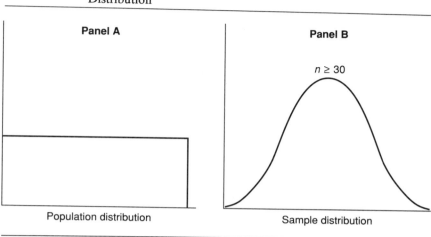

highly unlikely that a sample of 1,000 would contain predominantly small or large numbers. Instead, the numbers would tend to be uniformly distributed in accordance with the population distribution, resulting in an expected value of 50. Panel B of Figure 7.14 shows a series of CBS-generated sampling curves as a function of increasing sample size. It is this tendency of the sample means to move toward a normal curve as the sample size increases that characterizes the central limit theorem.

ESTIMATORS AND ESTIMATES

One of the basic objectives of this chapter is to introduce methods for estimating population parameters based on results obtained from a sample. As indicated in the introduction, two types of estimates can be made regarding a population parameter: point and interval. An **estimator** is a sample statistic used to estimate a specific population parameter. A specific numerical value generated from an estimator is called an **estimate**. Figure 7.15 presents several examples of populations, population parameters, estimators, and estimates.

Figure 7.16 reports population parameters and point estimates of absenteeism and the proportion of women employees at Wilcox Accounting. A point estimate (a single number used to estimate an unknown population parameter such as the mean) provides the most representative value of the population parameter of interest.

The population values obtained from an analysis of the entire employee

FIGURE 7.15 Examples of Populations, Population Parameters, Estimators, and Estimates

Population	Population Parameter	Estimator	Estimate
Mutual funds	Selling price	Mean selling price, 1988	$11.15 per share
Electrical demand	Peak residential demand	Mean peak demand for summer	200 kw
Women in top management	Proportion of women in top management	Proportion from survey of Fortune 500 firms	4% of top management positions are held by women
Machine-filling volume	Filling volume variance	Standard deviation from sample of 100 filling operations	2 oz.

population at Wilcox differ somewhat from the sample values. This result should not be surprising since the sample consisted of only 30 members of the population.

SELECTING A GOOD ESTIMATOR

Unfortunately, not all estimators of a population parameter are equally accurate. The four basic criteria for evaluating the quality of an estimator are summarized in Figure 7.17. Consider a normally distributed population in which the mean and median are equal. In this situation, both the mean and

FIGURE 7.16 Population and Sample Data for Wilcox Accounting

Population Parameter	Population Value	Point Estimator	Point Estimate
Mean absenteeism	42.5	Sample mean, absenteeism	44.20
Standard deviation of absenteeism	19.4	Sample standard deviation of absenteeism	21.08
Proportion of women employees	.4	Sample proportion of women employees	.43

FIGURE 7.17 Characteristics of a Good Estimator

Characteristic	Definition
1. **Unbiasedness**	The ability of an estimator to match the value of the population parameter on an expected-value basis
2. **Consistency**	The ability of an estimator to approach the value of the population parameter as the sample size increases
3. **Efficiency**	The ability of an estimator to show the smallest variance among the available estimators for a given population parameter
4. **Sufficiency**	The ability of an estimator to maximize the use of the available information regarding a population parameter

median are **unbiased** estimators of the population mean. Furthermore, both of these estimators are **consistent** since they both approach the value of the population mean as the sample size increases. On the other hand, the mean is a more **efficient** estimator of the population mean than is the median because in large samples its standard error is smaller. Additionally, the mean is a more **sufficient** estimator of the population mean since, unlike the median, it maximizes the use of the available information. Recall that the median is the centermost point in the data set. Half of the data fall above the median, and half fall below. Unlike the mean, however, the median is not based on the actual numerical values but on the frequency of occurrence of the data. It is therefore not surprising that the sample mean is used as the best estimator of the population mean.

INTERVAL ESTIMATES

Key idea 5

Although a point estimate provides the most representative value of a population parameter, it is unlikely ever to equal the true population value. Therefore, another type of estimate is needed to provide a more exact statement regarding the value of the population parameter. Typically, an **interval estimate** is used for this purpose. Because an interval estimate describes a range of values within which the population parameter most likely resides, it conveys the fact that the estimation process is probabilistic. Furthermore, interval estimates, unlike point estimates, provide a measure of confidence regarding the estimate. In fact, they are often called **confidence intervals**. In this regard, interval estimates are often considered more useful than point estimates.

The formulation of interval estimates, although computationally simple, is complicated by a large number of possible variations. For example, for an interval estimate or confidence interval to be developed, the population standard deviation must be specified. In cases where the population standard deviation is known, the interval can be determined based on the normal curve (i.e., Z statistic). In many business applications, however, the population standard deviation is unknown. In these cases, the sample standard deviation is used as a point estimate for the population standard deviation (i.e., $s \to \sigma$). The use of the sample standard deviation in developing a confidence interval requires several additional considerations as well, including sample size and the shape of the population distribution.

As the central limit theorem shows, the sampling distribution can be used to estimate population parameters without the need to know the shape of the population distribution. Operationally, this statement is appropriate for sample sizes of 30 or more. This means that the sampling distribution is assumed to be normal. However, when the sample size is less than 30, this assumption may not be valid. In fact, the use of the normal distribution (i.e., Z statistic) under these circumstances can yield misleading results. Instead, a different distribution, called the *t*-distribution, is used. The *t*-distribution is appropriate for sampling from a population that is normally distributed. However, when the sample size is small and the population distribution is unknown, neither the Z nor the *t*-distribution is appropriate. These cases require a totally different approach based on nonparametric analysis. An in-depth treatment of nonparametric statistics is presented in Chapter 14. Figure 7.18 identifies the recommended statistic as a function of sample size and the shape of the population distribution.

The difference between the sample mean and the population mean is called the **sampling error:**

$$\text{Sampling error} = \bar{X} - \mu$$

Unfortunately, this relationship cannot be directly used to determine the size of the sampling error because the population mean is unknown. Instead, the sampling distribution (i.e., the standard error of the mean or the standard error of the proportion) can be used to make probability statements about the size of the sampling error. The general form of an interval estimate is as follows:

$$\text{Lower and upper limits} = \text{Point estimate} \pm Z_{\alpha/2} \times \text{Standard error}$$

The expression $Z_{\alpha/2}$ refers to the number of standard deviations from the population mean for an area in the upper or lower tail of $\alpha/2$. Alpha is often referred to as the level of significance, which is the complement of the level of confidence. The value of Z used in developing an interval estimate is based on the desired level of confidence. For example, if $Z = 1$, then approximately 68% of the sample means will fall within the range of $\mu \pm \sigma_{\bar{x}}$.

The preceding statement about interval estimates can be expressed mathematically as follows:

$$\mu = \bar{X} \pm Z_{\alpha/2} \times \sigma_{\bar{x}} \quad \text{(population mean)}$$

$$p = \bar{p} \pm Z_{\alpha/2} \times \sigma_{\bar{p}} \quad \text{(population proportion)}$$

FIGURE 7.18 Relationship between Test Statistic, Sample Size, and Population Distribution When the Population Standard Deviation Is Unknown

Population Distribution	Sample Size	
	Small (less than 30)	Large (greater than or equal to 30)
Normal	t statistic	Z statistic
Unknown or nonnormal	Nonparametric statistic	Z statistic

These models are based on the assumption that the population standard deviation is known. However, in many situations σ is unknown. In these cases, the sample standard deviation is used as an estimate for the population standard deviation. Accordingly, the preceding models become:

$$\mu = \overline{X} \pm Z_{\alpha/2} \times s_{\overline{x}} \quad \text{(population mean)}$$

$$p = \overline{p} \pm Z_{\alpha/2} \times s_{\overline{p}} \quad \text{(population proportion)}$$

These interval models are appropriate when the sample size is large (i.e., $n \geq 30$) or when the population is normally distributed.

Mr. Barnes would like to develop a 99% confidence interval for both the sample mean and sample proportion obtained from the employee survey. These intervals will be helpful in characterizing the level of uncertainty in the sampling process. A 99% confidence interval corresponds to a level of significance (i.e., α) of .01. The Z-score for a .01 level of significance is 2.576; this value was obtained from the normal curve presented in Table A.1 in Appendix A for an area in the lower tail of .005 and an area in the upper tail of .005. Remember that a confidence interval is based on dividing the level of significance (i.e., α) equally between the lower and upper tails of the sampling distribution. Figure 7.19 shows a generalized sampling distribution and corresponding level-of-confidence limits. One can conclude from this graphic that there is a $(1 - \alpha)$ probability that the value of the sample mean will yield a sampling error of $Z_{\alpha/2}$ or less.

Based on our interval model, a 99% confidence interval for employee absenteeism at Wilcox can be determined as follows:

$$\mu = \overline{X} \pm Z_{\alpha/2} \times s_{\overline{x}} = 44.20 \pm 2.576 \times 3.85 = 44.20 \pm 9.92$$

Thus, Mr. Barnes can be 99% confident that the actual rate of absenteeism at Wilcox falls within the range 34.28 hours per year to 54.12 hours per year. The value 34.28 is called the *lower confidence limit*, and the value 54.12 is called the *upper confidence limit*.

It is *not* appropriate to conclude from the preceding results that the true population mean has a 99% chance of lying within the reported interval limits. Basically, probability statements about a population parameter are

FIGURE 7.19 Sampling Distribution and $(1 - \alpha)$ Confidence Limits

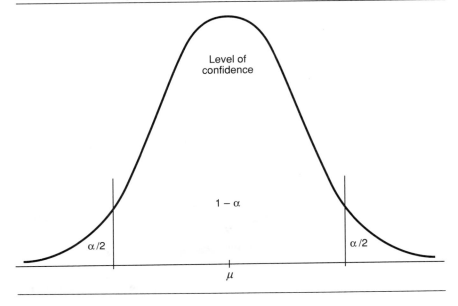

FIGURE 7.20 Confidence Intervals for the First Three Sample Means for the Wilcox Problem

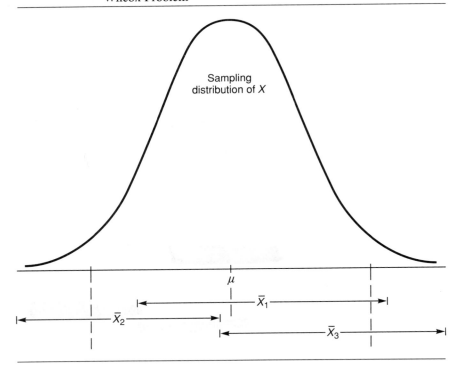

inappropriate. That is, the population parameter either lies within the interval limits or it does not. Instead, Mr. Barnes can observe that if he had calculated 99% confidence intervals for the 100 samples reported in Figure 7.6, 99 out of the 100 intervals would contain the true population mean. To illustrate this point graphically, Figure 7.20 presents 99% confidence intervals for the first three sample means reported in Figure 7.5.

Notice that in this example only two of the three intervals actually contain the population mean (i.e., the intervals for $\overline{X}_1$ and $\overline{X}_3$). This result should not be surprising. In general, for a given level of significance one would expect $100(1 - \alpha)$ out of 100 intervals developed to contain the true population mean.

This same general approach can be used to develop a confidence interval for a population proportion. Recall that Mr. Barnes wished to know the proportion of women employed at Wilcox. The summary results from the survey indicated that women compose approximately 43% of the total work force. The 99% confidence interval based on the survey results is given here:

$$p = \overline{p} \pm Z_{\alpha/2} \times s_{\overline{p}} = .43 \pm 2.576 \times .09 = .43 \pm .23$$

Mr. Barnes can conclude that the actual proportion of women in the work force lies between 20% and 66%. This result is not very helpful since the range is very large.

The Student *t*-Distribution

In cases where the sample size is small and the population is normally distributed, the *t*-distribution is used instead of the normal curve in developing confidence intervals. However, the *t*-distribution is not restricted to small-sample situations. In fact, the *t*-statistic can be used whenever the population is normal or near normal, and the sample standard deviation is used as an estimate for the population standard deviation.

The central limit theorem states that, as the sample size increases, the sampling distribution approaches normality regardless of the shape of the population distribution. Operationally, the standard normal curve can be used for estimating population parameters for sample sizes of 30 or more. Sample sizes below 30, however, require another distribution. This distribution, known as the *Student* t-*distribution,* was developed by W. S. Gossett in the early 1900s (see historical note). The use of the *t*-distribution requires the assumption that the population is normal or near normal. Furthermore, unlike the standard normal distribution, the *t*-distribution is influenced by the size of the sample — more specifically, by the **degrees of freedom,** which are equal to the sample size minus 1 (i.e., $n - 1$). The concept of degrees of freedom is important but somewhat difficult to understand. The following example should shed some light on it.

Consider the integer numbers 2, 3, and 7. The average of these three numbers is equal to 4. Suppose all three of these numbers are allowed to vary. Clearly, only two of the numbers can change if the average is to remain equal to 4. That is, one of the numbers becomes "fixed" based on the selection of the other two numbers. For example, if both the numbers 2 and 3 were increased by 1, the third number must be reduced by 2 (i.e., from 7 to 5) for the average of the three numbers to remain 4. Therefore, only $n - 1$ of the numbers are free to change. Thus, the degrees of freedom for a single random variable are equal to the sample size minus 1.

Although the t-distribution takes on a shape similar to the normal distribution, it is significantly influenced by the degrees of freedom associated with the sample. Figure 7.21 compares the normal distribution with three t-distributions of different sample sizes. In general, the tails of the t-distribution are more elongated than those of the normal distribution. That is, more of the area under the t-curve is found in the tails. This suggests that the interval widths for a t-based interval will be larger than those for a Z-based interval for the same level of confidence. Notice, however, that the shape of the t-distribution approaches the normal distribution as the sample size increases.

The models for developing interval estimates for means and proportions based on the t-distribution are similar to the ones used with the Z distribution. The interval relationships for means and proportions using the t-distribution are as follows:

$$\mu = \overline{X} \pm t_{\alpha/2} \times s_{\overline{x}} \quad \text{(Population mean)}$$

$$p = \overline{p} \pm t_{\alpha/2} \times s_{\overline{p}} \quad \text{(Population proportion)}$$

Notice that the only difference is that the term $Z_{\alpha/2}$ has been replaced by the

FIGURE 7.21 Comparison of the Normal Distribution with t-Distributions of Differing Sample Sizes

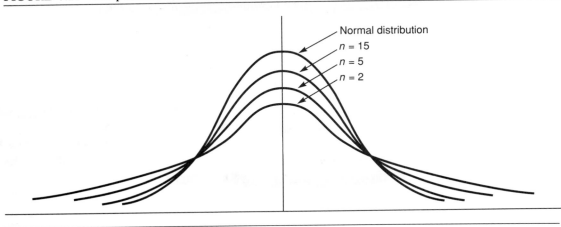

FIGURE 7.22 Selected *t*-Values Based on Degrees of Freedom and Level of Confidence

Degrees of Freedom	Area in Upper Tail			
	.05	.025	.01	.005
1	6.314	12.706	31.821	63.657
5	2.015	2.571	3.365	4.032
10	1.812	2.228	2.764	3.169
15	1.753	2.131	2.602	2.947
20	1.725	2.086	2.528	2.845
25	1.716	2.060	2.485	2.787
30	1.697	2.042	2.457	2.750
60	1.671	2.000	2.390	2.660
∞	1.645	1.960	2.326	2.576

term $t_{\alpha/2}$. Also note that the sampling error ($s_{\bar{x}}$ or $s_{\bar{p}}$) is used in lieu of the population standard error.

Figure 7.22 presents $t_{\alpha/2}$ values for selected degrees of freedom and levels of significance. The corresponding Z values are recorded in the final row (i.e., for x degrees of freedom). A cursory examination of this table shows the basis for the rule of thumb that the sampling distribution is assumed to be normal for sample sizes of 30 or more. The differences between the *t*-statistic and the Z-statistic for the probability values given in Figure 7.22 are less than 10% for a sample size greater than 30.

Application of *t*-Distribution to Wilcox Accounting Problem

Suppose that Mr. Barnes had selected a sample containing only 11 employees instead of 30. This sample has a mean absenteeism rate of 53.45 hours per year and a standard deviation of 25.27 hours per year. The sample proportion of women in the work force is .37. Because the sample size is less than 30, Mr. Barnes decided to use the *t*-distribution in developing a confidence interval. He is fairly confident that the absenteeism rate in the overall work force is normally distributed based on the findings of the previous study. The standard error of the mean for the new sample is 7.64 (i.e., $25.27/\sqrt{11}$). The *t*-value for 10 degrees of freedom ($11 - 1$) and a 99% confidence interval is 3.169. The 99% confidence interval is as follows:

$$\mu = \bar{X} \pm t_{\alpha/2} \times s_{\bar{x}} = 53.45 \pm 3.169 \times 7.64 = 53.45 \pm 24.21$$

Thus, Mr. Barnes can be 99% confident that the population mean absenteeism rate falls between 29.24 hours per year and 77.66 hours per year. Notice that this confidence interval is nearly twice the size of the interval estimate developed for the original sample of 30 employees. This is due to both the larger standard error and the larger *t*-statistic. Accordingly, Mr. Barnes should be less confident of the actual magnitude of absenteeism based on this interval compared with the previous interval.

The *t*-distribution can also be used to develop a confidence interval for a proportion. The standard error of the proportion based on a sample mean of .37 and sample size of 11 is .13. The corresponding 99% interval is:

$$p = \bar{p} \pm t_{\alpha/2} \times s_{\bar{p}} = .37 \pm 3.169 \times .13 = .37 \pm .41$$

This result suggests that Mr. Barnes can be 99% confident that the population proportion falls between $-.09$ to .78. This is an extremely wide range, which indicates that some caution should be exercised in drawing definite conclusions. Also, note that the lower confidence limit is actually below zero. Although technically correct, this is a physical impossibility. Therefore, the lower limit can be reported as zero. Finally, the use of the finite multiplier is appropriate for *t*-distributions involving finite populations.

DIFFERENCES BETWEEN POPULATIONS

In many applications the decision maker is interested in developing estimates for the difference between two population means. Examples include the difference in price between two competing VCR units, the difference in weight as a result of a new diet, and the difference in productivity under two sets of work rules. The process for developing estimates for the difference between two population means is very similar to the methods used to develop the estimate for a single population mean. Additionally, two types of difference intervals can be formulated based on the following methods of sampling: (1) unmatched pairs (i.e., independent samples), and (2) matched pairs (i.e., dependent samples).

Again, as in the case of an interval estimate for a single population, the size of the sample plays an important role. For small-sample applications (i.e., $n < 30$), a standard requirement is that the populations are assumed to be normally distributed and to have equal variances. Large samples require neither of these assumptions. In fact, two different procedures for estimating the sample error are used depending on whether the population variances are equal or not. The procedures yield near-equal interval widths except when both the sample sizes and sample standard deviations differ significantly. In these cases, the nonequal variance model should be used as a basis for developing an interval estimate.

Unmatched Pairs

The process of developing an unmatched pairs test involves selecting independent samples from the two populations. Consider that the manager of *Consumers' Buyers Guide* (CBG) is interested in reporting on the quality and

FIGURE 7.23 Results from *Consumers' Buyers Guide* Survey

	Sony	Hitachi
Mean	$325	$359
Standard deviation	24	32
Sample size	40	50

price of the two leading VCRs (Sony® and Hitachi®) being marketed in the Southern California area. The manager has recently conducted a random survey of retail outlets to determine the prices of the two VCR units. The results of the pricing survey are reported in Figure 7.23.

These results will be used to determine 95% confidence intervals for the difference between the sales prices of the two units. More specifically, two interval estimates will be developed: one based on equal variances and the other based on unequal variances. The main reason for developing estimates based on both approaches is to illustrate the computational procedures and the relative closeness of the results intervals. The model for computing the standard error for unequal variances is:

$$S_{x_1-x_2} = \sqrt{\frac{s_1^2}{n_1} + \frac{s_2^2}{n_2}} = \sqrt{\frac{24^2}{40} + \frac{32^2}{50}} = 5.91$$

The corresponding model for estimating the standard error based on equal variances is:

$$S_{x_1-x_2} = \sqrt{s^2\left(\frac{1}{n_1} + \frac{1}{n_2}\right)}$$

where s^2 is referred to as the *pooled estimator* of the population variance σ and is computed as follows:

$$s^2 = \frac{(n_1-1)s_1^2 + (n_2-1)s_2^2}{n_1 + n_2 - 2} = \frac{39 \times 24^2 + 49 \times 32^2}{40 + 50 - 2} = 825.45$$

Substituting the pooled estimate for the variance into the preceding model yields the desired standard deviation:

$$S_{\bar{x}_1-\bar{x}_2} = \sqrt{825.45 \times \left(\frac{1}{40} + \frac{1}{50}\right)} = 6.09$$

The interval estimator for the difference between two population means is:

$$\mu_a - \mu_b = \bar{X}_1 - \bar{X}_2 \pm Z_{\alpha/2} \times S_{\bar{x}_1-\bar{x}_2}$$

This estimator is appropriate when the sample sizes are large or when the two populations are normally distributed. The resultant 95% intervals (equal and unequal variances) for the *Consumers' Buyers Guide* survey are:

$$\mu_a - \mu_b = 325 - 359 + 1.96 \times 5.91 = -34 \pm 11.58 \quad \text{(Equal)}$$
$$\mu_a - \mu_b = 325 - 359 + 1.96 \times 6.09 = -34 \pm 11.94 \quad \text{(Unequal)}$$

These results suggest that the Sony® VCR is clearly the cheaper machine. The 95% confidence interval ranges from -45.58 to -22.42 based on the equal variance assumption and from -45.94 to -22.06 based on the assumption of unequal variance. Notice that even the upper limit is less than zero. These two intervals are very close in width, which suggests that the assumption regarding the equality of the variance is not important in this situation. Usually the interval based on equal variance will be smaller than the one based on unequal variance. This condition holds true for samples of approximately the same size. However, significant differences in interval width can occur whenever the sample sizes are substantially different. Consequently, the assumption of unequal variances is appropriate whenever the population variances are unknown and the sample sizes are significantly different.

Matched Pairs

A matched-pairs sample is based on exposing a sample member to two different situations (e.g., the same workers to two different training methods). Suppose that a production manager wishes to evaluate two different training procedures. One approach would be to randomly select two different sets of employees, one for each training program (i.e., independent samples). The other approach would be to expose the same set of employees to both training procedures (i.e., matched pairs). In this way, the variation in job training for the two programs due to employee variability could be significantly reduced. In general, matched-pairs samples can lead to a smaller sampling error than an independent sample design for a given sample size. This is mainly because the variation due to a particular factor (e.g.,

FIGURE 7.24 Results of Low-Fat Diet at Weber Health Spa

Patient Name	Weight (lbs.)		
	Before	After	Difference
1. Brooks	234	215	19
2. Jackson	190	185	5
3. Kelly	215	202	13
4. Truman	205	207	-2
5. O'Brian	170	161	9
6. Evans	242	220	12
7. Jones	185	179	6
8. Rogers	205	191	14

employee) can be significantly reduced or eliminated. This reduction in variability should, in turn, provide a clearer view of the actual performance of the two programs. However, it is not always possible or desirable to utilize matched-pairs samples.

The basic model for developing an interval estimate for a matched-pairs sample based on a large sample or taken from a normal population is as follows:

$$\mu_a - \mu_b = \overline{D} \pm Z \frac{s_d}{\sqrt{n}}$$

where $\overline{D}$ is the average difference and s_d is the sample standard deviation. Both of these statistics are computed as follows:

$$\overline{D} = \frac{\Sigma d_i}{n}$$

$$s_d = \sqrt{\frac{\Sigma(d_i - \overline{D})^2}{n-1}}$$

The corresponding model for small-sample problems is:

$$\mu_a - \mu_b = \overline{D} \pm t \times \frac{s_d}{\sqrt{n}}$$

Notice that the only difference between the two models is that t has replaced Z. The second model is appropriate only when the two populations can be assumed to be normally distributed and to have equal variances.

Suppose that Weber Health Spa's general manager, Ms. Schmitt, is interested in estimating the potential impact on weight loss of a new low-fat diet. She has decided to evaluate the new diet on current members of the spa and has randomly selected eight male members. Each member was weighed at the start of the diet and again after eight weeks. The results from this matched-pairs experiment are reported in Figure 7.24.

The summary statistics for these sample data are $\overline{D} = 9.5$ pounds and $S_d = 6.48$ pounds. These statistics were computed using the relationships presented above. Ms. Schmitt would like to develop a 95% confidence interval for the difference between the two means. She has assumed that the sampling distribution is normal and that the two populations have equal variances. The interval estimate for the difference in population means is

$$\mu_a - \mu_b = 9.5 \pm 2.365 \times \frac{6.48}{\sqrt{8}} = 9.5 \pm 5.4$$

This 95% interval ranges from 4.1 pounds to 14.9 pounds. More specifically, Ms. Schmitt can be 95% confident that the average weight loss due to the new diet lies within the approximate range of 4 to 15 pounds. Thus, she can be quite confident that the new diet actually results in weight loss for the population represented by the sample. The rather large interval width is the direct result of the relatively small sample size. Increasing the size of the sample would reduce the size of the interval. To reiterate a point made earlier, it is *not* appropriate to say that the actual population weight loss due to the diet has a 95% chance of falling within the specified interval limits.

DETERMINING THE SIZE OF THE SAMPLE

Key idea 6

A key aspect in designing a survey is determining the optimal sample size. Recall that Mr. Barnes had decided to select a sample of 30 employees. What factors guided Mr. Barnes in choosing a sample size of 30? A survey based on too small a sample, while reducing costs, can yield misleading results. On the other hand, a survey that uses an oversized sample can be very expensive and yet not yield more meaningful results than a survey based on a moderate sample size.

Recall that the Z transformation for a sample was based on the following model:

$$Z_{\alpha/2} = \frac{\bar{X} - \mu}{\sigma/\sqrt{n}}$$

This model provides a measure of the number of standard deviations between the population mean and the sample mean. The difference between the sample mean and the population mean is called the *sampling error*. Letting E equal the difference between the sample mean and the population and solving for n yields the following expression:

$$n = \frac{Z_{\alpha/2}^2 \times \sigma^2}{E^2}$$

Thus, the sample size is directly proportional to the square of the critical Z-value and the population variance and inversely proportional to the square of the sampling error. Thus, the sample size increases as the population variance increases or the sampling error decreases. The preceding model yields the minimum sample size required for a given level of confidence and a given sampling error. Notice that an infinite sample size is required (i.e., the entire population) if no sampling error is allowed.

Suppose that Mr. Barnes wishes to be 99% confident that the sample mean will provide a sample error of 10 hours per year or less. A 99% level of confidence corresponds to a $Z_{\alpha/2}$ values of 2.576. The preceding model can be used to determine the minimum sample size in this way:

$$n = \frac{2.576^2 \times 21.08^2}{10^2} = 29.5$$

This fractional result should be rounded up to the next integer value. Thus, the minimum sample size required to obtain the level of precision specified by Mr. Barnes is 30.

As another example, suppose that the U.S. Department of Labor's assistant secretary for personnel planning is interested in conducting a nationwide survey to determine the starting salaries of computer analysts entering the work force. The secretary wishes to be 95% confident that the sampling error will not exceed $100. A review of the historical records on computer analysts has revealed that the population standard deviation for starting salaries is $500. The secretary would like to determine the size of the sample required to meet the desired error and confidence specifications. The preceding model can be used to determine the minimum sample size required for the Department of Labor's salary survey. In this case, the critical Z value is 1.96, which corresponds to an error level of 2.5% on each side, or 5% overall. The resultant calculations are:

$$n = \frac{(1.96)^2 \times 500^2}{100^2} = 96$$

This result indicates that a sample size of 96 should be used to conduct the survey. Sometimes this model yields fractional results (e.g., 96.5). In these cases, the computed value should be rounded up to the next whole number. As indicated previously, the size of the sample is directly proportional to the square of $Z_{\alpha/2}$. The value of $Z_{\alpha/2}$ follows directly from a statement regarding the level of confidence for a given interval estimate. For example, in the Department of Labor problem the secretary wished to be 95% confident that the sampling error did not exceed $100. Clearly, a greater level of confidence would require a larger sample size. Figure 7.25 shows the relationship be-

FIGURE 7.25 Sample Size versus Level of Confidence for Department of Labor Problem

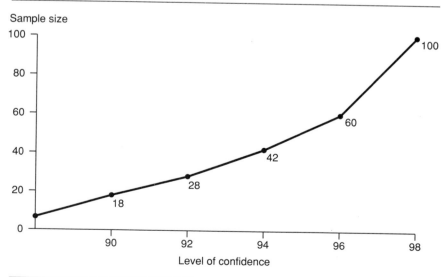

tween various levels of confidence and the required sample sizes for the Department of Labor problem. Notice that this relationship is highly nonlinear (i.e., nonproportional). In many instances, the value of the standard deviation may not be known. In these cases, a preliminary estimate can be developed from past data on similar studies or a pilot test.

The same general approach can be used for determining the sample size for a sample size for a population (p). The basic model is:

$$n = \frac{Z^2 \alpha/2 \times p(1-p)}{E^2}$$

where p is the population proportion and E is the acceptable error level. Typically, p is unknown, but it can be estimated from a "pilot" sample or by letting $p = 0.5$ (worst case). The latter approach will yield the largest sample size for a given error and alpha level.

7.5 COMPUTER ANALYSIS

Although the actual computations for developing point and interval estimates as well as sampling plans are relatively simple, the use of computer analysis provides a standard format for processing the relatively large number of estimating and sampling options. Furthermore, computer analysis reduces the potential for calculation errors. The following section presents CBS analyses for several of the problems introduced in the previous section.

FIGURE 7.26 CBS-Generated Interval for Estimating Absenteeism of Wilcox Accounting Service

```
CBS-Interval Estimation                              08-29-1990 - 08:03:33
                              Results

Standard Error of Mean:                              3.8487
Mean:                                                44.2000
Alpha Error:                                         0.0050
Degrees of Freedom:                                  29
Critical t (Test Statistic - alpha/2):               2.5800
Computed t:                                          11.4845
p value:                                             0.0002

              *** 99.5% interval:  44.2 +/- 9.9296 ***




                              press ↵
```

Wilcox Accounting Service (Interval Estimate for a Single Population Mean)

Figure 7.26 presents the required data input for computing an interval estimate for employee absenteeism at Wilcox Accounting (i.e., single population mean). The resultant 99% interval estimate of 44.20 ± 9.92 is exactly the same as the one computed in the previous section. The relative ease of this process underscores the efficiency of computer analysis in developing confidence intervals.

Wilcox Accounting Service (Interval Estimates for a Single Population Proportion)

Figure 7.27 presents a CBS-generated 99% interval estimate for the proportion of women in the work force at Wilcox Accounting. The large interval width (i.e., .20 to .66) suggests that some caution should be exercised in using these results. One potential remedy would be to increase the size of the sample.

Consumers' Buyers Guide (Interval Estimate for the Difference between Two Means from Independent Samples)

Shown in Figure 7.28 is a CBS analysis of the *Consumers' Buyers Guide* problem. The reported 95% interval of −34 ± 11.58 confirms the results

FIGURE 7.27 CBS-Generated Interval Estimate for Proportion of Women at Wilcox Accounting Service

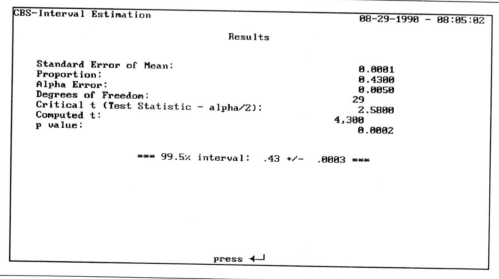

FIGURE 7.28 CBS-Generated Interval Estimate for the *Consumers' Buyers Guide* Problem

```
CBS-Interval Estimation                              08-29-1990 - 08:06:53
                              Results

         Standard Error of Mean (unequal variances):        5.9059
         Standard Error of Mean (equal variances):          6.0947
         Mean 1 - Mean 2:                                  -34
         Alpha Error:                                       0.0500
         Degrees of Freedom:                               88
         Critical Z (Test Statistic - alpha/2):             1.9600
         Computed Z (unequal variances):                   -5.7569
         p value:                                           1

              *** 95% interval: -34 +/-  11.5756 (unequal variances) ***

              *** 95% interval: -34 +/-  11.9456 (equal variances) ***



                              press ←┘
```

presented in the previous section. Again, one can conclude that the Sony® does sell for a lower price than the Hitachi®.

Weber Health Spa (Interval Estimates for the Difference between Two Means from Matched-Pairs Samples)

Figure 7.29 presents a CBS analysis of the Weber Health Spa problem. This problem involved a matched-pairs evaluation of the effects of a new low-fat diet on male spa members. The computer-generated results clearly reveal, at a 95% level of confidence, that the diet does work. The 95% confidence interval ranges between 4.1 pounds and 14.9 pounds, which is significantly above zero.

Department of Labor (Sample Size)

Recall that the assistant secretary at the Department of Labor wishes to know the size of the sample needed to estimate the mean monthly starting salary of computer analysts entering the work force, at a 95% level of confidence. The population standard deviation was estimated at 500, and the error was not to exceed 100. A CBS analysis of this problem is shown in Figure 7.30. These results, which replicate the ones presented in the previous section, show that the minimum sample size needed is 96.

FIGURE 7.29 CBS-Generated Interval Estimate for the Weber Health Spa Problem

```
CBS-Interval Estimation                       08-29-1990 - 08:09:08
                          Results

    Standard Error of Mean:                    3.8487
    Mean:                                     44.2000
    Alpha Error:                               0.0100
    Degrees of Freedom:                       29
    Critical t (Test Statistic - alpha/2):     2.7560
    Computed t:                               11.4845
    p value:                                   0.0002


            *** 99% interval:  44.2 +/- 10.6069 ***




                          press ↵
```

FIGURE 7.30 CBS Analysis of Department of Labor Problem

```
         C O M P U T E R I Z E D    B U S I N E S S    S T A T I S T I C S
                          Sample Size Analysis

                      Sample Size from Sampling Error

    Enter Standard Deviation, press ↵                      500

    Alpha error:  1=.1, 2=.05, 3=.025, 4=.01, 5=.005, 6=other
    Select alpha: enter 1-6, press ↵     3

    Critical Z for Alpha = .025 :          1.96


    Enter Sampling Error, press ↵                          100




              * * *    SAMPLE SIZE = 96     * * *
                          Send copy to printer? Y/N, press ↵
```

7.6 PRACTICAL APPLICATIONS

This section presents several business applications involving the use of sampling and estimating.

Von's Supermarket, Inc. (Estimation of Means)

The marketing director at Von's Supermarket, Inc., Ms. Sally O'Hare, is becoming increasingly concerned about the aggressive advertising campaigns being used by Von's major competitors. The supermarket business is very competitive, and this has resulted in a general industry strategy of lowering prices to increase customer volume. Several supermarkets have claimed to have the "lowest" prices in town. However, these claims are usually based on studies conducted by the supermarket sponsoring the ad. Ms. O'Hare wishes to overcome any potential biases by hiring an independent outside firm to conduct an industrywide survey. She believes that an independently conducted survey will improve the credibility of any future claims made by Von's in its advertising. Accordingly, she has retained a well-respected marketing firm to conduct an independent study of food prices for the three leading supermarkets. The firm, using a standardized one-week shopping plan, has made identical purchases at several of each chain's local stores over a period of 60 days. The costs of each shopping exercise are presented in Figure 7.31.

FIGURE 7.31 Comparison of Weekly Shopping Expenses for Leading Supermarket Chains

Observation	Vons	Lucky	Ralph's
1	$89.25	$77.76	$95.37
2	91.32	81.45	98.56
3	92.34	83.12	99.21
4	90.14	86.23	97.12
5	92.19	85.76	95.10
6	93.71	86.87	94.36
7	91.72	88.91	92.46
8	92.42	89.40	91.11
9	93.63	90.38	90.33
10	92.44	91.62	91.76
11	91.55	91.89	90.45
12	92.59	92.02	89.88
13	89.51	90.11	90.24
14	87.21	89.88	91.33
15	85.12	86.99	92.38

FIGURE 7.32 Point and Interval Estimates for Weekly Food Purchases

Chain	Average Weekly Food Purchase	
	Point	Interval
Vons	$91.01	$89.68–92.34
Lucky	87.49	85.20–89.78
Ralphs	93.31	91.56–95.06

The consulting firm developed point and 95% interval estimates for each of the three samples. These results are presented in Figure 7.32. These data show that Lucky's has the lowest overall prices for the survey (i.e., $87.49 for the standard weekly purchase). Furthermore, the developed interval estimate barely overlaps the interval for Von's (the second-place finisher). Clearly, Ms. O'Hare would be well-advised not to advertise that Von's has the lowest prices in light of these survey results.

Election Year 1988 (Estimation of Proportions)

Prior to the 1988 presidential election, a poll of 500 eligible voters indicated 235 for Bush, 205 for Dukakis, and 60 undecided. The news media wished to report these results as well as the margin of error for a 95% level of confidence. The proportion for each candidate, based on the sample, is shown in Figure 7.33.

The first step in determining the margin of error is to compute the standard error of the proportion based on $p = .5$ (the most conservative case). That is:

$$s_p = \sqrt{.5 \times .5/500} = .022$$

The next and final step is to multiply the standard error by the corresponding Z statistic, which yields the required error estimate. For a 95% confidence interval, the margin of error is $\pm(.022) \times (1.96)$, or $\pm 4\%$ (rounded off). Thus, the news media can report the survey's point estimates and indicate that these estimates are within ± 4 percentage points.

FIGURE 7.33 Survey Results of National Presidential Poll

Candidate	Percentage
Bush	47
Dukakis	41
Undecided	12

7.7 CASE STUDY: BOZART INVESTMENTS CORPORATION

The director of planning services at Bozart Investments, Mr. Bob Thomas, is interested in evaluating the performance of the mutual fund market. Currently, there are approximately 1,000 different active funds available in the U.S. market. The portfolio composition and operating performance of these funds can vary considerably depending on the investment philosophy of the management. Mr. Thomas has randomly selected a sample of 25 mutual funds. The sample results are presented in Figure 7.34. This table shows the fund, the net asset value per share, and the offer price. The symbol NL signifies "no load," which means there is no sales commission to buy into the fund.

Mr. Thomas has decided to determine point and interval estimates of the selling price for both the load and no-load funds. This information will provide an estimate of the average performance of the mutual fund market

FIGURE 7.34 Random Sample of 25 Mutual Funds' Net Assets and Asking Price

Fund	Net Asset Value per Share	Offer Price
1. AAL CapGrp	8.49	8.91
2. AARP Genbd	14.91	NL
3. Babson Gwth	11.10	NL
4. Benham Nitfl	10.59	NL
5. Conn Mutual Growth	10.35	11.04
6. Delware TF USI	11.02	11.57
7. Dean Witter Tax	10.69	11.14
8. Eaton Gvtob	11.32	11.88
9. FPA Funds	12.73	13.61
10. Gen Ele Effn	10.73	NL
11. HTins	10.20	10.68
12. Hummer	13.39	NL
13. IAI Apollo	11.14	NL
14. Kemper BluChp	8.20	8.59
15. Lexington Cldr	11.23	12.02
16. MFS Intbnd	11.64	12.55
17. Merrill Intrm	10.88	11.10
18. Penn Sqr	8.87	NL
19. Price Equin	13.25	NL
20. Putman Caprs	11.62	12.20
21. Scudder Cal	10.02	NL
22. Security Invest	8.23	8.99
23. Shearson Calmu	15.06	15.85
24. 20th Century Growth	11.75	NL
25. Vanguard Equinc	10.27	NL

FIGURE 7.35 CBS-Generated Point and Interval Estimates for Mutual Fund Case

Estimate	Load	No-Load	Total
Point	10.90	11.46	11.15
Interval			
90%	10.26–11.54	10.74–12.18	10.69–11.61
95%	10.06–11.74	10.50–12.42	10.55–11.75
DOF	14	10	24

as well as an indication of the variability in market performance. A CBS analysis of this problem is given in Figure 7.35. These data present point estimates for the load, no-load, and combined cases. These results reveal that the no-load funds are outperforming the load funds on the selling price. The point estimates provide the "best" single indicator of the population mean. Nevertheless, Mr. Thomas knows that the actual population mean may differ somewhat from these point estimates. The computed interval estimates are designed to specify the range most likely to contain the true population mean. Because of the relatively small sample sizes, Mr. Thomas will need to use the t-distribution to develop the required interval estimates. The developed interval estimates for 90% and 95% confidence levels are also shown in Figure 7.35. For example, Mr. Thomas can say that he is 90% confident that the interval of 10.74 to 12.18 contains the population mean for the no-load fund. Notice that increasing the confidence level from 90% to 95% in the case of the no-load fund enlarges the interval by a mere $0.24 per share. Armed with these point and interval estimates, Mr. Thomas can now provide his clients with a more detailed assessment of the performance of the mutual fund market.

7.8 SUMMARY

Drawing conclusions about an unknown population parameter from a sample is known as statistical inference. Sampling and estimation lie at the heart of the statistical inference process. This chapter has introduced a variety of sampling procedures, from simple random sampling to more complex methods (e.g., cluster and stratified sampling). A major aspect of the sampling process focuses on the central limit theorem. This rule states that the distribution of the sample means approaches normality regardless of the shape of the population distribution as the sample size increases. The central limit theorem is perhaps the single most important concept in statistical inference. A study of sampling and sampling distributions also showed that quite accurate estimates of an unknown population parameter can be obtained from a relatively small sample.

In addition, this chapter introduced two different types of statistical estimates: point and interval. Each of these estimates is designed to provide somewhat different information on a given population parameter. A point estimate provides a single value, while an interval estimate provides a range of values. The interval is based on the decision maker's desired level of confidence. Typically, most interval estimates are based on a 90% to 99% confidence level. Also identified were the characteristics that yield a "good" estimator (i.e., unbiasedness, efficiency, consistency, and sufficiency).

7.9 GLOSSARY

central limit theorem The proposition that the sampling distribution approaches a normal curve as the sample size increases, regardless of the shape of the population distribution being sampled.

cluster sampling A method of sampling in which the population is divided into heterogeneous groups and a random sample is drawn from each group.

confidence level The level of statistical confidence associated with statements regarding the value of a population parameter.

consistency The ability of an estimator to approach the value of the population parameter as the sample size increases.

degrees of freedom The number of values in the sample that are free to vary.

efficiency The ability of an estimator to show the smallest variance among the available estimators for a given population parameter.

estimate A numerical value generated from an estimator.

estimator A sample statistic used to estimate a population parameter.

interval estimator A confidence interval used to estimate an unknown population parameter.

parameter A population characteristic (e.g., mean or standard deviation).

point estimator A single numerical value used for estimating an unknown population parameter.

sampling distribution A probability distribution of the sampling means for a given population.

sampling error The variation among sample statistics due to chance.

simple random sampling A method of sampling in which every member of the population has an equal chance of being selected.

standard error of the mean The standard deviation of the sampling distribution of means.

standard error of the proportion The standard deviation of the sampling distribution of proportions.

statistical inference The process of making statements regarding a population parameter from data obtained from a sample(s).

stratified sampling A method of sampling in which the population is divided into homogeneous groups and a random sample is drawn from each group.

sufficiency The ability of an estimator to maximize the use of the available information regarding a population parameter.

systematic sampling A method of sampling in which population elements are selected on the basis of a uniform decision criterion.

unbiasedness The ability of an estimator to match the value of the population parameter on an expected-value basis.

7.10 BIBLIOGRAPHY

Aczel, A. D. *Complete Business Statistics.* Homewood, Ill.: Irwin, 1989.

Adams, W. J. *The Life and Times of the Central Limit Theorem.* New York: Kaedom, 1974.

Mason, R. D. *Statistical Techniques in Business and Economics.* 7th ed. Homewood, Ill.: Irwin, 1990.

Mendenghall, W.; L. Ott; and R. L. Schaffer. *Elementary Survey Sampling.* 2nd ed. Belmont, Calif.: Wadsworth, 1979.

Williams, B. *A Sampler on Sampling.* New York: Wiley, 1978.

7.11 PROBLEMS

1. Describe the following statistical concepts:
   a. Central limit theorem   b. Sampling distribution
   c. Point estimate   d. Interval estimate
2. What role does the central limit theorem play in statistics?
3. Illustrate the central limit theorem by collecting and plotting the means of 20 two-digit samples taken from the list of random numbers given in Figure 7.2 for sample sizes of 15, 30, and 45.
4. Why is sampling so fundamental to modern business practice?
5. Discuss the basic trade-offs in designing a sampling program.
6. Identify the advantages and disadvantages of stratified, cluster, systematic, and judgment sampling.
7. Identify whether each of the following business situations involves stratified, cluster, or systematic sampling.
   a. The director of personnel wishes to determine company morale by interviewing every 25th employee.
   b. The vice president of marketing plans to survey her friends on the market potential of a new product.
   c. A tax return specialist plans to audit a number of corporate accounts, which are divided into four groups according to income level.
   d. A market research firm wishes to determine the number of VCRs per household in Los Angeles by surveying three representative city blocks.
8. The following data present the financial credit rating (1–5) for the 10 firms making up the hard-disk-drive industry (i.e., the population).

Firm	Rating	Firm	Rating
A	3	F	4
B	2	G	3
C	2	H	1
D	3	I	2
E	5	J	4

    a. How many different samples of size 5 can be drawn?
    b. Randomly select 10 samples of size 5 and compute the mean of each sample.
    c. Compute the mean and variance of x.
    d. Compute the mean and variance of the population.
    e. Compare the results obtained in parts c and d.

9. Select a random sample of 20 firms listed in the Fortune 500. Compute a sample mean and variance for annual sales and net earnings. Compare these results with the average sales and earnings for the total population.

10. Brooks Company operates eight obsolete stamping processes. A recent survey on the operating status of the eight machines (i.e., population) revealed the following information:

Machine	Status	Machine	Status
1	OK	5	OK
2	Fix	6	Fix
3	OK	7	Fix
4	Fix	8	Fix

    a. What proportion of machines in the population need repair?
    b. How many samples of size 3 can be drawn (without replacement) from the population?
    c. Select six samples of size 3 and compute the proportion of each sample.
    d. Determine the average sample proportion.
    e. Compare these results with the population proportion from part a.

11. The Los Angeles Police Department maintains records on the interval between the time a crime is reported and the time a police officer arrives on the scene. A random sample of 20 reports revealed the following response times, in minutes:

Report	Time	Report	Time
1	7.5	11	8.9
2	11.4	12	11.3
3	15.2	13	14.1
4	10.1	14	17.5
5	19.3	15	24.1
6	14.3	16	19.8
7	8.8	17	20.3
8	7.2	18	12.1
9	12.6	19	15.1
10	22.1	20	10.6

a. Develop a point estimate for the mean response time.
b. Develop a point estimate for the standard deviation.
12. The vice president of operations at Ice Age Corporation is concerned about the filling variability associated with the company's standard 200 oz. ice cream carton. A random sample of 15 cartons produced the following results:

Sample	Quantity	Sample	Quantity
1	198	9	200
2	204	10	203
3	189	11	201
4	182	12	193
5	205	13	194
6	195	14	196
7	188	15	190
8	201		

a. Develop a point estimate for the mean volume.
b. Develop a point estimate for the standard deviation.
13. A random survey of 20 voting adults taken at a local shopping center revealed the following preference for the upcoming presidential election:

Voter	Party	Voter	Party
1	Rep	11	Dem
2	Rep	12	Rep
3	Dem	13	Dem
4	Dem	14	Rep
5	Dem	15	Rep
6	Dem	16	Dem
7	Rep	17	Rep
8	Rep	18	Dem
9	Dem	19	Rep
10	Rep	20	Rep

a. Develop a point estimate for the proportion of voters planning to support the Republican party.
b. Does the shopping center provide a representative sample of the voting electorate?
14. What role do confidence intervals play in statistical estimating?
15. The average monthly accounts receivable at Jade International total $100,000. The accounting manager estimates that the corresponding standard deviation is $10,000.
a. What is the probability that the monthly receivables will exceed $100,000?
b. What is the probability that the monthly receivables will range from $90,000 to $105,000?
c. Select a random sample of size 30 and compute a sample mean and variance.
d. Develop a 90% confidence interval for the developed sample.

16. What are the basic differences in developing interval estimates between large and small samples?
17. What basic assumptions are required to develop an interval estimate from a small sample?
18. What statement can be made about the relationship between a population parameter and an interval estimate?
19. Develop interval estimates when $\bar{X} = 500$, $s = 50$, and $n = 100$ for the following confidence levels:
    a. 90%   b. 95%
    c. 99%   d. 100%
20. Develop interval estimates when $\bar{X} = 500$, $s = 50$, and $n = 20$ for the following confidence levels:
    a. 90%   b. 95%
    c. 99%   d. 100%
21. Develop interval estimates for the following levels of confidence using the data given in problem 10:
    a. 90%   b. 95%
    c. 99%   d. 100%
22. Develop interval estimates for the following levels of confidence using the data given in problem 11:
    a. 90%   b. 95%
    c. 99%   d. 100%
23. Develop interval estimates for the following levels of confidence using the data given in problem 12:
    a. 90%   b. 95%
    c. 99%   d. 100%
24. Develop interval estimates for the following levels of confidence using the data given in problem 13:
    a. 90%   b. 95%
    c. 99%   d. 100%
25. A random sample of 20 accounts payable at Krick Corporation showed a sample mean of $50,000 and a sample standard deviation of $10,000.
    a. Develop a 90% confidence interval of the mean monthly accounts payable.
    b. How would the interval change if the sample size had been 50?
26. The Gallup Poll is planning to survey the public's preference in the upcoming presidential election. Typically, Gallup bases its estimates on a 95% confidence interval. How large a sample size would be required if Gallup wishes that the sampling error remain less than or equal to:
    a. 5%   b. 3%   c. 1%
27. The director of marketing at Clean Shave, Inc., is planning to test-market a new low-price razor via direct mail to a number of potential customers selected at random from the local telephone directory. The director wishes to determine the appropriate sample size based on a 95% confidence interval and a target sampling error of 2% or less for the following conditions:
    a. No population proportion estimates on customer demand are available.
    b. A preliminary estimate indicates that 15% of the targeted population will show interest in the new razor.
28. A recent survey of the voting electorate showed that 45% favored gun control, 42% opposed it, and 3% were undecided. The poll had a margin of error of 4% at a

95% level of confidence. What is the minimum sample size required to produce these results?

29. A random sample of 100 Los Angeles fire captains revealed that a firefighter spent 12.5 years on the job before being promoted to captain. The standard deviation for the sample was 2.2 years. Develop a 95% confidence interval for the population mean.

30. The engineering vice president at Bridgeman Manufacturing is interested in initiating a new employee training program. The program is designed to reduce the time required to test the company's CAD/CAM system. A pilot test of the new training procedure yielded a sample mean of 20 hours with a standard deviation of 4 hours. Develop a 95% confidence interval for the population mean.

31. A survey of 2,000 Californians revealed the 1,350 supported stricter gun control laws. Develop a 99% confidence interval for the population mean.

32. An auditing firm plans to survey 100 accounts receivable of a leading Fortune 500 company. Some of the receivables are very large (i.e., above $1 million) while others are quite small (i.e., below $10,000). What type of sampling plan should the auditing firm use?

33. Ampex Corporation, a market research firm, conducts telephone surveys for its clients. On the average, Ampex obtains a 30% response rate. The company is planning a new survey involving 500 calls. What is the probability that:
    a. At least 200 individuals contacted will participate in the survey?
    b. Fewer than 250 individuals contacted will participate in the survey?
    c. Between 300 to 400 individuals contacted will participate in the survey?

34. The accounting manager at Tenex Corporation earlier in the year surveyed the company's accounts receivable. The results of a random sample of 100 accounts revealed an average account age of 35 days with a standard deviation of 10 days. The manager recently initiated a pilot program whereby the company's vendors would receive a 5% discount if the account is paid within 40 days. The results from this study, which consisted of 50 accounts, yielded an average account age of 27 days with a standard deviation of 9 days. Construct a 95% confidence interval for the difference between the mean account age for the two samples.

35. The Department of Housing and Urban Development (HUD) is interested in determining the differences between housing rental costs throughout the country. A recent survey of the monthly rents for two of HUD's regions (East Coast and Midwest) revealed the following statistics:

Region	Location	Sample Size	Mean	Standard Deviation
I	East Coast	300	$1,150	$250
V	Midwest	250	750	175

Construct a 95% confidence interval for the difference between the two means.

36. The marketing manager at Seven-Day Tire Store is interested in developing a confidence interval for the difference in tire wear between the Mark 3000 (Seven-Day's leading radial) and the Eagle 400 (sold by J. C. Penney). The marketing manager has collected the following sample data on tire mileage:

	Mark 3000	Eagle 400
Mean	28,700	25,900
Standard deviation	2,500	2,800
Sample size	100	50

    a. Develop a 95% confidence interval for the difference between the two means.
    b. Is there sufficient evidence to recommend the Mark 3000 over the Eagle 400?

37. The R&D manager at Perkins Corporation wishes to determine the effectiveness of two new radar display systems. A recent test was conducted with 100 radar operators selected at random. The test involved measuring the reaction time needed on both systems for operators to detect an incoming flying object. The test results revealed an average difference of 15 seconds identification time between the two systems, with a standard deviation of 4 seconds. Develop a 95% confidence interval for the difference between the two means.

38. The operations manager at Talbert Computers is interested in determining the impact of a new packaging method. A sample of 12 employees was selected at random. The time required for each selected employee to package one of the company's computer systems using both the old and new methods is reported in the following table:

Worker	Completion Time (minutes)	
	Old Method	New Method
1	8.2	6.7
2	9.1	8.5
3	11.5	8.8
4	10.2	10.6
5	9.8	7.9
6	12.3	11.1
7	11.8	10.9
8	8.6	8.3
9	9.5	9.1
10	10.4	9.9
11	11.6	11.1
12	10.2	9.2

    a. Develop a 95% confidence interval for the difference between the means.
    b. Is there sufficient evidence to conclude that the new method is statistically better?

39. The managers at Healthwest, Inc., a diet and health clinic, are interested in measuring the impact of a new diet. Ten patients were selected at random and placed on the diet for a period of 12 weeks. The results of the 12-week program are reported in the following table:

Patient Name	Weight (pounds)	
	Before	After
Johnson	156	145
Grier	181	178
Chan	145	139
Wilcox	210	201
Jefferys	175	179
Thompson	255	242
Brown	115	114
Wo	151	146
Jackson	192	178
Pickens	167	161

   a. Develop a 95% confidence interval for the difference in the means.
   b. What conclusions can be drawn about the diet?
40. A Gallup poll has been commissioned by NBC News to estimate the difference between the support levels for nuclear power plants in the North and the South. A recent survey of 500 Northerners and 500 Southerners found that 200 and 300, respectively, supported nuclear power. Develop a 98% confidence interval for the difference between the proportions favoring nuclear power.
41. American Airlines wishes to determine whether the proportions of passengers using supersaver fares is on the increase (a passenger must purchase the ticket 30 days in advance in order to qualify for a supersaver fare). Last year a survey of 200 passengers found that 80 were able to purchase their ticket 30 days in advance. This year a similar survey revealed that 110 out of 200 passengers qualified for a supersaver fare. Construct a 90% confidence interval between the proportions using supersaver tickets.
42. The quality control manager at Briggs Wire and Cable Company is interested in developing a confidence interval for the difference between the mean number of defects produced from two wire-splicing machines. He has selected two large and independent samples of 100 from each machine. The proportion of defects found in the samples were .12 and .10, respectively.
   a. Develop a 98% confidence interval for the difference between the two means.
   b. What conclusions can be drawn about the reported differences?
43. The marketing manager at Weber Electronics wishes to determine the potential demand for the company's new optical scanner system. In a preliminary estimate, she found that 20% of the retail grocery population would be interested in the new scanner. Specifically, she wishes to know the minimum sample size required to provide a sampling error of .05 or less at a 95% level of confidence.
44. The vice president at Bridgeman (see problem 30) is concerned about the size of the developed confidence interval. She would prefer that the sampling error not exceed three hours. What is the minimum sample size required to yield a sampling error of three hours or less at a 95% level of confidence?
45. The manager at a local bank wishes to know the average time a teller spends with a customer. The population standard deviation for the entire banking system is six minutes. The manager wants the sample error to be less than two minutes.

Determine the minimum sample size based on:
a. A 90% confidence interval.
b. A 95% confidence interval.

46. The manager of Wallco Inc. would like to estimate the level of monthly operating expenses for the company's 150 retail outlets. The following data on monthly expenses were obtained from a stratified sample of the firm's three regional groups:

Stratum	Number of Stores	Sample Size	Monthly Expenses ($000)
Eastern	25	5	54, 51, 45, 48, 50
Central	75	15	38, 41, 43, 39, 40, 24, 27, 34, 37, 35, 27, 25, 31, 30, 34
Western	50	10	41, 45, 41, 47, 44, 39, 40, 44, 36, 39

a. Develop a point estimate for mean monthly expenditure.
b. Develop a 95% interval estimate for the sample.
c. Develop a statement regarding the confidence interval.
d. Discuss the rationale behind the use of stratified sampling for this problem.

47. A Japanese electronics manufacturer is planning to market its new high definition TV in the United States beginning in 1992. The firm's management wishes to estimate the proportion of households that might purchase the new product. The following summary data was obtained from a stratified sample of a "typical" community by level of education:

Educational Level	N_i	n_i	p
High school	8,250	198	0.37
Undergraduate	3,000	72	0.49
Graduate	1,250	30	0.62
	12,500	300	

a. Develop a point estimate for the proportion of households interested in the new product.
b. Develop a 90% interval estimate for the sample.
c. Develop a statement regarding the confidence interval.
d. Discuss the rationale behind the use of stratified sampling for this problem.

48. The marketing manager at Crimestop, Inc., is planning to launch a major advertising campaign on the firm's new automobile electronic beacon system. The

system is designed to alert police as to the location of a car in the event it is stolen. Prior to initiating the campaign, the manager has decided to conduct a household survey from 5 representative neighborhoods out of a total of 100 neighborhoods in the community. The primary objective of the survey is to develop an estimate of the price the motoring public would be willing to pay for the security system. The following summary data on unit price was obtained from a cluster random sample of the five neighborhoods:

	Neighborhood				
	1	2	3	4	5
Number of households	780	1,120	940	1,325	875
Total price per cluster ($ 000)	269	459	362	431	342

a. Develop a point estimate for the mean price.
b. Develop a 95% interval estimate for the sample.
c. Develop a statement regarding the confidence interval.
d. Discuss the rationale behind the use of cluster sampling for this problem.

49. The Los Angeles Department of Water and Power is interested in determining the average level of water consumption for its residential customers. A sample of 5 residential districts was randomly selected out of a total of 40. The following data on weekly water use, in gallons, was obtained from a clustered random sample of the five districts:

District				
1	2	3	4	5
147	179	232	156	218
152	121	179	201	209
176	148	161	189	101
198	215	141	177	154
156	245	193	222	183
212		110	188	
186		177		
		156		

a. Develop a point estimate for the mean weekly water use.
b. Develop a 90% interval estimate for the sample.
c. Develop a statement regarding the confidence interval.
d. Discuss the rationale behind the use of cluster sampling for this problem.

50. The operations manager at Wildwest Inc. is interested in measuring the level of absenteeism among the firm's 30 retail outlets scattered throughout Southern California. The following data on monthly absenteeism in hours per month were obtained from a cluster sample of four retail outlets.

	Retail Outlet		
1	2	3	4
12	3	2	7
2	6	9	11
8	3	4	2
1	5	0	8
4	10	11	4
6	8	7	5
5	7	10	
9	1		

    a. Develop a point estimate for the mean level of absenteeism.
    b. Develop a 95% interval estimate for the sample.
    c. Develop a statement regarding the confidence interval.
    d. Discuss the rationale behind the use of cluster sampling for this problem.

51. Develop a point estimate and 95% interval estimate for the IRS stratified sample data given on page 239. Develop a statement regarding the confidence interval.

52. Develop a point estimate and 90% interval estimate for the market research cluster sample data given on page 240. Develop a statement regarding the confidence interval.

53. The director of the local retail association is interested in estimating the average compensation of CEOs in the industry. A random sample of 50 CEOs yielded the following statistics: $\bar{X} = 225{,}000$ and $s = 27{,}000$. Construct a 95% confidence interval for CEO compensation.

54. The director of public housing for the city of Los Angeles would like to estimate the average room rental rate. A random sample of 100 facilities yielded the following statistics: $\bar{X} = \$17/\text{day}$, $s = \$2/\text{day}$. Construct a 95% confidence interval for room rates.

55. The operations manager at Palos Verdes Community Bank is interested in estimating the average loan level. A random sample of 20 loans yielded the following data ($000): 6, 8.5, 11, 12.5, 14, 17.5, 18, 20, 22.5, 24, 25, 28.5, 31, 34, 37, 39, 40, 42, 45, and 47. Construct a 95% confidence interval for the loan level.

56. What sample size is necessary to estimate the mean of a random variable, within two, where the minimum value is 80 and the maximum value is 100 at the 95% level?

57. For a population proportion, determine the minimum sample size required to yield an error of no more than 0.02, with $a = 0.05$, when:
    a. $p = 0.3$.
    b. p is unknown but $\bar{p} = 0.25$.
    c. p is unknown and there is no estimate.

58. For a population proportion, determine the minimum sample size required to yield an error of no more than 0.03, with $a = 0.01$, when:
    a. $p = 0.5$.
    b. p is unknown but $\bar{p} = 0.4$.
    c. p is unknown and there is no estimate.

59. The human resources manager at Goodwell Inc. is interested in estimating the proportion of employees that are currently engaged in a flextime work schedule. The manage wishes to estimate the population proportion within 0.05 with a 95% confidence. A preliminary estimate indicates that approximately 35% of al employees are on a flextime schedule. What sample size is required to estimate p within the required error specifications?

60. The director of the Public Opinion Institute is interested in estimating the public's support for President Bush's Persian Golf policy. The director wishes to estimate the population proportion within 0.02 with a 95% confidence. Currently, there is no indication of the public's support for the policy. What sample size is required to estimate p within the required error specification?

7.12 CASES

7.12.1 Southern California Edison

The director of short-term planning at Southern California Edison (SCE), Ms. Ivy Devesto, is concerned about the impact on electrical demand of a projected heat wave for next summer. Forecasts from the U.S. Weather Service have indicated that the Los Angeles area will experience unusually high temperatures between late June and early August. Some weather experts believe that the general trend in higher temperatures can be attributed to the so-called greenhouse effect.

SCE maintains sufficient generating reserves to meet most demand situations. However, when demand exceeds generating capacity, SCE can purchase power from other utilities at a premium price. Obviously, SCE would prefer to reduce demand through conservation rather than buy power from the outside. The other alternative would be for SCE to shed loads for periods of time until demand has subsided (i.e., brownouts). Obviously, this option would be very unpopular with SCE customers.

Ms. Devesto, who recently transferred from Potomac Power & Electric, knows that the use of air conditioning during the summer months on the East Coast is a major factor in the increased demand for electrical power. However, she is somewhat unsure of the situation in the Los Angeles area. Accordingly, Ms. Devesto has decided to research the use of air conditioning locally and assess its impact on power demand.

Table 1 presents power profiles for a typical weekday and weekend (measured in kilowatts) for two apartment facilities located on the west side. The apartment complexes are of approximately the same size (i.e., 30,000 square feet). These data were recorded over a 24-hour period during last year's peak summer heating season. One apartment building did not have central air conditioning, while the other was equipped with such a system.

Suggested Questions

1. Develop point estimates of hourly power demand for weekday and weekend applications.

TABLE 1 Weekday and Weekend Power Profile

Time (hours)	No Air Conditioning		Air Conditioning	
	Weekday (kw)	Weekend (kw)	Weekday (kw)	Weekend (kw)
1	9.6	11.5	10.1	10.6
2	12.0	11.5	10.1	9.6
3	7.2	9.6	9.6	10.1
4	9.1	10.1	8.2	9.6
5	9.6	9.1	8.6	8.2
6	8.1	8.2	7.2	8.2
7	9.1	8.6	9.6	11.0
8	9.6	9.1	7.7	8.6
9	8.6	10.6	7.7	9.6
10	8.2	10.6	7.7	9.6
11	9.6	11.0	7.7	8.2
12	8.2	10.1	8.2	9.6
13	11.0	11.5	7.7	8.6
14	8.6	11.5	9.1	7.2
15	11.5	10.6	10.1	11.5
16	8.6	12.5	11.5	10.1
17	9.1	10.6	11.0	9.1
18	8.6	13.0	12.5	10.6
19	13.4	14.0	7.7	10.1
20	14.4	12.5	11.0	16.3
21	16.3	12.5	14.9	15.4
22	13.9	12.5	10.1	16.3
23	11.0	12.5	9.1	14.9
24	12.5	11.5	7.7	12.0

2. Does there appear to be a difference in power demand between the non-A/C- and A/C-equipped apartments?
3. Develop interval estimates of hourly power demand at a 95% level of confidence.
4. What blocks of hours are associated with peak demand?
5. Are the sample data reported in Table 1 representative of power demand during the summer for apartments located in Southern California?
6. What factors, in addition to A/C, might contribute to the variation in power demand during the day?
7. What conclusions should Ms. Devesto draw from these results?

7.12.2 Brookline Robotics Corporation

Brookline Corporation manufactures and markets a robotic arm used in a variety of process control industries (e.g., newspapers). Mr. Al Hagen, Brookline's production

TABLE 1 Production Data for Brookline Robotics

Month	Output (units)	Production Quality (units)	Shipping Delays (units)	Cost Reduction (%)
Jan	946	17	98	0.0
Feb	923	19	105	−0.5
Mar	901	21	122	0.0
Apr	882	19	127	−1.0
May	863	16	145	−1.5
Jun	845	13	152	−0.5
Jul	911	16	126	0.0
Aug	931	17	110	0.5
Sep	904	15	116	−0.5
Oct	925	14	107	−0.5
Nov	956	17	95	0.0
Dec	973	19	82	0.0
Jan	942	21	78	−0.5
Feb	911	20	86	−0.5
Mar	934	18	80	0.0
Apr	945	19	72	0.0
May	915	17	77	0.5
Jun	907	16	84	0.0

manager, has been receiving increased attention from the general manager regarding the company's El Paso facility. Recent summary production reports on the El Paso plant have indicated an apparent problem in maintaining production schedules and quality control targets. Accordingly, Mr. Hagen had decided to investigate the situation by collecting and analyzing detailed production data from the last 18 months of plant operation. Table 1 indicates production output, production quality (measured in terms of the number of units rejected), the extent of any shipping delays (measured in the number of units scheduled for delivery but not delivered), and the level of cost reduction per month (measured as a percent of the previous month's costs).

The average production and quality control targets over the 18-month period are as follows:

- Production level: 925 units
- Quality level: 15 units
- Shipping delay: 80 units
- Cost reduction: 0.5% per month

Suggested Questions

1. Develop a point estimate for each of the production variables.
2. Develop a 90% confidence interval for each of the variables.

3. What effect would a larger sample size have on the interval estimates?
4. Compare the point and interval estimates with the production and quality control targets.
5. What statements can be made regarding the production and quality control targets relative to the interval estimates?
6. Should Mr. Hagen be concerned about these results?

Chapter 8

Hypothesis Testing

A scientific hypothesis transcends the facts that served as its basis.
V. I. Vernadsky

CHAPTER OUTLINE

- 8.1 Introduction
- 8.2 Example Management Problem: Iowa Department of Motor Vehicles
- 8.3 How to Recognize a Hypothesis Testing Problem
- 8.4 Statistical Hypothesis Testing
- 8.5 Computer Analysis
- 8.6 Practical Applications
- 8.7 Case Study: Heartwell Music Experiment
- 8.8 Summary
- 8.9 Glossary
- 8.10 Bibliography
- 8.11 Problems
- 8.12 Cases

CHAPTER OBJECTIVES

The primary objectives of this chapter are to develop an understanding of

1. The various testing procedures.
2. The hypothesis testing process.
3. The nature of testing errors.
4. How to establish decision criteria.
5. Statistical quality control.
6. How to perform hypothesis testing using computer analysis.

The problem of management compensation will be one of the critical issues facing corporate America in the 1990s. In the past, executives received automatic salary increases regardless of performance. This situation has now reached the point that the ratio of executives' salaries to workers' salaries in the United States is five times that of Japan. This has resulted in a growing resentment among American workers. However, these trends may be changing as a consequence of increased international competition.

A number of experts in both the public and private sectors have called for a reexamination of executive compensation, which many believe should be tied more directly to the firm's performance. In that regard, a recently completed study of 65 publicly held companies in California revealed that compensation for CEOs generally correlated with corporate performance.* However, the results tended to vary by industry group. It is interesting to note that companies with average growth paid above-average salaries, but those with below-average growth did not show below-average compensation.

8.1 INTRODUCTION

Hypothesis testing is fundamental to statistics. It is very closely related to estimation and confidence intervals, which were discussed in the previous chapter. The primary purpose of hypothesis testing is to support inferences, which are made every time one interprets an observation. For example, a manager who concludes that employee turnover is related to the company's restrictive compensation policies is inferentially relating turnover to policy. But does one observation support such an inference? How might imprecise measures of turnover affect the manager's conclusions? What is needed is a way of drawing conclusions with a small probability of error. This is what hypothesis testing is all about.

In reality most decision makers continually attempt to attribute to as large a population as possible conclusions reached from data obtained from as small a sample as possible. The process of inference thus starts at the sample level and proceeds to the population level; that is, it moves from the specific to the general. Unfortunately, estimating population characteristics from sample data can result in significant error. For example, a manager may infer a relationship between turnover and policy based on sample results when no such relationship exists. On the other hand, the manager may infer that no relationship exists when, in fact, one does. The control of these potential errors plays a significant part in the hypothesis testing process.

The first step in the hypothesis testing process is the specification of competing hypotheses (e.g., "there is a relationship between turnover and company policy" versus "there is no relationship between turnover and company policy"). The objective is to determine which hypothesis is probably true. But before evaluating a hypothesis, one must collect sample data and formulate a decision rule. Only then can a decision regarding the hypothesis be made.

---

* *CEO Compensation in the California 100 Industrials for 1985,* The Key Groups. 1986.

> **HISTORICAL NOTE**
>
> The modern notation of hypothesis testing can be traced to Jerzy Neyman (1894–1984), a Polish mathematician. Working with E. S. Pearson in the 1920s and 1930s, Neyman evolved the concept that the only valid reason for rejecting the null hypothesis was the existence of an alternative hypothesis with a higher probability of occurrence. The idea of focusing on the alternative hypothesis led to the discovery that two types of errors can occur in hypothesis testing.

8.2 EXAMPLE MANAGEMENT PROBLEM: IOWA DEPARTMENT OF MOTOR VEHICLES

The director of the Iowa Department of Motor Vehicles (IDMV) is interested in determining the average waiting time for service in the new Des Moines office. The current policy of the IDMV is that the average waiting time for service should not exceed 10 minutes. To evaluate the effectiveness of the new office, the director has requested that the staff collect a representative sample of actual waiting times over a 30-day period. The resultant data are provided in the following table:

Observation	Waiting Time (minutes)	Observation	Waiting Time (minutes)
1	11	16	14
2	21	17	7
3	15	18	10
4	8	19	11
5	12	20	16
6	14	21	9
7	8	22	18
8	7	23	5
9	22	24	11
10	10	25	14
11	9	26	7
12	12	27	10
13	6	28	11
14	9	29	8
15	14	30	12

What conclusion can the director draw from this sample? Does the director need to increase the staff at the new office to meet the policy guidelines?

8.3 HOW TO RECOGNIZE A HYPOTHESIS TESTING PROBLEM

The following list highlights the general characteristics of most hypothesis-testing problems:

1. The problem is stated in the form of a hypothesis (i.e., premise) regarding a population parameter.
2. An acceptable level of error is established.
3. A sample is selected from the population.
4. Descriptive statistics are computed for the sample.
5. A decision rule regarding the hypothesis is specified.
6. A decision is made to reject or not reject the hypothesis.

8.4 STATISTICAL HYPOTHESIS TESTING

Although the hypothesis testing process seems simple enough in theory, it is sometimes difficult to draw a sharp distinction between a population and a sample. Suppose, for example, the director of the Iowa Department of Motor Vehicles wishes to find out something about the department's 50 first-line supervisors. If all 50 are interviewed, do you have a sample or a population? They represent a population of the department's first-line supervisors. But in

INTERNATIONAL VIGNETTE

Mr. Audun Boerve, managing director of Crystalox Ltd., was concerned about the company's manufacturing effectiveness. The rapid growth in overhead coupled with the constant interaction between the project engineers and the production staff was causing serious production problems. In response to this situation, Mr. Boerve had taken several steps toward improving the production process. These included the implementation of a time and cost reporting system and having manufacturing report directly to him. A proposal from one of Crystalox's vendors had recently reached his desk suggesting a more radical approach. Basically, the proposal recommended forming a joint venture with the vendor and moving most of the manufacturing capability offsite.

Crystalox, which was founded in 1970 by Dr. David Hukin, a well-known expert in crystal growth technology, was acquired by Elkem AS of Norway in 1985. Crystalox's products are sold on a worldwide basis to a variety of research and development laboratories. The product line consists of a number of specialized machines for growing crystals to customer specifications. Mr. Boerve's basic hypothesis was that the current demand for the company's special machines and standard products would serve as a solid base for the proposed new venture. Additionally, there would be a significant amount of more traditional machine shop business that would develop. Based on these assumptions, Mr. Boerve was prepared to take the proposal to Elken's CEO for approval at the next board meeting.

a broader sense they are a small sample of the population of first-line supervisors in departments of motor vehicles throughout the country. Thus, it may be helpful to view the term *population* as mathematicians view infinity — you may approach it, but you can never get there. Generally speaking, a sample is a small portion of a population. Thus, the observations are made on samples and not on populations.

Key idea 1

The actual test procedures involved in hypothesis testing are often complex due to the large number of variations that can occur. The following list presents some of the more common situations encountered in hypothesis testing:

- A single population mean
- A single population proportion
- The difference between two population means
    Independent samples (unmatched pairs)
    Dependent samples (matched pairs)
- The difference between two population proportions
- Other population parameters (e.g., variance)

THE HYPOTHESIS TESTING PROCESS

Key idea 2

A **hypothesis** is a statement about a population parameter or relationship between populations. In general, it is an interpretation of the statement, "If X is true, then Y should be true." In classical statistics, X is considered to be the independent variable (the one manipulated) and Y the dependent variable (the result of the manipulation). Thus, one might hypothesize that productivity (Y) will increase if classical music (X) is played in the work area. Similarly, one can hypothesize that increased expenditures for advertising (X) will result in increased revenues (Y). Thus, the purpose of the hypothesis is to make a statement about reality that may be true or false. The purpose of statistical hypothesis testing is to determine whether the proposed hypothesis is probably true or probably false. Because statements regarding the population are normally based on a sample, one must examine and make some assumptions about the population from which the samples have been drawn.

Basically, there are three different approaches for testing hypotheses—classical method, standard scores, and p-values. Each of these approaches provides a somewhat different perspective of the testing process. In the case of the first two methods, however, the resultant conclusions will be the same. The hypothesis testing process can be characterized by the following five-step sequence:

1. Stating a null and an alternative hypothesis
2. Establishing the level of acceptable error
3. Formulating a decision rule
4. Collecting a sample from the target population
5. Making a decision based on the sample results

Each of these steps will be discussed in some detail below.

Null and Alternative Hypotheses

Central to an understanding of hypothesis testing is the concept of the null hypothesis. Generally speaking, the **null hypothesis** is a statement about some expectation. For example, the null hypothesis for the Iowa Department of Motor Vehicles might be that the average observed waiting time is no different from that of the stated policy (i.e., 10 minutes or less). It is the null hypothesis that is actually tested by statistical procedures. The existence of the null hypothesis implies the existence of an alternative hypothesis. Often called the **research hypothesis,** the **alternative hypothesis** is usually stated in "larger than," "smaller than," or "not equal to" terms. When the null hypothesis is rejected as unlikely, one can then accept the alternative. When the null hypothesis is not rejected, then the alternative can be rejected.

In general, there are three different options for testing a hypothesis about a single population mean. These options are given in Figure 8.1. The first

8.4 Statistical Hypothesis Testing

FIGURE 8.1 Hypothesis-Testing Options for a Single Population

Hypothesis	Option 1	Option 2	Option 3
Null (H_0)	$\mu = \mu_0$	$\mu \leq \mu_0$	$\mu \geq \mu_0$
Alternative (H_1)	$\mu \neq \mu_0$	$\mu > \mu_0$	$\mu < \mu_0$

option utilizes a hypothesis indicating a specific value for the unknown population parameter (e.g., the actual waiting time is exactly 10 minutes). The alternative hypothesis, in this case, suggests that the actual population parameter is either larger or smaller than that stipulated by μ_0. A hypothesis test that involves an exact hypothesis is often referred to as a **two-sided hypothesis**.

In the second option the null hypothesis states that the population parameter is less than or equal to a given value (e.g., the waiting time is less than or equal to 10 minutes). In the third option, the null hypothesis states that the population parameter is greater than or equal to a specified value (e.g., the average waiting time is greater than or equal to 10 minutes). In both of these cases, the test procedure is called a **one-sided hypothesis** because the range of values for the alternative hypothesis falls on only one side of the population value.

For the IDMV problem, the director would first state the null (H_0) and alternative hypotheses (H_1) as follows:

- H_0: The average waiting time is less than or equal to 10 minutes.
- H_1: The average waiting time is greater than 10 minutes.

Mathematically, this can be expressed as

$$H_0: \mu \leq 10 \qquad H_1: \mu > 10$$

where μ is the population waiting time.

This is a clear example of the second type of null hypothesis. If there is insufficient evidence to reject the null hypothesis, then the director can conclude that the average waiting time is 10 minutes or less. If the null hypothesis can be rejected, then the director can conclude that the average waiting time is greater than 10 minutes and that the sample mean is the "best" estimate of the actual waiting time.

Level of Acceptable Error

Key idea 3

Unfortunately, the process of hypothesis testing is plagued by uncertainty. This observation should not be surprising since the testing process uses a small sample to estimate the characteristics of what is usually a large population. Two basic errors are associated with hypothesis testing:

- **Type I error:** Rejecting the null hypothesis when it is true **(alpha)**.
- **Type II error:** Accepting the null hypothesis when it is false **(beta)**.

In the IDMV problem, a Type I error would occur if the director rejected the null hypothesis when in fact the actual waiting time was 10 minutes or less. A Type II error would occur if the director accepted the null hypothesis when the actual waiting time was greater than 10 minutes.

A court trial can be viewed, in general terms, as a hypothesis testing process. Typically, the null hypothesis is that the accused is not guilty, and the alternative is that the accused is guilty. Rejection of the null hypothesis of not guilty leads to the conclusion that the accused is guilty. Clearly, both types of errors are possible in the decision of the court. A Type I error would occur if the accused were actually not guilty but was found guilty. A Type II error would occur if the accused were actually guilty but was not convicted. Both of these types of errors have occurred throughout the history of jurisprudence. In a court trial, as in most other hypothesis-testing situations, committing a Type I error (placing a not guilty person in jail) is viewed as much more serious than committing a Type II error (letting a guilty person go free). The usual method for controlling Type I errors is to set alpha to a very small value (e.g., 1% to 5%). In the vernacular of the court system, the judge usually sets alpha to a value "beyond a reasonable doubt."

Figure 8.2 provides another view of the relationship of these errors to the decision process for the IDMV problem. The probability of committing a Type I (alpha) error is called the *level of significance,* and the probability of committing a Type II (beta) error is referred to as **power**. The primary objective in testing is to control the magnitude of these errors. Naturally, the ideal case would be to set the probability of both errors at zero. In practice, however, this is not feasible. For example, making alpha equal to zero would result in never rejecting the null hypothesis. Typically, an alpha value between 1% and 10% is used in testing the null hypothesis.

FIGURE 8.2 Error Comparison for Estimated versus Actual Situations

Actual \ Estimate	H_0 true ($\mu \leq 10$)	H_0 false ($\mu > 10$)
H_0 true ($\mu \leq 10$)	Correct decision (probability = $1 - \alpha$)	Type I error (probability = α)
H_0 false ($\mu > 10$)	Type II error (probability = β)	Correct decision (probability = $1 - \beta$)

Another important aspect of hypothesis testing involves the relationship between alpha and beta. From the preceding discussion, one might be tempted to conclude that $\alpha + \beta = 1$. However, this situation is true only under very special conditions. Typically, as alpha becomes smaller, beta becomes larger, and vice versa. This implies that reducing the probability of committing a Type I error increases the probability of committing a Type II error. The usual procedure is to set the value of alpha and then try to control the value of beta through the size of the sample. In general, dealing with beta errors is more difficult because the actual value of the population mean is unknown.

Decision Rules

Key idea 4

A **decision rule** is a statement used for making judgments regarding a given hypothesis (i.e., rejecting or not rejecting the null hypothesis). Decision rules are not symmetrical. That is, if the null hypothesis is rejected, then the logical conclusion is to accept the alternative hypothesis. Theoretically, however, *not* rejecting the null hypothesis is not equivalent to its acceptance. The proper conclusion, in this case, is to state that there is insufficient evidence to reject H_0. The trial court example discussed previously may help shed some light on this important principle. If the jury finds the defendant guilty, the null hypothesis of not guilty can be rejected. If, however, the jury finds the defendant not guilty, one cannot conclude that the defendant is innocent, only that there was insufficient evidence to support a guilty verdict.

Decision rules for hypothesis testing can be based on either means (classical method) or standard scores. Both approaches yield precisely the same results. In either case, the decision rule yields a **critical value** based on the specified level of significance (i.e., alpha). The critical value is the statistic used to determine whether to reject or not reject the null hypothesis. Typically, one of two probability distributions is used in formulating decision rules. The first is the Z-test (for large samples), and the second is the t-test (for small samples). The t-test is based on the assumption that the sample was drawn from a normal population. Both of these distributions were introduced in Chapter 7. Figure 8.3 summarizes the various test possibilities as a

FIGURE 8.3 Summary of Test Procedure Alternatives

Population Distribution	Sample Size	
	Small (<30)	Large (≥30)
Normal	t-test	Z-test
Unknown	Nonparametric test	Z-test

296 Chapter 8 Hypothesis Testing

FIGURE 8.4 Alternative Hypothesis Tests

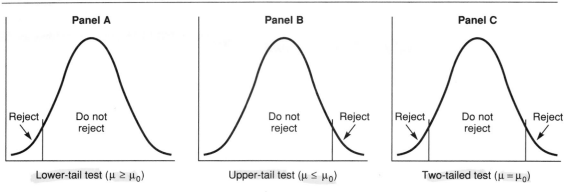

function of sample size. In the case of a small sample drawn from an unknown population, the usual procedure is to employ nonparametric analysis. This statistical procedure is covered in Chapter 14.

As indicated previously, the testing process can be either *unidirectional* or *bidirectional,* depending on the nature of the problem. A unidirectional test involves a null hypothesis in the form of $\mu \geq \mu_0$ or $\mu \leq \mu_0$. Figure 8.4 illustrates both unidirectional situations (panel A and panel B). The distribution represents the sampling error. In the case of $\mu \geq \mu_0$ a lower limit is used as a basis for rejecting or not rejecting the null hypothesis. If the sample statistic (mean value or standard value) appears to the left of the decision criterion line (shaded area), then the null hypothesis can be rejected. The shaded area corresponds to the assigned level of significance (i.e., alpha). When H_0 is of the form $\mu \leq \mu_0$, the decision rule is applied to the upper end of the distribution. In such a case, H_0 can be rejected if the sampling statistic falls to the right of the decision criterion value. A null hypothesis in the form $H_0: \mu = \mu_0$ calls for a two-tailed test. This is because the null hypothesis can be rejected for values of the test statistic located on either side of the sampling distribution. The nature of the two-tailed test is shown in Panel C of Figure 8.4.

The decision rule, based on means, for the three alternative test options is outlined in Figure 8.5. These relationships are appropriate when the sample size is 30 or more. Notice that they are similar to the interval estimate developed in Chapter 7. The equivalent decision rules based on a standard score are given in Figure 8.6. Because of the ease of use of this method, many managers find the standard Z score more convenient in establishing decision rules than one based on means. However, either approach yields exactly the same results.

The t-test is perhaps the most frequently used test procedure. It is primarily used for small samples of less than 30. When the sample size is 30 or more, the t-distribution is not significantly different from the normal curve. Developing a decision rule for a t-test is similar to the procedure for a Z-test with

FIGURE 8.5 Alternative Decision Rules Using Means (classical)

Decision Rule

$$\text{Reject } H_0 \text{ if } \overline{X} \geq \mu_0 + \frac{Z \times s}{\sqrt{n}} \quad \text{(Upper-tail test)}$$

$$\text{Reject } H_0 \text{ if } \overline{X} \leq \mu_0 - \frac{Z \times s}{\sqrt{n}} \quad \text{(Lower-tail test)}$$

$$\text{Reject } H_0 \text{ if } \overline{X} \leq \mu_0 - \frac{Z \times s}{\sqrt{n}}$$
$$\text{or } \overline{X} \geq \mu_0 + \frac{Z \times s}{\sqrt{n}} \quad \text{(Two-tailed test)}$$

where:

μ_0 = Hypothesized population mean
Z = Standard normal score based on given alpha
s = Estimated population standard deviation
n = Sample size

---

one important difference. Unlike the Z-test, the t-test depends on the size of the sample. More specifically, it depends on the *degrees of freedom* (df). For a one-population problem, the degrees of freedom are equal to the sample size minus 1 (i.e., $n - 1$). Thus, the critical t-value for a sample size of 10 (9 df) and a level of significance of 5% is 1.833 (see Table A.2, the standard t-table in Appendix A). In contrast, if the sample size is 15 (14 df), the corresponding critical t-value is 1.761. Notice that the critical t-value becomes smaller as the sample size grows. This suggests that for a given alpha the chances of rejecting the null hypothesis increase as the sample size

---

FIGURE 8.6 Alternative Decision Rules Using Standard Scores

$$Z = \frac{\overline{X} - \mu_0}{s/\sqrt{n}}$$

Reject H_0 if $Z \geq Z_c$ (Upper-tail test)

Reject H_0 if $Z \leq Z_c$ (Lower-tail test)

Reject H_0 if $Z \leq Z_c$
or $Z \geq Z_c$ (Two-tailed test)

where Z_c is the critical Z-statistic based on the specified level of significance.

increases. This conclusion should not be surprising since the availability of more data (in the form of a larger sample) reduces the standard error and thus more clearly reveals any differences between the population and sample means.

Data Collection and Analysis

One of the key steps in the hypothesis testing process is the collection and analysis of sample data. In formal terms, the collection of data requires the design of an experiment. Specific elements of the experiment include the size of the sample, the method for collecting the data, and the method of analysis. The general process for collecting data has already been discussed in Chapter 6. The basic steps in this procedure are given in Figure 8.7.

In the IDMV example, the director collected a representative sample of 30 observations. From these observations he calculated a variety of descriptive statistics including a mean and standard deviation. The director found that the sample mean was 11.37 with a standard deviation of 4.14. Can the director reject the null hypothesis since the sample mean is greater than the population mean (i.e., 11.37 > 10)? Obviously, this is a tempting conclusion. Unfortunately, the answer is no. In hypothesis testing, *both* the sample mean and the standard deviation must be used in making a decision about the null hypothesis. In general, the smaller the standard error, the smaller the differences must be to justify rejection of the null hypothesis. In contrast, when the standard deviation is large, large differences between the population and the sample means must exist to justify rejection.

Drawing a Conclusion

Drawing a conclusion about a stated hypothesis involves comparing the decision criterion with the sample statistic (e.g., sample mean). In general, H_0 can be rejected if the sample statistic falls outside the decision limits. The IDMV problem involves an upper-tail test in which the decision rule is in the form of $H_0: \mu \leq 10$. The director has selected an alpha of .05. This corresponds to a Z-value of 1.65 (from Table A.1 in Appendix A.). The upper limit for this problem is computed, using the relationships given in Figure 8.5, as follows:

$$\text{Upper limit} = 10 + \frac{1.65 \times 4.14}{\sqrt{30}} = 11.25$$

Since the sample mean is greater than the computed upper limit (i.e., 11.37 > 11.25), the director rejects the null hypothesis and accepts the alter-

8.4 Statistical Hypothesis Testing

FIGURE 8.7 Basic Steps in Data Collection and Analysis

Design Experiment	Collect Data
• Formulate survey instrument. • Determine sample size.	• Define measurement system. • Conduct survey.

Process Data	Analyze Data
• Select statistical model. • Compute descriptive statistics.	• Develop sampling distribution. • Check for errors.

native hypothesis at a 95% level of confidence. That is, he can conclude that the average customer waiting time at the new facility is greater than 10 minutes. The equivalent decision rule analysis based on the standard Z score is

$$Z = \frac{11.37 - 10}{4.14/\sqrt{30}} = 1.81$$

The null hypothesis can be rejected since the computed Z statistic is greater than the critical Z of 1.65. As indicated previously, this test procedure results in exactly the same conclusion as the test based on the means.

p-Values

Recall that the testing procedure just outlined was based on rejecting or not rejecting the null hypothesis using a preestablished level of significance (i.e., alpha error). This approach requires the decision maker to set the decision criterion before conducting the test. Although this testing procedure is decisive — reject or not reject H_0 — it does not provide a quantitative statement on the strength of the conclusion. This is particularly troublesome when the test statistic falls into the rejection region. One way around this problem is to report the observed significance level, or *p*-value, for the test. The *p*-value for a given test is the probability of observing a specific sample value that is different from that specified in the null hypothesis. The primary advantage in using *p*-values is that the analyst does not need to draw conclusions about the results of the hypothesis testing. Instead, the *p*-values are merely reported, thus allowing the decision maker to determine the significance of the test.

The *p*-value can be determined directly from the computed Z- or t- statistic via the appropriate probability table. For example, in the IDMV problem the computed Z-value was 1.81. The corresponding probability from the standard normal table (Table A.1 in Appendix A) is .4649. Therefore, the probability that the test statistic will be greater than or equal to this value, given that the null hypothesis is true,

$$p(Z \geq 1.81) = .5 - .4649 = 0.0351$$

Thus, the chance of rejecting the IDMV null hypothesis when it is true is approximately 3.5%. Another interpretation is that the null hypothesis can be rejected because this p-value is less than the alpha specified in the original problem (i.e., $\alpha = .05$). In general, the null hypothesis is rejected for alpha values greater than or equal to the computed p-value. Clearly, as the p-value becomes smaller, the chances of committing a Type I error also decrease. Notice that the p-value provides the decision maker with a quantitative indication of the chances of making a Type I error. This is, of course, in sharp contrast to the basic approach of either rejecting or not rejecting the null hypothesis based on a preset criterion.

The computational situation becomes somewhat more complicated for small-sample analysis since the t-distribution table usually consists only of selected probability values (e.g., .10, .05, .01). In these cases, the probability value is obtained through interpolation based on the degrees of freedom and computed t-value. Fortunately, most statistical software packages, including CBS, routinely provide p-values for each hypothesis test.

In summary, the p-value for a hypothesis test can be interpreted as follows:

- The value of α at which the hypothesis test procedure changes conclusions.
- The probability that the sample mean or test statistic will be at least as great (smaller or larger) as the observed sample mean on the test statistic, given that the null hypothesis is true.

OPERATING CHARACTERISTIC AND POWER CURVES

The primary focus of the discussion up to this point has been on controlling Type I or alpha errors. Recall, however, that there are two types of errors associated with the hypothesis testing process: Type I and Type II. A Type II or beta error occurs when a false null hypothesis is not rejected. When the null hypothesis is false, the purported relationship between μ and μ_0 is not true, e.g., if $H_0: \mu_1 = \mu_0$ and H_0 is false then $\mu_1 \neq \mu_0$. In these cases some other value for μ or relationship between μ and μ_o exists. A beta error reports the probability of not rejecting a false hypothesis for various values of μ. The general objective in hypothesis testing is to make beta as small as possible. A plot showing the probability of not rejecting the null hypothesis when it is false for various values of the population parameter is known as an operating curve. Figure 8.8 shows hypothetical operating curves for the three types of hypothesis tests—lower tail (Panel A), two tail (Panel B), and upper tail (Panel C). Notice that the probability of not rejecting the null hypothesis decreases exponentially as the difference between the true value of the popu-

FIGURE 8.8 OC and Power Curves for the Three Types of Hypothesis Tests

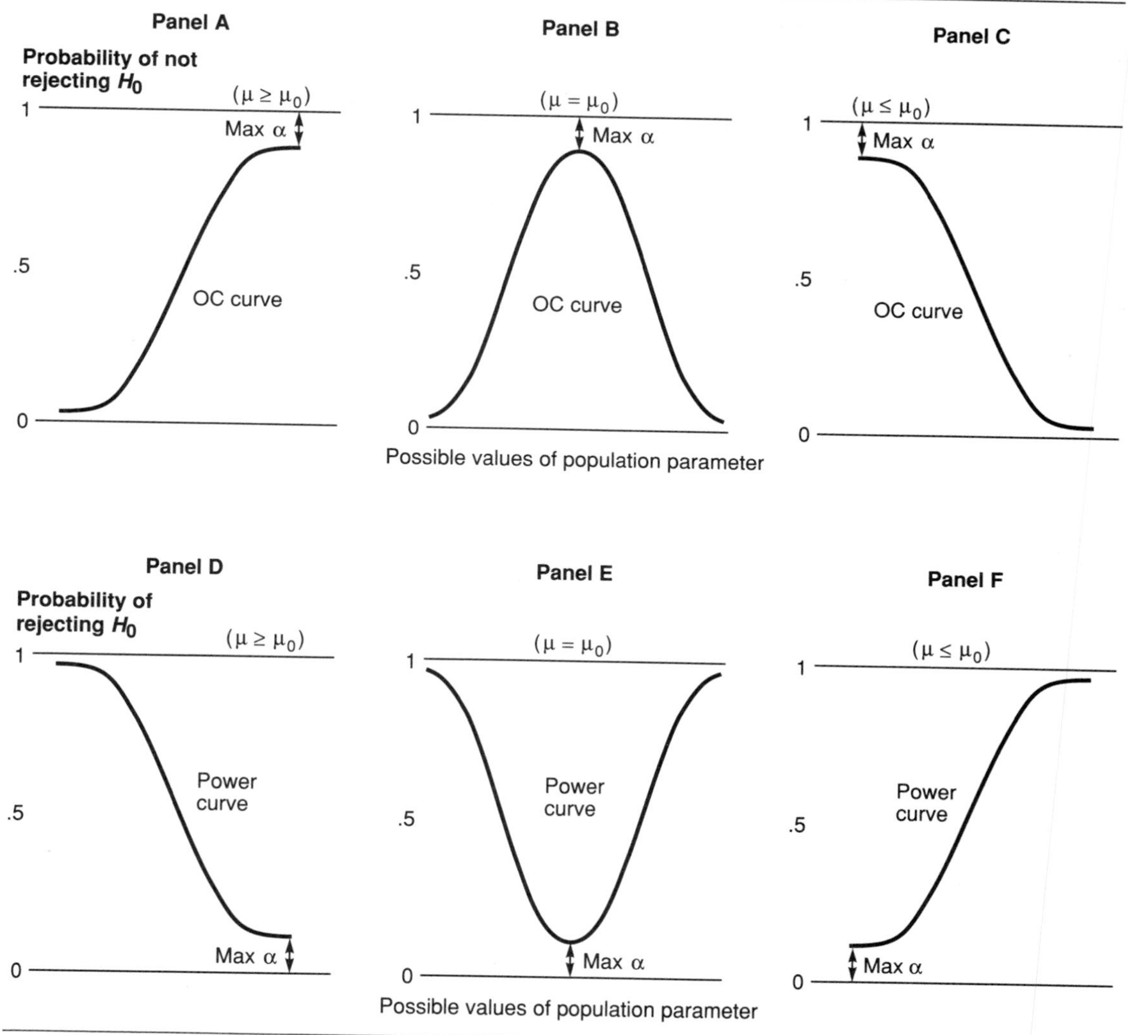

Panel A, Panel B, Panel C: OC curves (Probability of not rejecting H_0) for $(\mu \geq \mu_0)$, $(\mu = \mu_0)$, and $(\mu \leq \mu_0)$ respectively, each showing Max α.

Panel D, Panel E, Panel F: Power curves (Probability of rejecting H_0) for $(\mu \geq \mu_0)$, $(\mu = \mu_0)$, and $(\mu \leq \mu_0)$ respectively, each showing Max α.

Horizontal axis: Possible values of population parameter.

lation mean and the hypothesized population mean increases. Operating curves play an important role in statistical quality control. When used in this context they report the probability of not rejecting a batch of items with varying percents of defectives.

The probability of rejecting the null hypothesis when it is false is indicated by $1 - \beta$. Ideally, the goal is to make the value $1 - \beta$ approach one. The value of $1 - \beta$ provides a measure as to how well the test is performing. Consequently, it is known as the power of the test. A plot of the probability of rejecting the null hypothesis when it is false for various values of the population parameter is known as the power curve. Figure 8.8 also shows hypothetical power curves for the three types of hypothesis tests. As can be seen, the power curve is the complement to the OC curve. In this case, the probability of rejecting the null hypothesis when it is true decreases exponentially as the distance between the true value of the population mean and the hypothesized population mean decreases. This result should not be too surprising, since minor variations that form the true population mean are more difficult to detect than larger ones.

FIGURE 8.9 Illustration of the Probability Representing the Power of the Test for the IDMV Problem

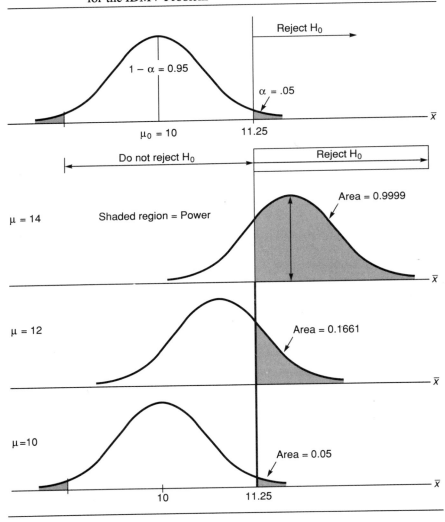

In the case of the IDMV problem, committing a Type II error would involve the director concluding that the average waiting time is 10 minutes or less, when the actual time is greater than 10 minutes. As indicated above, one approach for determining the likelihood of committing a Type II error is to compute β based on selected values of μ, i.e., develop an OC curve. The more common approach, however, is to develop a power curve since it provides a direct statement on the accuracy of the test. Suppose the director wishes to know the probability of rejecting the null hypothesis when it is false (i.e., making the correct decision) for various waiting time values. Figure 8.9 presents a power curve for selected values of the population mean. Notice that when $\mu = 14$ the probability of rejecting the null hypothesis is 0.9999. This result suggests that there is only a 0.0001 chance of accepting the null hypothesis when it is false (i.e., a Type II error). In contrast, when $\mu = 12$, the probability of rejecting the null has dropped to 0.1661. Thus the chances of making a Type II error have increased dramatically as a result of a two-minute reduction in the actual waiting time. As a final note, when $\mu \leq 10$ the probability of committing a Type II error is actually zero. That is, a Type II error can not be made when the actual value of the population mean equals the hypothesized value. However, when developing either an OC or power curve, this discontinuity is usually ignored, since the values around μ are very close to the specified level of significance.

TESTS OF PROPORTIONS

Some business situations require the testing of population proportions (p) instead of means. Examples include:

- The percentage of contracts awarded to minority firms.
- The percentage of defective parts.
- The popularity of a presidential candidate.
- The percentage of creditor customers who pay on time.

The standard hypotheses for a population proportion along with examples are given in Figure 8.10 (notice the parallel with the test for means). In general, the procedure for testing proportions follows the same process used in testing means. That is, either the classical method, standard scores, or p-values can be used.

To illustrate the use of this test, consider that the secretary of the Department of Health and Human Services (HHS) is interested in determining whether the department is in compliance with federal guidelines regarding minority contracts. Current department policy calls for awarding at least 25% of HHS contracts to minority firms. The secretary has asked the audit section to conduct a survey of contracts awarded over the past year. The head of the audit section selects at random 25 contracts and finds that 3 contracts were awarded to minority-owned firms. The secretary wishes to know whether or not this sample supports current departmental policy. This

Chapter 8 Hypothesis Testing

FIGURE 8.10 Summary of Proportion Hypothesis Forms

Hypothesis Form		Example
1. $H_0: p = p_0$	$H_1: p \neq p_0$	55% of the voting public approve of the president's policies
2. $H_0: p \geq p_0$	$H_1: p < p_0$	The proportion of minority contracts is 25% or greater
3. $H_0: p \leq p_0$	$H_1: p > p_0$	The proportion of defective parts is 5% or less

where:

p = Actual population proportion
p_0 = Purported population proportion

problem can be analyzed using a one-sided test of proportions because contract award levels above 25% would not violate department policy. Accordingly, the null hypothesis for this problem is $H_0: p \geq .25$. The alternative hypothesis is $p < .25$. The secretary has selected an alpha error of 5%. That is, the secretary does not want the chances of making a Type I error to exceed 5%. Figure 8.11 graphically shows the basic decision situation for this problem.

A t-test is appropriate for evaluating this problem because the sample size is less than 30. The critical t-value with 24 degrees of freedom and $\alpha = .05$ is 1.711. The standard error of the proportion is given by:

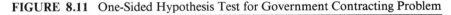

$$s_p = \sqrt{p(1-p)/n}$$

FIGURE 8.11 One-Sided Hypothesis Test for Government Contracting Problem

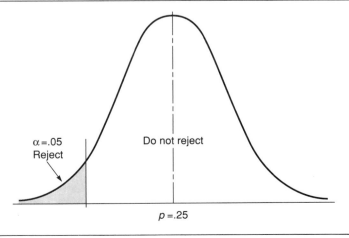

With $p = .25$ and $n = 25$, the standard error is .0866. The computed t-value using this estimate is:

$$t = \frac{\bar{p} - p}{s_p} = \frac{.12 - .25}{.0866} = -1.50$$

This result indicates that there is insufficient evidence to reject the null hypothesis, since its absolute value is less than the critical t-value (i.e., $1.50 < 1.711$). Thus, the secretary may conclude that the department is in conformance with policy.

TWO-POPULATION TESTS

Thus far, the discussion has focused on testing the significance of a sample drawn from one population. There are, however, many situations involving two populations. Examples include:

- Manufacturing costs for two different products.
- Two alternative methods of instruction for teaching a foreign language.
- Two different production methods.
- Differences in salary levels between male and female employees.

In each of these situations, the manager or researcher is often more interested in the difference between the two population parameters (e.g., means) than the actual values of the parameters. The *standard* null hypothesis for two populations is that there is no difference. The alternative or research hypothesis is that there is a difference. The basic hypothesis options for two populations are reported in Figure 8.12. Notice that these options are very similar to the ones presented in Figure 8.1 for the one-population case. For example, in a problem involving the differences in salary compensation between men and women, the null hypothesis is that men and women earn the same pay for the same work (i.e., $H_0 : \mu_1 = \mu_2$), while the alternative is that men and women do not earn the same pay for the same work (i.e., $H_1 : \mu_1 \neq \mu_2$).

The testing of two population means involves one of the following alternative designs:

- **Independent sample design (unmatched pairs):** Two independent random samples are obtained from two populations.

FIGURE 8.12 Hypothesis Testing Options for Two Populations

Hypothesis	Option 1	Option 2	Option 3
Null (H_0)	$\mu_1 = \mu_2$	$\mu_1 \leq \mu_2$	$\mu_1 \geq \mu_2$
Alternative (H_1)	$\mu_1 \neq \mu_2$	$\mu_1 > \mu_2$	$\mu_1 < \mu_2$

- **Dependent sample design (matched pairs):** Two dependent random samples are obtained from two populations.

The testing of two different methods of language instruction is an example of an independent sample design. In this case, one set of students is exposed to one method, and a second set is exposed to the other method. The evaluation of two different production techniques can be accomplished with a matched-pair design. Here the same set of workers is used to evaluate both production methods. In some circumstances, as will be outlined later, the use of matched designs can yield more accurate results for a given sample size.

Independent Sample Designs (Unmatched Pairs)

The test for two populations is very similar to the one-population test. Basically, they differ in the sampling distribution and the sampling variance. Recall that the sampling distribution for one population is assumed to be normal, based on the central limit theorem, when the sample size is 30 or more. The same conclusion can be reached regarding the sampling distribution of the difference between two means. The standard deviation of this distribution is called the *standard error of the difference.* Figure 8.13 presents a graphic of this distribution.

As in the case of a single population, the Z-test is used for large samples ($n \geq 30$), and the *t*-test is employed for small samples. The major assumptions for the application of the *t*-test for two populations are as follows:

1. Both samples are assumed to arise from a normal population.
2. The measurement system is linear (i.e., interval or ratio scale).

FIGURE 8.13 Theoretical Sampling Distribution for Two Populations

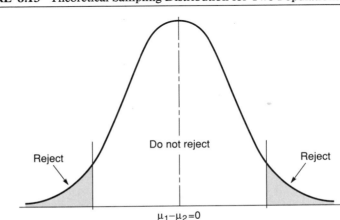

3. Each measure is independent and exclusive of other measures being used.
4. The variances are homogeneous.

Suppose that the director of human resources at Sky High Aircraft wishes to determine whether men and women are paid the same for equal positions of responsibility. The director is especially interested in potential compensation differences for department managers. The company currently has more than 800 department managers. The director has selected a random sample of 20 male managers and 20 female managers with approximately 10 years of experience. The data base is presented in Figure 8.14.

The director has established a null hypothesis that women (μ_2) make at least as much salary as men (μ_1); i.e., $H_0 : \mu_1 \leq \mu_2$. The director's alternative hypothesis is that men earn more than women (i.e., $H_1 : \mu_1 > \mu_2$.). The computations involved in a test of two populations are somewhat complex and lend themselves to computer analysis. A complete analysis of this problem can be found in Section 8.5.

Dependent Sample Designs (Matched Pairs)

The most common form of hypothesis testing for two populations involves independent samples. However, when the data can be grouped into pairs (e.g., "before" and "after"), a more efficient test can be implemented when the variation between the pairs is greater than the variation between data elements. This test procedure is called a *matched sample design*. The primary advantage of a matched design is that the variation between elements (e.g., workers) is eliminated as a source of sampling error. In such cases, the efficiency of the test or experiment is improved for a given sample size. Consider, for example, that an operations supervisor wishes to evaluate the

FIGURE 8.14 Sample of Managers' Salaries at Sky High Aircraft

Men		Women	
$45,500	$57,000	$41,500	$45,000
57,500	69,000	44,000	39,000
48,000	67,500	47,000	51,000
65,000	56,500	42,000	50,000
61,000	48,000	43,500	47,000
55,500	51,000	44,000	49,500
60,000	54,500	41,500	48,500
62,000	57,500	54,000	51,500
54,500	59,000	43,500	46,000
62,000	63,000	44,500	45,000
$\bar{X}_1 = \$57,700$		$\bar{X}_2 = \$49,900$	

effectiveness of two production methods. If the supervisor tests both methods using the same set of employees (i.e., elements), then the result is a paired experiment. Furthermore, if the differences in production times between the pairs (method 1 versus method 2) are larger than the differences in times among employees, then the test results will be more accurate. However, random pairing can yield misleading conclusions whenever the data are nearly homogeneous. Perhaps the most famous matched-pair design involves the use of twins in studying a variety of social problems.

A matched-pair test is conducted in exactly the same way as a group or independent test. The only difference involves the computation of the standard deviation. In a matched-pair test, the standard deviation is based on the sum of the differences between the average difference and each observed difference. Also, in a matched-pair design, the degrees of freedom are equal to $n - 1$, in contrast to $2(n - 1)$ for an independent design. Clearly, the reduction in degrees of freedom in a matched design is an important factor in selecting the appropriate design method.

The Dietfast Company is interested in mounting an advertising campaign to promote a newly developed diet. One of the key advertising elements is a forecast of potential weight loss. Before running the ad, the company's general manager has decided to conduct a trial program to determine the diet's actual effectiveness. A random sample of eight candidates was selected from a population of over 200 interested individuals. Each candidate was weighed before and after the 30-day diet. The results from the experiment are presented in Figure 8.15. The general manager wishes to avoid the possibility of false advertising and therefore has established a null hypothesis of no difference (i.e., the diet doesn't work). A CBS solution to this problem can be found in Section 8.5.

FIGURE 8.15 Survey Results from Diet Test Program

Candidate	Weight		Difference
	Before	After	
1	149	151	−2
2	172	167	5
3	216	210	6
4	185	178	7
5	137	136	1
6	115	112	3
7	155	149	6
8	108	106	2
		Total	28

$\overline{D} = 3.5$ pounds

FIGURE 8.16 Null and Alternative Hypothesis Options for Testing the Differences between Two Population Proportions

	Example
Option 1	
$H_0: p_1 - p_2 = 0$	The audit rate is the same for tax returns at two different IRS regional offices
$H_1: p_1 - p_2 \neq 0$	
Option 2	
$H_0: p_1 - p_2 \geq 0$	A larger proportion of male voters prefer the Republican candidate (P_1) to the Democratic candidate (P_2)
$H_1: p_1 - p_2 < 0$	
Option 3	
$H_0: p_1 - p_2 \leq 0$	The proportion of Americans living below the poverty line is lower this year (P_1) than three years ago (P_2)
$H_1: p_1 - p_2 > 0$	

Testing the Differences between Two Population Proportions

The process of testing the differences between two population proportions is similar to that of testing the difference between two means. The standard null and alternative hypothesis options for testing the difference between two proportions along with several examples are given in Figure. 8.16.

The theoretical sampling distribution for $p_1 - p_2$ is shown in Figure 8.17. Determination of the decision rule is based on the required level of significance. As in testing the difference between two means, if the sample statistic (i.e., $p_1 - p_2$) falls outside the critical limit(s), then the null hypothesis can be rejected. A CBS solution for this problem is given in Section 8.5.

FIGURE 8.17 Sampling Distribution of $\bar{p}_1 - \bar{p}_2$

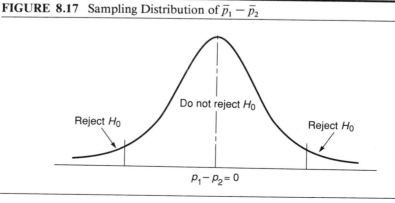

OPTIMIZING SAMPLE SIZE

The previous chapter introduced the idea of determining the size of the sample based on error considerations. That situation required the specification of alpha and the acceptable level of error (i.e., the difference between the population mean and the sample mean). This chapter has introduced a second kind of error (Type II or beta), which can be used along with the Type I error to determine the optimal sample size. As indicated in an earlier section, once the sample size has been established, any increase in alpha lowers beta, and any decrease in alpha raises beta. In this situation, the procedure is to specify maximum values for alpha and beta. In this way the sample size can be determined. The computational process for determining the sample size in this manner is somewhat complex and is best left for computer analysis. However, the following list outlines the basic steps involved in the computational process:

1. Specify the null hypothesis and alpha error level.
2. Specify the alternative hypothesis and beta error level.
3. Estimate the population standard deviation or proportion.

The CBS package contains a model for determining sample size. One word of caution, however: The resultant sample size may be outside the economic constraints associated with the given problem. That is, this approach may yield a sample size requirement that is too costly. In these cases, the maximum levels for alpha or beta may need to be increased depending upon which of the two errors is more critical.

Typically, time and money play an important role in determining the size of the sample. A balance should be found between the marginal benefits and the marginal costs of increasing sample size.

Figure 8.18 illustrates the relationship between sample size, alpha, and beta for the Iowa Department of Motor Vehicles problem. It clearly shows that the required sample size decreases for larger values of either alpha or beta and increases for smaller values of either alpha or beta.

TESTING FOR OTHER POPULATION PARAMETERS

The hypothesis testing procedures presented have involved the values of means and proportions. In some business situations, however, it may be more important to know the variability of a measure rather than the value of the population mean. Consider for example, a machine that is used for filling bags with granulated sugar. The filling process is designed to deliver 5 pounds of sugar per bag. Yet it is unlikely that each bag produced will contain exactly 5 pounds of sugar. In fact, if the filling process is not controlled properly, some bags may be overfilled, and others may be underfilled. The amount of

FIGURE 8.18 Analysis of Sample Size as a Function of Alpha and Beta

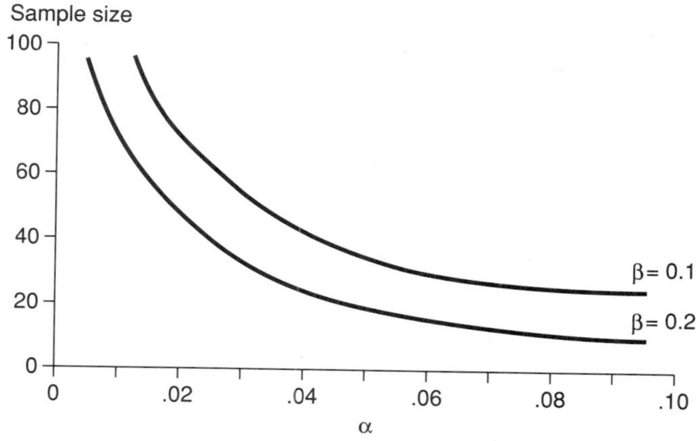

variability in the filling process can be determined by selecting a random sample of bags and measuring the actual amount of sugar in each bag. Guidelines for the amount of acceptable variance (i.e., decision rules) can be established for determining when the filling process requires adjustment. The process of performing hypothesis tests for variances requires a different type of analysis, called *chi-square*. A detailed presentation of the chi-square statistic and its application to testing variances is given in Chapter 12.

HYPOTHESIS TESTING LIMITATIONS

Like all other quantitative methods, the hypothesis testing process is not without its critics. The following discussion highlights several of the more significant ongoing criticisms of the use of hypothesis testing.

Insignificant Results

Researchers do not usually report insignificant results of hypothesis testing. But because most researchers are looking for significant results, it becomes quite possible — indeed probable — that they will report the results from any given study as significant (i.e., justifying rejection of the null hypothesis) even though the null hypothesis is true. This situation, of course, represents a Type I error. Suppose, for example, that six independent research teams conducted a similar experiment with a .05 level of significance. The chances that a single team would report statistical significance when it is not justified is 5%. On the other hand, the chances that at least one team would publish a false conclusion is approximately 27% [i.e., $1 - (.95)^6$]. It is not uncommon in today's complex world for several researchers to be studying the same phenomenon. The relatively large probability that a false conclusion will be reported while the correct conclusion from similar experiments goes unreported is the source of much concern among researchers and managers alike.

Violation of Assumptions

The process of hypothesis testing, as outlined in this chapter, is based on a set of assumptions. Specific examples include accurate identification of the target population, random sampling of the target population, and a normal population when the sample size is small. Whenever one or more of these assumptions is violated, the decision to accept or reject the null hypothesis can be erroneous.

Avoidance of Power

The concept of power in hypothesis testing is a difficult one to understand. Recall that power is associated with the probability of correctly rejecting the null hypothesis. Traditionally, Type II errors have not received the same level of attention as Type I errors. A quick review of the trial scenario presented earlier should underscore this point. (Generally speaking, society is much more concerned with avoiding a Type I error than a Type II error.) Consequently, power results are often neglected. This sometimes leads to an unfortunate outcome, particularly when the true population value is close to the value specified in the null hypothesis. In these cases, the null hypothesis may very well be accepted when the alternative hypothesis is actually true.

STATISTICAL QUALITY CONTROL

Key idea 5

Classical hypothesis testing usually involves decision making based on a single sample. In these cases, the analysis provides only a "snapshot" of the situation under investigation. This approach is usually adequate for processes that are not changing significantly in the short term, such as employee compensation or automotive fuel economy. Nevertheless, there are many operational situations which, in fact, can change dramatically in a short period of time. For example, a manufacturing line can be producing items that meet product specifications during one period (e.g., shift, week, month) and yet not meet them during another period. Clearly, making a determination on product quality based on a fixed single random sample will not reveal the full extent of any potential problems. What is needed, in these cases, is an ongoing sampling process. This process is generally known as statistical quality control (SQC) and can be viewed as complementary to classical hypothesis testing.

SQC involves continuous sampling to determine whether the process under investigation is operating within acceptable limits. Adjustments can be made to the process when the sampling results indicate that the process is operating outside these limits. The fact that a production or service process can rapidly "wander" in and out of control underscores the need for continuous sampling. Typically, control charts are used to monitor the quality status of the process. The concept behind the design of control charts was introduced in Chapter 3. Recall that control charts present a plot of the sample data, the center line (average), and the control limits. There are a variety of control charts used for monitoring product or service quality. Two of the most popular are attribute and variable charts. Presented in the bibliography at the end of this chapter are several texts on statistical quality control.

p-Charts

An attribute control chart, called a p-chart, is used when the measurements involve a specific characteristic such as color. Typically, these charts show the proportion of items sampled that are defective. To illustrate how to develop a p-chart, consider that the production manager at Wilcox Electric has collected the data shown in Panel A of Figure 8.19. These data report the number of defectives found in 15 samples taken over a five-day period. Each

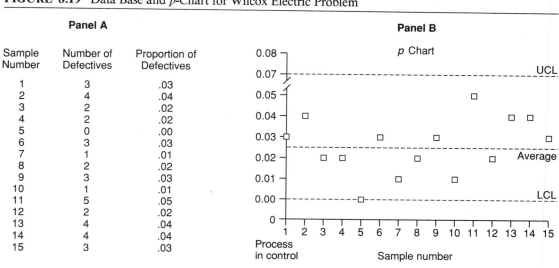

FIGURE 8.19 Data Base and p-Chart for Wilcox Electric Problem

Panel A

Sample Number	Number of Defectives	Proportion of Defectives
1	3	.03
2	4	.04
3	2	.02
4	2	.02
5	0	.00
6	3	.03
7	1	.01
8	2	.02
9	3	.03
10	1	.01
11	5	.05
12	2	.02
13	4	.04
14	4	.04
15	3	.03

sample contains 100 finished microchips. A *p*-chart featuring the individual sample results, the average defective level, and the control limits is shown in Panel B of Figure 8.19. The overall fraction defective *(p)*, or center line, is simply the total number of defectives divided by the product of the number of samples and the sample size. The lower and upper control limits are determined from the standard error of the proportion and the specified confidence level. The most common confidence limits used in statistical quality control are based on three standard errors or approximately 99.7%. These results show that the process is in control over the sample range. That is, each sample proportion falls within the control limits. However, this does not suggest that the process will always remain in control. This latter observation underscores why continuous sampling is essential for these types of applications.

$\bar{X}$ and R Charts

$\bar{X}$ and R charts are used to monitor processes involving variable measures. Examples include weight, length, and volume. More specifically, $\bar{X}$ charts are used to monitor the central tendency of the process, while R charts provide an indication of the process dispersion. As in the case of *p*-charts, control limits are determined based on the sampling distribution. Like classical hypothesis testing, concluding that a process is out of control when it is not is a Type I error, while concluding that the process is in control when it is not, is a Type II error. Usually, a Z value of 3 is used in constructing $\bar{X}$ and R charts. Suppose the production manager at Farey Foods is interested in determining whether the two-pound sugar filling machine is operating within limits. Panel A of Figure 8.20 presents the results of a random sample taken over a five-day period. Each sample, which reports the filling weight in pounds, contained four observations. $\bar{X}$ and R charts for this data are given in Panel B. Again, the center line for each chart represents the average of the mean and the range values. The lower and upper limits are based on $\pm$ three standard deviations and the computer standard error of the mean (range). These results show that the filling process is out of control, since the #8 sample on the $\bar{X}$ chart is above the upper control limit. Thus in standard hypothesis testing, the manager can reject the null hypothesis that the process is in control and accept the alternative hypothesis that it is out of control. Typically, if either the $\bar{X}$ chart or the R chart shows that one or more sample values fall outside the control limits, one can conclude that the process is out of control. At this point the process should be investigated to determine the probable causes for this situation.

8.5 COMPUTER ANALYSIS

Key idea 6

Although the actual computations for hypothesis testing are relatively straightforward, the use of computer analysis provides a standard format for processing the large number of possible testing options. Furthermore, com-

FIGURE 8.20 Data Base and $\bar{X}$ and R Charts for Farey Foods Problem

Panel A

Sample Number	Observation 1	2	3	4
1	1.98	1.98	1.98	1.98
2	1.97	1.97	1.97	1.97
3	1.96	1.96	1.96	1.96
4	2.02	2.02	2.02	2.02
5	1.95	1.95	1.95	1.95
6	2.02	2.02	2.02	2.02
7	1.94	1.94	1.94	1.94
8	2.04	2.04	2.04	2.04
9	1.96	1.96	1.96	1.96
10	2.01	2.01	2.01	2.01

Panel B

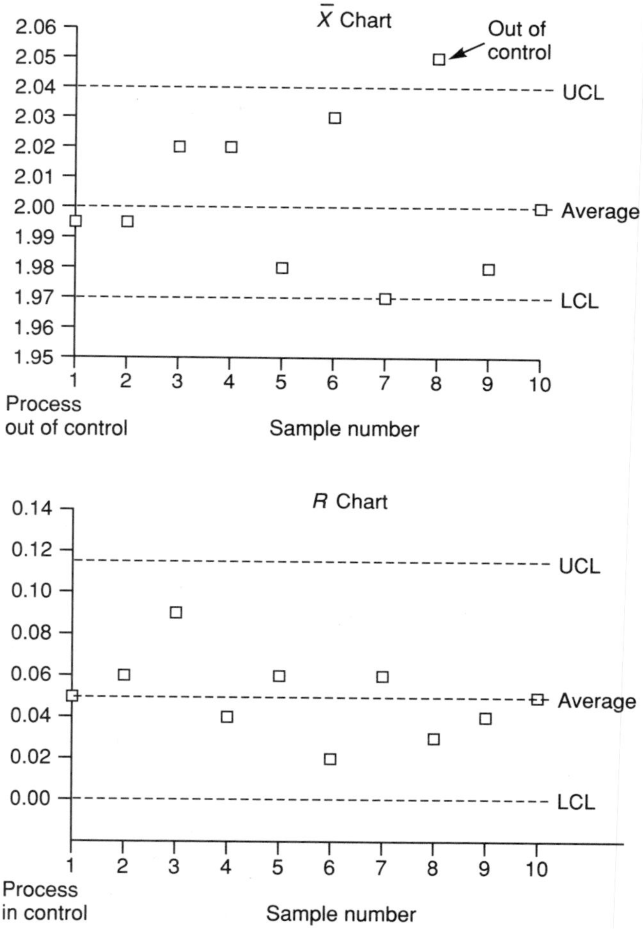

$\bar{X}$ Chart — Process out of control

R Chart — Process in control

puter analysis is particularly helpful in developing OC and power curves for a given application.

Iowa Department of Motor Vehicles (Testing One Population Mean)

A CBS analysis of the Iowa Department of Motor Vehicles problem is given in Figure 8.21. Recall that the director was interested in determining whether the average waiting time in the new Des Moines office exceeded 10 minutes. This problem involved an upper-tail one-sided test (null hypothesis, $\mu \leq 10$

FIGURE 8.21 CBS Hypothesis Testing of IDMV Problem

Information Entered

Test procedure	One-sided (U)
Alpha error	.0500
Test statistic (critical Z value)	1.6500
Null hypothesis	10
Sample size	30
Sample mean	11.3667
Standard deviation	4.1397

Results

Standard error of mean	.7558
Null hypothesis	10
Upper limit	11.2471
Sample mean	11.3667
Alpha error	.0500
Degrees of freedom	29
Test statistic (critical Z value)	1.6500
Computed Z value	1.8083
p-value	.0351
Conclusion	Reject null

FIGURE 8.22 CBS Hypothesis Testing for Government Contracting

Information Entered

Test procedure	One-sided (L)
Alpha error	.0500
Test statistic (critical t-value)	1.7110
Null hypothesis	.2500
Sample size	25
Sample proportion	.1200

Results

Standard error of mean	.0866
Lower limit	.1018
Null hypothesis	.2500
Sample proportion	.1200
Alpha error	.0500
Degrees of freedom	24
Test statistic (critical t-value)	1.7110
Computed t-value	−1.5011
p-value	.0750
Conclusion	Do not reject null

minutes). The analysis shows that the null hypothesis can be rejected. That is, the computed Z-value (1.81) exceeded the critical Z-value (1.65) at the .05 level. Therefore, the director can conclude that the average waiting time does exceed 10 minutes. Furthermore, notice that the reported p-value is smaller than α. Accordingly, the null hypothesis can also be rejected based on this observation.

HHS Department Contracting (Testing One Population Proportion)

A CBS analysis for the HHS Department contracting problem is given in Figure 8.22. These results show that the null hypothesis cannot be rejected even though the difference between the population and sample mean is 13 percentage points. This somewhat curious result can be attributed to the relatively small sample size, which yielded a relatively large standard error of the mean (.08). For example, if the sample size were doubled (i.e., $n = 50$), then the conclusion would be to reject the null hypothesis. However, with the current sample the secretary can only conclude that the contracting procedures used by HHS are meeting the minority guidelines. This conclusion is also supported by the fact that the reported p-value is larger than the specified α.

Sky High Aircraft (Testing Two Population Means)

A CBS analysis for the Sky High Aircraft problem is given in Figure 8.23. These results show that the null hypothesis that men and women are paid the same can be rejected. This conclusion is based on the fact that the computed t-value is greater than the critical t-value (7.00 versus 1.65). Therefore, the director can conclude that men and women are not paid the same for equal work since the difference in salary is statistically significant. In fact, the salary difference of $11,800 between the two sexes is the best estimate for the population difference within the company. The same conclusion can be drawn (i.e., reject H_0) when comparing the computed p-value with the given α.

Dietfast Company (Testing Matched and Unmatched Means)

A CBS analysis of the Dietfast problem for both matched and unmatched cases is given in Figure 8.24. These results show that the null hypothesis of no difference can be rejected for the matched-pair case; the computed t-value is greater than the critical t-value. However, the same conclusion cannot be drawn when the problem is analyzed as two independent groups (i.e., unmatched pairs). In that case there is insufficient evidence to reject the null hypothesis. The computed t-value of 0.199 is clearly less than the critical t-value of 2.145. These results should not be surprising based on a comparison of the two standard errors. Notice that the standard error for the group

FIGURE 8.23 CBS Hypothesis Testing for Sky High Aircraft

Information Entered

Test procedure (unequal variances)	One-sided (U)
Alpha error	.0500
Test statistic (critical *t*-value)	1.6800
Null hypothesis	0
Sample size for group 1	20
Sample size for group 2	20
Sample mean for group 1	57,700
Sample mean for group 2	45,900
Standard deviation for group 1	6,362.7202
Standard deviation for group 2	3,878.4153

Results

Standard error of mean	1,666.2280
Null hypothesis	0
Upper limit	2,799.2629
Mean 1 − Mean 2	11,800
Alpha error	.0500
Degrees of freedom	38
Test statistic (critical *t*-value)	1.6800
Computed *t*-value	7.0819
p-value	.0001
Conclusion	Reject null

FIGURE 8.24 CBS Analysis of Matched and Independent Tests for Dietfast

	Matched	Unmatched
Standard error of mean	1.0856	17.5859
Lower limit	−2.5675	−37.7217
Null hypothesis	0	0
Upper limit	2.5675	37.7317
Mean 1 − Mean 2	3.50	3.50
Alpha error	.05	.05
Degrees of freedom	7	14
Test statistic (critical *t*-value)	2.365	2.145
Computed *t*-value	3.224	0.199
p-value	.0075	.4200
Conclusion	Reject null	Do not reject null

case is approximately 17 times that for the matched-pair case. Which conclusion is correct? In this problem, the results for the matched case are more representative of the impact of the diet plan. Therefore, the company's manager can conclude that the diet does work.

8.6 PRACTICAL APPLICATIONS

As this chapter has indicated, there are many different hypothesis-testing situations (e.g., one or two populations, one- or two-sided testing, and small versus large sample size). This section presents several applications involving one or more of these cases.

Worldwide Insurance (Testing Two Population Means)

The vice president of human resources at Worldwide Insurance wishes to examine turnover as it relates to management style. More specifically, she is interested in determining the impact of both autocratic and democratic leadership styles on employee turnover. Following the standard approach to hypothesis testing, she first states the null (H_0) and research (H_1) hypotheses as follows:

- H_0: Employees working for autocratic leaders will show the same level of turnover as employees working for democratic leaders.
- H_1: Employees working for autocratic leaders will show a higher level of turnover than employees working for democratic leaders.

The vice president has decided to use a .05 level of significance in analyzing this problem. Next, she selects a random sample of employees that left within three years after joining the company. The sample records the length of service at termination (years) and the employee's last supervisor. These data are summarized in Figure 8.25.

A CBS analysis of this problem is given in Figure 8.26. The results show that there is a significant difference between the two leadership styles as measured by length of service. That is, employees tend to stay longer if they are supervised by managers with a democratic style. Of course, determining whether a manager is strictly democratic or autocratic represents a serious measurement challenge. Often the measurement issue is more complex than the actual experiment.

FIGURE 8.25 Results from Worldwide Ex-Employee Survey

	Leadership Style	
	Autocratic	*Democratic*
Mean	1.2	2.5
Standard deviation	0.5	0.7
Sample size	50	65

FIGURE 8.26 CBS Hypothesis Testing for Worldwide Insurance

Information Entered

Test procedure (unequal variances)	Two-sided
Alpha error	.0500
Test statistic (critical Z value)	1.9600
Null hypothesis	0
Sample size for group 1	50
Sample size for group 2	65
Sample mean for group 1	1.2000
Sample mean for group 2	2.5000
Standard deviation for group 1	0.5000
Standard deviation for group 2	0.7000

Results

Standard error of mean	0.1120
Lower limit	−0.2195
Null hypothesis	0
Upper limit	0.2195
Mean 1 − Mean 2	−1.3000
Alpha error	.0500
Degrees of freedom	113
Test statistic (critical Z value)	1.9600
Computed Z value	−11.6097
p-value	.0001
Conclusion	Reject null hypothesis

Internal Revenue Service (Testing Two Population Proportions)

The manager of the Internal Revenue Service's Western Division is interested in determining if there is a difference between the percentage of audits conducted by the division's Northern and Southern offices. The manager selected a random sample of 100 income tax returns from the Northern office and 100 tax returns from the Southern office. The manager's null and alternative hypotheses can be expressed as follows:

- H_0: The audit rate is the same for personal tax returns at the two offices.
- H_1: The audit rate is different for personal tax returns at the two offices.

This problem assumes that the variances from the two populations are the same. The sample data showed audit rates of 8% and 11%. A CBS analysis of these sample data, using a two-sided test procedure with a 5% level of significance, is given in Figure 8.27. These results show that the null hypothesis of

FIGURE 8.27 CBS Hypothesis Testing for IRS Audit

Information Entered

Test procedure (equal variances)	Two-sided
Alpha error	.0500
Test statistic (critical Z value)	1.9600
Null hypothesis	0
Sample size for group 1	100
Sample size for group 2	100
Sample proportion for group 1	.0800
Sample proportion for group 2	.1100

Results

Standard error of mean	0.0386
Lower limit	−0.0756
Null hypothesis	0
Upper limit	0.0756
Proportion 1 − Proportion 2	−.0300
Alpha error	.0500
Degrees of freedom	248
Test statistic (critical Z value)	1.9600
Computed Z value	−.7782
p-value	.2277
Conclusion	Do not reject null

no difference cannot be rejected even though there is a difference of 3 percentage points in audit rates. The manager can conclude that the two offices are auditing approximately the same percentage of tax returns.

8.7 CASE STUDY: HEARTWELL MUSIC EXPERIMENT

The Heartwell corporation manufactures and markets pacemakers. The operations vice president, Mr. Steve Growfaster, is concerned about increasing production so he calls a meeting of the production staff to discuss possible strategies for improving productivity. Steve opens the meeting by stating that product quality must remain at its current high level regardless of any new production strategies. Bill Heartwell, the son of the owner and part-time employee, suggests that Steve might wish to consider a somewhat nontraditional approach to the problem. He cites the results of the famous Hawthorne lighting experiment and indicates that a similar approach might help solve Steve's problem. Quite naturally, Steve is intrigued by Bill's suggestion. He asks Bill to prepare a plan and report back in one week.

FIGURE 8.28 Productivity Levels as a Function of Music Type

Observation	No Music	Classical Music	Popular Music
1	793	867	850
2	965	952	775
3	652	735	825
4	817	908	702
5	775	854	936
6	802	893	739
7	792	798	865
8	743	919	893
9	845	898	950
10	853	904	965

At the next meeting of the production staff, Mr. Growfaster calls on Bill to present his plan. Bill indicates that he believes production could be increased by providing music in the production bays. Ms. Susan Smith, Steve's assistant, is somewhat shocked by this proposal and asks Bill for details. He responds by stating that an experiment would be required to determine the effect of these strategies on production and that such an experiment would take about 20 working days. Steve intervenes in the discussion and authorizes Bill to proceed with his plan. Bill has collected first-shift production data over the past 10 weeks. This data base is to serve as the control group. He then exposes the same group to 10 days of classical music, followed by 10 days of popular music. The numerical results from this experiment are given in Figure 8.28.

Bill decides to analyze these data using the following hypothesis:

Classical music   Popular music

$H_0: \mu_c = \mu_n$ $H_0: \mu_p = \mu_n$

$H_1: \mu_c \neq \mu_n$ $H_1: \mu_p \neq \mu_n$

where:

μ_n = Population production with no music
μ_c = Population production with classical music
μ_p = Population production with popular music

A CBS analysis of the classical music hypothesis is presented in Figure 8.29. The results of this analysis indicate that classical music has an impact on production. That is, the computed t-value (2.1298) exceeds the critical t-value of 2.101, which leads to the rejection of the null hypothesis of no difference. Therefore, Bill can conclude that classical music does seem to have a positive effect on production. An analysis of the hypothesis for the

FIGURE 8.29 CBS Hypothesis Testing for Classical Music Option

Information Entered

Test procedure (unequal variances)	Two-sided
Alpha error	.0500
Test statistic (critical *t*-value)	2.1010
Null hypothesis	0
Sample size for group 1	10
Sample size for group 2	10
Sample mean for group 1	872.8000
Sample mean for group 2	803.7000
Standard deviation for group 1	63.7997
Standard deviation for group 2	80.3507

Results

Standard error of mean	32.4448
Lower limit	−68.1665
Null hypothesis	0
Upper limit	68.1665
Mean 1 − Mean 2	69.1000
Alpha error	.0500
Degrees of freedom	18
Test statistic (critical *t*-value)	2.1010
Computed *t*-value	2.1298
p-value	.0225
Conclusion	Reject null

popular music option is given in Figure 8.30. These results show that the null hypothesis cannot be rejected. The reported *p*-value indicates that there is a 12.25% probability that the sample mean occurred by chance given that the null hypothesis is true.

8.8 SUMMARY

Hypothesis testing is fundamental to statistical inference. It involves many different assumptions and test procedures. Fortunately the testing of hypotheses can usually be conducted by the following five-step process:

1. Stating a null and alternative hypothesis
2. Establishing the level of acceptable error
3. Formulating a decision rule
4. Collecting a sample from the target population
5. Making a decision based on the sample results

FIGURE 8.30 CBS Hypothesis Testing for Popular Music Option

Information Entered

Test procedure (unequal variances)	Two-sided
Alpha error	.0500
Test statistic (critical *t*-value)	2.1010
Null hypothesis	0
Sample size for group 1	10
Sample size for group 2	10
Sample mean for group 1	850
Sample mean for group 2	803.7000
Standard deviation for group 1	90.1665
Standard deviation for group 2	80.3507
Alpha error	.0500

Results

Standard error of mean	38.1919
Lower limit	−80.2412
Null hypothesis	0
Upper limit	80.2412
Mean 1 − Mean 2	46.3000
Alpha error	.0500
Degrees of freedom	18
Test statistic (critical *t*-value)	2.1010
Computed *t*-value	−1.2123
p-value	.1225
Conclusion	Do not reject null

Two types of errors can occur in hypothesis testing. A Type I error is the rejection of the null hypothesis when it is true. A Type II error is the acceptance of the null hypothesis when it is false. Type I errors can be controlled through the selection of the level of significance (i.e., alpha). Type II errors are usually controlled by increasing the sample size.

Sample size also plays a role in selecting the appropriate test procedure. For hypotheses involving large samples (i.e., $n \geq 30$) the standard Z-test analysis is used. The *t*-test is used for small sample ($n < 30$) applications. The hypothesis-testing procedures introduced in this chapter can be used for analyzing one or two populations. In the latter case, the *standard* null hypothesis is that the difference between the means of the two populations is equal to zero.

Although the actual computations are quite straightforward, selecting the appropriate test procedure can be somewhat confusing. The following list summarizes the basic options associated with hypothesis testing:

- One-sided or two-sided test procedure.
- Population or sample variance.
- Means or proportions.
- One or two populations.

Two alternative sample designs are available for tests involving two populations. One is based on two independent samples, and the other is based on a single matched sample.

This chapter also identified several limitations associated with hypothesis testing. These included insignificant results, violation of assumptions, and avoidance of power. Additionally, three different approaches to hypothesis testing were introduced: classical method, standard scores, and *p*-values.

8.9 GLOSSARY

alpha The probability of a Type I error.

alternative hypothesis A statement about a population parameter that is assumed to be true if the null hypothesis is proven false.

beta The probability of a Type II error.

critical value A statistic used to determine whether to reject or not reject the null hypothesis.

decision rule A statement used for making judgments regarding a given hypothesis.

dependent sample design (matched pairs) A testing procedure where two dependent samples are obtained from two populations.

hypothesis A statement about a population parameter.

independent sample design (unmatched pairs) A testing procedure where two independent samples are obtained from two populations.

null hypothesis A statement about a population parameter that is assumed to be true until proven false.

one-tailed test A test procedure involving either the lower or the upper end of the sampling distribution (i.e., one-sided).

operating characteristic (OC) curve A graph showing the probability of accepting the null hypothesis for various values of the population parameter.

power The probability of rejecting the null hypothesis.

power curve A graph showing the probability of rejecting the null hypothesis for various values of the population parameter.

***p*-value** The probability that the sample mean or test statistic will be at least as great (smaller or larger) as the observed mean or test statistic, given that the null hypothesis is true.

research hypothesis Equivalent to the alternative hypothesis.

two-tailed test A test procedure involving both the lower and upper ends of the sampling distribution (i.e., two-sided).

type I error Rejecting the null hypothesis when it is true.

type II error Accepting the null hypothesis when it is false.

8.10 BIBLIOGRAPHY

Enns, P. G. *Business Statistics: Methods and Applications.* Homewood, Ill.: Irwin, 1985.

Gitlow, H. *Tools and Methods for the Improvement of Quality.* Homewood, Ill.: Irwin: 1989.

Moore, D. S. *Statistics: Concepts and Controversies.* San Francisco: W. H. Freeman Co., 1979.

Peters, W. S. *Counting for Something.* New York: Springer-Verlag, 1987.

8.11 PROBLEMS

1. Test the following null and alternative hypotheses when the population standard deviation is 4 and the sample size is 50:

$$H_0: \mu = 20 \qquad H_1: \mu \neq 20$$

  a. For $\overline{X} = 18$
  b. For $\overline{X} = 16$
  c. For $\overline{X} = 21.5$

2. Test the following null and alternative hypotheses when the population standard deviation is 200 and the sample size is 25:

$$H_0: \mu \geq 1{,}000 \qquad H_1: \mu < 1{,}000$$

  a. For $\overline{X} = 1{,}100$
  b. For $\overline{X} = 750$
  c. For $\overline{X} = 1{,}350$

3. Formulate appropriate null and alternative hypotheses for each of the following business situations:
  a. The personnel manager at General Motors believes that a training program will reduce absenteeism below the current 2% of the work force per week.
  b. The vice president of marketing at Goodyear claims that the company's new radial tires will last for at least 30,000 miles.
  c. The quality control inspector at IBM will authorize shipment of a new batch of computer chips if the defective rate is less than 3%.
  d. The president of the AFL-CIO predicts that union membership will not fall below 12% of the total work force.

4. Characterize the Type I and Type II errors associated with each of the examples in problem 3.

5. Formulate appropriate null and alternative hypotheses for each of the following public policy situations:
  a. Government economists believe that the gross national product will not grow by more than 4% for 1989.
  b. The director of the National Safety Council believes that increasing the speed limit to 65 mph on certain rural routes will increase the traffic death rate.
  c. The director of the National Health Institute has indicated that a cure for AIDS will be discovered by 1995.
  d. The head of the Environmental Protection Agency stated that smog in Los

Angeles can be reduced by 30% through the introduction of subways and car pooling.

6. Characterize the Type I and Type II errors associated with each of the examples given in problem 5.
7. What roles do alpha and beta errors play in hypothesis testing? How can both of these errors be controlled?
8. Characterize a jury trial in terms of the potential for Type I and Type II errors.
9. Identify which of the following business situations involve one-sided testing and which involve two-sided testing. In the case of a one-sided test, indicate the direction.
    a. Ball bearings designed to fit into a housing
    b. The life of a light bulb
    c. The demand for a new product
    d. The mean time to failure for a computer system
10. Conduct a hypothesis test based on the following information:

    $H_0: \mu \geq 100$ $\bar{X} = 92.5$
    $H_1: \mu < 100$ $s = 3.5$
    $n = 40$ $\alpha = .05$

11. Conduct a hypothesis test based on the following information:

    $H_0: \mu = 40$ $\bar{X} = 55$
    $H_1: \mu \neq 40$ $s = 5.3$
    $n = 15$ $\alpha = .05$

12. Conduct a hypothesis test based on the following information:

    $H_0: p \geq 0.40$ $\bar{p} = 0.35$ $\alpha = 0.05$
    $H_1: p < 0.40$ $n = 50$

13. Conduct a hypothesis test based on the following information:

    $H_0: p = 0.50$ $\bar{p} = 0.55$ $\alpha = 0.01$
    $H_1: p \neq 0.50$ $n = 30$

14. Conduct a hypothesis test based on the following information:

    $H_0: p \leq 0.60$ $\bar{p} = 0.65$ $\alpha = 0.02$
    $H_1: p > 0.60$ $n = 25$

15. The Department of Housing and Urban Development estimates that the proportion of families with incomes below the poverty level, i.e., $12,000 per year, is 0.15. A recent survey of 40 families in the Denver area revealed that 8 fell below the $12,000 standard. Do the survey results support the HUD estimates at the 0.05 level?
16. The American Bar Association (ABA) estimates that the proportion of candidates that will pass the bar examination on the first try will not exceed 50%. A recent survey of 30 candidates from the Atlanta area revealed that 17 had passed the examination. Do the survey results support the ABA estimates at the 0.01 level?

17. The Environmental Protection Agency (EPA) estimates that at least 30% of the vehicles on the road exceed the EPA's emission standards. A recent survey of 40 vehicles from the New Jersey area revealed that 10 failed the EPA's emission test. Do the survey results support the EPA estimates at the 0.05 level?
18. The Surgeon General of the United States estimates that no more than 5% of the adult population is infected with the AIDS virus. A recent survey of 40 adults in the New York area revealed that 6 were infected with the virus. Do the survey results support the Surgeon General's estimates at the 0.02 level?
19. Discuss the role of p-values in hypothesis testing.
20. Identify which of the following cases would result in rejection of the null hypothesis.
    a. $p = .10, \alpha = .05$ b. $p = .001, \alpha = .10$
    c. $p = .05, \alpha = .05$ d. $p = .02, \alpha = .05$
21. Determine the p-values for the following cases:
    a. $H_0: \mu = 10, \bar{X} = 9, s = 1, n = 30, \alpha = .05$
    b. $H_0: p = .3, \bar{p} = .35, n = 25, \alpha = .05$
    c. $H_0: \mu = 100, \bar{X} = 90, s = 4, n = 20, \alpha = .01$
22. The national average annual income of physicians in this country is $100,000 with a standard deviation of $20,000. The AMA is interested in determining whether female physicians make less than the national average. The AMA wishes to use a 5% level of significance.
    a. State the null and alternative hypotheses.
    b. Characterize the nature of the Type I and Type II errors.
    c. State the decision rule.
    d. A sample of 100 female physicians was selected. The sample mean was computed at $82,000. What conclusions can be drawn?
23. The credit manager at Sears wishes to determine whether or not the mean unpaid balance has shifted from last year's level of $500. The manager selects 100 accounts at random and finds that the average account level is $515 with a standard deviation of $25. What conclusions can the manager make regarding the sample results at a .01 level of significance?
24. The director of admissions at Long Beach Hospital believes that patient residence time (i.e., the length of stay in the hospital) has been dropping over the last several years. Two years ago the average residence time was four days. The director examines the records of 40 recently released patients and finds that the average residence time was 3.5 days with a standard deviation of 1.2 days. What conclusions regarding patient residence time can be drawn from the survey results at a 90% level of confidence?
25. The program director at *TV Guide* wishes to determine if the average number of viewing hours has changed appreciably over time. The average number of television viewing hours five years ago was 25.5. The director undertakes a random survey of 300 families, which yields a mean viewing time of 32.7 hours with a standard deviation of 11.1 hours. What conclusions can the director make regarding viewing patterns at a .05 level of significance?
26. KNBC of Los Angeles advertises that its five o'clock news program is watched by more viewers than any other news show in town. Company managers have told their sponsors that 30% of the viewing audience tune in to the news program. However, the program's director is unsure of this estimate because of the recent introduction of cable TV news. Accordingly, the director has commissioned a

survey to estimate the actual viewing rate. A telephone survey of 350 viewers during the news broadcast reveals that 100 are watching the program.
   a. Formulate the null and alternative hypotheses for this problem.
   b. Describe the nature of both the Type I and Type II errors.
   c. If the director sets alpha at .05 and beta at .15, is the sample size adequate?
   d. Can KNBC's claim of 30% viewership be supported by the survey results?

27. The accounting firm of Smith & Cline has been retained by the Banc International to audit the bank's savings accounts. Smith's managing partner has sent out a questionnaire to 300 savings account customers. The response from the survey shows that 9% of those surveyed disagree with the balance reported by the bank (i.e., there was more than a $2 difference).
   a. Develop point and interval estimates on the proportion of customers who disagree with the bank (use a 95% level of confidence).
   b. Bank management claims that the actual amount of disagreement is less than 4%. Do the results from the survey support the bank's claim?

28. The owner of Rendondo Beach Ford has been told by her general manager that the average profit from new car sales has averaged $300 over the last two years. The owner is concerned that the average profit has been decreasing over the last year due to the impact of imports. She wishes to inspect the records of 10 sales selected randomly over the past six months. These data are given in the following table:

Vehicle	Net Profit	Vehicle	Net Profit
LTD	$325	Aerostar	$225
Taurus	175	Tempo	375
Thunderbird	225	Taurus	250
Taurus	190	Mustang	275
Mustang	255	LTD	400

What conclusions can the owner draw from this sample at a .05 level of significance?

29. The FDA has limited the dosage of a new cholesterol-reducing drug to 100 mg. A research analyst at the FDA has randomly selected 50 tablets of the new drug for quality testing. The test results revealed a sample mean of 100 mg with a standard deviation of 3 mg. Find the p-value for this quality assurance test.

30. The executive MBA program director at a local university believes that the average compensation for potential candidates is at least $100,000. The director interviewed 30 prospective students and found the sample mean to be $110,000 with a standard deviation of $15,000. Determine the corresponding p-value.

31. A preliminary assessment of the 1988 presidential campaign by the national media placed the race as dead even. The *Los Angeles Times* decided to undertake a survey to improve this subjective estimate. The results from a representative sample of 200 Californians indicated that 52% would vote Democratic and 48% would vote Republican. What conclusion regarding the outcome of the race should have been drawn by these survey results at the 0.05 level? Should the results from California have been used to estimate national trends?

32. The U.S. Department of Agriculture wishes to determine the impact of two new fertilizers on wheat production. Both fertilizers were tried on seven adjacent five-acre plots. The following production data (measured in bushels per acre) were obtained at the end of the growing season.

Fertilizer	1	2	3	4	5	6	7
A	54.5	56.7	50.3	55.5	61.2	59.8	53.2
B	49.8	46.7	51.3	52.1	48.7	44.9	50.2

  a. Develop appropriate null and alternative hypotheses.
  b. What conclusions can be drawn at the .05 level?
  c. What other factors might influence the observed differences in crop yield?

33. The following data were collected from two populations with equal variances.

Sample	Sample Size	Mean	Standard Deviation
1	22	101.5	8.9
2	27	96.5	10.1

  a. Formulate null and alternative hypotheses.
  b. Develop an appropriate decision rule at the .01 level.
  c. What conclusions can be drawn?

34. The following independent sample data were collected from two normal populations.

Sample	Sample Size	Mean	Standard Deviation
1	11	22.3	2.6
2	14	25.1	2.1

  a. Formulate null and alternative hypotheses.
  b. Develop an appropriate decision rule at the .05 level.
  c. What conclusions can be drawn?

35. What factors help to make a paired comparison more efficient than a group comparison?

36. The quality control manager at the Todd Lighting Company is concerned about the growing rate of defective light bulbs. Currently, the defective rate is approximately 6% of the production rate (i.e., 6 per 100 produced). The manager has decided to conduct a training program to improve product reliability among the production staff. The manager has randomly selected five production operators to attend a special two-week course on improving product quality. Subsequent to the course, the operator's performance was recorded over a 30-day period. The "before" and "after" data are given below:

| | Defect Rate (per 100) | |
Operator	Before	After
1	6	4
2	4	4
3	7	6
4	5	3
5	6	3

    a. Did the two-week course significantly reduce the defective rate at the .01 level?
    b. What size sample is needed to demonstrate significance at the .01 level?

37. The language department at Midwest University has recently designed a computer-aided course to help improve the teaching of language skills. A random sample of seven second-year Spanish language students were selected to measure the effectiveness of the new teaching aid. Each student was given a standardized test followed by a two-week exposure to the language instruction courseware. The students were then retested using a similar test instrument. The results from the two examinations are:

| | Test Performance | |
Student	Before	After
A	65	82
B	71	73
C	77	74
D	81	84
E	69	77
F	85	86
G	75	61

    a. Does the new courseware improve test performance at the .05 level?
    b. How would the results change if only the first five students had been used in the experiment?

38. The production supervisor at Browning Transmission is considering the adoption of a new transmission welding method. However, before proceeding with this project, the supervisor has decided to conduct a test to determine whether the new machine results in significantly lower production times. The supervisor has selected seven workers at random and has timed each worker in welding a transmission using both the current and new methods. The results of the test are reported in the following table. Determine if the new system is significantly different at the .01 level (assume that the samples were drawn from normally distributed populations).

	Completion Time (minutes)	
Worker	Current Method	Proposed Method
1	32	27
2	27	23
3	30	31
4	26	22
5	28	26
6	28	25
7	30	27

39. Develop and analyze an OC curve and a power curve for problem 10.
40. Develop and analyze an OC curve and a power curve for problem 11.
41. The managing partner at Wavel Executive Search is interested in analyzing the annual salary structure for regional vice presidents in the electronics industry. The partner randomly selects eight VPs currently placed in the South and eight VPs placed in the West. The salary data for the two groups are summarized in the following table:

Region	Candidate (annual salary)							
	1	2	3	4	5	6	7	8
South	79,500	63,250	67,500	81,250	94,000	73,750	84,500	92,000
West	89,400	94,500	105,000	98,000	110,500	89,250	94,500	118,500

Determine whether there is a difference between salary structures for the two groups of vice presidents at the .05 level under the following conditions:
  a. The population variances are equal.
  b. The population variances are unequal.

42. The head of the CIA has recently received some information on the test performance of a new Soviet intercontinental ballistic missile. The Soviets announced that the missile has a range of at least 6,000 miles. The data based on 15 test firings indicate a mean of 5,600 with a standard deviation of 800 miles. What conclusions can the CIA director draw at the .01 level?

43. A recent Pepsi versus Coke preference test conducted by an independent market research firm revealed average scores (on a scale of 0 to 100) of 66 and 57 and standard deviations of 15 and 8, respectively. The test involved two independent samples of 50 participants each. What did the test results show at the .05 level of significance?

44. The advertised delivery time for the Home Shopping Channel (HSC) is five working days. A recent sample of 40 orders yielded an average time of 6.5 days with a standard deviation of 1.2 days.
  a. What conclusions can be drawn with respect to the delivery claims of HSC at the .05 level of significance? At the .01 level?
  b. What is the probability of a Type II error if the actual delivery time equals the sample mean?
  c. What sample size is required to reduce the Type II error to 10% or less?

45. The assistant secretary of the U.S. Department of Transportation is interested in determining the potential impact on fuel economy of raising the maximum speed limit to 65 miles per hour. A recent pilot program in New Jersey measured the average fuel mileage of two fleets of 25 cars each. Both fleets used the standard urban driving cycle during the test program, with one fleet limited to a maximum speed of 55 mph and the other to a maximum speed of 65 mph. The results from the test program are reported in the following table:

Test Statistics	Maximum Speed Limit	
	55 mph	65 mph
Sample mean	25.6 mpg	23.1 mpg
Sample SD	1.6	1.8
Sample size	25	25

What conclusion can the assistant secretary draw at a .05 level of significance?

46. The superintendent of public instruction for the Los Angeles Unified School District recently received the following data on reading test scores for 12 randomly selected elementary schools within the district. The data show the average sixth-grade test scores for the 1986–1987 and 1987–1988 periods.

School	Reading Scores	
	1986–1987	1987–1988
1. 32nd Street	239	263
2. 36th Street	193	175
3. Toland Way	252	277
4. Trinity Way	193	166
5. 20th Street	189	190
6. 28th Street	190	189
7. 24th Street	202	204
8. Union Avenue	207	219
9. Utah Street	204	169
10. Vermont	232	213
11. Virginia Road	192	226
12. Westminster	282	278

a. What conclusions can the superintendent draw regarding the changes in test scores at the .01 level of significance?

b. What is the probability of committing a Type II error if the actual average reading score of the 1987–1988 period is 265?

47. California's statewide reading average for sixth-graders during the 1987–1988 period was 265. Do the sample data presented in problem 46 support this hypothesis at the .05 level of signficance?

48. The general manager at Henkiel Tool & Die has obtained the sample data presented below on total outgoing product quality (% defectives) from its two manufacturing plants in Bonn and Hannover, West Germany. The data were generated from observations taken over a 20-day period.

| | Plant | | | Plant | |
Observation	Bonn	Hannover	Observation	Bonn	Hannover
1	0.03	0.02	11	0.01	0.02
2	0.04	0.02	12	0.05	0.01
3	0.01	0.01	13	0.06	0.01
4	0.03	0.	14	0.01	0.02
5	0.02	0.03	15	0.01	0.
6	0.04	0.01	16	0.02	0.01
7	0.05	0.02	17	0.03	0.02
8	0.07	0.03	18	0.01	0.03
9	0.04	0.02	19	0.02	0.01
10	0.	0.01	20	0.04	0.01

   a. Formulate an appropriate null and alternative hypothesis.
   b. What conclusions can the general manager draw at the 0.05 level regarding the differences between the two plants?

49. The production manager at Los Negros Printing Co. has obtained the following summary sample data on outgoing product quality (% lot defectives) from its two printing shops in Sao Paulo, Brazil.

Plant	Sample Size	Percentage Defective
1	100	0.10
2	150	0.14

   a. Formulate an appropriate null and alternative hypothesis.
   b. What conclusions can the general manager draw at the 0.05 level regarding the differences between the two plants?

50. The health effects manager at Cornell Mines, located in southern Chile, has obtained the following sample data on air quality levels (CO %) from its two copper mines.

Observation	Mine 1	Mine 2
1	2.0	3.5
2	1.5	3.0
3	1.0	4.5
4	4.5	3.0
5	2.0	3.5
6	1.5	2.0
7	3.5	2.5
8	3.0	1.5
9	2.5	1.0
10	1.5	1.0
11	3.0	
12	3.5	

a. Formulate an appropriate null and alternative hypothesis.
b. What conclusions can the health effects manager draw at the 0.01 level regarding the differences between the two mines?

51. The Director of the FBI is interested in determining if the level of violent crimes has changed significantly between 1989 and 1990 on a nationwide basis. The following data have been obtained from the first 11 reporting states:

	Number of Homicides	
State	1989	1990
Alabama	421	463
Colorado	140	146
Connecticut	190	127
Indiana	353	388
Kentucky	293	308
Maryland	544	517
Massachusetts	254	310
Missouri	409	454
Ohio	652	652
Oregon	134	88
Tennessee	417	417

a. Formulate an appropriate null and alternative hypothesis.
b. What conclusions can the director draw at the 0.05 level regarding the differences in homicides for 1989 and 1990?

52. The vice president of auditing for Abbey National Bank, located in London, is interested in determining if there is a difference in the proportion of delinquent accounts between its international and domestic clients. The vice president has collected the following data from a random sample of the firm's accounting data base:

	Domestic	International
p	0.12	0.18
n	30	20

a. Formulate an appropriate null and alternative hypothesis.
b. What conclusions can the director draw regarding the differences in domestic and international accounts at the 0.05 level?

53. The baseball commissioner is interested in determining if there exists a difference between home and away attendance throughout both leagues. The following average attendance data was obtained from a random sample of 10 teams through the end of July.

| | Attendance | |
Team	Home	Away
Atlanta	14,558	26,902
Chicago (Cubs)	29,523	28,402
Montreal	18,309	24,880
Pittsburgh	23,933	25,058
San Diego	25,425	27,288
Baltimore	30,863	24,909
California	32,254	25,345
Minnesota	23,712	26,778
Oakland	35,958	32,123
Texas	26,021	27,028

a. Formulate an appropriate null and alternative hypothesis.
b. What conclusions can the commissioner draw at the 0.01 level regarding the differences between home and away attendence?

54. The managing partner at Union Securities is interested in analyzing regional bank performance. The manager has obtained the following sample data on earnings per share (EPS) for 15 regional banks for 1989 and 1990.

| | EPS | |
Company	1989	1990
Affiliated Bankshares	0.54	0.75
Constellation Bancorp	3.21	1.50
Corestates Financial	5.03	5.05
First City Bancorp	4.26	6.60
First Commerce Corp	2.26	2.65
First Fidelity	2.51	0.90
First Interstate	−3.89	6.30
First Wachovia	2.40	2.75
Mercantile Bancorp	3.82	4.15
NBD Bancorp	0.03	3.50
Shawnut National	3.46	3.80
Summercorp	2.01	2.45
SunTrust Banks	2.61	2.80
United Banks of Colorado	1.04	1.15

a. Formulate an appropriate null and alternative hypothesis.
b. What conclusions can the general partner make regarding changes in EPS performance at the 0.01 level?

55. The managing partner at Bronson & Smith Investments is interested in analyzing the changes in stock prices between 1989 and 1990 for the top international insurance corporations. The partner believes that the percentage of changes in stock prices in both U.S. dollars and the firm's local currency are similar. The

following data were obtained from a random sample of the top foreign insurance firms:

Firm	U.S. Dollars	Local Currency
1. Prudential (UK)	35	26
2. Union des Assurances (FR)	73	46
3. Assicurazioni (IT)	26	9
4. Alleanza (IT)	60	38
5. Tokyo Marine (JA)	−24	−19
6. Nissan Fire (JA)	52	62
7. Nationale (NE)	48	25
8. Skweiz (SW)	−3	−12
9. Skweiz (SZ)	52	25
10. Allianz (WG)	85	57
11. Munchener (WG)	42	20

a. Construct a null and alternative hypothesis.
b. Identify the most appropriate hypothesis test.
c. What conclusion can the partner make regarding changes in stock prices between the two currencies at the 0.05 level?
d. How would you interpret the results in light of the typical fluctuations between foreign exchange rates?

56. The director of the Los Angeles Department of Transportation is interested in determining whether the travel time to work during peak periods has changed between 1989 and 1990. The following sample data on average one-way trip times (minutes) were collected on 15 motorists during a 30-day period.

Motorist	1989	1990	Motorist	1989	1990
A	27	31	I	18	23
B	39	39	J	24	28
C	42	56	K	14	17
D	22	27	L	28	25
E	36	34	M	32	38
F	52	61	N	45	54
G	64	65	O	39	39
H	32	37			

a. Construct a null and alternative hypothesis.
b. Identify the most appropriate hypothesis test.
c. What conclusion can the director make regarding changes in travel times at the 0.05 level?
d. What assumptions regarding motorist behavior are needed to analyze this problem?

57. The production manager at Oxford United, a water valve manufacturing firm located in Birmingham, England, is interested in determining if a recent quality

control training program has had an impact on operator performance. The following sample data were collected over a one-week period prior and following the training program for 10 machine operators. The data represent the proportion of total defectives attributed to the operator.

Operator	Proportion Defective	
	Before	After
1	50	35
2	45	47
3	62	51
4	37	24
5	31	26
6	39	33
7	53	51
8	48	42
9	38	29
10	26	31

a. Construct a null and alternative hypothesis.
b. Identify the most appropriate hypothesis test.
c. What conclusion can the production manager make regarding the impact of the training program at the 0.05 level?
d. Comment on the size of the sample in terms of its impact on the conclusions.

58. The production manager at Vallourec Corporation, a manufacturing and engineering firm located in southern France, is interested in determining whether the firm's stamping press is operating within the current specifications of 10 cycles per minute. A random sample consisting of 25 observations was selected over a one-week period. Following is a CBS analysis of this problem.

Information Entered

Test procedure:	Two Sided
Alpha error:	0.025
Critical t (test statistic):	2.064
Hypothesis value:	10.
Sample size:	25
Mean:	7.0
Standard deviation:	2.5

Results

Standard error of mean:	0.5
Lower limit:	8.968
Hypothesis value:	10
Mean:	7
Alpha error:	0.025
Degrees of freedom:	
Critical t:	2.064
Computed t:	−6.
p value	0.0001
Conclusion: ?	

a. Describe the null and alternative hypothesis.
b. What is the number of degrees of freedom for this problem?
c. What conclusion should be drawn regarding this hypothesis test?
d. Discuss the significance of the reported p value.

59. The vice president of operations at Air Canada is interested in determining whether the current system is achieving the company's goal of at least a 95% on-time arrival rate. A random sample consisting of 35 observations was selected over a one-week period. A CBS analysis of this problem follows.

Information Entered

Test procedure:	One Sided
Alpha error:	0.01
Critical Z (test statistic):	2.326
Hypothesis value:	0.95
Sample size:	35
Proportion:	0.90

Results

Standard error of mean:	0.0368
Lower limit:	0.8643
Hypothesis value:	0.95
Proportion:	0.90
Alpha error:	0.05
Degrees of freedom:	
Critical Z:	2.326
Computed Z:	−1.3572
p value	0.5
Conclusion: ?	0.0900

a. Describe the null and alternative hypothesis.
b. What is the number of degrees of freedom for this problem?
c. What conclusion should be drawn regarding Air Canada's on-time claim?
d. Discuss the significance of the reported p value.

60. The director of training programs at Statoil Group, a Norwegian energy company, is interested in determining if the company's new safety program has had an impact on reducing accidents. A controlled experiment was conducted in which one group (group 1) of 25 employees undertook the training program and another group (group 2) of 30 employees did not. Sample data were collected on the number of accidents per month for both groups. Presented below is a CBS analysis of this problem.

Information Entered

Test procedure:	Two Sided
Alpha error:	.05
Critical Z (test statistic):	1.96
Hypothesis value:	0
Sample size for Group 1:	25
Sample size for Group 2:	30
Mean for Group 1:	10.50
Mean for Group 2:	11.75
Standard deviation for Group 1:	3.5
Standard deviation for Group 2:	4

Results

Standard error of mean (unequal variances):	1.0116
Lower limit:	−1.9827
Upper limit:	1.9827
Standard error of mean (equal variances):	1.0241
Lower limit:	−2.0073
Upper limit:	2.0073
Mean 1 − Mean 2:	−1.2500
Degrees of freedom:	
Critical Z:	1.9600
Computed Z:	−1.2357
p value	0.1075

Conclusion: ?

a. Describe the null and alternative hypothesis.
b. What is the number of degrees of freedom for this problem?
c. What conclusion should be drawn regarding the effectiveness of the training program?
d. Discuss the significance of the reported p value.

61. Use CBS to develop OC and power curves for problem 28.
62. Use CBS to develop OC and power curves for problem 32.
63. Use CBS to develop OC and power curves for problem 44.
64. Contrast the classical hypothesis testing procedure with statistical process control (SPC). In what ways can the SPC approach be viewed as an extension of the classical method?

65. The operations manager at Nippon Spic, a commodities firm located near Tokyo, is interested in determining whether the filling process for its pepper line meets the minimum standards of 125 Kg per bag. The following data was collected from 20 samples consisting of two observations per sample over a 20-day period.

Sample	Weight(kg)	Sample	Weight(kg)
1	126.1, 126.8	11	125.1, 125.9
2	126.5, 127.1	12	123.8, 125.1
3	123.4, 124.2	13	124.7, 125.3
4	123.9, 124.4	14	123.9, 124.7
5	126.2, 124.8	15	125.3, 124.1
6	126.3, 127.1	16	125.8, 127.2
7	124.1, 125.3	17	125.5, 125.3
8	125.6, 125.4	18	124.8, 123.9
9	128.0, 126.8	19	125.7, 124.7
10	127.6, 124.7	20	126.2, 125.6

a. Formulate a classical null and alternative hypothesis.
b. Develop $\bar{X}$ and R charts.
c. Determine the upper and lower control limits.
d. What conclusions can the manager make regarding the filling process?

66. The quality control manager at Drake Pulp & Paper is interested in analyzing the variation in paper thickness produced at the company's Dayton, Ohio, plant. The following data were obtained from sampling the production process over a 20-day period. Each of the 20 samples consisted of three observations.

Sample	Thickness (in)	Sample	Thickness (in)
1	0.12, 0.13, 0.13	11	0.09, 0.10, 0.11
2	0.11, 0.11, 0.11	12	0.10, 0.11, 0.12
3	0.10, 0.11, 0.12	13	0.09, 0.11, 0.13
4	0.09, 0.09, 0.10	14	0.10, 0.10, 0.09
5	0.10, 0.11, 0.12	15	0.13, 0.12, 0.10
6	0.12, 0.11, 0.13	16	0.10, 0.11, 0.13
7	0.13, 0.11, 0.12	17	0.12, 0.11, 0.09
8	0.09, 0.09, 0.10	18	0.13, 0.12, 0.08
9	0.10, 0.10, 0.09	19	0.12, 0.10, 0.09
10	0.11, 0.10, 0.08	20	0.10, 0.09, 0.10

a. Formulate a classical null and alternative hypothesis.
b. Develop an $\bar{X}$ and an R chart.
c. Determine the upper and lower control limits.
d. What conclusions can the manager make regarding paper thickness?

67. Develop $\overline{X}$ and R charts for the data presented in problem 48. Establish lower and upper control limits. What conclusions can be made regarding the process at the two plants?
68. Develop $\overline{X}$ and R charts for the data presented in problem 50. Establish lower and upper control limits. What conclusions can be made regarding the process at the two mines?
69. Explain which of the following statements are true or false:
   a. Decreasing α decreases β.
   b. The sume of α and β equals 1 (one).
   c. A type II error is controlled by increasing the sample size.
   d. A type I error is controlled by increasing the standard deviation.
70. Explain why the OC and power curves are useful in statistical quality control.
71. Calculate and discuss the p-value for problem 28. Using the p-value, would you reject H_0 at the 0.05 level?
72. Calculate and discuss the p-value for problem 31. Using the p-value, would you reject H_0 at the 0.05 level?
73. Conduct a standard hypothesis test on the data given in problem 65 at the 0.05 level.
74. Conduct a standard hypothesis test on the data given in problem 66 at the 0.01 level. Assume that the production standard calls for a paper thickness of 0.10 inches.

8.12 CASES

8.12.1 Forecasting Stock Growth

The price movement of equity stocks has long been the subject of general speculation. A frequently used indicator of future market performance is the price-to-earnings ratio (P/E). A recent study was undertaken to examine this relationship. Thirty-five stocks were chosen at random from those listed on the New York Stock Exchange (NYSE). The data base used for this study consisted of P/E and closing prices for June 30, 1976; June 30, 1981; and June 30, 1986. These data are presented in Table 1. Each of the selected companies has been in business for over 10 years.

Suggested Questions

1. Formulate an appropriate null and alternative hypothesis.
2. Identify the appropriate test procedure.
3. What data transformation process should be used?
4. Should separate tests be performed for the five- and ten-year periods?
5. What conclusions can be drawn at the 0.05 level regarding the use of P/E as a prediction of market performance?
6. What other data should be collected to support the analysis?

TABLE 1 Closing and P/E Quotes for Three Time Periods for Randomly Selected Stocks

	6/30/86		6/30/81		6/30/76	
Stock	Close	P/E	Close	P/E	Close	P/E
1 Ahmanso (HF) Co.	22.625	13	18.25	19	12.375	6
2 Ball Corp.	85	18	34.125	11	22	7
3 Bearing Corp.	39.5	51	32.125	11	25.375	11
4 Clorax	55.125	16	11.625	7	12.25	11
5 Delta Air	83	15	71.25	10	44.5	20
6 Dravo	15.875	57	20.625	12	23.75	9
7 Du Pont	249.75	15	160.5	12	139.5	18
8 Duquesne Light	13.25	6	12.625	7	18.25	8
9 Eastman Kodak	50.625	50	74.75	10	100.125	25
10 Exxon	121.75	9	68.5	6	104.625	9
11 General Dynamics	382.5	9	164.375	9	63.375	8
12 Greyhound	70.5	15	36.25	6	16.5	8
13 Georgia Pacific	31.375	23	27.375	12	51.5	19
14 Hammermill Paper	42.25	23	31.75	7	22.125	7
15 Hilton	71.75	19	46.875	12	17.5	12
16 Honeywell	75.75	13	85.375	7	49.625	14
17 Hutton, E.F.	203.125	16	156.25	6	20.5	5
18 IBM	586	14	231.5	9	276.75	20
19 Jostens	608.20312	20	23.90625	11	21.625	12
20 Lear Siegler	54.325	12	37	8	10.75	7
21 Lockheed	167.25	9	37.375	1	10.625	3
22 MCA	63.75	23	61.25	9	33.875	6
23 Marine Midland	52	8	21	5	11.125	9
24 Montana Power	38.875	7	32.875	9	24.5	8
25 Northrop	45	19	47.5	9	29.75	9
26 Ohio Mattress	150	11	43.375	8	43.25	9
27 Paine Weber	14.625	41	12.5	8	17.375	9
28 PPG	45.5	12	22.125	16	9	4
29 Pennzoil	204.375	15	70.5	5	54.875	10
30 Rockwell	80.0625	28	69.75	9	34.5	11

8.12.2 Evaluating the Economic Forecasters

Forecasting economic trends has reached a new level of technical sophistication over the last 10 years. Over this period, a growing number of institutions have been formed to provide regional, national, and international forecasts to both the public and private sectors. Economic models based on econometric principles represent the primary methodology used in developing quantitative forecasts. The basic role of econometrics is in estimating and testing these models. Most current econometric models contain hundreds of variables and hundreds of relationships (i.e., equations).

TABLE 1 Average Absolute Errors

Forecaster	\multicolumn{8}{c}{Forecast Horizon (quarters)}
	1	2	3	4	5	6	7	8
A. Gross national product (billions):								
Chase	27.4	32.1	46.8	56.9	65.8	76.0	91.8	109.2
DRI	25.4	26.5	32.7	39.4	43.3	48.8	46.2	59.4
WEFA	25.0	26.0	29.8	27.9	36.3	42.0	43.4	51.0
B. Consumer price index (1967 = 100):								
Chase	0.7	2.1	3.5	5.1	7.2	9.3	11.5	14.9
DRI	1.0	2.4	3.6	5.6	7.6	9.7	12.0	14.4
WEFA	0.6	1.6	2.6	4.1	5.7	7.7	9.6	11.1
C. Civilian employment (millions):								
Chase	0.3	0.6	1.0	1.2	1.4	1.8	2.2	2.5
DRI	0.3	0.6	0.8	0.9	1.0	1.2	1.4	1.6
WEFA	0.3	0.5	0.5	0.6	0.8	0.9	1.1	1.3

This level of complexity is required in order to adequately describe the economic process under investigation.

Economic forecasts, by their very nature, contain some level of error. Accordingly, there is strong interest in estimating the accuracy of developed forecasts before they can be used effectively. Typically, most forecasts are limited to a few quarters. This limitation is due in large part to the nature of the econometric estimating process. Basically, econometric models are estimated using a historical data base. Based on this approach, the accuracy of the model must decline over the forecast horizon (i.e., forecasting errors will increase over time).

To further explore the issue of forecasting errors, consider the error data presented in part A of Table 1. These data were compiled from forecasts of gross national product by three of the better-known economic forecasting institutions (Chase Econometrics; Decision Resources, Inc.; and the Wharton School). The error data represent the differences between the predicted economic indicator and the actual value. These forecasts were developed for an eight-quarter period during 1979–1980. As suggested earlier, the average absolute errors tend to increase over time. Similar trends are also seen in the error data given in parts B and C of Table 1.

Suggested Questions

1. What criteria would you use in evaluating the performance of each forecaster?
2. Describe both qualitatively and quantitatively the nature of the forecasting errors over time.
3. Beyond how many quarters do the forecasted estimates become unreliable?
4. Which forecaster would you select and why?

Chapter 9

Simple Linear Correlation and Regression

Science is nothing but perception.
Plato

CHAPTER OUTLINE

9.1 Introduction
9.2 Example Management Problem: Pacific Construction Company
9.3 How to Recognize a Simple Correlation and Regression Problem
9.4 Model Formulation
9.5 Computer Analysis
9.6 Practical Applications
9.7 Case Study: Pritikin Diet
9.8 Summary
9.9 Glossary
9.10 Bibliography
9.11 Problems
9.12 Cases

CHAPTER OBJECTIVES

The primary objectives of this chapter are to develop an understanding of

1. How to recognize simple linear correlation and regression problems.
2. How to determine and interpret the correlation coefficient and coefficient of determination.
3. How to develop a linear regression model.
4. How to perform residual analysis.
5. How to use the regression model for prediction and interval estimation.

The insurance industry represents the largest single sector of the U.S. economy. In 1986, total revenues for this industry exceeded $750 billion, or nearly 20% of the total GNP. Most major insurance companies rely heavily on accurate claim estimates in designing account portfolios and rate structures. This is especially true in the highly competitive automobile insurance business. Linear regression models are increasingly being used to forecast

> **HISTORICAL NOTE**
>
> Sir Francis Galton (1822–1911), in studying the implications of Darwinian theory in the 1870s, noted a peculiar phenomenon. He compared the heights of parents with those of their children. In converting each to standard scores, he noted that the offsprings' heights did not increase as rapidly as those of their parents. They tended to deviate less from their mean. This regression, or "falling backward," has become known as the *law of filial regression*. This law has since been shown to be merely a phenomenon of imperfect correlation or relationship of parent heights to offspring heights.

future claims and design pricing policies.* These analyses have shown that a variety of categorical variables (e.g., age, area, and vehicle type) have a statistically significant relationship to claim costs. Accordingly, these models can be employed to assist the manager in developing a better understanding of general claim trends as well as specific insight into particular market segments.

9.1 INTRODUCTION

Today's decision makers are constantly confronted by complex relationships. For example, a manager may wish to understand why revenues increase in some periods and decrease in others, why some workers get into accidents and others do not, or why some retail stores sell more than others. Simple linear correlation analysis and regression analysis are often used to investigate the potential relationship between two business factors (e.g., advertising expense and sales). Correlation and regression are actually parts of the same statistical process. **Correlation analysis** is used to measure the strength of the relationship between two variables, X and Y. (The actual degree of association is measured by the **correlation coefficient.**) **Regression analysis** involves the construction of an equation that best describes the linear relationship between X and Y and which can then be used to estimate values of Y based on selected values of X.

Often the scientist or researcher is more interested in correlation analysis, the manager in regression analysis. For example, a manager may be interested in projecting revenues based on a $100,000 advertising budget. In this case, the amount spent on advertising is the independent variable (X), and

---

*D. Samson and H. Thomas, "Linear Models as Aids in Insurance Decision Making: The Estimation of Automobile Insurance Rates," *Journal of Business Research* 15, no. 3 (1987).

the amount of revenues is the dependent variable (Y). The researcher, on the other hand, may be more concerned with the extent of the relationship between these two variables. In these cases, the correlation coefficient would be utilized.

9.2 EXAMPLE MANAGEMENT PROBLEM: PACIFIC CONSTRUCTION COMPANY

The management at Pacific Construction Company has become interested in determining the extent of the relationship between advertising and sales. The company's general manager has collected the following data on advertising expenditures and gross sales for the past 12 months:

Month	Advertising ($000)	Sales ($000)
JAN	15.25	212.18
FEB	14.15	218.27
MAR	17.35	221.45
APR	21.05	234.56
MAY	19.50	241.34
JUN	16.40	225.20
JUL	12.85	203.91
AUG	9.10	212.23
SEP	14.05	198.76
OCT	18.25	212.32
NOV	19.25	214.78
DEC	14.65	222.56

The general manager believes that the relationship between advertising and sales can best be explained with a simple linear regression model. If a relationship exists, the general manager wishes to use it to predict sales based on various levels of advertising.

9.3 HOW TO RECOGNIZE A SIMPLE CORRELATION AND REGRESSION PROBLEM

The following list highlights the basic characteristics of a simple (i.e., two-variable, or bivariate) regression problem:

Key idea 1
- There is a single dependent or response variable Y (e.g., sales in the example problem).
- There is a single independent or predictor variable X (e.g., advertising expense in the example problem).

348 *Chapter 9 Simple Linear Correlation and Regression*

INTERNATIONAL VIGNETTE

Mr. Vesa Kumpulainen, managing director of Kumera Oy, was vexed about the phone call he had received from one of his major customers. This was not the first time the problem of inadequate inventory control had come to his attention. Customer complaints were on the increase, operating costs were escalating, and disputes between departments were on the rise. He knew that something had to be done. A quick review of the income statements for the last two years showed that net earnings had declined approximately 80%.

Kumera Oy, founded in 1947 by the father of the current managing director, is located approximately 60 KM north of Helsinki. Its main product line is a series of high-quality worm gears and shaft-mounted speed reducers. The company has a customer base of some 200–300 firms, of which approximately 50% place an order in any given year. Typically, orders come directly from the customers by mail or phone or from the in-field sales engineers. Regular orders for stock accounts could take as long as six months to fill. Mr. Kumpulainen knew that better control over the ordering and scheduling systems could significantly improve customer response time. Therefore, he decided to incorporate a regression-based forecasting system into the inventory planning process.

- Measurement data are available for both variables.
- The objective is to determine the nature and extent of the linear relationship between X and Y.

The terms *dependent* and *independent variable* are standard in statistics; however they are somewhat misleading and can often cause confusion to the beginning student because they imply a cause-and-effect relationship. Typically, however, the interpretation of correlation and regression results should be limited to the relationships and not cause and effect. More specifically, the **independent variable** (X) is used to explain the variability in, or predict the value of, the **dependent variable** (Y).

9.4 MODEL FORMULATION

CORRELATION ANALYSIS

Data values for the two variables (i.e., X and Y) can be presented graphically using a scatter diagram. This graph shows the X values on the horizontal axis and the Y values on the vertical axis. A scatter diagram for the data for the Pacific Construction problem is shown in Figure 9.1. CBS can be used to generate scatter diagrams for up to three variables at a time.

The population **correlation coefficient** (ρ) is a numerical indicator of the degree of *linear association* between the independent and dependent variables in the population. As such, it describes the extent of the relationship

FIGURE 9.1 Scatter Diagram of Sales and Advertising Expenses for Pacific Construction

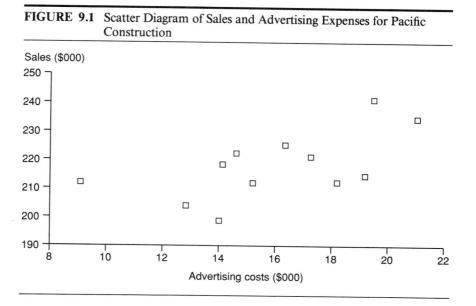

between X and Y. Usually, ρ is not known, and therefore the sample correlation coefficient (r) is used as a point estimate for ρ. This statistic is also known as the *Pearson correlation coefficient, correlation coefficient, first-order correlation coefficient,* and *bivariate correlation coefficient.*

The standard model for determining the Pearson correlation coefficient is:

$$r = \frac{\Sigma XY - n\overline{X}\,\overline{Y}}{\sqrt{(\Sigma X^2 - n\overline{X}^2)(\Sigma Y^2 - n\overline{Y}^2)}}$$

where n is the sample size and ΣX and ΣY are the sums of the individual values for the independent and dependent variables, and $\overline{X}$ and $\overline{Y}$ are the means of the independent and dependent variables.

In general, correlation coefficients vary from true positive ($+1.0$) to true negative (-1.0). These limit cases are graphically shown in Figures 9.2 and 9.3, respectively. The following list summarizes their characteristics:

- All the data points lie on a straight line.
- For $r = +1$, as X increases, Y increases proportionally.
- For $r = -1$, as X increases, Y decreases proportionally.
- For $r = +1$, as X decreases, Y decreases proportionally.
- For $r = -1$, as X decreases, Y increases proportionally.

Key idea 2

For relationships that are less than perfect, the correlation coefficient is a fraction between -1 and $+1$. This means that not all plotted points lie on a single, straight line. Figure 9.4 shows a partial positive relationship between

FIGURE 9.2 Perfect Positive Correlation

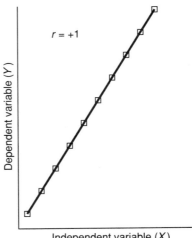

FIGURE 9.3 Perfect Negative Correlation

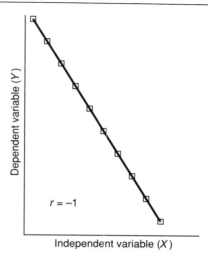

housing prices (Y) and floor space (X); Figure 9.5 demonstrates a partial negative relationship between new housing starts (Y) and interest rates (X).

In the case of a zero or near-zero correlation, an almost infinite number of lines can be computed for the data points. Therefore, no relationship can be inferred for such pairings. Figure 9.6 illustrates a case in which r is near zero.

FIGURE 9.4 Housing Prices (Y) versus Floor Space (X)

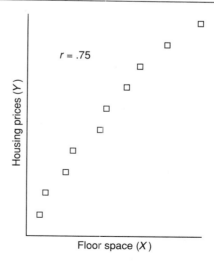

FIGURE 9.5 New Housing Starts (Y) versus Interest Rates (X)

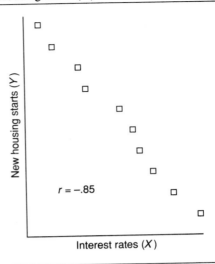

However, $r = 0$ does not necessarily mean that X and Y are not related. One can only conclude that X and Y are not *linearly* related.

The relationship between X and Y can also be plotted in the form of a curve instead of a straight line. Typically, such relationships are called *curvilinear* or *nonlinear*. Figure 9.7 illustrates a positive nonlinear relationship

FIGURE 9.6 Airline Passenger Complaints (Y) versus Jet Fuel Prices (X)

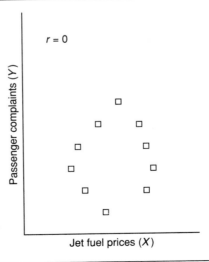

FIGURE 9.7 Probability of Being Served (Y) versus Waiting Time (X)

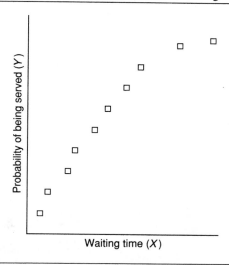

between the probability of being served (Y) and waiting time (X) at a local bank. Figure 9.8 shows a negative nonlinear relationship between the time required to produce an item (Y) and the number of units produced (X). This is an example of the so-called *learning curve*. Here, the time required to complete an item is reduced as more units are produced. However, the

FIGURE 9.8 Production Time per Unit (Y) versus Number of Units Produced (X)

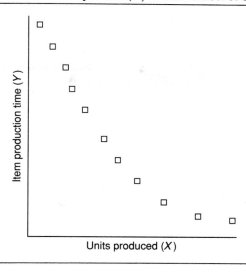

amount of improvement per unit drops as more units are produced. Thus, an initial sharp drop in the time to produce the first few units is followed by a gradual decline in the time required for subsequent units.

Interpretation

Two variables may be related for many different reasons. As indicated in Chapter 2, the assumption of causal relationship is more a logical matter than a statistical one. If the data are derived under controlled conditions, it is possible for the analyst to assume causation. Usually, however, data are collected as they exist in nature. Therefore, to assume causality would be highly presumptuous. For example, a high correlation between R&D expenses and sales does not mean that R&D *causes* sales. Thus, one of the most important things to remember in interpreting a correlation coefficient is that it *doesn't* imply cause and effect.

In general, the magnitude of r is a measure of the degree to which X and Y are related through some common variable. Expenditures on R&D and sales are related through the customer's perception of improved product quality. Sales of computers may be related to sales of TV sets because many computers require a TV-type screen. This interpretation is very useful because it sets the stage for multiple correlation, or the addition of more common measures (i.e., more independent variables) to increase the understanding of Y.

An accurate interpretation of the correlation coefficient requires an understanding of the numbering systems used for the variables and their probable implications. Figure 9.1 presented a plot of sales versus advertising expense for the Pacific Construction problem. Inspection of this graph appears to show that sales and advertising expense are related; that is, higher levels of advertising seem to result in larger sales. However, this does not necessarily mean that the two *are* related. To determine the actual relationship between the two variables, one must test for statistical significance using the standard hypothesis-testing procedure outlined in Chapter 8. The null hypothesis for the Pacific Construction problem is that there is no relationship between advertising expense (X) and sales (Y) in the population. The alternate or research hypothesis is that there is a relationship between X and Y in the population. In the terminology of Chapter 8, the null and alternative hypotheses can be expressed as follows:

$$H_0: \rho = 0 \text{ (null)}$$

$$H_1: \rho \neq 0 \text{ (alternative)}$$

Thus, the null hypothesis of no relationship can be tested for a given level of significance (alpha). If the null hypothesis is rejected, the logical response is to accept the alternative hypothesis that X and Y are linearly related.

9.4 Model Formulation

Usually the test for significance for the correlation coefficient is conducted concurrently with the test for significance for the regression model.

The interpretation of r is fairly simple. Two variables are either linearly related or are not. It is obvious that the difficulties in interpretation are not related to the statistic but to the logical processes of developing meaningful independent and dependent variables. Additional difficulties thus occur in deriving the numbers necessary to apply the model. The following examples illustrate some of the more common problems encountered.

Earthquake Real Estate (Lagged Effects)

Suppose that the manager of the Earthquake Real Estate agency wishes to determine whether or not a relationship exists between the mortgage lending rate (X) and housing starts (Y) in the local area. Figure 9.9 presents monthly lending rates and housing starts for 1985. An analysis of these data shows a very small negative correlation of $-.15$. Is it appropriate to conclude that there is no relationship between lending rates and housing starts? The economic literature clearly indicates that a relationship should exist, and so does our intuition. What went wrong? Maybe the January interest rate doesn't affect the builder until March or later. Thus, the proper pairing could be Jan-Mar, Feb-Apr, or Mar-May. Figure 9.10 depicts lending rates versus housing starts on a month-by-month basis. This graphic supports the low reported r value. A similar analysis in which housing starts were lagged by three months yielded a much stronger relationship of $r = -.55$. Thus, selection of the proper pairing can be crucial in assessing the extent of the relationship.

FIGURE 9.9 Lending Rates and Housing Starts Data

Month	Lending Rate (%)	Housing Starts (000)
JAN	10.6	225
FEB	11.2	300
MAR	11.1	290
APR	10.8	270
MAY	10.5	255
JUN	10.3	265
JUL	10.7	250
AUG	11.3	225
SEP	11.7	230
OCT	12.1	240
NOV	11.8	215
DEC	10.6	190

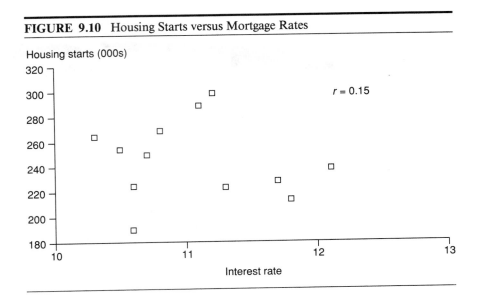

FIGURE 9.10 Housing Starts versus Mortgage Rates

Budget Computer, Inc. (Missing Data)

The sales manager at Budget Computer, Inc., wishes to determine if a relationship exists between the budgeted advertising costs (X) and the resultant revenues (Y). The collected data are reported in Figure 9.11. These data show two quarters of budget expenditures and six months of revenues. Since r requires pairings, what is paired with what? One approach is to assume equal monthly expenditures, divide the $15,000 for the first quarter by 3, and state that the budget amounts are $5,000 for January, February, and March. In this manner six monthly budget amounts are paired with six revenue amounts. However, that reduces the amount of variance in X and therefore may rule out the opportunity of finding an important relationship.

FIGURE 9.11 Advertising Expenditures and Revenue Data

	Quarterly Budget Amounts ($000)	Monthly Revenues ($000)
JAN		270
FEB	15	290
MAR		230
APR		220
MAY	20	210
JUN		250

FIGURE 9.12 Productivity versus Test Scores

Employee	Test Score	Productivity (units/hr)
1	85	50
2	80	55
3	75	57
4	89	38
5	84	41
6	88	68
7	72	45
8	82	70
9	71	52
10	78	55

Another approach is to compute mean values for revenues for January, February, and March and use those as the value of Y. The practice of pairing disparate periods must be given a great deal of thought.

Amex Machine Shop (Restricted Sample)

The managers of Amex Machine Shop are considering hiring only those applicants that score above 70% on an aptitude examination. But first they wish to know if there is a relationship between test scores and productivity. A random sample of 10 current employees who scored 70 or above on the aptitude test was selected. The collected data on employee test performance and employee productivity are given in Figure 9.12. The computed r for the data is .10, which indicates that test scores are not a very good indicator of work performance. It should be noted, however, that this analysis is restrictive because the data were limited to the top 20%. The selected sample has limited the variance in both X and Y and therefore cannot effectively explain the remaining variance in the sample. A full range of aptitude scores would have increased the probability of finding the desired relationship.

REGRESSION ANALYSIS

The ability to predict sales, performance, or other events from present data is obviously an important asset to the manager. Regression analysis, the complement of correlation, is used extensively in business for both estimating and predicting. In fact, it is the principal analytical tool used in business forecasting. Before using regression analysis, however, one must understand the assumptions underlying this model. The following discussion outlines some of the basic elements and assumptions associated with the regression model.

The Model

Key idea 3

Simply stated, the bivariate regression model is a straight line of best fit through paired data (i.e., X versus Y). The normal procedure is to determine the best fit using the **least squares method,** a way of developing estimates for the regression model coefficients by minimizing the sum of the squares of the residuals. The population regression equation for one dependent and one independent variable is given as:

$$Y = B_0 + B_1 X + e$$

where:

Y = Dependent variable
B_0 = Y intercept when X is equal to zero (population)
B_1 = Slope for the regression line (population)
X = Independent variable
e = Error term (i.e., difference between the actual value and estimated value of Y)

The following list identifies the basic assumptions regarding the development of the regression model and the nature of the error term:

1. A linear relationship exists between X and Y in the population.
2. The error term is a normally distributed random variable with a mean of zero [$E(e) = 0$].
3. The values of the error term are independent.
4. The variance of the error term is constant for the values of the independent variable (no space heteroscedasticity).

One simple approach for determining if the error terms are normally distributed is to construct a histogram of the sample residuals (i.e., errors). Such a plot would reveal whether the error terms are normally distributed or highly skewed. Similarly, a plot of the error terms as a function of the independent variable should indicate whether assumptions 3 or 4 have been violated. The basic idea is to look for a random pattern of residuals on the graph. If the plot shows that most of the residuals are either positive or negative for small values of X, and the opposite is seen for larger values of X, then this would suggest that the error terms are not independent. Additionally, if the values of the residuals tend to increase or decrease as X becomes larger, this would suggest a violation of assumption 4.

In most cases, an estimate of the population regression model is based on sample data. That is, the intercept and slope are computed from the sample

FIGURE 9.13 Scatter Diagram and Regression Line for Pacific Construction Company

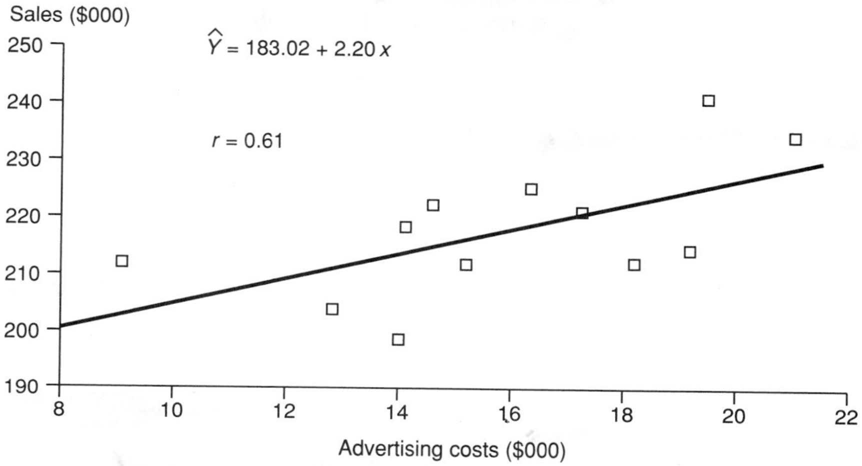

Sales ($000)
250
240
230
220
210
200
190
Ŷ = 183.02 + 2.20 x
r = 0.61
8
10
12
14
16
18
20
22
Advertising costs ($000)

data and used to estimate the population intercept and slope as follows:

$$b_0 \text{ (sample)} \rightarrow B_0 \text{ (population)}$$

$$b_1 \text{ (sample)} \rightarrow B_1 \text{ (population)}$$

In practice, therefore, the actual regression model takes on the following form:

$$\hat{Y} = b_0 + b_1 X$$

where $\hat{Y}$ is the regression line extention of Y for a given value of X.

Figure 9.13 shows a scatter plot and regression line for the Pacific Construction Company example discussed earlier in this chapter. The regression equation was determined according to the least squares principle. The least squares procedure involves determining values for b_0 and b_1 in such a way that the total error (i.e., Σe_i) is minimized. The error term is determined by squaring the differences between the estimated Y values (as computed from the regression equation) and the actual Y values and summing the total. The difference between the estimated and actual Y values is called the **residual.** Any other straight line used in fitting the data will result in a larger error term or residual (this point will be illustrated later).

As Figure 9.13 shows, the data points are not all on the regression line. For example, a marketing expense of $19,500 (May's value) yields estimated sales revenues of approximately $225,920. This figure is approximately $15,420 lower than the actual value. Clearly, there is no assurance that the regression equation will estimate the dependent variable values (in this case, total sales) with 100% accuracy. Accordingly, some measure of dispersion

Coefficient of Determination

Key idea 2

Because the regression line in Figure 9.13 does not perfectly fit the points on the scatter diagram, use of the model will result in some level of error. The extent of the error is often stated in terms of the sum of squares:

$$\text{Sum of squares due to error} = \text{SSE} = \Sigma(Y_i - \hat{Y})^2$$

The data given in Figure 9.14 illustrate the process for computing the sum of squares due to error for the Pacific Construction Company problem.

If one were asked to estimate total sales without knowing the marketing expense, the best estimate would be to use the value of the sample mean for Y (i.e., $\bar{Y}$). A measure of the variability of Y can be made by computing the total sum of the squares about the mean. This process is illustrated here:

$$\text{Sum of squares about the mean} = \text{SST} = \Sigma(Y_i - \bar{Y})^2$$

The sum of the squares about the mean is usually referred to as the sum of the squares of the total, or SST. The computed SST for the Pacific Construction Company problem is 1,581.90. The difference between SST and SSE represents the regression sum of squares (i.e., SSR). If the regression model gener-

---

FIGURE 9.14 Computation of Sum of Squares Error

Month	Expense	Sales Actual	Sales Predicted	Residual	Squared Difference
JAN	15.25	212.18	216.51	−4.33	18.75
FEB	14.15	218.27	214.09	4.18	17.44
MAR	17.35	221.45	221.12	0.33	0.11
APR	21.05	234.56	229.25	5.31	28.20
MAY	19.50	241.34	225.85	15.495	240.10
JUN	16.40	225.20	219.04	6.164	38.00
JUL	12.85	203.91	211.24	−7.329	53.71
AUG	9.10	212.23	203.00	9.227	85.14
SEP	14.05	198.76	213.88	−15.115	228.46
OCT	18.25	212.32	223.10	−10.779	116.19
NOV	19.25	214.78	225.30	−10.516	110.59
DEC	14.65	222.56	215.19	7.368	54.29
			Totals	0	990.99

ated a perfect fit for the data, then SSE would be zero and SSR would equal SST. The ratio of SSR to SST is often used as a measure of the accuracy of the regression model. This ratio measure is called the **coefficient of determination.**

1. *Coefficient of determination* (R^2) — The amount of variance in Y as explained by the regression model.
2. *Coefficient of nondetermination* — The amount of unexplained variance or error variance.

The computational method for estimating the coefficient of determination is given by:

$$R^2 = \frac{SSR}{SST} \quad \text{or} \quad R^2 = \frac{(SST - SSE)}{SST}$$

The coefficient of determination for the Pacific Construction Company problem is computed as follows:

$$R^2 = \frac{(1{,}581.90 - 990.99)}{1{,}581.90} = .37$$

Thus, 37% of the variance in sales is explained by the variance in advertising expense. This result suggests that other factors in addition to advertising contribute to the fluctuations in company sales. The incorporation of additional predictor variables will be addressed in the next chapter.

As indicated above, the regression line produces the best fit of the data based on the principle of least squares. The graphic shown in Figure 9.15

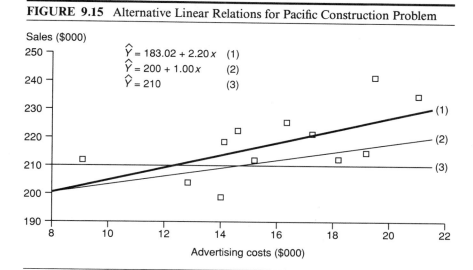

FIGURE 9.15 Alternative Linear Relations for Pacific Construction Problem

illustrates this point. Here, two other lines, which were selected on the basis of an inspection of the data, are plotted along with the developed regression line. The obvious question is which one best fits the data.

Figure 9.16 presents a comparison of the SSE terms associated with the three different linear functions. Clearly, model 1, which is based on the least squares method, yields the lowest SSE. On the other hand, model 3, which is based on a constant value of 210, results in the largest SSE.

Testing for Significance

Before using the developed model for estimating or forecasting, one must evaluate the basic relationship for statistical significance. Recall that the developed model is based on information contained in the sample. The objective, then, is to estimate the population parameters [i.e., intercept (B_0) and slope (B_1)] based on the sample data. The standard hypothesis-testing procedure outlined in Chapter 8 can be used to assess the statistical significance of the model. If the relationship is not significant, then the two variables are not linearly related. Conversely, if the relationship is significant, then the two variables are most likely linearly related. A conclusion regarding the statistical significance of the developed model can be made based on the following hypotheses:

$$H_0: b_1 = 0 \text{ (null)}$$
$$H_1: b_1 \neq 0 \text{ (alternative)}$$

FIGURE 9.16 Comparison of SSEs for Candidate Linear Functions

Actual Sales	Model 1		Model 2		Model 3	
	Predicted Sales	Squared Difference	Predicted Sales	Squared Difference	Predicted Sales	Squared Difference
212.18	216.51	18.75	215.15	8.82	210.00	4.75
218.27	214.09	17.44	214.15	16.97	210.00	69.39
221.45	221.12	0.11	217.35	16.81	210.00	131.10
234.56	229.25	28.21	221.05	182.52	210.00	603.19
241.34	225.85	240.10	219.50	476.99	210.00	982.20
225.20	219.04	38.00	216.40	77.44	210.00	231.04
203.91	211.24	53.71	212.85	79.92	210.00	37.09
212.23	203.00	85.14	209.10	9.80	210.00	4.97
198.76	213.88	228.46	214.05	233.78	210.00	1.80
212.32	223.10	116.19	218.25	35.16	210.00	5.38
214.78	225.30	110.59	219.25	19.98	210.00	22.85
222.56	215.19	54.29	214.65	62.57	210.00	157.75
SSE totals		990.99		1,220.76		2,251.59

This hypothesis model is similar to the one used in testing for significance of the correlation coefficient. Notice that the null hypothesis states that X and Y are not linearly related; that is, the slope b_1 is zero. Rejecting the null hypothesis results in adopting the alternative hypothesis that X and Y are linearly related. CBS makes the testing procedure fairly straightforward. The computer model uses the standard error of the slope (SES) and the b_1 regression coefficient to compute a t-value (i.e., $t = b_1/\text{SES}$). This value is compared to the critical t, which is based on a given confidence level and the degrees of freedom (i.e., $N - 2$) for the problem. If the computed t-value is greater than the critical t-value, then the null hypothesis of no significance can be rejected. There is a direct relationship between the significance of the regression and correlation coefficients in the bivariate model. Either both are significant or neither is significant for a given problem.

Most computer-based regression models also provide p-values. These values are used for testing the significance of the regression coefficient (b_1). Recall from the previous chapter that p-values are reported so that the user can determine the significance of the results independent of a preestablished level of significance. The same p-value can be used for interpreting the significance of both the regression coefficient and the correlation coefficient.

Nonlinear Effects

Perhaps the single most important assumption introduced in this chapter is that X and Y are linearly related in the population. As indicated, this is not always the case. The following list identifies the two standard methods for addressing situations in which the assumption of linearity does not hold:

- Approximating the nonlinear relationship with a set of linear regression lines.
- Transforming the data into a linear relationship.

For example, the scheduling manager at Vatco Electronics is interested in estimating the number of labor hours required to produce the company's Mark 8 radar system. An estimate of the number of labor hours needed to manufacture a system is given in Figure 9.17. This is an example of the learning curve in which the number of labor hours declines in a nonlinear fashion as a function of the number of units produced. A scatter diagram of this relationship is given in Figure 9.18. Also shown are two linear regression lines developed over segments of the total curve. The R^2 values for these two lines are .83 and .97, respectively. This is in comparison to an R^2 of approximately .56 when a single linear regression model is used to estimate the entire data base. In practice, these two lines could be used to estimate the number of labor hours, depending on the number of units produced.

Another approach for addressing curvilinearity involves transforming one or both of the variables. For example, the values of the dependent variable

FIGURE 9.17 Units Produced and Direct Labor Hours for the Vatco Problem along with Natural Logarithms

Number of Units	Direct Labor Hours
1 (0)	2,500 (7.82)
5 (1.61)	1,485 (7.31)
10 (2.30)	1,096 (7.00)
15 (2.71)	948 (6.85)
20 (3.00)	856 (6.73)
25 (3.22)	790 (6.67)
30 (3.40)	740 (6.61)
35 (3.56)	700 (6.55)
40 (3.69)	668 (6.52)
45 (3.81)	640 (6.46)
50 (3.9)	616 (6.42)

can be transformed using logarithms or the square root function. Insight into the selection of the most appropriate transforming method can be gained by developing a scatter plot of X and Y. In the case of Vatco Electronics, an examination of the scatter plot indicates a logarithmic relationship. More specifically, experience has indicated that the learning curve tends to become linear when plotted on log-log paper. In this case, both the dependent and independent variables are transformed by taking the natural logarithm of each variable value (e.g., $\log_e 2{,}500 = 7.84$). A plot of the transformed data

FIGURE 9.18 Plot of Direct Labor Hours and Units Produced for the Vatco Problem

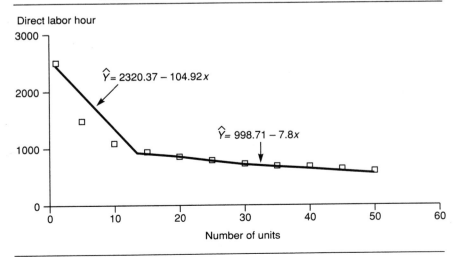

FIGURE 9.19 Plot of Transformed Direct Labor Hours versus Units Produced for the Vatco Problem

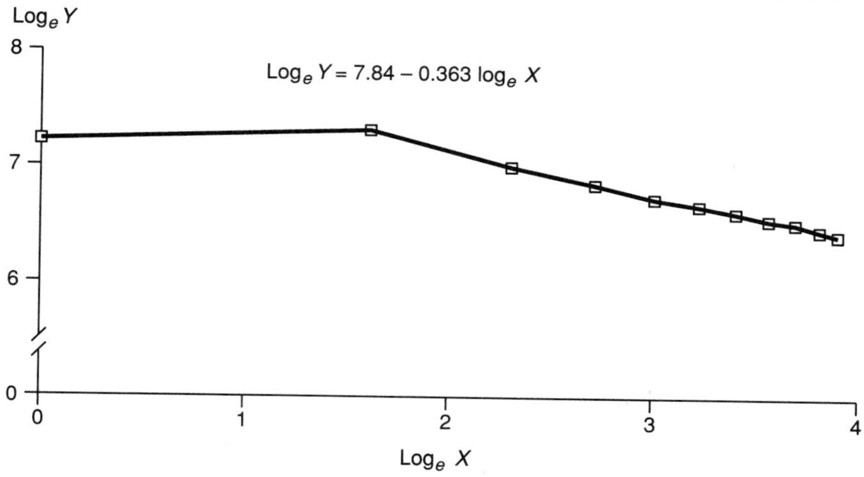

$Log_e Y = 7.84 - 0.363 \, log_e X$

base is shown in Figure 9.19. The resultant linear regression model also appears in Figure 9.19. This model explains nearly 99% of the variance in the natural logarithm of Y.

The following example illustrates the use of the model for estimating the number of direct labor hours needed to produce 20 radar systems:

$$\log_e Y = 7.84 - 0.363 \log_e 20 = 6.75$$

$$\hat{Y} = \text{antilog } 6.75 = 854 \text{ hours}$$

The resultant estimate compares very closely with the actual time required to produce the 20th system. CBS contains a number of data transformation options including logarithms and power functions. (See Appendix B for details.)

Model Limitations

Some caution must be exercised in the use of regression analysis, especially when the model is used as a basis of forecasting. Consider, for example, the Pacific Construction Company. If a reliable relationship exists between advertising expenditures and revenues, then the model can be used to predict revenues for various advertising expenditure values. For example, an advertising budget of $15,000 should result in sales of approximately $215,000. This result is obtained by substituting the value 15,000 into the regression model. However, what if the company wishes to spend $30,000 on advertising? In this case, the estimated sales figure is projected as $244,000. Although such a conclusion may be mathematically correct, the actual results might be quite different. This is because the data are outside the range of the data set. Beyond the original data range the straight line might become curvilinear. The result could then be substantially more or less revenue than the projected $244,000. This is especially true in problems where the coefficient of determination is relatively small. The analyst must exercise care when going beyond the original data range. The assumption of continued linearity outside the range of the data (i.e., extrapolation) can yield misleading results.

Caution must also be exercised in inferring causal relationships. Regression does not mean causation. Even though variables relate to one another in a natural fashion, this relationship may not hold when one of them is artificially derived. Thus, the analyst must be very careful in interpreting the logical status of an application of correlation and regression to the data.

A third reason for caution involves the general nature of forecasting. Forecasts are often used by managers as a strategy and do not necessarily reflect the logical status of the regression model. For example, if the U.S. Department of Labor states that there will be a critical shortage of computer programmers, the probability is high that that shortage will not occur. The mere hint of a shortage seems to set a compensating mechanism in motion. Thus, reactions to a forecast can affect the validity of the forecasting methodology.

9.5 COMPUTER ANALYSIS

Today, most correlation and regression problems are solved by computer analysis. The computational requirements for even a small data set are time-consuming when performed manually. Further, the possibility of computational errors is ever-present. Figure 9.20 presents a CBS summary output for the Pacific Construction problem. Notice that both advertising expense and sales were input in units of thousands of dollars. This is a common practice when the numerical values are very large. The first two values represent the Y intercept (1) and slope (2) of the regression line. The next two values report the averages for the dependent (3) and independent (4) variables. These are followed by three values showing the sum of the squares for the regression line (5), the error term (6), and the total (7).

The coefficient of determination (8) and correlation coefficient (9) are given next. These results, as indicated earlier, reveal that the model explains approximately 37% of the variance in Y. The standard error of the estimate (10) indicates the variability of the regression line, and the standard error for the b_1 coefficient (11) indicates the level of variability around the slope estimate. More specifically, the standard error of the estimate measures the dispersion of the values of Y observed around the developed regression model. The computed (12) and critical (13) t-values can be used to determine whether or not the regression model is statistically significant. For example,

FIGURE 9.20 CBS Regression Analysis of Pacific Construction Company Problem

```
CBS-Simple Correlation & Regressi               03-15-1990 - 09:09:59
                        Results
    B0 Coefficient:                             183.0159   (1)
    B1 Coefficient:                               2.1963   (2)

    Mean of X:                                   15.9875   (3)
    Mean of Y:                                  218.1300   (4)
    Sum of Squares Regression:                  590.9362   (5)
    Sum of Squares Error:                       990.9610   (6)
    Sum of Squares Total:                     1,581.8972   (7)

    Coefficient of Determination:                 0.3736   (8)
    Correlation Coefficient:                      0.6112   (9)
    Standard Error Estimate:                      9.9547  (10)
    Standard Error B1:                            0.8994  (11)

    Computed t:                                   2.4420  (12)
    Critical t:                                   2.2280  (13)
    p value:                                      0.0348  (14)

    Conclusion: B1 is statistically significant
                        press ↵
```

the critical t-value for 10 degrees of freedom (i.e., $12 - 2$) and $\alpha = .05$ is 2.228. Based on these results, the null hypothesis of no relationship can be rejected. This conclusion is substantiated by the computed p-value (14) of .040. The reported p-value suggests that there is only a 4% chance that the variables are not linearly related.

Residual Analysis

Key idea 4

An analysis of the errors or residual can also provide some important insights into the accuracy of the developed model. Recall that the residual is the difference between the actual Y value and its predicted value. The fifth column in Figure 9.14 presents residual data for the Pacific Construction Company problem. Note that the residuals sum to zero. This is because the mean of Y represents the mathematical center for the distribution. Panel A of Figure 9.21 shows a relative frequency distribution of the error terms. Notice that the general shape appears somewhat normal, i.e., lower frequency counts in the tails and higher frequency counts in the middle. The fact that the shape is not completely symmetrical can be attributed to the relatively small sample size associated with this problem. Panel B of Figure 9.21 presents a plot of the residuals as a function of the independent variable (X). This plot suggests that the error terms are independent (assumption 3) and that the error variance is constant (assumption 4). These conclusions are based on the observations that the pattern of errors appears random as a function of X and that the magnitude of the error values appears neither to

FIGURE 9.21 CBS Residual Analysis for Pacific Construction Company Problem

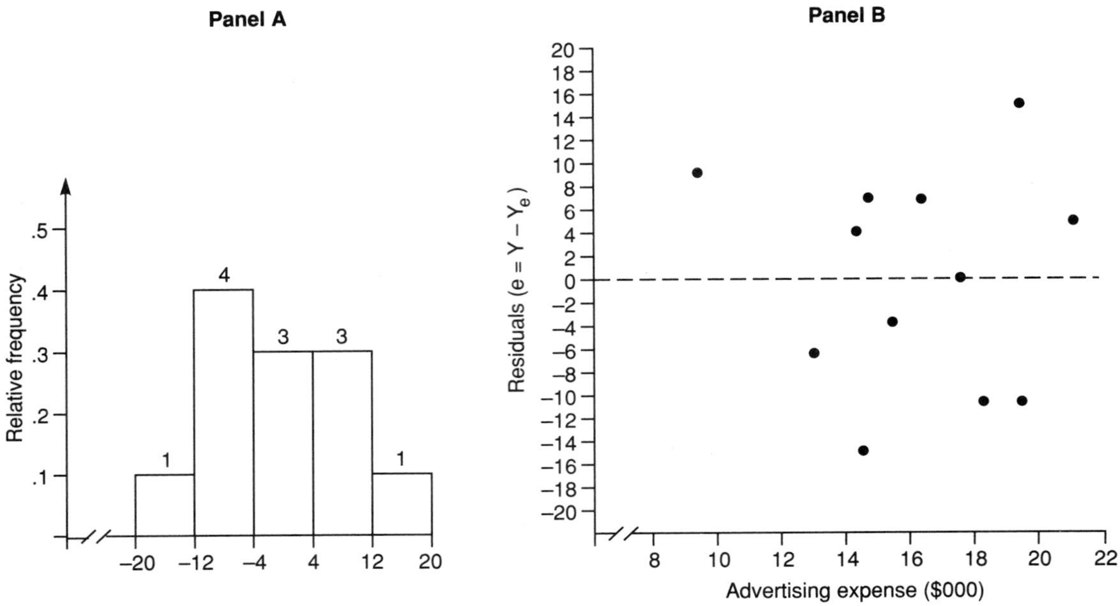

increase or decrease as X increases in value. In contrast, Figure 9.22 illustrates the presence of a trend (panel A) and heteroscedasticity (panel B) in the residuals as a function of X. The first condition is a clear violation of assumption 3 (independence of error terms), while the second condition violates assumption 4 (constant variance). Typically, when violations of these assumptions do occur, the normal approach is to employ more advanced techniques, several of which will be presented in Chapter 10.

Forecasting and Interval Estimation

Key idea 5

One of the primary uses of the regression model is in forecasting or predicting. For example, Pacific's managers can use the regression model to help optimize corporate performance. Based on the developed regression model, a planned advertising expense of $18,000 yields a sales forecast of approximately $222,620. In this way, management can set the level of advertising that helps achieve a specific corporate objective. Generally speaking, whenever management has control over the independent variable, they may be able to exert at least partial indirect control over the dependent variable.

The use of sample data in developing the regression model suggests the potential for some uncertainty in a given forecast, even though the relation-

FIGURE 9.22 Residual Plots Illustrating Dependent Error Terms (Panel A) and Heteroscedasticity (Panel B)

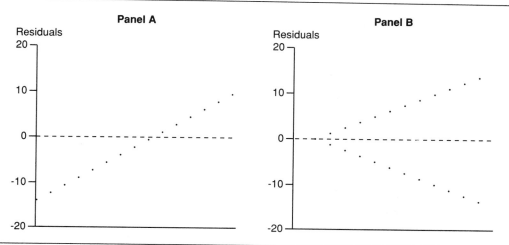

ship is statistically significant. Specifically, the selection of another sample will result in a different regression model, which in turn will yield a somewhat different forecast. Developing an interval estimate represents one approach to this problem. The concept of an interval estimate was introduced in Chapter 7. Basically, the idea is to establish a confidence limit around a specific forecast. Unfortunately, this process is somewhat complicated. In regression forecasting, two types of interval estimates can be developed:

- An estimate of the *mean* value of Y for a given value of X.
- An estimate of an *individual* value of Y for a given value of X.

Although the point estimates for both types of intervals will be the same, the interval estimates themselves will differ. Recall that an advertising budget of $18,000 for the Pacific Construction Company yielded a sales estimate of $222,620. How should this point estimate be viewed? Is it the mean value of all sales estimates based on an advertising budget of $18,000, or is it an individual estimate? The answer, of course, depends on the objective of the analysis. Intuitively, the amount of error associated with predicting a mean value should be somewhat less than the error associated with predicting an individual value. This is because a calculation of the mean makes high and low values of Y tend to cancel each other out. On the other hand, a prediction of an individual value of Y is much more likely to obtain a large or small value. This is because there is one source of error in predicting a mean value of Y and two sources of error in predicting an individual value of Y.

The confidence interval for the dependent variable can be estimated according to the following relationship:

$$\mu_{y \cdot x} = \hat{Y} \pm Z \cdot \sigma_{Y_x}$$

where $\mu_{y \cdot x}$ is the true population value of Y given a value of X, and $\hat{Y}$ is the forecast value of Y. The term σ_{Y_x} is called the standard error of prediction. When the sample size is large ($n > 30$), the standard error of prediction can be *approximated* using the standard error of the estimate (SE). Figure 9.23 presents 95% (i.e., ±1.96 standard deviations) prediction intervals using the regression line developed for the Pacific Construction Company. For example, the 95% upper and lower limits for $\hat{Y}$ for $\overline{X} = 16$ is shown below:

$$\hat{Y} \pm 2 \cdot SE = \underset{\text{Intercept}}{183.02} + \underset{\substack{\text{Slope} \\ }}{(2.2)} \underset{\substack{X \\ \text{value}}}{(16)} \pm \underset{\substack{95\% \\ \text{level}}}{(1.96)} \underset{\substack{\text{Standard} \\ \text{error of} \\ \text{estimate}}}{(10)}$$

$$\hat{Y} \pm 2 \cdot SE = 218.22 \pm 19.6(\$000)$$

The prediction intervals for other X values are determined in a similar manner.

This discussion has focused on developing confidence intervals by approximating the standard error of prediction. In reality, the confidence interval is a function of both the sample size and the distance of the given value of X from $\overline{X}$. When the sample size is large and the distance between X and $\overline{X}$

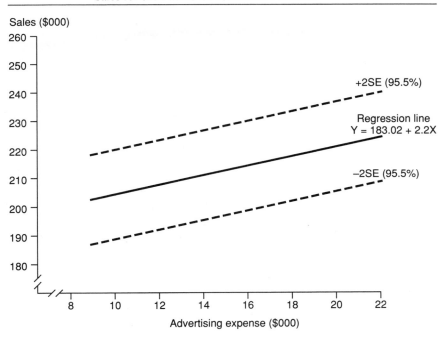

FIGURE 9.23 Approximation of 95% Interval Estimates for Individual and Mean Sales for the Pacific Construction Problem

FIGURE 9.24 95% Interval Estimates for Individual and Mean Sales for the Pacific Construction Company Problem

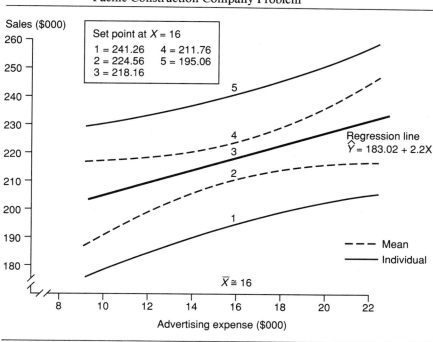

is small, these approximations are adequate. Otherwise, a more complicated set of formulas is required to produce the desired intervals. Fortunately, computer analysis is available to handle the computations. Figure 9.24 shows a CBS-generated set of interval estimates for both mean and individual values of $\hat{Y}$. Again notice that the individual interval is much wider than the interval for the mean value. Equally important, however, is the curvilinear nature of these intervals. For example, the 95% confidence interval for an individual value of $\hat{Y}$ when X equals $\bar{X}$ is approximately ±$23,000. This is in contrast to an interval width of approximately ±$26,000 when X equals $10,000. This result underscores the need for caution whenever X is outside the range of the sample values of X. In fact, use of the model outside the range of the data base amounts to little more than extrapolation where, in some cases, the error term can exceed the value of the point estimate.

9.6 PRACTICAL APPLICATIONS

Typically, regression models involve more than one independent variable because of the complex nature of most business relationships (the multiple-variable regression model is presented in the next chapter). Nevertheless, simple correlation and regression models, because of their straightforward nature, are used in supporting the decision-making process. This section presents several examples involving simple correlation and regression.

Research and Development Expenditures (Lagged Variables)

R&D is playing an ever-increasing role in shaping the future direction of most firms. When a company spends money on R&D, it expects these expenditures to increase future profits. Presented in Figure 9.25 is a random sample of eight firms taken from the Fortune 500. A CBS analysis of these data is shown in Figure 9.26. The independent variable is R&D expenditures for 1983, and the dependent variable is earnings for 1985. A preliminary inspection of the computer results indicates that R&D and earnings are not linearly related. That is, the computed t-value is less than the critical t-value at a .05 level of significance. Accordingly, there is insufficient evidence to reject the null hypothesis of no linear relationship.

Jacobs Software Corporation (Forecasting and Interval Estimation)

The Jacobs Corporation designs and markets a variety of analytical software packages for commercial applications. Testing represents one of the most important elements in the development of new computer programs. However, the levels of effort and costs associated with testing are often significantly underestimated. This can result in cost overruns and lost profits. The chief designers at Jacobs collected the data presented in Figure 9.27 on 15 recent software projects.

FIGURE 9.25 R&D Expenditures and Earnings for a Random Sample of Fortune 500 Firms

	R&D Expenditures (% of sales)			Earnings (% of total assets)		
	1983	1984	1985	1983	1984	1985
Boeing	3.9	4.9	3.0	4.8	9.3	6.1
Litton Ind	4.9	5.0	5.3	5.8	7.1	6.4
Lockheed	4.4	4.7	4.5	9.3	10.9	9.6
McDonnell	3.7	3.8	3.7	5.7	5.2	4.8
Raytheon	3.7	3.9	4.1	8.0	6.7	10.9
RCA	2.4	2.4	2.8	3.1	4.1	5.5
Rockwell	3.1	3.2	3.2	7.4	8.5	8.1
Sperry	7.4	8.3	7.9	2.2	3.9	5.0

FIGURE 9.26 CBS Regression Analysis of R&D Expenditures

Results	
B_0 coefficient	8.4764
B_1 coefficient	−.3406
Mean of X	4.1875
Mean of Y	7.0500
Sum of squares regression	1.8343
Sum of squares error	33.5857
Sum of squares total	35.4200
Coefficient of determination	.0518
Correlation coefficient	−.2276
Standard error estimate	2.3659
Standard error B_1	0.5950
Computed t-value	0.5724
Critical t-value	2.447
p-value	.4
Conclusion	Do not reject null

A CBS analysis of these data is given in Figure 9.28. These results indicate a statistically significant relationship between the size of the program (lines of code) and the number of hours required for testing at the 5% level of significance. The developed regression model, however, is not entirely satisfactory since the coefficient of determination is less than .5. This suggests

FIGURE 9.27 Data Base for Jacobs Corporation Problem

Lines of Code	Hours of Testing
4,500	135.0
750	52.5
250	8.5
750	48.5
1,000	22.0
250	5.0
6,500	105.5
4,000	131.0
2,000	178.0
500	27.5
1,500	70.5
1,250	70.0
3,750	275.0
3,250	195.0
500	12.0

FIGURE 9.28 CBS Regression Analysis of Jacobs Problem

Results

B_0 coefficient	31.8539
B_1 coefficient	0.0279
Mean of X	2,050
Mean of Y	89.0667
Sum of squares regression	39,451.0230
Sum of squares error	49,866.4100
Sum of squares total	89,317.4300
Coefficient of determination	.4417
Correlation coefficient	.6646
Standard error estimate	61.9345
Standard error B_1	0.0087
Computed t-value	3.2070
Critical t-value	2.160
p-value	.005
Conclusion	Reject null

Interval Estimate

Model: ht = 31.85386 + 2.790869E-02 lc

Value of lc: 1000

Interval: 59.767 ± 36.84

that, although the size of the program is important in estimating testing time, it is not the only factor. Figure 9.28 also features a forecast and interval estimate for a program with 1,000 lines of code. The results report a point estimate of 59.77 hours with a 95% confidence interval of ±36.84 hours. This very large interval with respect to the point estimate underscores the fact that the model explains less than half of the variance in the dependent variable. Accordingly, extreme care should be exercised in using the model for estimating or forecasting.

9.7 CASE STUDY: PRITIKIN DIET

Nathan Pritikin, founder of the Pritikin Longevity Center and a major proponent of the Pritikin diet, has stated that atherosclerosis (hardening of the arteries) may be prevented, its progress may be halted, and certain of its pathological processes may even regress if risk factors (e.g., obesity) are eliminated or controlled for a protracted period by a change in lifestyle.

9.7 Case Study: Pritikin Diet

The managers at a local medical center wish to determine the extent of correlation between cholesterol and triglyceride levels and atherosclerosis. For this study they have decided to use maximum heart rate and maximum systolic blood pressure as proxies for the level of atherosclerosis. (The only definite way to measure the reduction in atherosclerosis of the heart is to perform a cardiac angiogram — an invasive procedure that involves some risk — before and after the diet program.) Data were collected on 23 patients with heart disease. This group, consisting of 16 men and 7 women, undertook the Pritikin diet over a 30-day period. The results from this experiment are presented in Figure 9.29.

Computer analysis of the data yielded the correlation coefficients given in Figure 9.30. The correlation coefficients show that changes in triglyceride levels do not have a significant impact on either heart rate or systolic pressure. The results for cholesterol levels appear a little more promising. However, even in these cases the amount of explained variance is less than 20%. Other results from this analysis show:

FIGURE 9.29 Data Base for Pritikin Diet Case

Observation	Change in Triglyceride Level	Change in Cholesterol Level	Change in Heart Rate	Change in Systolic Pressure
1	87	−71	33	22
2	−150	−89	20	−22
3	−138	−24	18	−4
4	−78	−101	1	2
5	−62	−45	−7	10
6	−23	−86	6	−28
7	−65	−121	1	−6
8	−117	−64	−1	−20
9	−1	9	8	−10
10	−88	−47	37	10
11	−137	−20	27	4
12	−12	−42	42	26
13	−42	−78	20	14
14	−73	−91	4	0
15	−101	−96	−13	−48
16	−39	−81	1	−34
17	−60	−83	−20	−16
18	61	−80	19	10
19	−42	−14	17	40
20	−136	−79	4	−38
21	−325	−31	14	−20
22	−142	−25	34	18
23	−48	−87	28	−18

FIGURE 9.30 Correlation Coefficients for Pritikin Diet Problem

Independent Variable	Dependent Variable: Change in	
	Maximum Heart Rate	Maximum Systolic Pressure
Triglyceride	−.054	.258
Cholesterol	.400	.413

- The maximum systolic blood pressure decreased by an average of 4.69 mmHg as a result of the diet.
- The maximum heart rate increased by 12 beats per minute.

9.8 SUMMARY

Correlation and regression analysis are used extensively throughout the business and scientific communities. The scientist is often more interested in determining the extent or magnitude of relationships among variables. The manager, on the other hand, is generally more interested in establishing specific relationships and forecasting events. In correlation analysis, the primary objective is determining the magnitude of the relationship between +1.0, a perfect positive relationship, to −1.0, a perfect negative relationship. In regression analysis, a least squares line is used to describe the relationship between the dependent variable (Y) and the independent variable (X). The resultant equation can be used to forecast values of Y from values of X (or vice versa). The coefficient of determination and computed t-statistic are among the primary measures used to describe the explanatory power of the developed equation model. The standard error of the estimate is used in formulating interval estimates for specific forecast values. The interpretation of both correlation and regression is partially based on a logical analysis of the data. The numbers should be meaningfully assigned, they should be interval- or ratio-scaled, and they should contain enough variance to be explained. Interpretations should be made cautiously, considering all of these criteria and not just the coefficients and forecasts. In general, forecasting should not go beyond the data base used in developing the model. Furthermore, care should be exercised in assuming causality between two variables even though they exhibit a high degree of statistical correlation.

9.9 GLOSSARY

coefficient of determination (R^2) A numerical measure of the amount of variance explained by the regression model.

correlation analysis The process of determining the extent of the relationship between two variables.

correlation coefficient (r) The degree of linear association between two variables (also known as Pearson's r).

dependent variable (Y) The variable that is being explained or predicted.

independent variable (X) The variable used to explain the variability in, or to predict the values of, the dependent variable.

least squares method A procedure for developing estimates for the regression model coefficients based on minimizing the sum of the squares of the residuals.

regression analysis The process of developing a statistically based linear model between two variables.

residual The difference between the actual values of the dependent variable and its predicted values from the regression model.

scatter diagram A plot of the independent (horizontal axis) versus dependent (vertical axis) variable values.

standard error of the estimate An estimate of the standard deviation of the dependent variable (Y) for any given value of the independent variable (X).

9.10 BIBLIOGRAPHY

Draper, N., and H. Smith. *Applied Regression Analysis.* 2nd ed. New York: Wiley, 1981.

Hanke, J., and A. Reitsch. *Business Forecasting.* Boston: Allyn & Bacon, 1981.

Meek, G. E., and S. J. Turner. *Statistical Analysis for Business Decisions.* Boston: Houghton Mifflin, 1982.

Neter, J., W. Wasserman, and M. H. Kutner, *Applied Linear Regression* Analysis, 2nd ed. Homewood, Ill.: Irwin, 1989.

9.11 PROBLEMS

1. Describe the differences between correlation and regression analysis (use examples).
2. Does a high positive or high negative correlation coefficient indicate that two variables are causally related?
3. Can two variables be causally related yet yield a low correlation coefficient? If the answer is yes, explain with examples.
4. Rank-order the following plots in terms of the highest correlation coefficient:

5. Explain the difference between linear and curvilinear relationships.
6. Develop several business examples involving curvilinear relationships.
7. Describe the four basic regression model assumptions.
8. A regression analysis of average fuel economy (Y = mpg) versus vehicle weight (X = 000 pounds) from data collected on 50 late-model cars yielded the following model:

$$\hat{Y} = 24.5 - 0.04X$$

   a. What is the significance of the intercept?
   b. If one car weighs 1,000 pounds more than another car, what reduction in fuel economy should be expected?
   c. Estimate the fuel economy for a car weighing 10,000 pounds.
   d. Does the answer to part c make any sense? Why or why not?
   e. How might this model be used?

9. A regression analysis of executive compensation (Y = $000) versus corporate earnings (X = $000) for the CEOs of 100 Fortune 1,000 companies yielded the following model:

$$\hat{Y} = 285.4 + 0.0035X$$

   a. Would you expect the relationship between compensation and earnings to be linear?
   b. What would be the impact of negative corporate earnings on executive compensation?
   c. Provide an interpretation of the slope and intercept.

10. Why do the model results presented in problems 8 and 9 underscore the need to limit the value for X to the range defined by the original data? Give examples.

11. The following data summarize monthly marketing expenditures and sales revenues for the Pulmory Irrigation Company:

Marketing Expenditures	Sales Revenues
$15,000	$105,000
19,000	155,000
17,000	137,000
22,000	148,000
18,000	142,000
19,000	149,000
24,000	167,000
27,000	172,000
23,000	151,000
21,000	135,000
16,000	121,000
14,000	110,000

   a. Develop a scatter diagram of market expenses (horizontal axis) versus sales revenues (vertical axis).
   b. What preliminary conclusions can be drawn regarding the possible relationship between marketing expenses and sales revenues?
   c. Estimate a value for the correlation coefficient (i.e., positive or negative).
   d. Draw a straight line through the data (use your own judgment).

e. Develop computer-based estimates for both the correlation coefficient and the regression line.

f. Compare the average error associated with your straight-line estimate with the computer-generated results.

12. The Atwater Oil Company is interested in determining the relationship, if any, between the price of gasoline and the amount of gasoline sold. The following data show the price versus volume for the past 10 weeks:

Price ($/gallon)	Volume (gallons)
1.25	85,000
1.20	92,000
1.20	87,000
1.10	101,000
1.05	112,000
1.00	117,000
0.99	115,000
0.95	120,000
0.95	124,000
0.90	127,000

a. Develop a scatter diagram of price (horizontal axis) versus volume sold (vertical axis).

b. What preliminary conclusions can be drawn regarding the possible relationship between price and volume?

c. Estimate a value for the correlation coefficient (i.e., positive or negative).

d. Draw a straight line through the data (use your own judgment).

e. Develop computer-based estimates for both the correlation coefficient and the regression line.

13. The managers at the Bindright Bookstore wish to know the extent of any relationship between number of floor personnel at their three retail outlets and net sales. They believe that the payroll should be related to sales. The following is a summary of the relevant data for the past year:

Month	Full-Time Equivalent	Net Sales
1	7.8	16,000
2	8.9	20,200
3	8.5	8,900
4	8.4	11,200
5	23.4	125,900
6	26.0	170,060
7	10.7	50,200
8	7.2	18,200
9	7.8	12,900
10	7.8	13,100
11	6.8	15,200
12	6.8	14,700

a. What is the nature of the relationship between staffing and gross sales?
b. What other factors might be influencing sales?
c. What is the estimated monthly sales level if the firm uses 12 full-time equivalents in the three stores?

14. The Pumping Iron Health Club has decided to initiate a drive for new members. The managers are aware that a relationship exists between the amount spent on advertising and the number of new members. The following table summarizes historical data for the past 20 months:

Month	Advertising	New Members
1	$3,028	61
2	1,586	53
3	1,023	55
4	654	43
5	767	45
6	305	36
7	201	55
8	215	38
9	121	45
10	161	37
11	319	60
12	310	31
13	441	58
14	202	75
15	54	24
16	1,831	47
17	1,068	131
18	1,123	137
19	949	49
20	1,574	34

a. Is there a relationship between advertising and new members at the 95% level?
b. What is your interpretation?
c. How many new members would the firm expect if management spent $1,500 on advertising?
d. If management wishes to gain 50 new members each month, how much should be spent on advertising?

15. The Los Angeles Fire Department would like to determine the nature of the relationship between the age of a standard fire engine and the annual maintenance cost. A sample of 12 fire engines yielded the following data:

Age (years)	Maintenance Cost ($/month)
1	425
1	375
2	450
2	460
3	515
3	490
3	540
4	590
4	610
5	825
5	790
6	890

a. Determine the sample correlation coefficient for these data.
b. Prepare an interpretation of your results.
c. Develop a regression model for these data.
d. Is the model statistically significant at the 5% level?
e. What is the estimated cost for a four-year-old fire engine?
f. Develop a 90% confidence interval estimate for a three-year-old fire engine.

16. At B. S. Manufacturing, foremen are assigned to supervise from 10 to 12 unskilled laborers. The union has filed a report with the management stating that employee complaints are increasing with respect to several of the foremen. In response, the company's vice president for human relations has decided to test each foreman to determine, among other things, his or her propensity for authoritarian management. A standard industrywide index for measuring management style (0 = Democratic, 100 = Authoritarian) was used. The results of the testing along with the level of complaints are summarized for the company's eight foremen:

Foreman	Authoritarian Index	Complaints (per 1,000 hours)
1	75	.12
2	63	.08
3	62	.07
4	70	.11
5	35	.06
6	38	.06
7	64	.10
8	20	.04

a. What is the extent of the relationship between test scores and the level of complaints?

b. If management decides to limit the foreman position to individuals who test below 50, what would be the maximum expected number of complaints?
c. What would you advise the management to tell the union?

17. The general manager of the Computer Vision Supply Company is concerned about the number of malfunctions reported in the company's deluxe tape drives. Since the hours of operation are the same, it should be possible to compare the age of the various drives with their corresponding number of malfunctions. The manager has decided to investigate this situation by performing regression analysis on the following data sample:

Age (years)	Errors (no./1,000 hrs)
1	5
1	6
2	8
4	12
4	16
5	23
6	25

a. Is there a relationship between age and machine errors?
b. How many errors should be observed for a three-year-old machine?
c. Develop a 90% confidence interval estimate for a two-year-old machine.
d. How should the manager use the regression model?

18. The chief engineer at HighPump, a manufacturer of high-pressure pump seals, is concerned about seal wear rate in high-pressure applications. The following data were provided by the company's testing lab:

Pressure (psi)	Wear (mils/khw)
3200	2.4
3200	4.9
1600	5.4
2400	6.7
2400	10.2
4400	7.8
5000	8.7
5000	7.9
5000	8.3
5000	8.7
125	0.7
125	0.6
125	0.6
190	0.6

a. What is the extent of the relationship between pressure and seal wear?
b. Is the regression model significant at a 95% level of confidence?
c. The chief engineer has recently received an order for a 3,000 psi application. What amount of wear should be expected?
d. If wear above 10 mils/kwh is unacceptable, what is the maximum pressure that should be used in future applications?

19. Describe the differences between an interval estimate for the mean of Y and one for an individual value of Y. Which interval estimate is larger and why?

20. The Protein Food Company has become interested in reducing inventory costs. Normally, inventory is taken at least monthly because of the volatility of the stock. The current method of determining inventory levels involves counting the amount of stock on the shelves and adding the total to the amount of stock still in cases. This had been fully accomplished on the last day of the month. Evaluating the product on the shelves takes a three-person crew eight hours to complete. Counting the number of boxes in the storeroom takes one person two hours to complete. For the last year the stockroom values and total inventory values were:

Month	Stockroom Value ($000)	Total Value ($000)
JAN	55.64	275.47
FEB	39.31	231.64
MAR	59.17	271.79
APR	46.21	239.00
MAY	33.24	218.83
JUN	60.40	301.78
JUL	31.49	243.91
AUG	32.88	206.99
SEP	35.59	226.28
OCT	57.92	273.11
NOV	54.27	247.99
DEC	61.23	293.54

The managers at Protein wish to know if there might be a more effective method for estimating inventory levels. Develop an appropriate regression model for this problem.

21. Using the model developed in problem 20, forecast the total inventory value based on a stockroom value of $50,000 and prepare a 95% confidence interval for the forecast.

22. Develop a 95% confidence band for predicting individual values of Y for selected values of X for problem 20.

23. The managers at Smith & Johnson are interested in stabilizing the company's sales force because of high customer demand for after-sales service and support. Accordingly, management are interested in retaining experienced sales representatives as a sure way of increasing business. Some of the yardsticks used to measure performance are sales volume, gross profit, sales growth, and number of invoices produced. The data presented here were collected on the company's 212 sales representatives over a one-year period:

Length of Service (years)	Invoices Generated	Sales Volume ($000)
2	656	147.5
4	511	121.7
4	317	119.7
5	856	181.5
6	962	174.
6	798	202.5
8	1178	288.3
9	976	287.7
12	1585	334.8
13	1321	355.9
15	1002	303.2
18	892	189.5

    a. Is there a linear relationship between years of service and either invoices generated or sales volume?
    b. Describe the correlation coefficient for part *a*.
    c. Develop an invoice forecast and confidence interval ($\alpha = .05$) for an employee with seven years of service.

24. Give three or four reasons for not predicting values outside the range of the data base.

25. The following data summarize the United States GNP (in 1982 dollars) and passenger car sales for the period 1977 to 1984.

Year	GNP ($ millions)	Passenger Car Retail Sales (000)
1977	2,959	11,185
1978	3,115	11,312
1979	3,192	10,671
1980	3,187	8,979
1981	3,249	8,536
1982	3,166	7,982
1983	3,279	9,182
1984	3,490	10,391

    a. Is there a linear relationship between GNP and the number of passenger cars sold in the United States?
    b. Describe the correlation coefficient for part *a*.
    c. Develop a unit sales forecast and confidence interval ($\alpha = .05$) using the 1985 GNP value of $3,585 (millions).
    d. Actual passenger car sales for 1985 were 11,042,000. Compare the forecast value and interval estimate developed in part *c* with this value.

e. What other factors might be used in explaining passenger car sales?
26. Develop a 95% confidence band for predicting individual values of Y for selected values of X for problem 25.
27. The following data summarize, for 15 selected geographical areas throughout the United States, the average daily maximum temperature and the average percentage of sunshine for the month of May. These data are based on a 30-year period, 1951 to 1980.

Location	Temperature (degrees Fahrenheit)	Sunshine (%)
Mobile	85	65
Phoenix	92	93
Los Angeles	69	66
San Francisco	66	72
Denver	71	64
Hartford	71	58
Miami	85	70
Chicago	70	60
Des Moines	73	61
New Orleans	85	62
Portland	63	55
Baltimore	74	56
Kansas City	75	64
New York	72	61
El Paso	87	89

a. Is there a linear relationship between average maximum temperature and the average percentage of sunshine in the United States?
b. Describe the correlation coefficient developed in part a.
c. Develop a forecast for percentage of sunshine and a confidence interval ($\alpha = .05$) for an average maximum temperature of 75°.
d. Would you advise using the developed model for predicting sunshine percentage for the month of June? December? Explain.
e. What other factors might be used to explain average sunshine percentage?
28. Develop a 95% confidence band for predicting the mean value of Y for selected values of X for problem 27.
29. The following data summarize 1984 employment in manufacturing and value added by manufacture for nine Northeast states.

State	Employment (000)	Value Added (millions)
ME	108	$ 4,812
NH	114	4,804
VT	44	2,077
MA	657	31,486
RI	114	4,241
CT	419	19,809
NY	1,347	72,361
NJ	738	36,543
PA	1,126	51,725

a. Is there a linear relationship between employment level and value added?
b. Describe the correlation coefficient developed in part a.
c. What other factors might be used to explain value added by manufacture?

30. The data presented here show the winning percentage and points scored for the 14 teams composing the American Conference of the National Football League (midseason 1988):

Team	Winning Percentage	Points Scored
Seattle	556	158
Los Angeles	444	191
Denver	444	190
San Diego	222	116
Kansas City	167	112
Cincinnati	778	252
Houston	667	215
Cleveland	667	153
Pittsburgh	222	189
Buffalo	889	199
New York	611	206
Miami	556	176
Indianapolis	444	205
New England	444	155

a. Is there a linear relationship between points scored and winning percentage?
b. Describe the correlation coefficient developed in part a.
c. What other factors might be used in explaining why a team wins?

31. The director of marketing for the U.S. Linseed Oil Association is interested in determining the extent of the relationship between wholesale price and consumption in the refining industry for linseed oil. The following data present price and consumption figures for the period from 1980 to 1986. What conclusions can be drawn at a .05 level of significance?

Year	Wholesale Price (cents/lb)	Consumption (000 lbs)
1980	29.9	184,200
1981	30.5	177,500
1982	29.5	131,500
1983	24.9	87,000
1984	31.1	129,000
1985	31.6	78,200
1986	30.7	142,000

32. Develop a 90% confidence band for estimating individual wholesale prices in problem 31.
33. The vice president of production for Dewitt Manufacturing is interested in determining the extent of the relationship between production and marginal production costs based on the following sample data:

Production (units)	Marginal Costs ($)
5,000	12
10,000	10
15,000	9
20,000	8
25,000	12
30,000	15
35,000	18
40,000	24

a. Using a scatter diagram, describe the shape of the apparent relationship between production and marginal costs.
b. Develop a linear regression model for this problem.
c. How well does the linear model match the data?
d. Select a method for transforming the dependent variable.
e. Formulate a revised regression model based on the transformed data.
f. Compare the results obtained in b and e.
g. Which model would you recommend and why?

34. Using the model developed in problem 33, forecast the total production costs based on a production level of 30,000 units and prepare a 95% confidence interval for the forecast.
35. Develop a 95% confidence band for predicting individual values of Y for selected values of X for problem 33.
36. The vice president of marketing for Gaylord International is interested in determining the extent of the relationship between advertising expense and market share. The following sample data show quarterly advertising expense and market share over the past two years:

Advertising Expense ($000)	Market Share (%)
15	10
21	12
32	14
55	18
67	21
81	24
102	26
127	27

    a. Using a scatter diagram, describe the shape of the apparent relationship between advertising expense and market share.
    b. Develop a linear regression model for this problem.
    c. How well does the linear model match the data?
    d. Select a method for transforming the dependent variable.
    e. Formulate a revised regression model based on the transformed data.
    f. Compare the results obtained in b and e.
    g. Which model would you recommend and why?

37. Presented in the following table are sample data on the average maximum temperature and the crime index for the city of Los Angeles for 1987:

Temperature (degrees Fahrenheit)	Crime Index
65	88
70	92
72	98
77	102
83	105
87	117
91	135
95	160
99	190
104	205

    a. Using a scatter diagram, describe the shape of the apparent relationship between temperature and the crime index.
    b. Develop a linear regression model for this problem.
    c. How well does the linear model match the data.
    d. Select a method for transforming the dependent variable.
    e. Formulate a revised regression model based on the transformed data.
    f. Compare the results obtained in b and e.
    g. Which model would you recommend and why?

38. Formulate a set of piecewise linear regression lines to estimate the data given in problem 36. Compare the results with those developed in problem 35.

39. Formulate a set of piecewise linear regression lines to estimate the data given in problem 36. Compare the results with those developed in problem 36.
40. Presented below is a CBS printout for a simple regression analysis ($n = 22$). Determine the coefficient of determination, the correlation coefficient, and the total sum of squares from the information presented. Is the regression model statistically significant at the .05 level?

	Results
B_0 coefficient	2,020.5726
B_1 coefficient	−59.5847
Mean of X	12.6667
Mean of Y	1,265.8334
Sum of squares regression	1,467,471.2500
Sum of squares error	599,305.5600
Coefficient of determination	
Correlation coefficient	
Standard error estimate	387.0741
Standard error B_1	19.0390
Computed t-value	3.1296
Critical t-value	2.086
p-value	.001

41. Develop an approximate 95% confidence interval for the model given in problem 40 using the mean value of X.
42. Presented below is a CBS printout for a simple regression analysis ($n = 25$). Determine the coefficient of determination, the correlation coefficient, and the total sum of squares from the information presented. Is the regression model statistically significant at the .05 level?

	Results
B_0 coefficient	919
B_1 coefficient	−6.1600
Mean of X	40
Mean of Y	672.6000
Sum of squares regression	9,486.4004
Sum of squares error	104.8000
Coefficient of determination	
Correlation coefficient	
Standard error estimate	5.9104
Standard error B_1	0.3738
Computed t-value	16.4790
Critical t-value	2.069
p-value	.000

43. Comment on the statistical significance of the printout results given in problem 42. Develop a forecast and an approximate 95% interval estimate for the mean value of X.
44. Presented below is a CBS printout for a simple regression analysis ($n = 15$).

Results	
B_0 coefficient	80.5386
B_1 coefficient	2.4703
Mean of X (X_1)	18.1333
Mean of Y (Y)	125.3333
Sum of squares regression	3,903.8977
Sum of squares error	1,811.4357
Sum of squares total	5,715.3335
Coefficient of determination	0.6831
Correlation coefficient	0.8265
Standard error estimate	11.8043
Standard error B_1	0.4667
Computed t-value	5.2931
Critical t-value	2.1600
p-value	0.0001

    a. Is the model statistically significant?
    b. Discuss the meaning of the p-value for this problem.
    c. Develop a forecast of Y when $X = 10$.
    d. Approximate a 95% confidence interval for the developed model.

45. Presented below are the computed residuals for problem 44.

	Residual Analysis		
Number	Y-Actual	Y-Predicted	Residual
1	123	110.1822	12.8178
2	147	117.5931	29.4069
3	98	107.7119	−9.7119
4	89	102.7713	−13.7713
5	118	122.5337	−4.5337
6	125	127.4743	−2.4743
7	138	134.8852	3.1148
8	135	142.2961	−7.2961
9	119	120.0634	−1.0634
10	128	125.0040	2.9960
11	117	115.1228	1.8772
12	151	147.2367	3.7633
13	164	157.1179	6.8821
14	127	144.7664	−17.7664
15	101	105.2416	−4.2416

a. Develop a histogram of the residuals.
b. Describe the resultant distribution.
c. Develop a plot of the residuals as a function of $\hat{Y}$.
d. Does this plot appear to violate any of the error term assumptions?

46. Using CBS and the data given in problem 27, develop a 95% interval estimate for both mean and individual Y values.

47. Using CBS and the data given in problem 30, develop a 95% interval estimate for both mean and individual Y values using the mean value of X.

48. The production manager at Clayton Can Company is interested in determining the nature of the relationship between training and productivity. The following data were collected over a six-month period on 20 employees.

Training Level (hrs)	Productivity (units/hr)	Training Level (hrs)	Productivity (units/hr)
10	115	0	118
8	121	8	114
15	132	12	98
4	108	16	128
0	110	10	114
20	141	0	100
16	119	6	114
0	110	20	136
12	102	10	110
24	137	0	96

a. Formulate a simple regression model.
b. Is the model statistically significant at the 95% level?
c. Develop a 95% interval estimate for both mean and individual Y values.

49. Presented below are the foreign exchange rates between the U.S. dollar and both one British pound and one Canadian dollar for the period January through December 1986.

U.S. Dollar/ British Pound	U.S. Dollar/ Canadian Dollar
0.7140	1.43
0.7140	1.43
0.7138	1.47
0.7213	1.50
0.7255	1.53
0.7196	1.51
0.7252	1.52
0.7209	1.49
0.7207	1.48
0.7204	1.43
0.7215	1.42
0.7248	1.44

a. Determine the correlation coefficient between these two variables.
b. Comment on the extent of the relationship.
c. Formulate a linear regression model between the two variables.
d. Is the model statistically significant at the 90% level?
e. If the model is significant, develop a point and 90% interval estimate for the Canadian dollar, based on a U.S.-to-British pound exchange rate of 0.72.

50. The following table highlights key economic indicators for the Eastern bloc countries for 1985:

Country	GNP per Capita ($)	Real Growth (%)	Inflation (%)	Exports ($ billions)
Bulgaria	6,420	−0.8	3.3	12.8
Czechoslovakia	8,750	1.7	2.2	17.8
East Germany	10,440	2.4	0.	24.0
Hungary	7,560	−0.9	8.6	13.5
Poland	6,470	1.6	16.2	17.8
Rumania	5,450	1.8	1.0	12.2
Soviet Union	7,400	1.2	1.5	87.0
Yugoslavia	5,600	0.2	11.9	31.3

a. Determine the extent of the relationship between GNP per capita and:
- Real growth
- Inflation
- Exports

b. Develop a regression model where real growth is the independent variable and inflation is the dependent variable.
c. Is the developed model statistically significant at the 90% level?
d. What is the impact of the Yugoslavia data on this model?

51. The following table highlights financial data on the top 15 worldwide commercial banking corporations for 1985:

Bank	Assets ($ billions)	Pretax Profit on Assets (%)	Pretax Profit on Capital (%)
1. Citcorp	142.7	1.15	25.32
2. Dailchi	119.0	0.48	18.66
3. Fuji	115.1	0.55	18.96
4. BankAmerica	113.7	0.40	8.84
5. Mitsubishi	110.7	0.53	18.54
6. Sumitomo	107.6	0.61	26.62
7. Banque Paris	98.9	0.35	22.84
8. Sanwa	96.4	0.51	18.78
9. Credit Lyonnais	90.4	0.37	31.37
10. Society Generale	87.1	0.25	18.09
11. Barclays	85.1	0.95	23.56
12. Westminster	82.7	1.02	24.35
13. Chase	81.6	0.80	16.54
14. Tokai	76.0	0.45	17.54
15. Deutsche	73.3	0.87	26.47

a. Use CBS to develop a scatter plot between corporate assets (X) and pretax profits on assets (Y) and pretax profits on capital (Y).
b. Does either of the two plots appear linear? If so, which one?
c. Develop for each dependent variable a simple regression model using corporate assets as the independent variable.
d. Which model is statistically significant at the 95% level?
e. What type of transformation might improve the accuracy of the model?

52. The following table highlights financial data on the largest 20 non-U.S. multinationals for 1985:

Name	Annual Sales ($ billions)	Profits/ Assets	Assets per Employee
1. Dutch Shell (E)*	84.87	0.071	463.56
2. British Petroleum (E)	50.66	0.037	302.51
3. ENI (E)	25.80	−0.002	183.26
4. Toyota	24.11	0.081	25.79
5. Unilever (E)	21.60	0.057	35.16
6. Elf-Aquitaine (E)	20.66	0.044	221.66
7. Matsushita	19.99	0.059	127.37
8. Pemex	19.41	0.000	212.39
9. Hitachi	18.49	0.034	127.01
10. Cie (E)	18.16	0.014	244.80
11. Philips (E)	17.84	0.023	44.47
12. Nissan	17.51	0.021	135.24
13. Petrobras	17.09	0.046	235.83
14. Siemens (E)	16.64	0.025	48.17
15. Volkswagenwerk (E)	16.03	0.008	43.79
16. Damler-Benz (E)	15.27	0.044	45.39
17. Bayer (E)	15.11	0.035	57.32
18. Kuwait Petroleum	16.69	0.057	1,140.16
19. Nippon Steel	14.79	0.008	666.54
20. Hoechst (E)	14.55	0.042	50.20

*(E) = European Community.

a. Use CBS to develop a scatter plot between corporate annual sales (X) and profits/assets on assets (Y) and pretax profits on capital (Y).
b. Does either of the two plots appear linear? If so, which one?
c. Develop for each dependent variable a simple regression model using corporate assets as the independent variable.
d. Which model is statistically significant at the 95% level?

53. Redo problem 52 using the subset of companies identified with the letter E (European Community).

54. The following table reports foreign investment position in the United States by the top seven countries for the banking and real estate industries for 1986:

Country	Investment ($ million)	
	Banking	Real Estate
United Kingdom	2,539	4,623
Netherlands	1,570	2,325
Japan	2,176	1,054
Canada	1,332	2,580
Germany	222	1,049
Switzerland	88	444
France	483	28

a. Describe the extent of the relationship between the amount invested in banking and the amount invested in real estate holdings by foreign countries.
b. Develop a linear regression model for predicting the level of investment in real estate as a function of the level of investment in banking.
c. Is the model significant at the 90% level?
d. Comment on the impact of the relatively small size of the data base.

55. The following data report the average annual growth rate by country and the estimated growth rate due to the integration effects of the European Common Market (ECM) on the member countries for the period 1974–1981:

Country	Actual Annual Growth Rate (%)	Growth Due to ECM (%)
1. Germany	2.65	0.91
2. France	2.66	1.57
3. Italy	2.74	0.42
4. Netherlands	1.99	0.53
5. Belgium	2.03	0.71
6. United Kingdom	1.24	0.37
7. Ireland	3.84	0.31
8. Denmark	1.98	−0.64

a. Determine the extent of the relationship between the effects of growth due to the ECM and the actual growth rate.
b. Develop a regression model for this purpose.
c. For what purposes could this model be used?

9.12 CASES

9.12.1 Analyzing Stock Market Trends

Many stock market analysts believe that stocks in an industry group tend to show similar sensitivity to the movement of the general market over time. One parameter often used as a measure of sensitivity is called a *beta factor*. Beta factors are computed as the relationship of the change in a stock's price to the change in the general market (in either direction) over a multiyear period. Another parameter used in measuring a stock's performance is called an *alpha factor*. The alpha factor expresses how much the stock price would have moved, on the average, every month over time, assuming that the general market remained unchanged during that period. An alpha factor is expressed as a percentage of growth per month. An analysis of computed alpha and beta factors could provide important insights into stock performance for both industry groups and specific companies.

Alpha and beta factors for the insurance industry (16 companies) and steel industry (13 companies) are given in Table 1. The period of performance is from 1980 to 1985. Each of these companies is listed on the New York Stock Exchange.

TABLE 1 Alpha and Beta Factors for the Insurance and Steel Industries, 1980–85

Insurance Industry			Steel Industry		
NYSE Symbol	Alpha	Beta	NYSE Symbol	Alpha	Beta
ACIG	4.73	0.33	AS	−0.31	0.99
AET	0.01	0.76	BS	−0.34	1.22
AML	1.30	0.57	BNY	0.71	0.66
BWD	−1.73	1.16	CRS	0.56	1.12
CI	−0.45	1.17	COS	0.70	0.99
CHUB	1.15	0.44	IK	0.56	0.77
GEC	2.33	1.44	KSC	0.71	0.71
GRN	1.15	0.90	LTV	0.42	1.91
FBH	−0.35	0.52	NII	−0.57	1.32
LNC	0.33	0.87	NX	−0.24	1.51
MLC	1.56	0.59	RS	−0.20	0.83
MMC	0.24	0.67	X	−0.23	1.14
TMK	0.57	0.66	WHX	−0.47	1.05
TIC	0.45	0.80			
FG	0.45	0.72			

Suggested Questions

1. Is there a linear relationship between alpha and beta for the insurance group? For the steel group?
2. Discuss the strength of any significant relationship.
3. Do insurance stocks show a similar sensitivity to changes in the general market?
4. How could this relationship be used in forecasting the future direction of an industry group?
5. Is there a linear relationship between the alpha factors for the two industry groups?

9.12.2 Measuring R&D Performance

Research and development (R&D) is becoming an increasingly important element in most corporations' business plans. Unfortunately, most general managers lack specific insight into the technical dimensions of R&D. Nevertheless, they are called upon to make important decisions regarding long-term goals, budgets, and milestones. Beyond these general issues, however, the general manager usually leaves the actual day-to-day operations to the technical staff until the new product appears ready for the market. This management style often leads to the misappropriation of corporate resources and false expectations.

One approach for improving the general manager's involvement in the management of R&D projects is to evaluate the R&D portfolio in terms of its strength, timing, and consistency. Two specific measures of R&D performance are project achievability and potential commercial value. Developing a model that relates achievability to commercial value could assist the general manager in identifying promising and not-so-promising technologies early in the R&D process. The use of simple analytical models could significantly improve the overall management of these often expensive and complex projects. Achievability and commercial value ratings for 15 projects recently funded by the Electric Power Research Institute (EPRI) are shown in Table 1. EPRI is the R&D management agency for the U.S. electric utility industry. Both ratings are measured on a 1-to-10 (best) scale.

TABLE 1 Ratings for 15 EPRI Projects

Project	Project Achievability	Potential Commercial Value
1	3	4
2	4	1
3	5	2
4	5	3
5	5	4
6	5	6
7	5	10
8	6	3
9	6	4
10	6	5
11	6	10
12	7	4
13	7	7
14	8	8
15	8	10

Suggested Questions

1. Is there a linear relationship between a project's achievability and its commercial value?
2. Discuss the meaning of the correlation coefficient.
3. Develop a 95% confidence interval for commercial value using an achievability level of 5.
4. What other factors might help to explain the potential value of a new product?
5. Will the type of measurement scale used influence the results?

Chapter 10

Multiple Regression Analysis

In the kingdom of the blind the one-eyed man is king.
Michael Apostolis

CHAPTER OUTLINE

10.1 Introduction
10.2 Example Management Problem: Far Filtration Company
10.3 How to Recognize a Multiple Regression Analysis Problem
10.4 Model Formulation
10.5 Computer Analysis
10.6 Practical Applications
10.7 Case Study: National Baseball League
10.8 Summary
10.9 Glossary
10.10 Bibliography
10.11 Problems
10.12 Cases

CHAPTER OBJECTIVES

The primary objectives of this chapter are to develop an understanding of

1. How to recognize multiple regression problems.
2. The basic measures used in interpreting the regression model.
3. The different methods for solving regression problems using computer analysis.
4. How to interpret the solution output.
5. How to use multiple regression models for developing point and interval estimates.

Corporate insolvency and business failures represent a major challenge facing this country's lending institutions. The decision to extend credit or to expand current credit levels is a complex issue that is influenced by a number of financial and operating factors. Recent evidence has indicated that the use of multiple regression analysis offers lending institutions an effective

> **HISTORICAL NOTE**
>
> The basic multiple regression model was developed by Karl Pearson in the early 1900s. Multiple regression analysis, however, did not receive widespread attention in either business or economics until after World War I. A significant contribution to the further development of this methodology was made by Henry Moore, who is regarded as the father of modern econometrics.

methodology for evaluating the creditworthiness of current or prospective borrowers.*

Many studies have shown that the use of financial ratios has great potential for predicting business failure. Specific ratios that have proved effective include profitability, liquidity, and leverage. This analytical approach has potential beyond mere evaluation of credit applications. Indeed, it can be applied to evaluating the entire loan portfolio as well as the performance of the portfolio manager.

10.1 INTRODUCTION

The previous chapter focused on relationships between one independent variable and one dependent or criterion variable. Before proceeding further, it should be reemphasized that labeling one variable *independent* and one variable *dependent* may not reflect the logical status of the two variables as "causer" and "caused." Rather, this is a matter of communication convenience, and it is the interest in explaining a specific factor (e.g., sales) that may dictate which is the dependent variable for a given problem. It should be no surprise that a bivariate relationship (i.e., a model with one dependent and one independent variable) may not adequately describe the behavior of the dependent variable. For example, in the Pacific Construction Company problem encountered in the last chapter, the developed regression model explained only 37% of the variability in sales volume.

In most business decision-making situations, many variables need to be considered. For example, the tendency for the Dow Jones averages to move up or down may partially depend on interest rates, but it may also depend on the bond market, the price of gold, consumer "confidence," and other eco-

---

* A. Rushinek and S. F. Rushinek, "Using Financial Ratios to Predict Insolvency," *Journal of Business Research* 15, no. 1 (February 1987).

nomic and political variables. Thus, a Dow Jones average may be said to be multiply determined or explained by various independent variables. The strength of the relationship will depend on the manner in which each factor relates to the criterion, as well as on how the factors relate to each other. This chapter focuses on the formulation and solution of multiple-variable regression models.

10.2 EXAMPLE MANAGEMENT PROBLEM: FAR FILTRATION COMPANY

The Far Filtration Company manufactures filtration elements and equipment for industrial and sewage plant applications. A major part of the company's sales revenue is generated as a result of competitive bidding. Typically, customer bid requests require specification data on filtration efficiency (measured as a percentage), filter strength (measured as the percentage of polyester), and filter operating life (measured in months). The company's chief engineer has collected the following performance and sales data over the past 14 months.

Month	Efficiency (%)	Composition (%)	Life (months)	Sales ($000)
Sep	95	12.0	27	192.360
Oct	84	12.2	24	116.560
Nov	92	12.6	29	214.200
Dec	90	12.0	22	136.830
Jan	88	12.8	27	152.230
Feb	91	12.3	26	168.400
Mar	86	13.0	29	179.400
Apr	85	12.5	29	187.210
May	84	12.2	23	145.230
Jun	93	13.5	25	165.540
Jul	89	11.8	24	161.330
Aug	88	13.3	21	138.780
Sep	92	11.9	30	172.500
Oct	93	13.7	27	186.110

The chief engineer wishes to demonstrate to management that product quality is an important factor in sales. The chief engineer plans to develop a multiple regression analysis to evaluate the impact of each quality factor (e.g., filter life) on sales.

INTERNATIONAL VIGNETTE

"How can we make products that were not forecast?" asked Mr. Pescarmona, manager of Olivetti's Distributed Processing and Office Automation Systems plant at Scarmagno, 9 kilometers from the company's administrative headquarters in Ivrea, Italy. Olivetti had been caught off guard by the huge surge in demand following the economic recovery in the early 1980s, combined with a general increase in the use of microprocessors in many market segments. Mr. Pescarmona continued, "Failure to make long-term commitments can cause future shortages when the market gets tight and delivery lead time for microprocessor chips is eight months. Last year, due to economic conditions, we were not experiencing component shortages. Today, it is a different story." Mr. Bonanate, the director of marketing support for direct sales, observed, "The supply planning process itself has become overbureaucratic. Everyone knows that the reliability of the detailed forecasts is poor. Eliminating much of the detail might help." As a short-term measure for stabilizing production, the planning staff have "pulled" some orders forward. However, the need for a more sophisticated forecasting system has become apparent to everyone on the management team.

10.3 HOW TO RECOGNIZE A MULTIPLE REGRESSION ANALYSIS PROBLEM

The main characteristics of the **multiple regression model** are as follows:

Key idea 1

1. A dependent or criterion variable Y (e.g., sales in the Far Filtration problem).
2. Two or more independent variables that can be used to explain the dependent variable (e.g., filter efficiency).
3. An objective to determine the nature and extent of the relationship between the dependent variable and the independent variables.

The basic assumptions associated with the multiple regression model are the same as those outlined in Chapter 9 for the bivariate model.

10.4 MODEL FORMULATION

The general form of the multiple regression model for the population is as follows:

$$Y = \beta_0 + \beta_1 X_1 + \beta_2 X_2 + \beta_3 X_3 + \cdots + \beta_i X_i + e$$

where:

    Y = Dependent variable
    X_i = ith independent variable
    β_i = ith population variable regression coefficient
    e = Error term

Since sample data are usually used in characterizing the coefficients, the actual form of the model is as follows:

$$Y = b_0 + b_1 X_1 + b_2 X_2 + b_3 X_3 + \cdots + b_n X_n$$

where the b_i terms are linear and unbiased estimators of the population parameters β_i.

The multiple-variable model, like the bivariate model, involves the construction of a least squares equation, the basic difference being that the former is multidimensional. In the Far Filtration example outlined above, one could predict company sales for any given value of filter efficiency, filter composition, and filter life. It should be noted that, as in any other regression model, each independent variable must be separately weighted to bring its units into common measure with the units of the criterion (Y).

The regression model for the Far Filtration problem takes on the following form:

$$Y_r = \underset{\text{Intercept}}{b_0} + \underset{\text{Efficiency}}{b_1 X_1} + \underset{\text{Composition}}{b_2 X_2} + \underset{\text{Life}}{b_3 X_3}$$

(Sales: Y_r)

where:

Y_r = Predicted value of Y (sales)
b_0 = Intercept
b_1 = Coefficient for filter efficiency
b_2 = Coefficient for filter composition
b_3 = Coefficient for filter life

The variable regression coefficients, when interpreted singularly, show the amount of change in Y for a given change in X with the effects of all other independent variables held constant. This is a generalized form of the model where the effect of each X and Y value may be viewed separately.

A more specific multiple regression model can be developed that includes the **beta coefficient,** also called the standard partial regression coefficient. Betas (as distinguished from the b coefficients) are called standard because they are in standard or Z measure. Beta coefficients are often used as a means of overcoming the problem of differing units of the independent variables. They are partial regression coefficients because all the independent variables are held constant, or "nullified."

The basic advantage in using the beta coefficients is that they indicate directly the relative contribution of each independent variable. For example, a change of one standard deviation in X_1 will result in a β_1 standard deviation change in Y. The independent variables can be ranked by the absolute value of their beta coefficients to indicate the relative sensitivity of Y to changes in each.

One of the primary tasks in multiple regression analysis is selecting the appropriate independent or predictor variables. The selection of candidate predictor variables should be based, in large part, on an understanding of the basic dynamics governing the problem. For example, an analysis of the factors impacting new housing starts would clearly focus on interest rates and regional weather conditions. The ideal situation in multiple regression analysis is that the candidate predictor variables are independent of each other. That is, they each explain a unique proportion of the total variance in the dependent variable. Panel A of Figure 10.1 illustrates this condition for a regression analysis involving three independent variables. Notice that each of the three variables explains a unique amount of the variability in Y. That is, none of the variables overlaps any of the other variables. This is not the case in Panel B of Figure 10.1. Here, variables X_2 and X_3 *do* overlap, which suggests that X_2 and X_3 are related. The phenomena of variable overlapping, technically called multicollinearity, can cause serious problems in developing accurate forecasts of Y whenever all of the overlapping variables are

FIGURE 10.1 Illustration of Ideal (panel A) and Imperfect (panel B) Relationship between Predictor Variables in Explaining Variability in Y

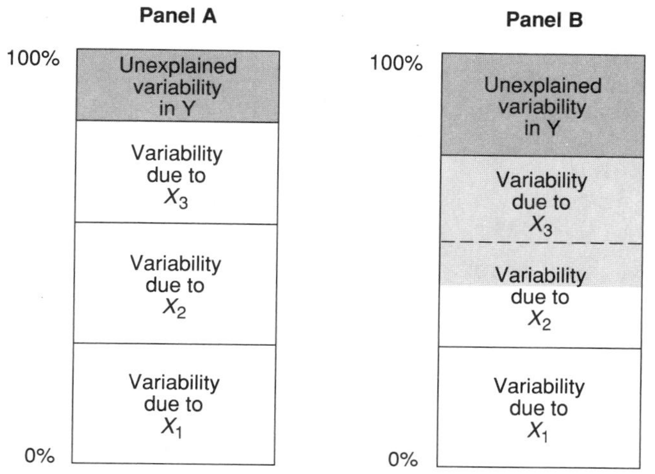

included in the model. This following section provides some additional insight into this problem along with other measures for evaluating the performance of the developed model.

MODEL MEASUREMENTS

Key idea 2

A number of measures are used in interpreting the developed regression model. Some of the more common measures are:

- *Coefficient of determination (R^2):* A measure of the amount of variance explained by the regression model, the **coefficient of determination** ranges in value between 0 and 1. For example, $R^2 = .36$ indicates that the regression model has explained 36% of the variability in the dependent variable. Recall from Chapter 9 that R^2 is computed by dividing the sum of the squares of the regression by the total sum of the squares. Care should be exercised in relying too heavily on R^2 for interpreting model results. This is especially the case for problems involving small samples and a large number of independent variables.
- *Adjusted coefficient of determination:* For problems involving small samples, the standard R^2 often "over" estimates the degree of explained variance. In fact, in small sample problems, it is possible to develop a model with an R^2 of one by merely adding new variables. To account for this "over" estimating, R^2 is adjusted downward to reflect the actual sample

> **THEORETICAL NOTE**
>
> The coefficient of determination is one of the most important measures in the evaluation of the regression model. Often, the objective is to generate large values for R^2 instead of using R^2 as an indicator of how well the regression analysis supports a given business model. Large R^2 values can be obtained from applications involving small samples or additional independent variables. Adding independent variables, particularly ones that are not statistically significant, can result in a condition known as *overfitting*. The problem of overfitting can be somewhat offset by including only statistically significant variables or by using the adjusted coefficient of determination instead of R^2 in interpreting the developed regression model.

  size. This revised measure is called the **adjusted coefficient of determination**.
- *Multiple correlation coefficient (R):* A measure of the strength of association between the dependent variable and the independent variables, the **multiple correlation coefficient** ranges in value between 0 and 1. An R value of .9 indicates a strong association, and a value of .1 indicates a weak association between the criterion and the explanatory variables.
- *Standard error of estimate:* The **standard error of estimate** is used to develop probability statements about the likely departure of predicted Y values around the regression line. For example, one standard error means that 68% of the values of Y will lie within one unit of the standard error from the regression line. The standard error of the estimate is thus interpreted in the same way as a standard deviation.
- *Multicollinearity:* In many problem situations, high correlations may exist between some of the independent variables. In multiple regression analysis this results in some degree of unreliability. In fact, as the amount of correlation between the independent variables (i.e., intercorrelation) increases, the regression coefficients become less reliable. The term for this phenomenon is **multicollinearity**. The following list outlines some of the effects of multicollinearity:
  1. Variables that are significant on a first-order basis may become collectively significant but lose individual significance in the multiple model.
  2. Individual contributions to explaining Y may be somewhat distorted.
  3. The standard error may be high for the regression coefficients, which widens the confidence intervals for the prediction model.
  4. The resultant R^2 may be spuriously high.
- *Autocorrelation:* **Autocorrelation** measures the extent of correlation between successive values of the dependent variable. For example, the demand for electrical power this year is related to the previous year's de-

FIGURE 10.2 Selected Residual Values for Pacific Construction Company Sales

Observation	Value Actual	Value Predicted	Residual
1	$212,180	$216,633	−$4,530
2	218,200	214,280	3,929
3	221,450	221,125	320
4	234,560	229,039	5,521

mand. This phenomenon is primarily observed in times series data and is more fully developed in Chapter 11.

- *Residuals:* A **residual** is the algebraic difference between an actual Y value and the predicted Y value from the regression model. Figure 10.2 shows a partial set of residuals for the Pacific Construction Company model detailed in Chapter 9. In general, the residuals should be studied very carefully for patterns. A random pattern suggests that the current model is viable. However, a negative or positive pattern could indicate that other variables should be considered to improve the fit. Another possible solution to a nonrandom residual pattern is to transform the current variables. For example, one might try to square each data point and use a quadratic form for the basic quantification. In general, one should carefully analyze the residuals prior to using the developed model.

As a final note on sample size, it is recommended that the data base contain at least 5 times as many data points (i.e., unique observations) per predictor variable. For example, a problem with a total of 5 predictor variables should have about 25 unique observations.

MODEL EXTENSIONS

This section highlights several important extensions of the basic multiple regression model. These extensions include dummy or qualitative variables, supression variables, nonlineal effects, interactive terms, and discriminant analysis.

Dummy or Qualitative Variables

Dummy variables are used to help clarify the relationship between the dependent variable and specific *qualitative* independent measures. In the standard case, a dummy variable is assigned one of two values—typically, 0 or 1. Perhaps the simplest example is classifying individuals based on sex, where males could be assigned a value of 1 and females a value of 0.

It is also possible for a dummy variable to take on more than two values. For example, an individual's highest level of education could be classified as follows: 0 = Grade school, 1 = High school, 2 = Undergraduate, 3 = Graduate. Unfortunately, it is difficult to understand the statistics associated with a discrete variable that can take on more than two values. For example, how would one interpret an average of 2.4 associated with the preceding classification scheme? Traditionally, there are two approaches for addressing problems involving qualitative independent variables that can take on more than two values. One approach involves the use of analysis of variance (ANOVA) techniques (see Chapter 13). Unfortunately, ANOVA is significantly more restrictive than multiple regression in analyzing many problems. The other approach incorporates the use of multiple dummy variables. In the educational example, one dummy variable could be used to characterize whether or not an individual had completed high school (1 = Yes, 0 = No). A second dummy variable could be used to characterize whether or not an individual had received an undergraduate degree (1 = Yes, 0 = No). A third dummy variable would be used to signify if the individual had received a graduate degree (1 = Yes, 0 = No). If all three variables were 0, this would indicate that the individual had only a grade school education. Obviously, the number of variables required to characterize several dummy factors, each possessing multiple categories, greatly expands the size of the overall variable set. This adds considerable complexity to the modeling process.

Suppose the human resources manager at Brookline Consulting Group, Ms. Betty Roberts, wishes to develop a model to help predict employee absenteeism Y based on the following employee attributes: salary, tenure, job satisfaction, and sex. Ms. Roberts is particulary interested in assessing the impact of sex. Accordingly, she collects data on 15 randomly selected employees, which are given in Figure 10.3.

One approach for assessing the impact of sex on absenteeism would be to segregate the data into two groups based on sex. A separate multiple regression analysis could then be performed on each data set. If the two resultant models showed differences, one could attribute the differences to sex. These two models can be combined into one through the use of a dummy variable that takes on the value of 1 for male and 0 for female. The resultant model would provide Ms. Roberts with the capability of directly assessing the impact of sex as well as the other independent variables on absenteeism. It should be noted that the assignment of 1 for male and 0 for female is completely arbitrary. The opposite assignment would yield exactly the same results.

This example illustrates the essence of the regression discontinuity problem. That is, the values of at least one independent variable are restricted to integers, more specifically, to 0 and 1. The interpretation of the statistics associated with dummy variables is relatively straightforward. The computed mean for the sex variable of .6 indicates that 60% of the individuals sampled were males. A solution and interpretation of the Brookline Group problem is given in Section 10.6.

FIGURE 10.3 Employee Survey Data for Brookline Consulting Group

Salary ($000)	Tenure (years)	Job Satisfaction (0–99%)	Sex (M = 1)	Absenteeism (days per year)
42	9	82	1	14
34	7	55	0	26
38	4	88	1	11
29	2	65	1	18
44	10	72	1	30
40	9	90	1	8
27	5	75	0	25
31	6	72	0	27
25	4	68	1	19
45	10	77	1	13
31	5	82	0	24
37	8	60	1	15
22	1	70	0	31
30	3	82	1	15
33	5	85	0	20

Suppression Variables

In some cases involving correlation between two independent variables, X_1 and X_2, there may remain some variance that has either zero or slightly negative correlation with the dependent variable. This variance, sometimes called *error variance*, may prevent a larger first-order correlation between X_1 and X_2. When a new variable, X_3 is added, X_2 and X_3 might correlate highly because of shared variance that is not common to X_1. In this case, X_3 may acquire a negative regression weight even though it may not correlate negatively with the criterion. This variable is called a **suppression variable** because it suppresses the variance present in other independent variables but not in the criterion variable.

Nonlinear Effects

In many cases, not all of the selected independent variables may be linearly related to the dependent variable. This does not necessarily mean, however, that the variables are not related. Recall that in Chapter 9 the methodology of data transformation was introduced, in which variable values are changed using one of several candidate data transformation schemes (e.g., logarithms). The objective behind data transformation is to improve the predictive power of the model. Interestingly, the introduction of nonlinear variables into a model via data transformation does not violate the assumption of linearity. Therefore, it is possible to include terms such as X_1^2, $\log X_2$, and

$X_1 X_2$. This development is surprising to many students. Actually, the assumption of linearity in regression pertains to the model coefficients. Accordingly, terms such as $B_2^2 X_2$ or $B_1 B_2 X_3$, which are nonlinear with regard to the coefficients, are not permitted. On the other hand, terms such as $B_4 X_1^2$, $B_5 \log X_2$, and $B_6 X_1 X_2$, which are nonlinear in X but linear in the coefficients, are acceptable. Thus, **nonlinear effects** can be incorporated directly into the model by treating them as new variables (e.g., $B_4 X_1^2$ would simply become $B_4 X_3$).

One the most difficult tasks in incorporating nonlinear variables is deciding which effects are indeed nonlinear. Suppose the marketing manager at United Electronics wishes to develop a modeling system for estimating monthly sales. The manager believes that corporate sales are influenced by the amount spent on marketing (X_1) and by interest rates (X_2). Figure 10.4 shows data on these three variables for the last 12 months. The manager, however, is unsure as to the nature of the relationship between the dependent variable and the candidate independent variables. Perhaps the simplest way to address this problem is to observe the general relationship between the dependent variable and each independent variable via a scatter diagram.

Figure 10.5 shows scatter diagrams for market expenses (X_1) and interest rates (X_2) plotted against the dependent variable. The dependent variable appears somewhat nonlinear with respect to X_1 (panel A) and linear with respect to X_2 (panel B). The apparent nonlinear relationship between Y and X_1 should not be too surprising since, in a competitive market, the rate at which sales increase as marketing expense increases should level off beyond a certain point. These scatter diagrams provide the manager with insight into the basic relationship between Y and the Xs. It should be noted that apparent nonlinear effects may not require the incorporation of nonlinear variables;

FIGURE 10.4 United Electronics Sales Data

Month	Sales ($000)	Marketing Expense ($000)	Regional Disposable Income ($billions)
1	495	85	19.5
2	507	90	16.25
3	432	87	21.5
4	389	63	20.75
5	512	125	17.5
6	435	72	22.5
7	492	96	19
8	532	118	17.25
9	447	67	22.25
10	467	90	20.25
11	501	108	17.75
12	420	82	23.50

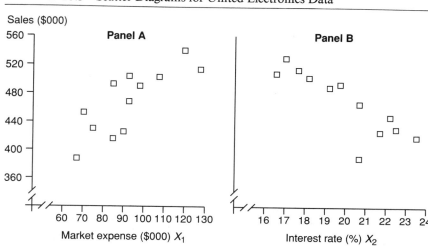

FIGURE 10.5 Scatter Diagrams for United Electronics Data

the combination effects of multiple linear variables may be sufficient to describe the variability in Y. However, in most situations it is advisable to evaluate various nonlinear relationships based on an assessment of the scatter diagrams.

The use of computer analysis greatly reduces the time required for evaluating alternative variable transformations (e.g., CBS has a built-in data transformation function for facilitating this task). A solution and interpretation of the United Electronics problem is given in Section 10.6.

Interactive Terms

A specialized case of the general nonlinear model occurs when two or more of the independent variables interact. That is, the use of **interactive terms** results in an effect on the dependent variable that cannot be described by combining the effects of the separate variables. Consider that the production manager at Wainwright Forge Company is interested in determining the extent of the relationship between production output (number of forgings per hour), the experience of the operator (years), and the downtime of the equipment. The manager collects a random sample of production data for 15 operators over a three-week period. Figure 10.6 reports the average results for each operator.

Suppose the manager believes that output (Y) is related to years of experience (X_1) and machine downtime (X_2) in the following manner:

$$Y = 290 + 11X_1 - 4X_2$$

FIGURE 10.6 Wainwright Forge Production Data

Operator	Average Production (units per hour)	Experience (years)	Downtime (%)
1	356	7	5
2	301	4	8
3	255	3	15
4	211	1	21
5	395	8	2
6	196	1	29
7	201	1	26
8	307	5	8
9	276	5	13
10	237	3	20
11	206	3	24
12	222	1	21
13	322	6	7
14	287	4	11
15	244	2	18

This relationship suggests that output should increase based on the experience of the machine operator and should decrease as machine downtime increases. Based on this relationship, if X_1 or X_2 is held constant, then Y is linearly related to the other. For example, if $X_1 = 5$, the model becomes:

$$Y = 345 - 4X_2$$

Notice that the only difference between the two models is the value of their intercepts (i.e., 290 versus 345). That is, the slope of X_2 is the same in both cases. This suggests, of course, that X_1 and X_2 are independent of each other. Suppose, however, that the manager believes the Y intercept as well as the slope of X_2 depends on the value of X_1. This situation can be addressed by incorporating a new "interactive" variable X_1X_2 into the model. The term X_1X_2 would be treated as a new variable and the estimation performed in the standard manner. Suppose the regression analysis yielded the following model:

$$Y = 280 + 15X_1 - 3X_2 - 0.5X_1X_2$$

To illustrate the principle of variable interaction, this model can be rewritten as follows:

$$Y = (280 + 15X_1) - (0.5X_1 + 3)X_2$$

In this situation, both the Y intercept and the slope of X_2 are influenced by the value of X_1. For example:

$$X_1 = 5: \quad Y = 355 - 5.5X_2$$
$$X_1 = 1: \quad Y = 295 - 3.5X_2$$

A general interpretation of these results is that output decreases as machine downtime increases. Furthermore, the rate of increase is greater for employees with more experience.

Usually, it is difficult to determine the presence of interactive effects even using scatter diagrams. Because of computational and interpretational complexities, interactive terms should be introduced only if there is some prior evidence of interaction between the independent variables. A complete solution and interpretation of the Wainwright Forge problem is given in Section 10.6

Discriminant Analysis

Thus far, multiple regression model formulation has involved a dependent variable that can be measured on a continuous scale (e.g., sales). Some business problems, however, involve situations in which the dependent variable takes on only discrete values. Specific examples include:

- Credit reference (granted or denied)
- Brand selection (brand A, B, or C)
- Sales territory performance (good, average, or poor)
- Bid proposal outcome (win or lose)

Recall that one of the basic assumptions associated with the regression model is that the dependent variable Y is normally distributed. Obviously, this assumption does not hold when Y can take on only discrete values. In these cases, a technique called **discriminant analysis** is used to develop a model for estimating a nominally scaled dependent variable.

The primary objective of discriminant analysis is to identify which of the independent variables discriminates most effectively. One approach is to develop a set of bivariate models, one for each independent variable, and then select the model that yields the highest predictive power. This approach, although useful, suffers from the fact that several variables may improve the discriminatory capability of the model. In these cases, multiple regression analysis is often used, especially when the number of categories for the dependent variable is limited to two (e.g., win or lose). More advanced techniques are required for applications involving more than two categories for the dependent variable. In any event, great caution must be exercised in interpreting the results from a discriminant analysis.

The following example illustrates the application of discriminant analysis in a business setting. Suppose the vice president at Second Federal Bank wishes to develop a mathematical system for determining the creditworthiness (Y) of loan applicants and randomly selects 12 previously evaluated applications. The financial data from the survey are given in Figure 10.7.

For this problem, the dependent variable takes on the value of either 1 (granted) or 0 (denied). Such problems can be solved directly using the

FIGURE 10.7 Second Federal Bank Financial Data

Firm	Debt/Equity	Inventory Turns	Profits/Sales	Application Status
A	45%	10	12%	Granted
B	100	4	4	Denied
C	60	6	5	Denied
D	50	8	10	Granted
E	70	7	11	Granted
F	80	9	14	Granted
G	95	7	8	Denied
H	75	8	10	Granted
I	55	7	9	Granted
J	105	8	8	Denied
K	80	7	15	Granted
L	90	6	8	Denied

multiple regression model. A solution and interpretation of the Second Federal Bank problem is given in Section 10.6.

SOLUTION METHODS

Unlike the bivariate model, multiple regression problems can be solved using several different methods. The following discussion summarizes some of the more popular options. The first three solution methods are contained in CBS.

1. *Full regression:* This method develops a relationship between the dependent variable and all of the independent variables in a nonselective manner. This approach often incorporates a number of independent variables that are statistically insignificant, thus reducing the explanatory power of the model. Furthermore, the full method does not attempt to control the degree of collinearity between the independent variables.

Key idea 3

2. *Self-stepwise:* In this approach, the user identifies the order of entry for the independent variables. This method permits the user to incorporate specific knowledge of the problem in order to develop an efficient model. But, as with the full method, the resultant model may contain a number of insignificant variables. Further, the user may not discover the most significant variables.

3. *Auto-stepwise:* This computational procedure determines the order of entry based on the variable sequence that accounts for the most amount of unique variance. The procedure terminates either when all of the independent variables have been introduced or when the remaining inde-

> **THEORETICAL NOTE**
>
> In the previous chapter, bivariate correlation was described as not implying causal relationship. The strength of a relationship was said to indicate the degree to which two variables measure the same thing. For example, the $r = .37$ relationship between sales and advertising expenditures for the Pacific Construction Company indicated that these two variables were partial measures of some underlying phenomenon. Therefore, in selecting independent variables to form a multiple-variable model, one could search for those variables with a high bivariate relationship to the dependent variable but low correlations with one another. Adding an independent variable that has a high correlation with one or more of the independent variables already included in the model would thus serve no useful purpose.

pendent variables are not statistically significant. This method allows variables added at an early stage of the computation process to be dropped as additional variables are added. More importantly, this method ensures that the degree of collinearity between the independent variables incorporated into the model is minimized.

4. *Backwards elimination:* This approach starts by incorporating all of the independent variables, as in the full method. Next, each variable in turn is dropped from the model. The significance of each dropped variable can then be measured. This approach, however, is used more for evaluating variables than regression models. The self-stepwise option in CBS can be used to simulate the backwards method.

MODEL SIGNIFICANCE

Typically, two measures are used in evaluating the statistical significance of the developed regression model. One is F, which deals with the statistical significance of the overall model. The other is t, which is used to evaluate the significance of specific variables. These two statistics are summarized below:

1. The F statistic for regression analysis is the ratio of the **mean square of the regression** (numerator) to the **mean square of the error** (denominator). The degrees of freedom for the numerator equal the number of independent variables included. The degrees of freedom for the denominator equal the number of observations minus the total number of variables in the model. For example, if there are four independent variables and 20 data points, then there are 4 degrees of freedom for the numerator and 15 degrees of freedom (i.e., $20 - 5$) for the denominator. The critical F value is obtained by using the F distribution table (see Table A.4, Appendix A) or CBS for a given level of significance and the computed degrees of

freedom (e.g., with 4 df and 15 df and a .05 level of significance, the critical F statistic is 3.06). If the computed F is greater than the critical value, the null hypothesis of no overall relationship can be rejected. In these cases, the research hypothesis that the variables operating together have a significant relationship to Y can be confirmed.

2. The t-statistic has already been described in the two preceding chapters. It is used in evaluating the statistical significance of each independent variable in relation to the dependent variable. The number of degrees of freedom is equal to the sample size minus the total number of included variables. The corresponding critical t is obtained from a t-table based on the desired level of significance (Table A.2 in Appendix A) or from CBS. The null hypothesis can be rejected when the computed t is larger than the critical t.

Rejection of the null hypothesis using either statistic implies that the relationship being tested is likely nonzero. In most computer analyses, p-values are reported for each variable, and this statistic can also be used as an indicator of statistical significance. Furthermore, unlike the t-statistic, the p-value provides the user with a direct indication of the degree of significance of the results.

MODEL FORECASTING

The developed regression model can be used for forecasting values for the dependent variables based on selected values of the independent variables. As in the case of simple regression, the forcast consists of both a point estimate and an interval estimate. The interval estimate provides a specified confidence level (i.e., 95%) around the point estimate. Typically, two different interval estimates are developed. The first is for estimating the **mean** or **average** value of Y. The second is for estimating individual values of Y. The interval estimate for the latter case, often called prediction intervals, will be somewhat larger than the former case. CBS generates both types of intervals when the forecasting option is selected. Again, as in simple regression, forecasting should be limited to values of the independent variables within the database. Estimating Y for X values outside the range of the database can yield misleading results. Typically, the confidence intervals for X values outside the database are extremely large, which tends to invalidate the point estimate for Y.

10.5 COMPUTER ANALYSIS

Most multivariable regression problems are solved using computer analysis, which allows for the rapid solution of large problems while reducing the potential for calculation errors. It is not uncommon to find business models

Key idea 4

with more than 10 independent variables. Although the input for multiple regression models is more complex than for the bivariate model, the input process is relatively straightforward. CBS uses a spreadsheet input format, which greatly simplifies the input and editing process. The basic CBS prompts are highlighted in Figure 10.8.

The Far Filtration Problem

The critical F and t statistics are derived directly from the specified level of significance (i.e., alpha) and the corresponding degrees of freedom. A commonly used alpha in business problems is .05. Unlike the bivariate model, however, the degrees of freedom will vary as a function of the number of independent variables included in the model. Thus, in a stepwise analysis the degrees of freedom will change as new independent variables are added to the model.

Figure 10.9 presents CBS-generated output for the Far Filtration problem using the full regression option for a .05 level of significance. The variable mnemonics used for this problem are: effy = Filter efficiency, comp = Filter composition, life = Filter life, and sale = Company monthly sales. The key elements of the printout are explained in some detail in the following discussion. The annotation on the printout refers to a specific discussion section.

1. B *coefficients:* The coefficients used in the multiple regression model.
2. *Standard errors:* The standard deviations associated with each of the regression coefficients.

FIGURE 10.8 CBS Input Prompts for Multiple Regression Analysis

Prompt A: Raw (R) or correlation (C) data

Prompt B: Number of variables (1–10)

Prompt C: Dependent variable number

Prompt D: Regression type:
  F = Full
  S = Self-stepwise
  A = Auto-stepwise

Prompt E: Alpha error

Prompt F: Number of data points

Prompt G: Variable names

Prompt H: Data values (raw or correlation matrix)

FIGURE 10.9 Full Regression Analysis for the Far Filtration Problem

CBS Multiple Variable Regression
Results

Variable	(1) B-Coeff	(2) Std Err	(3) Beta	(4) t-value	(5) p-value
effy	2.629	1.258	0.363	2.090	.090
comp	3.930	7.094	0.093	0.554	.600
life	6.163	1.574	0.674	3.915	.000
B_0 intercept:		−278.437			(6)
Critical t-value:		2.228			(4)
Sum squares regression:			6,321.0820		
Sum squares error:			2,454.1838		
Sum squares total:			8,775.2656		
Mean square regression:			2,107.0273		
Mean square residual:			245.4184		
C.O.D. (R^2):			.7203		(7)
Adjusted C.O.D. (R^2):			.6364		(8)
Multiple correlation coefficient (R):			.8487		(9)
df regression:			3		(10)
df error:			10		(10)
Standard error estimate:			15.6658		(11)
Computed F:			8.5855		(12)
Critical F:			3.71		(12)
Durbin-Watson statistic			2.4		(13)

Residual Analysis (14)

Actual	Predicted	Residual
192.360	184.916	7.444
116.560	138.293	−21.729
214.200	191.713	22.487
136.830	140.953	−4.123
152.230	169.655	−17.425
168.400	169.414	−1.014
179.400	177.509	1.891
187.210	172.914	14.296
145.230	132.126	13.104
165.540	173.225	−7.685
161.330	149.864	11.466
138.780	134.639	4.141
172.500	195.125	−22.625
186.110	186.338	−0.228

FIGURE 10.9 *(concluded)*

Correlation Matrix (15)

	effy	comp	life	sale
effy	1	.1242	.2364	.5335
comp		1	.0157	.1490
life			1	.7611
sale				1

Variable Statistics (16)

Variable	Mean Value	Standard Deviation
effy	89.286	3.583
comp	12.557	0.617
life	25.929	2.841
sale	165.477	25.981

3. *Beta coefficients:* The partial standard regression coefficients. They indicate the relative contribution of each variable in the model based on a one-standard-deviation change in each independent variable. For example, a one-standard-deviation increase in filter efficiency will result in a 0.363 standard deviation increase in sales.
4. *t-statistics:* The computed *t*-values for the variable coefficients. They are determined by dividing the regression coefficients by their corresponding standard errors (e.g., $t_{effy} = 2.629/1.258 = 2.090$). Both filter efficiency and filter life yield computed *t*s greater than the critical *t* value of 2.228. Therefore, both of these variables can be viewed as statistically significant.
5. *p-value:* Reported for each independent variable, they indicate the probability of observing a *t*-statistic at least as extreme as the computed *t*-statistic when the null hypothesis is true. Typically, the null hypothesis is rejected when the *p*-value is smaller than the given alpha. Notice that the *p*-value for the third variable is .000. This result suggests that the probability, as measured to three places, that the computed *t*-statistic occurred by chance is zero. In many studies *p*-values are reported instead of the *t*-values.
6. *Intercept:* The value of the dependent variable when each of the independent variables is equal to zero.
7. *Coefficient of determination (R^2):* Indicates that this model explains approximately 72% of the variability in *Y*.
8. *Adjusted coefficient of determination:* A measure that takes into account the impact of including additional independent variables in the model. The adjusted coefficient of approximately 64% can be viewed as a more

conservative estimate of the amount of explained variance. The adjusted coefficient approaches the value of R^2 as the sample size becomes large.
9. *Multiple correlation coefficient (R):* Indicates a high degree of association (i.e., .85) between sales and the three independent variables.
10. *Degrees of freedom (df) for regression and error:* The df for the regression term is equal to the number of independent variables in the model (3). The df for the term error is equal to the sample size minus the total number of variables in the model (10).
11. *Standard error of estimate:* The computed multiple standard error of estimate.
12. *F statistics:* The critical F is specified by the user from the F table, and the computed F is output from the program. In this case, the computed F statistic of 8.6 is greater than the critical F statistic of 3.71, which leads to the conclusion that the overall model is significant.
13. *Durbin-Watson statistic:* Used to indicate the presence of autocorrela-

---

FIGURE 10.10 Stepwise Regression Analysis of Far Filtration Problem

CBS Multiple Variable Regression

Results: Iteration 1

Variable	B-Coeff	Std Err	Beta	t-value	p-value
life	6.961	1.712	0.761	4.065	.000

B_0 intercept: −15.004
Critical t: 2.179

Sum squares regression:	5,083.9419
Sum squares error:	3,691.3240
Sum squares total:	8,775.2656
Mean square regression:	5,083.9419
Mean square residual:	307.6103
C.O.D. (R^2):	.5793
Adjusted C.O.D. (R^2):	.5443
Multiple correlation coefficient (R):	.7611
df regression:	1
df error:	12
Standard error estimate:	17.5388
Computed F:	16.5272
Critical F:	4.75
Durbin-Watson statistic:	2.4

FIGURE 10.10 *(concluded)*

Results: Iteration 2

Variable	B-Coeff	Std Err	Beta	t-value	p-value
life	6.151	1.524	0.673	4.037	.000
effy	2.716	1.208	0.374	2.248	.035

B_0 intercept: −236.487
Critical t: 2.201

Sum squares regression:	6,245.7729
Sum squares error:	2,529.4924
Sum squares total:	8,775.2656
Mean square regression:	3,122.8865
Mean square residual:	229.9539
C.O.D. (R^2):	.7117
Adjusted C.O.D. (R^2):	.6593
Multiple correlation coefficient (R):	.8437
df regression:	2
df error:	11
Standard error estimate:	15.1642
Computed F:	13.5805
Critical F:	3.98
Durbin-Watson statistic:	2.4

Remaining variables are insignificant

---

tion associated with time series data. A more detailed discussion of this statistic can be found in Chapter 11.
14. *Residual analysis:* Indicates the difference between the actual Y value and the estimated value from the regression model. For example, the residual for the first data point is 7.444.
15. *First-order correlation coefficients:* A table that presents the Pearson correlation (r) coefficient between each pair of variables. This table is an intercorrelational matrix. Notice that the diagonal shows a value of 1.00 for each variable correlated with itself.
16. *Variable means and standard deviations:* The mean and standard deviation for each of the variables.

This problem was also evaluated using the stepwise solution procedure. The standard CBS output is given in Figure 10.10. Notice that this printout consists of two iterations. The first iteration introduces the variable *life* into the model. This variable was selected first because it possesses the largest correlation coefficient with respect to the dependent variable (i.e., .76). In this case, the computed t is larger than the critical t, which leads to the conclusion that *life* is statistically significant. This bivariate model explains approximately 57% of the variance in sales. The issue at this point is whether either of the two remaining independent variables can further improve the predictive power of the model.

Recall the correlation coefficient matrix presented in Figure 10.9. This matrix showed that *effy* had the second largest correlation coefficient with respect to sales. In general, one cannot always conclude that the variable with the next largest correlation coefficient will be selected for inclusion. In this case, however, the second iteration has selected *effy* as the next explanatory variable. The revised R^2 is 71%. That is, the incorporation of both variables has increased the explanatory power of the model from 58% to 71%. The third iteration shows that the remaining variable, *comp*, is statistically insignificant.

Thus, the resultant stepwise model contains two variables instead of the three used in the full regression model. The R^2 value for the two-variable model has dropped 1% (i.e., from 72% to 71%) compared to the full regression model. Recall, however, that only two of the three variables in the full model were statistically significant. Therefore, one can conclude that the two-variable model is more appropriate than the three-variable model. An example of using this model for forecasting appears in Figure 10.11. The presented forecast is based on an efficiency value of 90 and a life value of 26 estimated sales value of $167,865, with a 95% interval of $9,105 based on the mean valve of Y. Notice that both the critical F and critical t vary from iteration to iteration. This is the direct result of the change in the degrees of freedom.

10.6 PRACTICAL APPLICATIONS

Key idea 5

Ongoing surveys of the business community indicate that multiple regression continues to be one of the most used quantitative tools. Examples of several different multiple regression applications are presented in the following sections.

Home Real Estate (Correlation Analysis)

The manager at Home Real Estate, Ms. Alice Stroud, is interested in better understanding the dynamics of residential home prices. Accordingly, she has collected data from the real estate multiple-listing board for the Santa Barbara area. The data is a listing base of sales price, dwelling size (square footage), location (1–5 scale), and lot acreage (square footage) for 30 single-

FIGURE 10.11 Forecast of Sales for Far Filtration Company

Forecasting

Variable	b Coefficient	Significant	Value
life	6.1510	Yes	26
effy	2.7158	Yes	90

sale est. = 167.865 ± 9.105

family homes. A preliminary analysis of the data base generated the correlation matrix and statistics that are shown in Figure 10.12.

The correlation matrix shows that a strong positive correlation exists between home size and acreage (i.e., .70). This should not be surprising since larger homes usually require a larger lot. The statistical data also shows that the average sales price for the 30 homes is $210,000 and that the average size is 2,500 square feet. This data set was evaluated using stepwise regression analysis at a .05 level of significance. The results are shown in Figure 10.13. The stepwise procedure selected location (*loca*) as the first entering variable; this was found to explain 56% of the variability in sales price. The next iteration selected dwelling size (*size*) for incorporation into the model. The combination of these two explanatory variables yields an R^2 of nearly 96%. The remaining independent variables are deemed not statistically significant and are therefore not incorporated into the model.

Ms. Stroud wishes to use this model to estimate the price for a home recently placed on the market. This home has 2,000 square feet of living area

FIGURE 10.12 Correlation Matrix and Statistics for Home Real Estate

	size	loca	acre	pric
size	1	.1000	.7000	.7000
loca		1	.1500	.7500
acre			1	.5000
pric				1

Variable	Mean Value	Standard Deviation
size	2,500	300
loca	3	1
acre	6,500	500
pric	210	40

and an average location (i.e., loca = 3). The developed forecast is given in Figure 10.14. Ms. Stroud can conclude that an appropriate asking price for this home is on the order of $167,917, with a 95% interval of $6,357 based on the mean valve of Y. Naturally, other factors such as the physical condition and specific amenities may substantially alter this estimate.

Brookline Consulting Group (Dummy Variables)

Recall that the human resources manager at the Brookline Consulting Group, Ms. Betty Roberts, was interested in developing a model for predicting employee absenteeism based on a number of employee attributes, including sex (see Section 10.4). The results of a stepwise analysis of the collected data at the .05 level are given in Figure 10.15. The analysis shows that sex, a dummy variable, was selected first. This variable explains approximately 47% of the variability in absenteeism. The other significant variable was job satisfaction. The remaining two independent variables were determined not to be statistically significant. This two-variable model explains

FIGURE 10.13 CBS Stepwise Regression Analysis of Home Real Estate Problem

Results: Iteration 1

Variable	B-Coeff	Std Err	Beta	t-value	p-value
loca	30	5	0.750	6	.000

B_0 intercept: 120
Critical t-value: 2.048

Sum squares regression:	26,100
Sum squares error:	20,300
Sum squares total:	46,400
Mean square regression:	26,100
Mean square residual:	725
C.O.D. (R^2):	.5625
Adjusted C.O.D. (R^2):	.5469
Multiple correlation coefficient (R):	.7500
df regression:	1
df error:	28
Standard error estimate:	26.9258
Computed F:	36
Critical F:	4.18
Durbin-Watson statistic:	2.3

FIGURE 10.13 (concluded)

Iteration 2

Variable	B-Coeff	Std Err	Beta	t-value	p-value
loca	27.475	1.603	0.687	17.139	.000
size	0.084	0.005	0.631	15.753	.000

B_0 Intercept: −82.862
Critical *t*-value: 2.052

Sum squares regression:	44,408.0820
Sum squares error:	1,991.9192
Sum squares total:	46,400
Mean square regression:	22,204.0410
Mean square residual:	73.7748
C.O.D. (R^2):	.9571
Adjusted C.O.D. (R^2):	.9539
Multiple correlation coefficient (*r*):	.9783
df regression:	2
df error:	27
Standard error estimate:	8.5892
Computed *F*:	300.9706
Critical *F*:	3.34
Durbin-Watson statistic:	2.3

Remaining variables are insignificant

---

approximately 61% of the variability in the dependent variable. Notice that both regression coefficients are negative. This suggests, for example, that male individuals with higher job satisfaction will have lower rates of absenteeism. This trend is illustrated in Figure 10.16, which shows a forecast of absenteeism based on a male employee with a satisfaction rating of 80.

FIGURE 10.14 Forecast of Housing Price for Home Real Estate

Variable	b-Coefficient	Significant	Value
loca	27.475	Yes	3
size	0.084	Yes	2000

pric est. = 167.917 ± 6.357

United Electronics (Nonlinear Effects)

Recall that the marketing manager of United Electronics was interested in developing a model for forecasting monthly sales (see Section 10.4). Furthermore, the manager believed that sales and marketing expenses were related in a nonlinear manner. Two candidate multiple regression models that relate sales to marketing expense (ME) and interest rates (IR) are presented on the next page. The first model represents a standard regression analysis where both independent variables are linearly related to sales. The numbers in parentheses represent the computed t-values. This model explains 80% of the variation in the dependent variable. In this model, only interest rates are statistically significant at the .05 level. The second model transforms the marketing expense data using logarithms. In this case, R^2 is increased to 84%, and both of the explanatory variables are significant at the .05 level. Clearly, other possible transformations could be used. One approach for determining the "best" transformation is to develop a scatter diagram of the dependent variable values plotted against each of the independent variables. In most cases, the general shape of the relationship will become apparent (see Figure 10.5).

FIGURE 10.15 CBS Stepwise Regression Analysis of Brookline Problem

Variable	B-Coeff	Std Err	Beta	t-value	p-value
sex	−9.611	2.843	−0.684	−3.380	.002

B_0 intercept: 25.500
Critical t-value: 2.160

Sum squares regression:	332.5444
Sum squares error:	378.3889
Sum squares total:	710.9334
Mean square regression:	332.5444
Mean square residual:	29.1068
C.O.D. (R^2):	.4678
Adjusted C.O.D. (R^2):	.4268
Multiple correlation coefficient (R):	.6839
df regression:	1
df error:	13
Standard error estimate:	5.3951
Computed F:	11.4250
Critical F:	4.67
Durbin-Watson statistic:	2.3

FIGURE 10.15 *(concluded)*

Iteration 2

Variable	B-Coeff	Std Err	Beta	t-value	p-value
sex	−8.847	2.544	−0.630	−3.477	.001
sat	−0.270	0.126	−0.387	−2.136	.030

B_0 intercept: 45.236
Critical *t*-value: 2.179

Sum squares regression:	436.8010
Sum squares error:	274.1324
Sum squares total:	710.9334
Mean square regression:	218.4005
Mean square residual:	22.8444
C.O.D. (R^2):	.6144
Adjusted C.O.D. (R^2):	.5501
Multiple correlation coefficient (R):	.7838
df regression:	2
df error:	12
Standard error estimate:	4.7796
Computed *F*:	9.5604
Critical *F*:	3.88
Durbin-Watson statistic:	2.3

Remaining variables are insignificant

$$\text{Sales} = 568.94 + 1.028 \times \text{ME} - 9.712 \times \text{IR}$$
$$\phantom{\text{Sales} = 568.94 + 1.0}(2.090) \phantom{\times \text{ME} - } (-2.409)$$
$$R^2 = .80$$

$$\text{Sales} = 209.94 + 228.26 \log \text{ME} - 9.310 \times \text{IR}$$
$$\phantom{\text{Sales} = 209.94 + 228.}(2.320) \phantom{\log \text{ME} - } (-2.400)$$
$$R^2 = .84$$

Wainwright Forge Company (Interactive Effects)

Recall that the production manager at Wainwright Forge Company was interested in estimating the extent of the relationship between production (Units) and the variables operator experience (Years) and machine down-

FIGURE 10.16 Forecast of Absenteeism for Brookline Consulting Group

Variable	b-Coefficient	Significant	Forecasting Value
sex	−8.8468	Yes	1
sat	−0.2697	Yes	80
	abs est. = 14.819		

time (Dtime). Based on a random sample of 15 operators, the following multiple regression model was generated:

$$\text{Units} = 293.821 + 10.988 \times \text{Years} - 4.367 \times \text{Dtime}$$
$$(2.944) \qquad\qquad (-4.405)$$
$$R^2 = .96 \qquad F = 152.37$$

Notice that both independent variables are significant at the .05 level, as indicated by their respective t-values (in parentheses). Furthermore, this model explains approximately 96% of the variability in the dependent variable. This model, not surprisingly, suggests that production will increase for operators with more experience and will decrease as machine downtime increases.

However, the manager felt that there may be some interaction between operator experience and machine downtime. Therefore, a second model was estimated, which incorporated an interactive term. The resultant model, where *prod* represents the interactive term, is:

$$\text{Units} = 286.211 + 15.547 \times \text{Years} - 3.034 \times \text{Dtime} - 0.728 \times \text{Prod}$$
$$(8.482) \qquad\qquad (-6.146) \qquad\qquad (-6.813)$$
$$R^2 = .99 \qquad F = 501.47$$

This model explains approximately 99% of the variability in Y. Notice that all three terms are statistically significant at the .05 level. More specifically, this model suggests that an interactive effect does exist. It would appear that output (units) decreases as machine downtime increases. Moreover, the rate of decrease is greater for operators with more experience, as indicated by the negative coefficient of the interactive term (−0.728). This means that although the output for more experienced operators is greater than for less experienced operators, as measured by the positive coefficient for the variable *Years*, the rate of production dropoff as downtime increases will also be greater. This interactive effect yields a result that cannot be predicted by either of the independent variables taken separately.

Second Federal Bank (Discriminant Analysis)

Recall that the vice president of Second Federal Bank wishes to develop a system for evaluating the creditworthiness of loan applicants (see Section 10.4). This problem requires a slightly different approach due to the fact that the dependent variable is measured on a nominal scale (i.e., approve or deny credit). A stepwise analysis yielded the following model at the .05 level:

$$\text{Credit} = 0.503 + 0.1034(p/s) - 0.0119(d/e)$$
$$(4.333) \quad (-3.047)$$

These results show that both the profit-to-sales ratio (p/s) and the debt-to-equity ratio (d/e) are statistically significant at the .05 level. These two variables explain approximately 81% of the variance in the dependent variable.

Consider that a firm has applied for a credit line at Second Federal. The company reports a profit-to-sales ratio of 10 and a debt-to-equity ratio of 60. Substituting these values into the developed discriminant function yields an estimate of .828. Notice that this value is neither 0 nor 1 (theoretically the only two possibilities). However, since this value is larger than the average (i.e., .5) the vice president should consider granting the credit line. In general, means can be used for assigning a forecast outcome to a particular category.

At this point the vice president might wish to know the statistical accuracy of the developed model. As mentioned earlier, using the computed R^2 as a guide can be somewhat misleading. An alternative approach is to formulate an error analysis table, as shown in Figure 10.17. This table looks very similar to the error table used in hypothesis testing. The discriminant model was used to evaluate the creditworthiness of 12 previous applicants. For this problem, the model was 100% accurate. That is, the model correctly classified the bank's credit decisions in each case. The error table approach provides the user with a relatively straightforward methodology for evaluating the statistical significance of the decision-making process.

FIGURE 10.17 Error Analysis Table for Second Federal Predictions

Actual	Decision Made		Total
	Deny	Approve	
Approve	0	7	7
Deny	5	0	5

10.7 CASE STUDY: NATIONAL BASEBALL LEAGUE

Professional baseball became increasingly competitive in the 1980s. Between the period 1980 to 1986, no team repeated as world champion, and only three teams repeated as league champion. Many factors contributed to this situation, including the advent of free agency. Because of this competitive environment, team management has placed even more emphasis on identifying those factors that contribute to winning baseball. Figure 10.18 presents the National League final standings for the 1983 season (the teams are listed in order of finish). The data show the winning percentage (WP), the batting average (BA), the number of runs scored (RS), the number of home runs produced (HR), and the earned run average (ERA).

The management of the "lowly" New York Mets wishes to improve the team's performance for the next season. The management has hired a consulting firm to undertake a multivariate regression analysis of this problem. The primary objective of the analysis is to identify the most significant areas that contribute to winning baseball. The firm elected to use winning percentage as the dependent variable (another possibility would be final standing). A CBS computer analysis of the data yielded the results given in Figure 10.19.

These first-order correlations reveal that of the independent variables ERA has the largest correlation with the dependent variable, WP. This supports the old baseball adage that "winning baseball is 80% pitching." Notice that the correlation coefficient is negative. This indicates that as the ERA increases (i.e., more runs allowed), the chances of winning are reduced.

FIGURE 10.18 1983 Final Standings of National Baseball League

| Team | Wins (%) | Batting | | | Pitching (ERA) |
		Average (%)	Runs	Home Runs	
Philadelphia	566	249	696	125	3.34
Los Angeles	562	250	654	146	3.10
Atlanta	543	272	746	130	3.67
Houston	525	257	643	97	3.45
Pittsburgh	519	264	659	121	3.55
Montreal	506	264	677	102	3.58
San Diego	500	250	653	93	3.62
San Francisco	488	247	687	142	3.70
St. Louis	488	270	679	83	3.80
Cincinnati	457	239	623	107	3.98
Chicago	438	261	701	140	4.08
New York	420	241	575	112	3.68

FIGURE 10.19 CBS-Generated Correlation Matrix for Baseball Standings

Correlation Matrix

	WP	BA	RS	HR	ERA
WP	1				
BA	.296	1			
RS	.485	.676	1		
HR	.179	−.133	.336	1	
ERA	−.79	.05	.018	−.212	1

10.8 SUMMARY

Multiple correlation and regression analysis are natural extensions of the bivariate model developed in Chapter 9. Both models are based on the principle that the degree to which two variables are correlated represents the degree to which they are measures of something common (i.e., they share a common variance). Several different computational methods are available for solving multiple regression problems. These include: full, self-stepwise, auto-stepwise, and backwards elimination. The auto-stepwise is perhaps the most frequently used method since it is designed to minimize the degree of colinearity in the developed model. The stepwise computational process for multiple regression involves adding new independent variables until the variable set has been exhausted, until all of the variance has been explained, or until the remaining variables are statistically insignificant. The developed model can be used to estimate values of Y based on given values of the independent variables. The resulting values of Y are then treated as any other predictions, with the standard error of estimate determining the range around any given point on the line.

In addition to the basic multiple regression model, this chapter introduced several important extensions. These included dummy variables, nonlinear effects, and discriminant analysis. Furthermore, considerable attention was given to interpreting the output of computer analysis. In conclusion, it should be restated that multivariate relationships do not imply multiple causation. A high R^2 simply means that Y varies with a number of Xs operating together. Only a logical analysis can shed light on multiple causation. However, the idea that one can increase the efficiency of a forecast by increasing the amount of data on which it is based is intuitively appealing. In interpreting model results, one should exercise care, however, that multicollinearity, autocorrelation, or some randomness doesn't produce spuriously high relationships.

10.9 GLOSSARY

adjusted coefficient of determination The proportion of total variance in the dependent variable explained by the independent variables adjusted for the sample size.

autocorrelation Measures the extent of correlation between successive values of the dependent variable.

auto-stepwise This solution method determines the order entry of the independent variables, based on the amount of unique variance.

backward elimination This solution method starts with the full model and then eliminates variables one at a time.

beta coefficient The standard normal regression coefficient expressed in terms of the standard deviation of the corresponding variable.

coefficient of determination (R^2) The proportion of total variance in the dependent variable explained by the independent variables.

discriminant analysis A special form of the regression model where the dependent variable is measured on a nominal scale.

dummy variables Variables used to characterize qualitative factors (e.g., sex).

full regression This solution method develops a relationship between the dependent variable and all the independent variables in a nonselective manner.

interactive terms A condition where the combination of two or more independent variables has an effect on the dependent variable that cannot be accounted for by their separate contributions.

mean square regression (MSR) The sum of squares of the regression divided by the degrees of freedom for the regression.

mean square regression error (MSE) The sum of squares of the error divided by the degrees of freedom associated with the error.

multicollinearity A situation where some of the independent variables are interrelated.

multiple correlation coefficient (R) A measure of the association between the dependent variable and the set of independent variables.

multiple regression model A statistically based model containing a single dependent or criterion variable and two or more independent or explanatory variables.

nonlinear effects A condition where the relationship between one or more of the independent variables and the dependent variable can best be described with a nonlinear function.

residual The difference between the actual Y value and the predicted Y value from the regression model.

self-stepwise This solution method allows the user to identify the order entry for the independent variables.

standard error of the estimate A measure of the variability of Y about the regression line.

suppression variables A variable that suppresses or reduces the variance present in other independent variables but not in the criterion variable.

10.10 BIBLIOGRAPHY

Affi, A. A., and V. Clark. *Computer-Aided Multivariate Analysis.* Belmont, Calif.: Wadsworth, 1984.

Hanke, J.; A. Reitsch; and J. P. Dickson. *Statistical Decision Models for Management.* Newton, Mass.: Allyn & Bacon, 1984.

Meek, G. E., and S. J. Turner. *Statistical Analysis for Business Decisions.* Boston: Houghton Mifflin, 1983.

Neter, J., W. Wasserman, and M. H. Kutner. *Applied Linear Regression Models,* 2nd ed. Homewood, Ill.: Irwin, 1989.

————. *Applied Linear Statistical Models,* 3rd ed. Homewood, Ill.: Irwin, 1990.

10.11 PROBLEMS

1. Define and give examples for each of the following terms:
   a. Stepwise regression analysis
   b. Full regression analysis
   c. Collinearity
   d. Dummy variable
   e. Nonlinear effects
2. Identify two or more independent variables that might be useful in explaining the following situations:
   a. Executive compensation
   b. Corn production
   c. Television sales
   d. Oil prices
3. What role does the correlation matrix play in multiple regression analysis?
4. See the following correlation matrix, where Y is the dependent variable:

	Y	X_2	X_3	X_4
Y	1			
X_1	.345	1		
X_2	−.545	.72	1	
X_3	.85	.10	−.35	1

   a. Why are only half of the entries presented?
   b. Why are the entries all 1 along the main diagonal?
   c. Which independent variable will be selected first and why?
   d. Which set of independent variables has the highest level of collinearity?
   e. What is the significance of the negative signs in the matrix?
5. The management at Micro Computer Company wish to evaluate the impact of advertising on monthly gross sales. Currently, the company advertises in several trade journals, on the radio, and in the newspaper. The following data present sales and advertising expenses over the past 15 months.

Trade Journals	Radio	Newspaper	Sales
$2,500	$4,250	$1,250	$64,500
2,500	5,200	1,000	57,500
2,000	4,900	1,250	62,000
1,500	4,500	1,500	55,200
1,800	4,500	1,200	60,500
2,200	5,500	1,000	65,700
2,000	4,000	750	52,300
1,750	3,800	900	54,250
1,500	3,500	600	49,800
1,900	3,900	1,000	56,300
2,000	3,700	1,250	59,500
2,300	4,000	1,300	67,900
2,100	4,200	1,100	68,500
2,000	4,400	1,200	70,200
2,000	4,500	1,250	73,400

    a. Which variable best explains the variability in sales?
    b. Which combination of variables yields the highest coefficient of determination?
    c. Prepare a forecast, including a 90% interval estimate, using the mean values for the significant predictor variables.
    d. What other factors might contribute to explaining sales variability?
6. Presented in the following table are wheat acreage, yield, production, and price statistics from the major wheat production states for 1985.
    a. What factor should serve as the dependent variable?
    b. What factor best explains the variance in the dependent variable?
    c. What model do you recommend?
    d. What is the computed R^2 associated with part *c*?
    e. What other factors might be used in improving the predictive power of the model?

State	Acres Harvested (000s)	Yield per Acre (bu)	Production (bu)	Price ($/bushel)
CA	630	83	68.9	3.44
CO	3,522	40	139.3	2.91
GA	825	31	25.6	2.85
ID	1,350	53	72.0	3.26
IL	750	49	36.8	3.05
IN	700	53	37.1	2.90
KS	11,400	38	433.2	3.05
MI	750	60	45.0	2.95
MN	2,683	53	142.4	3.23
MO	1,280	39	49.9	2.95
MT	3,960	13	50.2	3.55
NE	2,300	39	89.7	3.00
NC	760	29	22.0	3.35
ND	8,870	36	323.3	3.32
OH	950	62	58.9	3.00
OK	5,500	30	165.0	3.20
OR	1,065	53	56.0	3.40
SD	3,755	30	11.2	3.36
TX	5,860	32	187.2	3.10
WA	2,690	48	128.3	3.40

7. Develop a 95% confidence band for predicting the mean value of Y for selected values of X for problem 6.
8. The management of Offshore Realty is interested in determining the extent of relationship between the dependent variable sales price ($) and site location, lot size (ft^2), and the number of bedrooms for single-family dwellings. A survey of 50 recently sold homes yielded the following correlation matrix along with means and standard deviations for the four variables:

	Correlation Matrix					
	Site	Size	Bedrooms	Price	Mean	SD
Site	1	.3	.1	.8	3	1
Size	.3	1	.6	.7	8,000	2,000
Bedrooms	.1	.6	1	.5	3	1
Price	.8	.7	.5	1	150,000	20,000

a. Based on examination of the correlation matrix, which independent variable will be selected first?
b. What is the final form of the regression model?
c. Does the stepwise approach yield the same model as the full regression procedure?

d. Prepare a forecast and 90% confidence interval for the developed model.
e. Discuss the significance of the beta coefficients for this model.

9. Develop a 95% confidence band for predicting the mean value of Y for selected values of X for problem 8.

10. Following is a correlation matrix of cotton yield (lb/acre), cotton production (000 bales), and cotton price (cents/lb) for the major cotton-producing states for 1985. Also shown are means and standard deviations for the three variables.

Correlation Matrix

	yield	prod	price
yield	1	−.0022	.9003
prod		1	.0346
price			1

Variable	Units	Mean Value	Standard Deviation
yield	lb/acre	760.500	243.386
prod	bales (000)	1,273.400	1,271.692
price	cents/lb	54.780	1.903

a. Which of the two independent variables seems to best explain the variability in price?
b. Does the correlation matrix indicate the presence of multicollinearity?
c. Why is the correlation coefficient between production and price so small?
d. What is the value of R^2?
e. What other variables might be added to the model to improve R^2?

11. The following CBS computer output is from a problem consisting of five variables and eight data points that was analyzed using the full regression method:

Multiple Variable Regression

Variable	B-coeff	Std Err	Beta	t-value	p-value
M	1.411	0.773	0.943	1.825	.06
R	2.123	2.567	0.759	0.827	.4
A	.86	1.193	0.31	0.721	.600
Q	−5.473	5.173	−0.915	−1.058	.320

B_0 intercept: 89.727

C.O.D. (R^2):	.847
Adjusted C.O.D. (R^2):	.643
Multiple correlation coefficient (R):	.92
df Regression:	4
df Error:	3
Computed F:	4.173
Critical F:	28.7
Durbin-Watson statistic:	1.2

a. Identify the independent variables that are statistically significant.
b. Is the overall model statistically significant?
c. If the answer to part b is no, then explain why R^2 is so large.
d. Formulate the regression model in standard form.
e. What modifications to this model would you recommend?
f. Comment on the size of the data base relative to the number of variables.

12. Provide an interpretation of the p-values presented in problem 11.
13. The following CBS computer output presents an analysis of hospital costs for the 50 states (1984) using the full regression method. The data base consisted of the number of hospitals (hosp), the average daily census (cens), the average occupancy rate (rate), the average number of employees (empl) and the average cost per day (cost). The occupancy rate is the ratio of the average daily census to every 100 beds. The average daily census and employment levels are measured in thousands.

CBS Multiple Variable Regression

Information Entered

Number of variables:	5
Number of data points:	50
Alpha error:	.1
Critical t-value:	2
Critical F value:	5
Dependent variable:	5(cost)

Results

Variable	B-coeff	Std Err	Beta	t-value	p-value
hosp	−0.528	0.278	−0.769	−2.541	.050
cens	−15.011	3.412	−3.867	−4.400	.000
rate	−0.821	0.969	−0.111	−0.847	.300
empl	4.974	1.039	4.785	4.788	.000

B_0 Intercept:	455.083
Critical t-value	2

Results

Sum squares regression:	118,191.2580
Sum squares error:	202,091.1410
Sum squares total:	320,282.4100
Mean square regression:	29,547.8145
Mean square residual:	4,490.9146
C.O.D. (R^2):	.3690
Adjusted C.O.D. (R^2):	.3129
Multiple correlation coefficient (r):	.6075
df regression:	4
df error:	45
Standard error estimate:	67.0143
Computed F:	6.5795
Critical F:	3.79
Durbin-Watson statistic:	1.3

Correlation Matrix

	hosp	cens	rate	empl	cost
hosp	1	.8634	.0994	.8987	.1814
cens		1	.2968	.9890	.1684
rate			1	.2625	−.0793
empl				1	.2405
cost					1

Variable	Mean Value	Standard Deviation
hosp	137.100	117.711
cens	19.266	20.827
rate	68.968	10.936
empl	72.120	77.780
cost	395.540	80.848

a. Identify the independent variables that are statistically significant.
b. Is the overall model statistically significant?
c. Comment on the magnitude and sign of the beta coefficients.
d. Comment on the relatively small value of R^2.
e. Explain why the first-order correlation coefficients are relatively small with respect to the dependent variable, yet several of the independent variables are statistically significant.
f. Is multicollinearity present in this model?
g. What modifications to this model would you recommend?

14. Provide an interpretation of the p-values presented in problem 14.
15. Refer to the data given in Section 10.7 on the National Baseball League. Instead of using winning percentage as the dependent variable, consider one of the following alternatives:
    - The actual ordinal ranking of the team (e.g., New York ranked 12th).
    - Assigning a 2 to the top four finishers, a 1 to the next four finishers, and a 0 to the last four teams.

    a. How does a change in the dependent variable alter the model?
    b. Which independent variable is most important in predicting winning performance?
16. The purchasing manager at Jade Electronics, a manufacturer of electronic counter systems, is interested in reducing the average requisition backlog due to extremely tight production schedules. The manager has collected the following parts requisition data for 16 months of operations.

Month	Total Requisition	Total Buyers	Average Backlog/Buyer
1	230	13	0.64
2	224	12	0.71
3	225	13	0.71
4	290	13	0.36
5	236	13	0.63
6	271	13	0.48
7	278	14	0.36
8	234	14	0.38
9	215	12	0.29
10	218	12	0.51
11	162	11	0.49
12	142	10	0.78
13	149	9	0.90
14	165	8	0.97
15	180	8	0.83
16	138	8	0.90

a. Is there a relationship between the average backlog per buyer and the number of requisitions?
b. Is there a relationship between the average backlog per buyer and both the number of requisitions and the number of buyers?
c. How might the developed model be used by the purchasing manager?

17. Following are regional manufacturing summary statistics for 1984. The *value added* column represents the difference between sales value and the cost of merchandise sold.

Region	Total Employment (000)	Payroll ($millions)	Value Added ($millions)	Value of Shipments ($millions)
NE	1,456	31,790	67,229	124,563
MA	3,211	73,080	160,629	324,071
ENC	4,236	107,086	231,601	536,372
WNC	1,308	29,433	73,319	180,843
SA	3,011	56,778	140,581	311,713
ESC	1,276	23,214	60,391	145,581
WSC	1,531	33,579	86,403	278,257
Mt	572	12,728	28,708	65,609
Pac	2,538	60,944	134,705	286,739

   a. In what ways might the data base be inadequate for developing a multiple regression model?
   b. Is there a relationship between total employment and payroll?
   c. Which of these factors might serve as the dependent variable?
   d. Develop a multiple regression model using this data base.
   e. Prepare a forecast based on selected values of the independent variables.

18. A recent report from the U.S. Environmental Protection Agency (EPA) indicates that more new models than ever are paying the special tax on low-fuel-efficiency cars. In most cases, these are the expensive European models. The following table presents engine, transmission, and mileage data for a random sample of 25 compact vehicles (1988).

	Engine Size (CI)	Number of Cylinders	Type of Transmission	Mileage	
Vehicle				City (MPG)	Highway (MPG)
Volvo 780	174	6	Auto	17	21
Peugeot 505	174	6	Manual	18	22
Chry LeBaron	152	4	Auto	18	23
GM Beretta	173	6	Manual	18	29
GM Cavalier	173	6	Manual	18	29
Saab 900	121	4	Auto	19	23
Ford Tempo	140	4	Auto	19	23
Austin Rover	163	6	Manual	19	24
Chry Shadow	152	4	Auto	19	24
Chry Sundance	152	4	Auto	19	24
Toyota Camry	153	6	Manual	19	24
Nissan Maxima	181	6	Auto	23	25
Peugeot 405	116	4	Auto	20	24
GM Skylark	206	6	Auto	20	27
GM Cutlass	204	6	Auto	20	27
GM Sunbird	122	4	Auto	20	27
Mazda 323	98	4	Manual	27	29
Ford Probe	133	4	Manual	21	27
GM Grand Am	122	4	Auto	21	28
Nissan Stanza	120	4	Manual	22	28
Honda Accord	119	4	Auto	22	28
Chry Omni	135	4	Manual	26	35
Nissan Sentra	98	4	Manual	28	36
Ford Escort	113	4	Manual	27	36
GM LeMans	98	4	Manual	30	39

a. What type of variable would be used to describe the type of transmission?
b. Develop a multiple regression model to estimate both city and highway fuel economy.
c. Which variables are statistically significant?
d. Estimate the difference in mileage based on manual versus automatic transmission.
e. Why might a manual transmission provide better fuel economy?
f. Estimate the impact of foreign versus domestic cars on fuel economy. (*Hint:* Incorporate a new dummy variable, where 1 = Domestic and 0 = Foreign.)

19. The following personnel data were obtained from a random sample of 12 office employees at the Clair Corporation.

Employee	Annual Compensation ($000)	Experience (years)	Sex (M = 1)
1	26.5	4	1
2	17.8	3	0
3	22.3	4	0
4	32.5	5	1
5	41.3	6	1
6	37.6	7	1
7	44.3	5	1
8	17.8	3	0
9	31.1	4	1
10	15.4	2	0
11	52.3	10	1
12	25.5	3	0

   a. Which of these factors might serve as the dependent variable?
   b. What type of variable might be used to describe sex?
   c. Formulate a multiple regression model using these data.
   d. Which variables are statistically significant at the .05 level?
   e. Estimate the value of Y for selected values of the independent variables.

20. The following table presents financial data for a random sample of 25 taxable money market funds for the week ending September 20, 1988. The fund's average maturity was 34 days for the reporting period. The yields represent the total returns to the shareholders for the past 7- and 30-day periods.

Fund	Assets ($millions)	Average Maturity (days)	7-Day Yield (%)	30-Day Yield (%)
Alex Brothers	252.4	28	7.2	7.5
Alliance Capital	1,786.5	39	7.1	7.4
Bayshore Cash	322.4	35	7.7	8.0
Capital Preservation	2,191.0	36	6.7	6.9
Cash Assets Trust	204.1	31	7.7	8.0
Columbia Daily	503.8	30	7.4	7.7
DBL Cash Fund	1,522.2	53	7.4	7.7
Dean Witter Active	2,566.2	44	7.6	7.9
Delaware Group	997.1	24	7.2	7.5
Dreyfus Assets	7,257.8	43	7.4	7.6
Franklin Money	1,326.5	25	7.4	7.7
John Hancock	249.5	23	7.3	7.6
Kemper Money	5,021.8	29	7.9	8.2
Liberty Gov't	1,312.9	24	7.1	7.3
Merrill Lynch	19,849.9	44	7.9	8.2
NLR Cash	1,336.9	32	7.6	7.9
Oppenheimer	724.4	25	7.7	7.9
Pacific Horizon	925.8	26	7.6	7.9
Paine Webber	3,847.3	31	7.6	7.9
Prudential-Bache	5,039.8	40	7.5	7.8
Shearson Temp	5,063.7	42	8.0	8.3
T. Rowe Price	3,575.9	26	7.5	7.8
Twentieth Century	484.9	47	7.5	7.7
USAA Money	680.4	20	7.6	7.9
Vanguard	6,000.8	42	7.9	8.3

a. Does the incorporation of the Merrill Lynch fund bias the data base?
b. Develop a regression model using assets and maturity date as the independent variables.
c. Which of the two dependent variables does the model predict with more precision?
d. What other variables might be incorporated to improve model performance?

21. Consider the following regression model, which predicts juvenile delinquency (expressed as a percentage of the 13–21-year-old population) as a function of the unemployment rate (X_1), the percentage of single-parent homes (X_2), and percentage of high school dropouts (X_3).

$$Y \text{ (delinquency)} = b_0 + b_1X_1 + b_2X_2 + b_3X_3$$

a. Should local, regional, or national data be used in this analysis?
b. What signs would you anticipate for the regression coefficients?
c. Discuss the potential for multicollinearity.
d. What other variables might be used to estimate juvenile delinquency?

22. Prepare a forecast and 90% confidence interval for the regression model developed in problem 18 using the average values of the independent variables.

23. Prepare a forecast and 90% confidence interval for the regression model developed in problem 19 for a female with four years of experience.

24. A study on factors governing the demand for part-time workers, involving more than 5,000 firms (Mark Montgomery, "On the Determinants of Employer Demand for Part-Time Workers," *Review of Economics and Statistics,* February 1988), yielded the following empirical results:

Variable	Definition	Regression Coefficient	t-value
1. Wage	Ratio of earnings of part-timers to full-timers	−0.43	−3.33
2. Benefit	Ratio of fringe benefits to payroll	1.24	2.99
3. Size	Number of full-time workers	−7.88	−14.55
4. Union	Percentage of nonsupervisory workers covered by collective bargaining agreements	−0.27	−0.92
5. Office	Percentage of total hours worked by office staff	0.07	1.84

The dependent variable is the percentage of part-timers in the work force. The resultant R^2 is .385.

a. Which variables are statistically significant at the .10 level?
b. Interpret the coefficients for the wage and union variables.
c. Interpret the coefficient of determination.
d. What other variables might be included in the model?

25. A recent study was undertaken to analyze the factors governing chief executive officer (CEO) compensation using data from more than 200 large manufacturing firms (D. N. Winn and J. D. Shoenhair, "Compensation-Based Incentives for Revenue-Maximizing Behavior," *Review of Economics and Statistics,* February 1988). Each executive was classified as either a veteran (six or more years of service) or a rookie (two years of service or less). The empirical results from the study are reported below:

Variable	Definition	Regression Veteran	Regression Rookie	t-statistic Veteran	t-statistic Rookie
1. RG	Revenue growth (%)	−6.067	−5.125	1.63	1.16
2. RP	Profit growth (%)	5.465	2.033	2.87	0.88
3. As	Log of assets	146.3	110.9	7.43	5.87
	Constant	−548.1	−334.6	—	—

The dependent variable is total CEO annual compensation ($000). The coefficients of determination are .329 and .443, respectively.

a. Which variables are statistically significant at the .10 level?
b. Interpret the coefficients for each of the significant variables.

c. Estimate CEO compensation for a firm with assets of $1 billion that experienced no growth in revenues or profits during the year.
d. What other variables might be included in the model?
26. Is it possible for all of the independent variables in a regression model not to be statistically significant (measured by the t-statistic) yet for the overall model to be statistically significant (as measured by the F statistic)? Explain.
27. Explain why it is possible to develop a regression model with a large R^2 but for which none of the independent variables are statistically significant.
28. Develop a regression model using the following data on 21 of the top Fortune 500 (1988) firms. Use earnings per share and investor return as two candidate variables. Which of the explanatory variables are significant at the 0.05 level? Provide an interpretation of the Beta coefficients. Discuss the overall significance of the developed model.

Company	Sales ($ millions)	Profits/ Assets	Profits/ Equity	Earnings per Share ($)	Investor Return (%)
Philip Morris	25,860	6.3	30.4	10.0	24.4
Chevron	25,196	5.2	12.0	5.2	21.9
Amoco	21,150	6.9	15.5	4.0	13.8
Occidental	19,417	1.5	4.7	1.2	14.6
Procter & Gamble	19,336	6.9	16.1	6.0	5.6
UTC	18,088	5.2	13.7	5.1	26.3
Atlantic Richfield	17,626	7.4	25.3	8.9	22.8
Eastman Kodak	17,034	6.1	20.6	4.3	−4.1
RJR Nabisco	16,956	7.8	24.2	5.9	109.8
Boeing	16,692	4.9	11.4	4.0	68.4
Dow Chemical	16,682	14.8	33.1	12.8	0.4
Xerox	16,441	1.5	6.8	3.5	8.8
USX	15,792	3.9	13.1	2.6	2.5
Tenneco	15,707	4.7	25.2	5.5	31.2
McDonnell Douglas	15,072	2.9	11.0	9.1	31.5
Pepsico	13,007	6.8	24.1	2.9	20.8
Westinghouse	12,500	4.9	21.7	5.7	9.8
Rockwell	11,946	8.8	22.0	3.0	18.6
Allied-Signal	11,909	4.6	14.2	3.1	21.4
Digital Equipment	11,475	12.9	17.4	9.9	−27.1
Philips	11,304	5.4	30.2	2.7	44.7

29. Develop a forecast of earnings per share and investor return for the model developed in problem 28. Using the mean values for the independent variables, construct 95% prediction intervals for the regression model.

30. The following data represent absenteeism, years of experience, and sex for a random sample of 12 engineers at Yardely Corporation.

Absenteeism (hr/yr)	Experience (years)	Sex (M = 1)
36.5	4	1
27.7	1	0
32.3	3	0
42.4	5	1
47.6	7	1
54.3	5	1
27.8	2	0
41.1	4	1
35.4	3	0
62.3	9	1
38.9	4	1

   a. Which variables are statistically significant at the .10 level?
   b. What role do dummy variables play in the model?
   c. Interpret the coefficients for each of the variables.
   d. Estimate absenteeism for a male engineer with five years of experience.
   e. What other variables might be included in the model?
   f. Construct a 95% confidence interval for the forecast for both mean and individual Y. Provide an interpretation of the developed intervals.
31. Explain why, in a three-variable model where the significance level for testing each variable is 0.01, one cannot claim that the significance of the overall model is 0.01.
32. The following data on annual income, level of education (0 = Grade school, 1 = High school, 2 = Undergraduate degree, and 3 = Graduate degree), and sex were obtained from a random telephone survey of 20 heads of households between 30 and 40 years of age.

Household	Annual Income ($000)	Level of Education	Sex (M = 1)
1	27.5	1	1
2	33.8	2	0
3	18.5	0	0
4	25.4	1	1
5	19.9	0	1
6	45.5	3	1
7	42.4	2	0
8	33.4	2	1
9	29.8	1	1
10	37.8	1	1
11	44.1	3	0
12	22.8	1	0
13	54.5	2	1
14	53.1	3	0
15	24.2	0	1
16	19.6	1	0
17	29.8	0	1
18	37.2	2	0
19	35.1	1	1
20	26.2	2	0

a. Identify the dependent and independent variables.
b. Which factors should be characterized as dummy variables?
c. What are the difficulties associated with the use of dummy variables in this problem?
d. Develop a multiple regression model at the .05 level.
e. Which variables are statistically significant?
f. Estimate the value of the dependent variable for a male head of household with a high school education.
g. Construct a 95% prediction interval for the forecast for both mean and individual Y. Provide an interpretation of the developed intervals.

33. Consider a multiple regression analysis involving compensation (Y) versus level of education (X_1) and race (X_2). Define five (5) categories for each predictor variable. How many dummy variables are needed per factor?

34. Transform the data given in problem 20 using logarithms. Formulate a new model. Does this model better explain the variance in the dependent variable?

35. Transform the data given in problem 28 using logarithms. Formulate a new model. Does this model better explain the variance in the dependent variable?

36. The following two models are designed to predict the profit per acre for corn as a function of price per bushel (X_1), yield (X_2), and cost (X_3):

$$Y = b_0 + b_1 X_1 + b_2 X_2 + b_3 X_3$$
$$Y = b_0 + b_1 X_1 X_2 - b_2 X_3$$

a. Discuss the logic behind each of these two models.
b. Which of these two models should provide a better fit?
c. Identify other possible model formulations for this problem.

37. Use the following data to analyze the two models presented in problem 36 at the 0.05 level.

Profit ($/acre)	Price ($/bu)	Yield (bu/acre)	Cost ($/acre)
275	2.49	110.4	35.5
145	2.75	92.0	38.3
320	3.44	108.8	41.1
198	2.63	113.5	44.8
195	2.98	81.0	40.2
290	3.46	106.3	47.3
256	2.79	119.0	48.2
279	2.35	121.2	51.2
167	2.11	109.2	52.4
211	2.43	104.5	53.6
234	2.65	112.1	54.2

38. The following two models are designed to predict productivity as a function of employee aptitude (X_1), employee experience (X_2), and sex (X_3).

$$Y = b_0 + b_1 X_1 + b_2 X_2 + b_3 X_3$$

$$Y = b_0 + b_1 X_1 + b_2 \log X_2 + b_3 X_1 X_3$$

a. Discuss the logic behind each of these two models.
b. Which of these two models should provide a better fit?
c. Identify other possible model formulations for this problem.

39. Use the following data to analyze the two models presented in problem 38 at the 0.05 level.

Production (units/hr)	Aptitude (%)	Experience (months)	Sex ($M = 1$)
123	77	12	1
89	71	6	0
101	82	9	0
115	81	10	1
145	87	14	1
77	65	3	0
114	75	11	1
117	74	10	1
126	78	9	1
88	63	4	0
109	67	8	1
121	84	11	1
92	69	6	0

40. The operations vice president at Palos Verdes Community Bank is interested in predicting quarterly earnings (Y, $000) as a function of average quarterly deposits (X_1, $000), change in number of quarterly transactions (X_2, %), and average quarterly loan balance (X_3, $000). The data base consists of 5 years of quarterly data. The developed regression model is as follows, where the numbers in parentheses are the computed t-values. The error sum of squares is 10,915, and the total sum of squares is 98,765.

$$Y = -1{,}245.34 + 0.0345X_1 + 124.23X_2 + 0.01923X_3$$
$$(3.245) \quad (1.112) \quad (4.156)$$

a. Compute R^2 and discuss its significance.
b. Discuss the significance of the regression coefficients.
c. Comment on the overall significance of this model.

41. The surgeon general of the United States is interested in determining the state of rural health care. The study consisted of analyzing data selected from a random sample of 10 rural communities. Specifically, data were collected on the use of public hospitals (Y, %), the average family income (X_1, $000), the proportion of families with children (X_2, %), and the doctor-to-patient ratio (X_3, %). The developed regression model is reported below. The numbers in parentheses are the computed t-values. The regression sum of the squares was 421.25 and the error sum of the squares was 85.67.

$$Y = 8.345 + 0.6032X_1 + 17.54X_2 - 0.4565X_3$$
$$(-1.612) \quad (2.456) \quad (-1.023)$$

a. Compute R^2 and discuss its significance.
b. Discuss the significance of the regression coefficients.
c. Comment on the overall significance of this model.

42. Develop a regression model using the data from problem 18 to evaluate possible interactive effects between engine size and the number of cylinders.

43. Develop a regression model using the data from problem 19 to evaluate possible interactive effects between experience and sex.

44. Develop a regression model using the data from problem 32 to evaluate possible interactive effects between level of education and sex.

45. Develop a regression model using the data from problem 39 to evaluate possible interactive effects between:
    a. Experience and aptitude.
    b. Experience and sex.
    c. Experience, aptitude, and sex.

46. Discuss the implications of forecasting using values for the independent variables that are outside the range of the data base.

47. Develop a regression model for the data given in problem 16 by transforming the dependent variable using the square function. Develop a forecast and 95% prediction interval for using the mean values of the significant variables. Compare the computed R^2 with the one developed in problem 16.

48. The vice president of finance at Southeast Gas is interested in raising construction capital for a new gas pipeline through the issuance of a new bond. The vice president has collected the following data on 25 previous bond offerings throughout the industry. The data included, at the time of issuance, the bond yield, the Moody's rating of the utility (AAA, AA, or A), the U.S. T-bill rate, the anual earnings per share, and the bond maturity level (0 = 15 years or 1 = 30 years).

Bond Yield (%)	Moody's Rating		U.S. T-Bill Rate (%)	Earnings per Share ($)	Bond Maturity
	AAA	AA			
11.1	1	0	8.8	16.2	30
11.4	1	0	8.6	13.5	30
10.6	0	1	9.1	17.2	30
9.8	0	0	7.8	12.3	15
10.3	0	1	8.3	15.5	30
10.5	1	0	8.8	16.1	15
9.4	0	1	8.0	14.3	15
8.9	0	0	7.7	12.4	30
11.3	1	0	9.1	16.8	30
10.6	1	0	9.3	18.2	30
10.4	1	0	9.4	16.2	15
9.5	0	1	7.4	13.5	30
10.0	1	0	8.8	16.5	30
9.2	0	0	7.7	11.3	15
8.8	0	0	7.1	10.8	30
10.7	1	0	9.1	13.4	30
10.9	0	1	9.0	14.5	30
11.3	1	0	9.7	16.9	15
11.1	1	0	9.9	15.8	30
9.7	0	1	7.3	13.6	30
12.3	1	0	10.2	17.8	30
11.6	1	0	9.4	18.1	15
9.9	0	1	8.1	14.7	15
10.2	0	1	9.1	15.1	30

a. Develop a regression model using the stepwise procedure at the 0.05 level where bond yield is the dependent variable.

b. Which of the variables should be treated as dummy variables? How should each dummy variable be characterized?
c. Which of the explanatory variables are statistically significant?
d. Discuss the statistical significance of the developed model in terms of the standard measures.
e. Develop a point and interval forecast of Y using the mean values for the independent variables.
f. What additional explanatory variables might help improve the model's predictive capability?

49. The following table presents trade and investment data between the United States and Taiwan for the period 1976–1984.

Year	Balance of Trade (U.S. $ million)	Taiwan Tariff Reductions (number)	U.S. Foreign Investment (U.S. $ million)
1976	−1,241	59	21.8
1977	−1,672	139	24.2
1978	−2,634	999	69.8
1979	−2,271	453	80.4
1980	−2,086	111	110.1
1981	−3,393	0	203.2
1982	−4,195	111	79.6
1983	−6,688	0	93.3
1984	−9,972	283	231.2

a. Develop a regression model to predict U.S. balance of trade with Taiwan as a function of the number of tariff reductions and U.S. foreign investment in Taiwan using the stepwise method at a 0.05 level of significance.
b. Are either of the two explanatory variables significant?
c. Develop scatter diagrams for each of the independent variables.
c. If the answer to part b is yes, construct a point and prediction estimate using the mean values for the significant independent variables.
d. What additional variables might be used to explain the balance of trade?

50. The following table presents population, gross national product (GNP), gross investment (GDI), consumer price index (CPI, 1975 = 100), and foreign exchange rate (FER) data on Brazil for the period 1970 to 1981 (GNP and GDI are in constant 1970 prices).

Period	Population (000)	GNP ($ million)	GDI ($ million)	CPI	FER (per $ U.S.)
1970	95,847	208.3	47.7	38.5	4.59
1971	98,169	233.2	58.0	46.3	5.29
1972	100,547	259.3	65.4	53.9	5.93
1973	102,980	295.6	82.2	60.8	6.13
1974	105,477	323.8	105.3	77.6	6.79
1975	108,032	340.2	100.1	100.0	8.13
1976	110,018	372.6	104.6	141.9	10.67
1977	112,040	392.4	104.1	203.9	14.14
1978	114,099	408.5	101.3	282.8	18.07
1979	116,196	434.2	104.4	431.8	26.95
1980	118,332	466.1	113.2	789.4	52.71
1981	120,507	448.3	98.4	1,622.6	93.13

a. Develop a stepwise regression model at the 0.05 level using FER as the dependent variable.
b. Which if any of the independent variables are statistically significant?
c. Develop scatter diagrams for each of the independent variables.
d. How much of the variability in Y is explained by the developed model?
e. Develop a point and prediction estimate using the mean values of the significant variables.
f. What additional variables might be used to explain FER?

51. Develop a regression model using the data presented in problem 50 where GNP is the dependent variable and population and CPI are the independent variables. How much of the variability in GNP is explained by the model? Use alpha = 0.05.

52. The following table presents population, gross national product (GNP), gross domestic investment (GDI), consumer price index (CPI, 1975 = 100), and foreign exchange rate (FER) data on India for the period 1970 to 1981 (GNP and GDI are at constant 1970 prices):

Period	Population (000)	GNP ($ million)	GDI ($ million)	CPI	FER (per $ U.S.)
1970	547,569	400.1	69.6	59.4	7.5
1971	560,119	408.9	79.6	51.8	7.4
1972	571,850	405.9	74.8	66.1	7.7
1973	583,826	421.3	87.4	79.9	7.8
1974	596,053	423.1	89.5	101.3	8.0
1975	608,534	465.0	88.0	100.0	8.7
1976	621,280	472.7	92.7	96.2	8.9
1977	634,292	511.1	99.5	103.5	8.6
1978	647,576	543.3	117.3	105.8	8.2
1979	661,138	517.7	104.1	115.3	8.1
1980	674,984	554.0	118.4	128.1	7.9
1981	690,183	584.6	127.4	144.1	8.9

a. Develop a stepwise regression model at the 0.05 level using FER as the dependent variable.
b. Which if any of the independent variables are statistically significant?
c. Develop scatter diagrams for each of the independent variables.
d. How much of the variability in Y is explained by the developed model?
e. Develop a point and prediction estimate using the mean values of the significant variables.
f. What additional variables might be used to explain FER?

53. Develop a regression model using the data presented in problem 52 where GNP is the dependent variable and population and gross domestic investment are the independent variables. How much of the variability in GNP is explained by the model? Use alpha $= 0.10$.

54. The following table presents economic growth data (%) for selected European countries for the period 1970–1981:

Country	Population	Gross Domestic Product	Gross Production Manufacturing
Austria	0.1	3.5	3.6
Belgium	0.2	3.0	2.1
Denmark	0.3	2.1	1.8
Finland	0.4	3.1	3.9
France	0.5	3.3	2.2
W. Germany	0.	2.6	1.7
Ireland	1.3	4.0	4.9
Italy	0.4	2.9	3.2
Netherlands	0.8	2.7	2.2
Norway	0.5	4.5	0.8
Spain	1.1	3.4	4.8
Sweden	0.3	1.8	0.8
United Kingdom	0.1	1.7	−0.6

a. Develop a regression model where gross production manufacturing is the dependent variable.
b. Which of the two independent variables is statistically significant at the 0.05 level?
c. Develop a scatter diagram for each of the independent variables.
d. How much of the variability in Y is explained by the developed model?
e. Develop a point and prediction estimate using the mean values for the independent variables.

55. The following table presents key 1988 financial data on 25 U.S. corporations taken from a sample of the 100 to 250 Fortune 500 firms:

Company	Sales ($ million)	Profits ($ million)	Assets ($ million)	Earnings per share	Investor Return (%)
Time Inc.	4,507	298.3	4,913	5.01	31.5
Teledyne	4,401	391.8	5,125	34.03	10.6
Warner	4,206	423.2	4,598	2.65	34.7
Boise Cascade	4,095	289.1	3,610	6.34	4.0
Whitman	3,915	233.5	3,486	2.19	19.4
Stone Container	3,743	341.8	2,395	5.69	32.1
Control Data	3,628	1.7	2,534	0.03	−9.2
Gillette	3,581	268.5	2,868	2.45	25.4
Interco	3,341	145.0	1,986	3.50	−65.6
FMC	3,287	129.2	2,749	3.60	−5.2
Paccar	3,112	175.8	2,832	4.90	51.0
Intel	2,875	452.9	3,550	2.51	−10.4
USG	2,811	189.0	1,596	2.38	−62.9
Pitney Bowes	2,665	243.3	4,788	3.08	14.1
B. F. Goodrich	2,520	195.7	2,073	7.72	32.0
Olin	2,308	98.0	1,940	4.63	25.7
Black & Decker	2,281	97.1	1,825	1.65	24.8
Knight-Rider	2,194	156.4	2,357	2.76	16.2
Fruehauf	2,143	−56.6	1,583	−6.50	−16.7
Westvaco	2,134	200.4	2,513	3.10	7.6
Compaq	2,066	255.2	1,590	6.30	7.7
Asarco	1,988	207.2	2,223	2.22	−5.9
Maytag	1,909	158.6	1,330	2.07	−9.5
Cabot	1,677	60.4	1,543	2.24	20.2
Holly Farms	1,583	36.2	757	2.06	113.5

a. Develop a stepwise regression model where earnings per share and investor return are two candidate dependent variables.
b. Which of the three candidate independent variables is statistically significant at the 0.05 level?
c. Develop a scatter diagram for each of the independent variables.
d. How much of the variability in Y is explained by the developed model?
e. Develop a point and prediction estimate using the mean values for the independent variables.

56. The following table presents key 1988 financial data on 14 of the top U.S. aerospace companies:

Company	Sales ($ Mil)	Profits/ Assets (%)	Profits/ Equity (%)	Earnings per Share (%)
United Tech	18,088	5	7	5.1
Boeing	16,962	5	11	4.0
McDonnell Douglas	15,072	3	11	9.1
Rockwell	11,946	9	22	5.7
Allied-Signal	11,909	5	14	3.0
Lockheed	10,667	9	2	10.4
General Dynamics	9,551	6	20	9.0
Textron	7,111	2	10	2.7
Northrop	5,797	3	10	2.2
Martin Marietta	5,727	11	30	6.8
Grumman	3,591	3	10	2.5
Sequa	1,948	4	10	6.0
Rohr	907	4	9	1.9
Kaman	767	6	14	1.4

a. Develop a stepwise regression model where earnings per share is the dependent variable.
b. Which of the three candidate independent variables is statistically significant at the 0.05 level?
c. Develop a scatter diagram for each of the independent variables.
d. How much of the variability in Y is explained by the developed model?
e. Develop a point and prediction estimate using the mean values for the independent variables.
f. What additional variables might be used to explain earnings per share?

57. The following table presents key 1988 financial data on 23 of the top U.S. computer companies:

Company	Sales ($ million)	Profits/Assets (%)	Profits/Equity (%)	Earnings per Share (%)
Digital	11,475	13	17	9.9
Unisys	9,902	6	14	2.7
Hewlett	9,831	11	18	3.6
NCR	5,990	9	20	5.3
Apple	4,071	19	40	3.1
Control Data	3,628	0	0	0
Wang Labs	3,068	3	6	0.6
Zenith	2,686	1	2	0.5
Pitney Bowes	2,665	5	19	3.1
Compaq	2,066	16	31	6.3
Amdahl	1,802	12	22	2.1
Prime	1,595	1	4	0.4
Data General	1,365	−1	−3	−0.6
Tandem	1,315	7	11	1.0
Seagate	1,266	7	18	1.5
Sun Micro	1,052	9	18	1.8
Storage Tech	874	5	16	0.2
Intergraph	800	11	13	1.6
SCI	774	4	13	0.9
Cray	756	16	23	5.0
Atari	705	−23	−102	−1.5
Apollo	654	0	1	0.1
Miniscribe	603	5	20	0.7

a. Develop a stepwise regression model where earnings per share is the dependent variable.
b. Which of the three candidate independent variables is statistically significant at the 0.05 level?
c. Develop a scatter diagram for each of the independent variables.
d. How much of the variability in Y is explained by the developed model?
e. Develop a point and prediction estimate using the mean values for the independent variables.
f. What additional variables might be used to explain earnings per share?

58. The following table presents key 1988 financial data on 24 of the top U.S. food companies:

Company	Sales ($ million)	Employees (000)	Profits/Assets (%)	Investor Return (%)
Sara Lee	10,424	85,700	6	22.9
Conagra	9,475	42,993	5	17.1
Borden	7,244	46,300	7	23.1
Archer Daniels	6,798	9,006	8	3.7
Pillsbury	6,191	101,800	2	92.8
Ralston	6,176	56,734	10	30.7
General Mills	5,778	74,500	11	8.3
Quaker Oats	5,330	31,300	9	30.5
Heinz	5,224	39,000	11	19.3
Campbell	4,869	48,389	8	16.4
CPC Int	4,700	32,000	9	32.1
Kellogg	4,349	17,461	15	26.0
Whitman	3,915	25,396	7	19.4
United Brands	3,503	42,000	4	9.0
Hershey	2,561	12,100	12	8.9
Hormel	2,293	7,994	9	4.7
Tyson	1,936	26,000	9	34.3
Int. Multifoods	1,698	9,048	4	7.8
Holly Farms	1,583	15,000	5	113.5
Dean Foods	1,552	7,100	9	23.8
Wilson Foods	1,324	5,100	3	88.5
McCormick	1,184	7,626	6	56.8
Gerber	1,166	14,658	3	75.0
Savannah	917	1,665	5	34.2

a. Develop a stepwise regression model where investor return is the dependent variable.
b. Which of the three candidate independent variables is statistically significant at the 0.05 level?
c. Develop a scatter diagram for each of the independent variables.
d. How much of the variability in Y is explained by the developed model?
e. Develop a point and prediction estimate using the mean values for the independent variables.
f. What additional variables might be used to explain investor return?

59. Presented below is a CBS stepwise analysis (alpha = 0.05) of key 1988 financial data for 22 of the top U.S. chemical companies. The dependent variable for this problem is the annual growth in earnings per share (EPS). The candidate independent variables are sales in billions (sale), profits per assest (PPA), and profits per equity (PPE).

Iteration 2

Variable	B-coeff	std err	Beta	t value	p value
ppe	0.1555	0.0279	0.7040	5.5683	0
sale	0.3891	0.1229	0.4004	3.1655	0.0051
B_0 Intercept:		−0.1949			
Critical t:		2.0930			

Results for Iteration 2

Sum Squares Regression:	163.5225
Sum Squares Error:	70.7451
Sum Squares Total:	234.2676
Mean Square Regression:	81.7613
Mean Square Residual:	3.7234
C. O. D. (R^2):	0.6980
Adjusted C. O. D. (R^2):	0.6662
Multiple Correlation Coefficient:	0.8355
Standard Error Estimate:	1.9296
df Regression:	2
df Error:	19
Critical F:	3.5200
Computed F:	21.9586
F (p value):	0.0001
Durbin-Watson Statistic:	2.5742

Remaining variables are insignificant

	sale	ppa	ppe	eps
sale	1	.017	0.075	.453
ppa		1	0.699	.642
ppe			1	0.734
eps				1

Variable	Mean Value	Standard Deviation
sale	3.482	3.435
ppa	8.920	5.979
ppe	21.760	15.120
eps	4.544	3.340

a. Which of the candidate independent variables is statistically significant?
b. How much of the variability in Y is explained by the model?
c. Contrast the difference between R^2 and the adjusted R^2.
d. Describe the developed model.
e. Analyze the extent of collinearity between the independent variables.
f. What additional variables might be used to explain earnings per share?
g. Forecast EPS using the mean values of the significant predictor variables.

60. Presented below is a CBS stepwise analysis (alpha = 0.5) of quality control data for 25 industrial firms. The dependent variable is the average outgoing quality (AOQ) measured in the number of defects per 100 units produced. The candidate independent variables are the average years of exprience per employee (AYE), the average number of hours of quality control training per employee (ANT), the existence of quality circles (1 = yes, 0 = no) and the average quality control budget (AQB).

			Iteration 2			
Variable	B-coeff	std err		Beta	t value	p value
AQB	−0.0217	0.0065		−0.4581	−3.3358	0.0025
QC	−1.9713	0.5968		−0.4536	−3.3029	0.0027
B_0 Intercept:		7.1982				
Critical t:		2.0520				

460 Chapter 10 Multiple Regression Analysis

Results for Iteration 2

Sum Squares Regression:	87.2953
Sum Squares Error:	48.7047
Sum Squares Total:	136
Mean Square Regression:	43.6476
Mean Square Residual:	1.8039
C. O. D. (R-Squared):	0.6419
Adjusted C. O. D. (R-Squared):	0.6153
Multiple Correlation Coefficient:	0.8012
Standard Error Estimate:	1.3431
dOf Regression:	2
dOf Error:	27
Critical F:	3.3500
Computed F:	24.1966
F(p value):	0.0001
Durbin-Watson Statistic:	1.9321

Remaining variables are insignificant

	AYE	*ANT*	*QC*	*AQB*	*AOQ*
AYE	1	−0.1692	0.2639	0.0524	−0.1250
ANT		1	0.6323	0.5430	−0.6412
QC			1	0.5446	−0.7031
AQB				1	−0.7051
AOQ					1

Variable	*Mean Value*	*Standard Deviation*
AYE	6.900	1.783
ANT	16.533	10.332
QC	0.600	0.498
AQB	138.667	45.617
AOQ	3	2.166

a. Which of the candidate independent variables is statistically significant?
b. How much of the variability in Y is explained by the model?
c. Contrast the difference between R^2 and the adjusted R^2.
d. Describe the developed model.
e. Analyze the extent of collinearity between the independent variables.
f. What additional variables might be used to explain average outgoing quality?

61. The marketing manager at the International Automotive Association is interested in determining whether sex, age, or income (low, medium, or high) influences the decision to purchase a domestic (A) or foreign-made (B) car. A random sample of 12 recent buyers yielded the following data:

Sex (m = 1)	Age	Income	Type of car
1	18	L	A
1	22	L	A
0	27	M	B
0	33	M	B
0	41	H	A
1	24	M	B
1	33	L	A
0	37	M	A
1	21	L	B
0	28	M	A
1	36	H	B
1	44	H	B

a. What type of analysis is most appropriate for the problem?
b. What values does the dependent variable take on?
c. Which of the predictor variables are significant at the 0.05 level?
d. Construct an error analysis based on the mean value of Y.

62. The sales manager at Western Distributors, a wholesale electronics firm, is interested in determining whether sex, experience, or sales district (A, B, C) has an impact on sales performance (good or poor). A random sample of 11 employees yielded the following data:

Sex (m = 1)	Experience (years)	District	Sales Performance
1	5	A	Good
0	3	B	Good
1	2	C	Poor
0	4	B	Good
0	2	A	Poor
1	4	C	Poor
0	1	A	Good
0	2	A	Poor
1	4	B	Good
1	2	B	Good
1	3	C	Poor

a. What type of analysis is most appropriate for the problem?
b. What values does the dependent variable take on?
c. Which of the predictor variables are significant at the 0.05 level?
d. Construct an error analysis based on the mean value of Y.

10.12 CASES

10.12.1 Plumbing West Inc.

"The only thing plumbers need to know is that effluent runs downhill and payday is every other Friday," said Gordon Long, vice president of operations at Plumbing West. "Wrong," said the company's chief engineer. "Heavy stuff also sinks to the bottom." At that point, the company's president interjected, "We have received a letter from the county regarding their concern about potential infiltration of rain water into the sewer lines. The county's board of supervisors has requested our help in solving this problem. Does anyone have an idea on how we might proceed?"

"Let's use multiple regression to analyze this problem," said the chief engineer. "What variables should we consider?" responded the vice president. The chief engineer moved to the blackboard and listed the following candidates:

Y = Amount of infiltration into the sewer lines (millions of gallons)
X_1 = Total area (acres)
X_2 = Total length of pipe (feet)
X_3 = Number of faulty manhole covers
X_4 = Number of defective joints
X_5 = Number of faulty connections

Following a few minutes of discussion, the group agreed that this list was satisfactory. The chief engineer was instructed to collect the appropriate data and report the findings by the end of the week. After the meeting ended, the chief engineer placed a call to the county engineer requesting information on the latest infiltration data. The next day the chief engineer received via FAX the infiltration data given in Table 1. These data consist of measurements taken from a representative sample of 13 sites located throughout the county after the last major rainstorm. This put the chief engineer in a position to determine if any of the identified variables seemed related to infiltration.

TABLE 1 Infiltration Data

Flow (millions of gals)	Area (acres)	Length (ft)	Manhole Defects	Joint Defects	Faulty Connections
0.86	619	97,400	0	0	36
3.87	2,619	204,160	50	7,860	74
3.87	732	102,380	32	4,658	43
3.04	748	9,700	4	446	5
20.7	1,029	243,480	74	10,226	114
4.59	530	58,160	18	2,646	27
10.38	1,413	218,160	64	8,399	91
2.1	209	37,730	10	1,001	17
2.07	646	88,580	18	1,860	37
4.71	980	144,710	33	4,558	60
22.21	4,030	618,590	99	15,155	291
8.46	2,559	283,280	64	8,923	118
1.2	1,651	202,420	32	4,959	74

Suggested Questions

1. Which regression method would you recommend that the chief engineer use?
2. Which of the variables are statistically significant at the .05 level?
3. How much of the variability in the dependent variable is explained by the model?
4. Forecast the amount of infiltration using selected values for the significant independent variables. Provide an interpretation of these results.
5. What other variables might be used in this analysis?
6. Discuss the extent of collinearity in the model.
7. Would transforming the variables improve the explanatory power of the model?
8. Does the county appear to have a problem?

10.12.2 Pronet Tennis Company

The owners of Pronet Tennis Company were discussing methods of training gifted amateurs in the art of becoming seriously competitive players. Among the variables they wished to stress in the training were the following:

- Attitude
- Concentration
- Service
- Frame shots

"Obviously, we don't really know how these affect outcomes on the professional circuit," said the director, "but we sure could pontificate about them like everyone else."

"It's simple to show that a good attitude affects the results by just observing some of the players who win a lot," said a well-known coach. "Super," joined in the others. "We can teach concentration and frame shots using the same philosophy. But what can we do about the effect of the service on the end results? That is certainly not clear from watching the pros."

"We should study the results of a tournament and see what happens," said a coach with a degree in physical education. "We'll count the serves and the sets won and then put the data together." And so it came to pass. The tournament was held, and the results in Table 1 were obtained for 40 professional players.

Suggested Questions

1. Do players with more experience (i.e., more serves) have a higher win percentage?
2. Do players with a higher first-serve percentage win more frequently?
3. Does the number of aces or double faults significantly impact winning?
4. What additional factors should the school concentrate on to improve winning?

TABLE 1 Grand Prix Tennis Tournament

Player Name	Successful First Serves (%)	Aces/Set	Double Faults/Set	Winning Percentage
1 Alexander	63.79	2.1	1.3	.655
2 Amaya	61.85	5.3	1.8	.483
3 Borg	68.50	1.0	3.7	.667
4 Borowiak	69.37	1.9	1.9	.500
5 Buehning	65.49	4.5	1.3	.465
6 Clerc	57.30	2.6	1.5	.786
7 Connors	71.92	1.0	1.0	.885
8 Curren	56.59	3.8	1.4	.750
9 Dent	56.98	2.1	2.1	.419
10 Denton	57.94	7.8	7.9	.649
11 Dibbs	69.06	0.4	1.0	.630
12 Dupre	58.38	2.2	2.8	.484
13 Edmonson	67.13	2.0	1.9	.537
14 Fibak	69.96	0.8	0.8	.656
15 Fleming	56.94	4.4	7.9	.250
16 Gerulaitis	61.36	3.8	2.8	.823
17 Glickstein	66.91	1.0	1.1	.591
18 Gomez	63.56	3.0	1.6	.683
19 Gonzalez	60.66	4.3	2.6	.429
20 Gottfried	63.03	1.5	2.3	.662
21 Gullikson, Ti	63.07	2.6	1.6	.400
22 Gullikson, To	65.66	1.6	1.1	.514
23 Higueras	58.80	0.8	1.0	.731
24 Hooper	62.22	5.9	3.9	.614
25 Kirmayr	64.55	0.4	0.8	.483
26 Kriek	53.80	4.0	4.6	.667
27 Lendl	54.72	3.6	1.2	.868
28 Manson	61.95	2.1	2.9	.429
29 Mayer, Gene	63.78	1.9	0.7	.769
30 Mayer, Sandy	59.51	1.8	2.0	.647
31 Mayotte, Tim	61.89	3.2	2.4	.633
32 McEnroe	57.05	3.8	1.7	.884
33 McNamara	54.16	2.1	2.3	.646
34 Moor	56.28	2.9	4.0	.375
35 Mottram	64.18	1.5	0.6	.686
36 Nastase	57.70	4.3	3.1	.424
37 Noah	65.13	4.6	1.1	.773
38 Pecci	61.71	1.2	1.8	.500
39 Pfister	65.14	3.6	3.5	.590
40 Purcell	61.08	2.0	1.0	.619

Chapter 11

Time Series and Forecasting

When the time is run, and the future becomes history, it will become clear how little of it we today foresaw or could foresee.

                                            J. R. Oppenheimer

CHAPTER OUTLINE

11.1 Introduction
11.2 Example Management Problem: Dialnet Telephone Exchange
11.3 How to Recognize a Time Series Problem
11.4 Classical Decomposition Model
11.5 Forecasting Models
11.6 Forecast Validation
11.7 Computer Analysis
11.8 Practical Applications
11.9 Case Study: Thermhouse Insulation Corporation
11.10 Summary
11.11 Glossary
11.12 Bibliography
11.13 Problems
11.14 Cases

CHAPTER OBJECTIVES

The primary objectives of this chapter are to develop an understanding of

1. How to recognize time series problems.
2. The classical decomposition model.
3. The basic forecasting models.
4. How to validate a forecast.
5. How computer models can be used to develop forecasts.

The Civil Rights Act of 1964 was designed to give nonwhites equal access to public accommodations and to ban discrimination in employment. One major objective of the act was to improve the economic conditions of blacks throughout the country. A comprehensive study was undertaken, using time series analysis, to determine the impact of the act on the standard of living of blacks over an extended time period.* The dependent variable used in this analysis was the ratio of nonwhite to white median income.

The developed model explained 86% of the variation in the dependent variable (not extraordinarily high for time series analysis). The analysis showed that black/white income differentials decreased by 1.66% per year after the act, compared to 0.29% prior to the act. The results also revealed that the only significant change in the black/white income ratio occurred in the South. The impact of the act in other areas of the country was found to be insignificant.

11.1 INTRODUCTION

Typically, historical data can be classified into two categories: time series data and cross-sectional data. With cross-sectional data, values are recorded for a variety of variables (e.g., market share for the top five firms in 1982) at the same point in time. A **time series,** on the other hand, is a chronologically ordered sequence of data values for one variable (e.g., annual GNP values for the period 1975 to 1985). Publications such as the *Statistical Abstract of the United States* and *Business Statistics,* both published by the U.S. Department of Commerce, provide the reader with extensive time series data on a variety of economic, business, and demographic factors. The time periods used for most times series are years, quarters, months, or weeks, although periods as short as hours or days are sometimes encountered, such as recording the level of quality on a production line. A considerable amount of data used in business and government are quarterly. A basic objective of this chapter is to introduce the reader to a variety of methods of analyzing time series data. Typically, time series analysis consists of decomposing the time series into its basic parts in order to uncover any specific patterns (e.g., seasonal effects) in the data. This detailed analysis of a time series represents a first step in developing an accurate forecast of future events.

A primary responsibility of any organization is to plan for the future. These planning activities involve the generation of a series of predictions regarding future events. Predicting future revenues, market share, and inventory requirements are but a few well-known examples. The prediction of future events is called **forecasting.** Forecasts are developed to help the decision maker evaluate alternative business strategies. Chapters 9 and 10 have already introduced regression-based forecasting. This chapter expands on

---

* D. J. McCone and R. J. Hardy, "Civil Rights Policy and the Achievement of Racial Economic Equality, 1948–1975," *American Journal of Political Science* 22 (February 1978).

> **HISTORICAL NOTE**
>
> Simon Kunznets (1902–1985) was one of the pioneers in the development of national income measurements (i.e., time series). In 1933 the Secretary of Commerce appointed Dr. Kunznets as the first director of the newly formed group for the study of national income measures. In 1971 he received the Nobel prize in economics for his efforts in the analysis of economic growth.

these concepts by presenting a number of additional forecasting methods. Past experience indicates that no one forecasting method is appropriate for all situations. Generally speaking, forecasting techniques can be classified into four basic categories: (1) qualitative (judgmental), (2) time series (historical), (3) causal (regression), and (4) simulation. The primary focus of this chapter is on time series methods. The basic characteristic of a time series forecasting model is its reliance on historical data. For example, the forecast of a firm's sales revenues for the coming year should take into account the sales history from previous years. The ability of a forecasting model to describe past history is often taken as an indicator of its capability to predict the future.

Several different analytical approaches are available for formulating time series models. The two most popular are the classical decomposition model and the multiple regression model. Both of these models will be discussed in some detail in this chapter. In addition, the issue of model validation will be introduced along with several parameters used in measuring the accuracy of time series models.

11.2 EXAMPLE MANAGEMENT PROBLEM: DIALNET TELEPHONE EXCHANGE

Dialnet Telephone Exchange is a regional communications company serving the southwestern United States since 1976. The company has experienced significant growth in recent years as a result of population shifts to the "sunbelt" states. Dialnet's general manager has been asked by the company's board of directors to prepare a growth plan for the coming decade. A key aspect of this plan involves a forecast of new phone line installations and quarterly sales. In response to the request from the board, the general manager has acquired the following data on yearly phone installations and quarterly sales from the company's computer data base:

| | Installations | Quarterly Sales ($ millions) | | | |
Year	(000)	1	2	3	4
1976	45.5	7.2	9.2	4.5	5.6
1977	79.2	10.3	12.8	8.2	9.8
1978	77.8	10.0	13.4	7.5	10.3
1979	81.4	11.1	16.3	7.7	9.5
1980	102.2	15.9	20.4	9.9	11.2
1981	124.8	18.4	22.8	14.8	14.8
1982	132.6	19.7	23.1	15.4	14.5
1983	156.3	22.2	24.3	16.5	15.4
1984	161.2	21.1	26.7	17.2	16.9
1985	168.6	24.3	28.2	16.5	18.1
1986	174.2	25.2	27.3	18.5	20.2

The general manager is interested in formulating a forecast of new phone installations for the next three years. These estimates will be used to develop a master plan for improving phone service for the coming decade. Because of the uncertainty in the future demand for phone installations, the general manager wishes to use several different forecasting methods.

> **INTERNATIONAL VIGNETTE**
>
> Jacob Diddens, director of distribution at Vanderbruk Spice, was going over his notes after the board meeting. He had received approval to spend up to 2.5 million guilders on the installation of a new distribution information system for Japanese operations. The proposal originated with Vanderbruk's Japanese subsidiary, Nippon Spice. In conjunction with a consultant from a leading Japanese university, the management team at Nippon developed a set of planning requirements to manage the flow of products through the distribution system. Nippon Spice was mainly responsible for packaging the bulk product, which was ordered from Vanderbruk in Holland. In addition to the packing capabilities, however, Nippon did produce a few products with mostly local appeal. Some of these local products had been incorporated in the Vanderbruk product line for worldwide distribution.
>
> Diddens realized that the proposed system was the most advanced in the company and could serve as a model for other corporate applications. In making his presentation to the board, Mr. Diddens outlined three benefits from the system: improved inventory control, better information-customer linkages, and a pilot test system with companywide potential. There was some skepticism from the board, and their reaction was split. Even though he had received the budget for carrying out the project, Diddens felt uneasy. He didn't want to let the board down, yet he couldn't put his finger on any one item that really was a trouble point. As he was musing, his secretary came in with a telex from Nippon Spice saying that the team was waiting to know the results of the board meeting.

11.3 HOW TO RECOGNIZE A TIME SERIES PROBLEM

Key idea 1

Following are the primary characteristics of the classical time series problem:

1. The values of one or more variables are measured over time (e.g., phone installations).
2. The time series can be described using one or more of the following components: trend, seasonal variation, cyclical variation, and irregular variation.
3. One primary objective is to forecast values for the variables using historical data patterns.

11.4 CLASSICAL DECOMPOSITION MODEL

Key idea 2

The classical decomposition model is one of the few forecasting systems that can be effectively used for relatively long-range forecasting (i.e., 3 to 5 years). The decomposition approach involves segmenting the data into four basic components: trend, cyclic, seasonal, and irregular. Several different decomposition models are employed for this purpose. Perhaps the most frequently used is the so-called multiplicative model,

$$Y_t = T_t \times C_t \times S_t \times I_t$$

where:

Y = Dependent variable
T = Trend component
C = Cyclic component
S = Seasonal component
I = Irregular component

This model assumes that the product of the four components will produce the desired time series. Typically, the trend component is measured in the same units as the variable being forecast. The other three components (C, S, and I) are measured as percentages. The trend component is then adjusted to reflect the impact of these variations. A C, S, or I value of 100% indicates that no adjustment of the trend line is necessary.

Occasionally, a time series can be better described using models that allow for additive or mixed component effects. The additive time series model takes on the following form:

$$Y_t = T_t + C_t + S_t + I_t$$

This model requires all of the components to be expressed in the same units.

In some specialized situations, the time series model can consist of both multiplicative as well as additive components. Following are two examples of a mixed-components model:

$$Y_t = T_t \times C_t \times I_t + S_t$$
$$Y_t = T_t \times C_t + I_t \times S_t$$

Again, the units for these models must be consistent (e.g., T and S in the first example are in the units of Y, whereas the units of C and I are in percentage terms). Selection of the most appropriate model form is usually based on the characteristics of the specific problem under study.

TREND COMPONENT

The **trend component** represents the long-term direction (i.e., increase or decrease) in the time series. Long-term shifts in a time series are usually due to factors such as technological developments, changes in demographic patterns, or consumer preferences. These shifts are called trends. Understanding trends is very important since trends provide the management with a

11.4 *Classical Decomposition Model* 473

FIGURE 11.1 Estimated Trend Line for U.S. Federal Budget Outlays ($ billions)

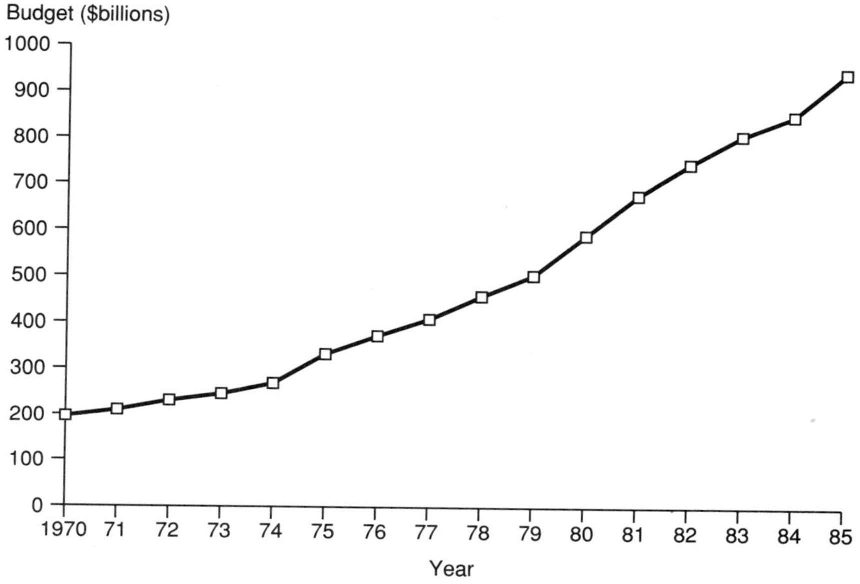

basic indication of the general direction of specific business parameters (e.g., sales). Figure 11.1 illustrates the trend line for U.S. federal budget outlays for the period 1970 to 1985.

CYCLIC COMPONENT

The **cyclic component** represents the variations around the trend line due to changes in business conditions. A cycle is measured from one peak to the next and can last from 2 to 10 years. Cycles may differ in amplitude and length based on such factors as buying patterns, government policies, or international incidents. Figure 11.2 demonstrates the cyclic variation in the annual percentage change in the consumer price index for the period 1969 to 1985.

SEASONAL COMPONENT

The **seasonal component** represents a pattern of change that is completed within one year and repeats itself regularly over the time series. Seasonal

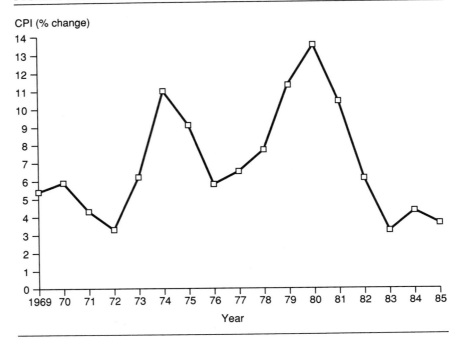

FIGURE 11.2 Demonstration of Cyclic Variation

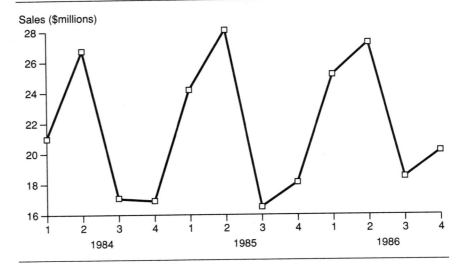

FIGURE 11.3 Demonstration of Seasonal Variation for Dialnet Sales

influences are often the result of annual weather patterns (e.g., the increased demand for air conditioning in the summer) or calendar events (e.g., Christmas). The considerable increase in tax preparation activities during the first quarter of each year and the reduction in demand of airline traffic over the same period are other examples of seasonal effects. The uneven demand for goods and services during the year must be taken into account in forecasting future demand. Figure 11.3 demonstrates the seasonal variation in Dialnet sales, by quarter, during the period 1984 to 1986.

IRREGULAR COMPONENT

The **irregular component** represents fluctuations in the time series that are the result of random or unpredictable events (e.g., presidential elections or natural disasters). The irregular component accounts for the variation in the time series that is not explained by the trend, cyclic, and seasonal components. Because the irregular component is random, its effect on the time series is very difficult to predict in advance. Figure 11.4 demonstrates the irregular variation for crude oil prices during the period 1973–1985. The dramatic rise in prices in 1975 and 1980 were the result of increasing tensions in the Middle East.

FIGURE 11.4 Demonstration of Irregular Component for Crude Oil Prices ($ per barrel)

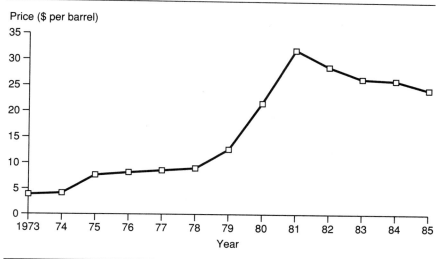

11.5 FORECASTING MODELS

Key idea 3

A variety of models are available for forecasting. These range from the very simple naive model to the very complex multivariable regression model. Each model possesses certain characteristics and limitations that should be understood. One key factor in selecting a forecasting technique is the required time horizon. Typically, forecasts can be classified as either short-term ($6 <$ months), medium-term ($6-18$ months), or long-term ($18 >$ months). Short-term forecasts are generally used by first-line and middle managers to address an immediate need. Examples of short-term forecasting include inventory requirements and the development of operating schedules. Medium-term forecasts, the general domain of middle managers, tend to focus on capacity planning and financial operations. Long-term forecasts, on the other hand, are usually the primary responsibility of top management. Capital requirements and technology development are specific examples of activities requiring long-term forecasts. Figure 11.5 presents a brief summary of the standard forecasting methods along with an indication of the appropriate time horizon. This section discusses several of the models presented in Figure 11.5. The moving averages and simple exponential smoothing models are generally limited to analyzing *stationary* time series; that is, a time series where the process average does not change over time. The more advanced models, such as smoothing with trend and autoregression can be used in analyzing *nonstationary* time series.

FIGURE 11.5 Overview of Forecasting Methods

Model	Description	Horizon
Naive	The forecast value for the next period is equal to the current actual value	Short
Simple moving average	The forecast value is an average based on two or more data values of equal importance	Short
Weighted moving average	The forecast value is an average based on two or more data values where each value can be weighted differently	Short
Exponential smoothing	The forecast value is an average based upon the previous period's actual value, the previous forecast value, and a smoothing coefficient	Short
Decomposition methods	The forecast value is based on an analysis of the following components: trend, cyclic, seasonal, and irregular	Long
Bivariate regression	The forecast value is based on the relationship between the dependent variable and time	Medium
Multivariate regression analysis	Similar to bivariate analysis except that the model contains two or more predictor variables	Medium
Autoregressive analysis	A special form of the multivariate regressive model where some of the predictor variables are past (i.e., lagged) values of the dependent variable	Medium

MOVING AVERAGES

The moving average is one of several standard techniques used in processing time series data. The concept of averages was introduced in Chapter 3. The term *mean* or *average* simply denotes the center point of a distribution of numbers. Typically, two types of moving averages are used in developing trend estimates for times series data: simple and weighted. The **simple moving average** is determined by averaging a specified number of contiguous data points. The actual number of data points used in computing the average is usually determined by the analyst. The computed average for a given data set is then used to forecast the next time period. This process is then repeated by dropping the oldest data value and adding the next value, which results in a new moving average. The computational process continues until all of the data values have been processed.

Simply stated, the use of moving averages generates a set of "smoothed" estimates, in contrast to the actual data. These smoothed estimates form the basis for forecasting beyond the existing data base. Further, any discernible trends are assumed to continue; recent data are more indicative of those trends than data from more remote time periods, which accounts for keeping the recent data and dropping the earliest. The general expression for computing a moving average is as follows:

$$M_t = \frac{A_{t-1} + A_{t-2} + A_{t-3} + A_{t-N+1}}{N}$$

where:

M_t = Moving average at period t
A_{t-1} = Actual value at period $t - 1$
N = Number of time periods used in computing the average

To illustrate the computational process, consider the installation data given in the Dialnet problem. The simple moving average (SMA) estimate for 1979, using three periods, is determined as follows:

$$\text{SMA}\,(1979) = \frac{45.5 + 79.2 + 77.8}{3} = 67.5$$

A complete set of CBS-generated moving average estimates for a three- and a five-period average are presented in Figure 11.6. Notice that the first three positions for the three-period average and the first five positions for the five-period average are not reported. In general, a moving average analysis based on n time periods will omit the first n positions; a moving average based on all of the time values will result in a single estimate that, of course, is the mean of the time series. An examination of the two moving average estimates shows that, in this case, the three-period model is more responsive than the five-period model. This result should not be surprising. Time series that involve rapid changes (such as that of Dialnet problem) are usually

FIGURE 11.6 Simple Three- and Five-Period Moving Averages for the Dialnet Company

Year	Actual Installations (000)	Three-Period Moving Average (000)	Five-Period Moving Average (000)
1976	45.5	—	—
1977	79.2	—	—
1978	77.8	—	—
1979	81.4	67.5	—
1980	102.2	79.5	—
1981	124.8	87.13	77.22
1982	132.6	102.80	93.08
1983	156.3	119.87	103.76
1984	161.2	137.90	119.46
1985	168.6	150.03	135.42
1986	174.2	162.03	148.70

better described using a smaller number of time periods. Figure 11.7 presents a plot of the actual number as well as the three- and five-period moving average forecasts of phone installations for the Dialnet Company.

An important variation on the simple moving average model is called the **weighted moving average** (WMA). The chief advantage of this method rela-

FIGURE 11.7 Comparison of Actual Data with Three- and Five-Month Simple Moving Averages for Dialnet

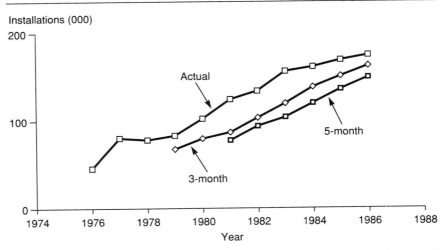

tive to the simple model is that more recent values are given more emphasis than older values. As the name implies, weights are assigned to each of the data values used in computing the average. Typically, weights are assigned in ascending order. That is, the smallest weight is assigned to the oldest data point, and the largest weight goes to the most recent data value. Often, the sum-of-digits method is used in computing weighting values. For example, the relative weights for a three-period moving average using this method would be one sixth, two sixths, and three sixths. The weighted moving averages along with the simple moving averages for the Dialnet problem (using a three-period average) are presented in Figure 11.8. The WMA estimate for 1979 was obtained as follows:

$$\text{WMA }(1979) = (1/6)45.5 + (2/6)79.2 + (3/6)77.8 = 72.88$$

One can see that the weighting technique yields a different fit for the same time series data. Selection of the number of periods and their corresponding weights is something of a mystery. How many periods should be used? What should the weights be, and how should they be applied? One approach to developing answers to these questions is to evaluate various combinations of periods and weights. The use of computer analysis greatly facilitates this process. Those combinations that significantly reduce the error of fit may then be applied to the new period for which a forecast is needed. Nevertheless, the user must remember that applying "backward validation" techniques is based on the assumption that the future will be similar to the past. Both moving average models can be used to forecast future values; unfortunately, they are technically limited to predicting only one period beyond the

FIGURE 11.8 Comparison of Simple and Weighted Moving Averages for the Dialnet Company

Year	Actual Installations (000)	Three-Period Moving Average (000)	Three-Period Weighted Moving Average (000)
1976	45.5	—	—
1977	79.2	—	—
1978	77.8	—	—
1979	81.4	67.5	72.88
1980	102.2	79.5	79.83
1981	124.8	87.13	91.20
1982	132.6	102.80	110.03
1983	156.3	119.87	124.93
1984	161.2	137.90	143.15
1985	168.6	150.03	154.80
1986	174.2	162.03	164.08

current data base. This limitation restricts the use of these methods for many applications. In general, these methods lend themselves to applications involving short time horizons.

To illustrate the forecasting process using moving averages, consider the two predictions that follow. The first one is based on simple moving averages, and the second one uses weighted moving averages. Both forecasts are for Dialnet phone installations (in thousands) for 1987. The weighted model yields a slightly higher result. This should not be surprising since larger weights are applied to larger data values. The weighted moving average model will always yield higher forecast values vis-à-vis the simple moving average model when the trend in the data base is upward and larger weights are applied to more recent time values.

$$\text{SMA (1987)} = \frac{161.2 + 168.6 + 174.2}{3} = 168$$

$$\text{WMA (1987)} = (1/6)161.2 + (2/6)168.6 + (3/6)174.2 = 170.16$$

EXPONENTIAL SMOOTHING

Exponential smoothing is a technique for continuously updating a forecast with new data. It is the most widely used of all forecasting models. Some characteristics of the exponential smoothing model are:

- Ease of understanding
- Relative accuracy over short time periods
- Straightforward formulation
- Minimal data storage requirements

Actually, there is a family of exponential smoothing models. The basic options are as follows:

- Single-parameter model
- Two-parameter model (includes trend effects)
- Three-parameter model (includes trend and seasonal effects)

The simple or single-parameter exponential model requires only three basic inputs: the actual and forecast variable values from the previous period and a smoothing coefficient. The standard form of the single-parameter exponential smoothing model is

$$F_{t+1} = F_t + \alpha(A_t - F_t)$$

where:

F_{t+1} = Forecast value for period $t + 1$
F_t = Forecast value for period t
A_t = Actual value for period t
α = Smoothing coefficient ($0 < \alpha < 1$)

The smoothing coefficient (alpha) is defined over the range of 0 to 1. Using an alpha of 0 results in a forecast equal to the previous forecast value. On the other hand, an alpha of 1 results in the decision to use all of the residual to correct the forecast. This produces results equivalent to the naive model (where the forecast value is equal to the current actual value). The most appropriate alpha value depends on the nature of the time series data. Generally speaking, larger alpha values result in a quicker response in the forecast values, and smaller alpha values yield a slower response. Typically, a small alpha value is appropriate for volatile data, and larger alpha values can be used for stable data. Determining the "best" alpha value can be done on a trial-and-error basis or through computer analysis. CBS has an option for determining the value of alpha that minimizes the total error associated with the forecast. As a closing note, some statisticians recommend an upper alpha limit of .3 regardless of the level of error. This is due to the fact that changes in the forecast may be the result of random fluctuations or real effects that cannot be isolated.

Figure 11.9 presents a graphic comparing the original data and forecast values based on two smoothing coefficients, $\alpha = .2$ and $\alpha = .5$. The computational process involves multiplying the residual or error term by the smoothing constant and adding the value to the forecast for the previous period to arrive at the forecast for the next period. This requires the estab-

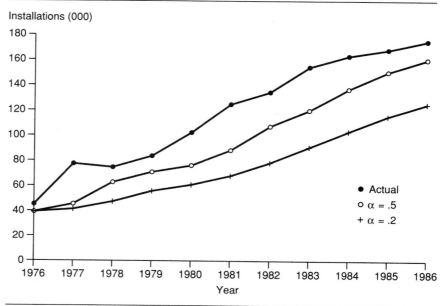

FIGURE 11.9 Single-Parameter Exponential Smoothing for Dialnet Company Showing Actual versus Forecast Demand for Selected Alpha Values

lishment of an initial starting value. Typically, one of the following two approaches can be used:

- Utilize the first data point as the starting value, and reduce the length of the time series by 1.
- Estimate a starting value based on the historical trend, and maintain the original length of the time series.

In most cases the two approaches yield nearly identical forecasts. For this analysis the computational process is based on the second approach with an initial value of 40.

An examination of the forecasts in Figure 11.9 shows that an alpha of .5 produces the better fit. This can be explained by the fact that the actual data values for the time series are increasing rapidly, which would require a larger alpha value.

The developed exponential smoothing model can now be used to forecast phone installations for 1987. The computational process using an alpha of .5 is

$$F(1987) = F(1986) + .5[A(1986) - F(1986)]$$
$$= 159.11 + 0.5(174.2 - 159.11) = 166.65$$

Thus, the single-parameter exponential smoothing model with an alpha of .5 yields a nearly identical forecast to the two previous moving average models.*

Single-parameter smoothed averages tend to lag behind a general trend in the data. The general upward direction of the Dialnet data suggests the presence of a positive trend. One approach for ameliorating this problem is to incorporate a trend factor into the model. The *Holt two-parameter exponential smoothing model* is often used in this regard. The basic form of the model is:

$$F_t = \alpha Y_t + (1 - \alpha)(F_{t-1} + T_{t-1})$$
$$T_t = \beta(F_t - F_{t-1}) + (1 - \beta)T_{t-1}$$
$$F_{t+1} = Y_t + T_t$$

where:

Y_t = Actual value for period t
Y_{t-1} = Actual value for period $t - 1$
T_t = Trend estimate for period t
T_{t-1} = Trend estimate for period $t - 1$
α = Smoothing coefficient
β = Smoothing coefficient for trend

The first equation is nearly identical to the single-parameter model, except that a term has been added to account for the trend. The second equation

---

* In fact, both models can be viewed as equivalent whenever $\alpha = 2/n + 1$.

utilizes a second smoothing constant, β, which is known as the trend coefficient. This coefficient is used to smooth the trend and ranges in value from 0 to 1. The final equation combines the trend estimate with the smoothed forecast for the current period to produce the desired forecast for the next period. The two-parameter model requires five basic inputs: the alpha value, the beta value, the initial trend value, the smoothed value, and the initial actual value. The initial trend is either estimated from the slope of past data or assigned a value of 0. Numerical values for alpha and beta can be estimated based on the nature of the times series.

Figure 11.10 shows a plot of the actual time series versus a two-parameter forecast with $\alpha = .5$, $\beta = .2$, an initial trend value of 10, and an initial starting value of 40. This graphic reveals that the two-parameter model more closely matches the actual data than the one-parameter model. Furthermore, the lag encountered in the earlier model has been significantly reduced. Again, this should not be too surprising because of the presence of the general upward trend in the time series data. Naturally, the selection of different values for the smoothing and trend coefficients will produce different forecast results. Determining the optimal values for these parameters can be accomplished using the CBS courseware.

An even more complex smoothing model can be used when a seasonal

FIGURE 11.10 Two-Parameter Exponential Smoothing for Dialnet Company Showing Actual versus Forecast Demand for $\alpha = .5$ and $\beta = .2$

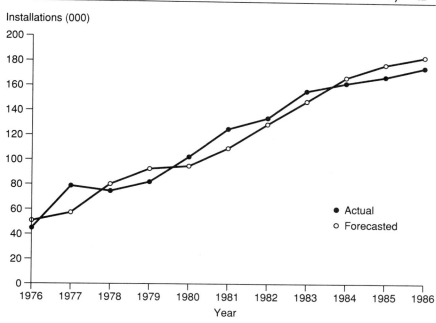

pattern appears to exist in the time series. This model, often known as *Winter's three-parameter model*, incorporates a smoothing coefficient, a trend coefficient, and a seasonal coefficient. The CBS software package contains an option for developing forecasts based on this three-parameter model.

Each of the time series forecasting methods discussed so far is particularly useful when the time series data vary considerably from period to period (i.e., large data scatter). However, as already stated, the effective range of forecasting is usually limited to the next period.

REGRESSION ANALYSIS

Regression analysis represents an important alternative to the moving averages and exponential smoothing models. It is particularly useful for problems requiring forecasts beyond the next period. The general structure of the regression model was presented in Chapters 9 and 10. Figure 11.11 highlights some of the major advantages and disadvantages of using regression modeling to analyze time series data.

As was seen in the previous two chapters, regression models (which are based on the principle of least squares) require extensive data, particularly when statistical significance is a requirement. Furthermore, regression models often do not match the data as effectively as some other forecasting models. However, these disadvantages are, in many applications, offset by the fact that the regression model can be used for forecasting beyond one period. Recall that the moving averages and exponential smoothing models, by and large, are restricted to forecasting one time period beyond the current data base. The multiple regression model, moreover, is based on specific explanatory variables instead of being based on time. The incorporation of such variables significantly enhances the power of the modeling approach.

The following time dependent model illustrates the use of regression analysis in forecasting:

$$Y = B_0 + B_1 \times T$$

This model is equivalent to the standard bivariate regression model presented in Chapter 9, except that the independent variable in this case is time (i.e., T). This model can also be used for estimating the trend component, in a similar way as for moving averages and exponential smoothing. The model specification based on the Dialnet phone installation data is:

$$Y \text{ (installations)} = 40.45 + 13.01 \times T$$

The coefficient of determination for this model is approximately .96. This result should not be surprising, given the near-linear growth experienced over the time horizon of the problem (see data in Section 11.2). One can

FIGURE 11.11 Characteristics of Regression Modeling for Analyzing Time Series Data

Advantages	Disadvantages
• Extended forecasts	• Autocorrelation
• Causal theory	• Extensive data base
• Theory-based	• Accuracy levels

conclude that a simple linear time-dependent model is a very good estimator for phone installations. However, will this simple model stand the test of predicting future sales? Probably not, since the use of a single variable to predict the future is generally ineffective.

Often there exists a strong correlation between successive observations in a time series. In these cases, it is possible to predict future values of the time series using the previous values. This forecasting method is known as **autoregression,** which literally means regressing the time series on itself. Autoregressive models tend to perform well when the time series does not fluctuate excessively and the forecast period does not exceed two years.

One of the basic assumptions associated with the regression model is that the residuals are uncorrelated. Unfortunately, this assumption is often violated in autoregressive models. A condition called **autocorrelation** occurs, in which the residuals are related to the previous values. The presence of autocorrelation in a time series model can have the following consequences:

- The regression coefficients may not be reliable.
- The standard error of the estimate may significantly underestimate error term variability.
- The hypothesis tests will report statistical significance when, in fact, no logical relationship exists.
- The developed confidence limits may be inaccurate.

Positive autocorrelation occurs when a positive residual ($e > 0$) follows another positive residual, or when a negative residual ($e < 0$) follows another negative residual. For example, the price of gold may remain below the norm for a period of several weeks, followed by a swing to a level above the norm, where it remains for another period of several weeks; the residuals for each period represent a form of positive autocorrelation. When a positive residual ($e > 0$) follows a negative residual ($e < 0$), or when a negative residual ($e < 0$) follows a positive residual ($e > 0$), *negative* autocorrelation exists. The history of domestic oil prices provides an example of negative correlation. Lower oil prices discourage production from marginal wells, which leads to shortages, prompting a move toward higher prices. Panels A and B of Figure 11.12 illustrate the nature of positive and negative autocorrelation, respectively.

FIGURE 11.12 Illustration of Positive (Panel A) and Negative (Panel B) Autocorrelation

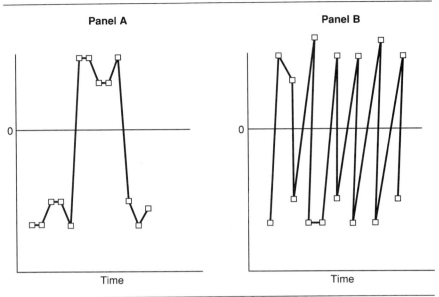

Several tests are available for determining the presence of autocorrelation in a time series. Perhaps the most useful is the Durbin-Watson (DW) test statistic. This statistic ranges in value between 0 and 4, with values around 2 indicating the absence of autocorrelation.

Normally, the existence of positive autocorrelation is of primary concern. The DW test procedure can be structured to test for positive autocorrelation using the following hypothesis:

H_0: No autocorrelation exists

H_1: Positive autocorrelation exists

The DW test is somewhat unique in that there is a certain range over which the test is inconclusive. The decision rule

DW > du: Do not reject H_0

dl ≤ DW ≤ du: Test is inconclusive

DW < dl: Reject H_0

where dl and du represent the critical lower and upper limits, respectively. These values are obtained from the DW table in Appendix A (Table A.7) based on the sample size, number, predictor variables, and alpha level.

There are several ways to address the problem of autocorrelation. Among them are:

- Incorporate additional explanatory variables (i.e., identify new independent variables).
- Redefine the variable set in terms of percentage changes (i.e., compute the percentage change from the previous year).

Finding additional explanatory variables seems the most useful method for reducing the impact of autocorrelation. Unfortunately, this is not possible in all situations. For example, a key factor in the continued growth of telephone service demand may be changing requirements in local building ordinances. This factor may be very difficult to quantify for incorporation into the model. In general, the problems of model specification and autocorrelation are difficult to resolve completely. In fact, a whole branch of statistics, called econometrics, has been established to address these important issues.

To illustrate the use of autoregression, consider replacing the current independent variable (i.e., time) in the bivariate model with two new variables, gross regional product (GRP) and installations lagged by one period. The object, of course, is to reduce the extent of autocorrelation on the one hand and to improve the model's accuracy on the other. GRP for the Southwest has been selected as a good indicator of economic activity in the region, and the previous number of installations would seem to indicate future demand. Presented below is the proposed model:

$$Y_t = B_0 + B_1 \text{GRP}_t + B_2 Y_{t-1}$$

Notice how the second independent variable is simply the dependent variable lagged one period. This model would be specified and statistically tested in the way outlined in Chapter 10. The extent of autocorrelation could also be measured using the DW test statistic. Based on these results, a decision could be made to replace the bivariate forecasting model with this new model.

CYCLIC AND SEASONAL VARIATIONS

In many situations the developed trend does not adequately fit the historical data. In these cases the trend (T) can be adjusted by incorporating the cyclic (C), seasonal (S), or irregular (I) component. Recall that the generalized classical decomposition model has the form:

$$Y_t = T_t \times C_t \times S_t \times I_t$$

Each component of interest can be isolated by either factoring out or eliminating the remaining components. This factoring process generally starts with the trend, since that is usually the first component to be estimated. The trend can be factored out by dividing both sides by T. This results in the following expression:

$$Y/T = C \times S \times I$$

Cyclic Analysis

When the data are yearly, as in the case of the Dialnet phone installation problem, the seasonal component can be eliminated (i.e., $S = 1$). This is because the seasonal effects are indeterminate on an annual basis. The use of annual data does not permit the isolation of any short-term irregular fluctuations, and any long-term irregular effects are confounded with the cyclic component. Accordingly, no attempt is made to isolate the irregular component when the problem consists of annual data. Thus, the CI component can be computed as follows:

$$CI = Y/T$$

Figure 11.13 shows the computation of the CI component using the simple time-dependent regression model (i.e., $Y = 40.45 + 13.01T$) for estimating the trend component. A plot of the CI component for the Dialnet problem appears in Figure 11.14.

These cyclic components can then be used to adjust or modify the basic trend estimate. An inspection of Figure 11.14 indicates approximately a three-year cycle. Therefore, a set of average adjustments can be computed that will be more accurate than any one value. Figure 11.15 depicts the average three-year cyclic factors for this problem. These averages were com-

FIGURE 11.13 *CI* Analysis of Dialnet Installation Data

Period	Y, Actual	Trend Component	CI Component
1	45.50	53.46	85.11
2	79.20	66.47	119.15
3	77.80	79.48	97.88
4	81.40	92.50	88.00
5	102.20	105.51	96.86
6	124.80	118.52	105.30
7	132.60	131.54	100.81
8	156.30	144.55	108.13
9	161.20	157.57	102.30
10	168.60	170.58	98.84
11	174.20	183.59	94.89

FIGURE 11.14 *CI* Component for the Dialnet Installation Data

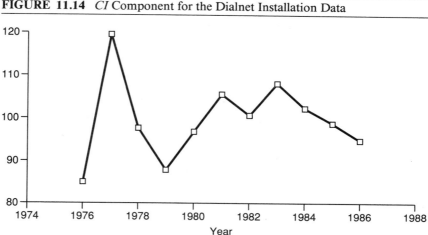

puted simply by adding the corresponding cyclic components for a three-year cycle and dividing by the number of values. For example, the average for the first year was computed by adding the first, fourth, seventh, and tenth values and dividing by four: $(85.11 + 96.86 + 100.81 + 98.84)/4 = 93.19$. A more detailed analysis involving the cyclic component can be found in Section 11.7.

Seasonal Analysis

Whenever the data are based on a time interval of less than one year (e.g., monthly or quarterly), the issue of seasonality becomes relevant. Recall that seasonal variation refers to the fluctuations that occur within a one-year period on a regular basis, year after year. In many forecasting exercises it becomes important to remove the effects of seasonality to gain insight into the overall business pattern. In effect, the removal of seasonal fluctuations

FIGURE 11.15 Computed Average *CI* Component for the Dialnet Installation Data

Year	Average Cyclic Component
1	93.19
2	104.76
3	101.83

often helps the manager identify specific operational strengths and weaknesses. For example, a 10% increase in the demand for home heating oil last winter may or may not signal an overall trend in the demand for heating oil.

The process of isolating the seasonal effects (i.e., deseasonalization) of the time series is very similar to that used for the cyclic component. In this case, the basic model is as follows:

$$S = Y/TCI$$

This model suggests that the seasonal component of a time series can be determined by dividing by the *TCI* component. The technique most often used to estimate the *TCI* component in these cases is known as the *method of centered moving averages* (CMA). The CMA method is very similar to the moving average model presented earlier. The process for deseasonalizing a time series will be illustrated using Dialnet's quarterly sales data. The computational process is shown in Figure 11.16 for the period 1983 to 1986. In this example the seasonal indexes were determined by dividing the actual time series by the corresponding CMA estimates.

A plot of the unadjusted and adjusted sales is shown in Figure 11.17. Notice that the seasonal factors are well above the norm (i.e., 100) for the first two quarters and below the norm for quarters 3 and 4. This suggests that the trend line, without adjustment, would significantly underestimate sales for

FIGURE 11.16 *SI* Factors for Dialnet Problem Using Four-Quarter Centered Moving Average Model

Year	Quarter	Actual	Centered Moving Average	Seasonal Component
1983	1	22.2	—	—
	2	24.3	—	—
	3	16.5	19.4	85.1
	4	15.4	19.0	81.1
1984	1	21.1	18.8	112.2
	2	21.7	19.1	109.9
	3	17.2	19.6	87.8
	4	16.9	20.9	80.9
1985	1	24.3	21.6	112.5
	2	28.2	21.7	129.9
	3	16.5	21.9	75.3
	4	18.1	21.9	82.6
1986	1	25.2	22.0	114.5
	2	27.3	20.8	131.3
	3	18.5	—	—
	4	20.2	—	—

FIGURE 11.17 Unadjusted and Adjusted Sales for Dialnet Company

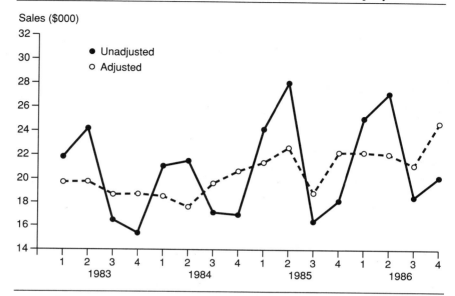

the first two quarters and overestimate sales for the last two quarters. This observation underscores the need to adjust the trend line for seasonal fluctuations.

An average set of seasonal factors can be computed by averaging the quarterly *SI* estimates over the four-year period. This process has the effect of eliminating the irregular component, since the year-to-year fluctuations in the *SI* factor can be attributed to irregular influences. The average quarterly seasonal components along with the seasonal indexes are shown in Figure 11.18. The seasonal index estimates have been obtained by normalizing the average seasonal components so that they sum to 400. This is accomplished by dividing 400 by the sum of the average seasonal components and multi-

FIGURE 11.18 Seasonal Index Computations for Dialnet Sales Data

Quarter	Average Seasonal Component	Seasonal Index
1	113.1	112.8
2	123.7	123.4
3	82.6	82.4
4	81.6	81.4
Total	401.0	400.0

FIGURE 11.19 Deseasonalized Data Base for Dialnet Sales

Year	Quarter	Actual	Seasonal Index	Adjusted Sales
1983	1	22.2	112.8	19.7
	2	24.3	123.4	19.7
	3	16.5	87.4	18.9
	4	15.4	81.4	18.9
1984	1	21.1	112.8	18.7
	2	21.7	123.4	17.6
	3	17.2	87.4	19.7
	4	16.9	81.4	20.8
1985	1	24.3	112.8	21.5
	2	28.2	123.4	22.9
	3	16.5	87.4	18.9
	4	18.1	81.4	22.2
1986	1	25.2	112.8	22.3
	2	27.3	123.4	22.1
	3	18.5	87.4	21.2
	4	20.2	81.4	24.8

plying that ratio by each of the quarterly average seasonal components (e.g., Seasonal index for quarter 1 = 400/401 × 113.1 = 112.8).

The computed seasonal indexes can now be used to adjust or "deseasonalize" the time series. This process is illustrated in Figure 11.19. The deseasonalized sales estimates are computed by simply dividing the actual sales by the seasonal index and multiplying the quotient by 100 (e.g., First-quarter 1983 adjusted sales = 22.2/112.8 × 100 = 19.7).

A more representative trend estimate of Dialnet sales can now be developed using the seasonally adjusted data base. The normal procedure is to develop a linear regression line.* A CBS analysis of the adjusted sales yields the following simple regression model:

$$Y \text{ (Sales)} = 18.08 + 0.299 \times T$$

where T represents time, and $T = 1$ is the first quarter of 1983.

An estimate of the trend using the regression model for the period 1983 to 1986 is given in Figure 11.20. Next, the cyclic variation around the trend line can be found by dividing the adjusted sales figures by the trend line. This result is also shown in Figure 11.20. This computation assumes that the irregular variation is relatively insignificant (i.e., $I = 1$). Notice that the cyclic estimates, like the seasonal indexes, are based on a norm of 100.

---

* Another approach is to incorporate any seasonal effects directly into the regression model through the use of dummy variables. These dummy variables would take on values of either 0 or 1 depending on the specific period.

FIGURE 11.20 Estimate of Trend and Cyclic Components for Dialnet Sales

Year	Quarter	Adjusted Sales	Trend	Cyclic Component
1983	1	19.7	18.4	107.1
	2	19.7	18.7	105.3
	3	18.9	19.0	99.5
	4	18.9	19.3	97.9
1984	1	21.1	19.6	107.7
	2	21.7	19.9	109.0
	3	17.2	20.2	85.1
	4	16.9	20.5	82.4
1985	1	24.3	20.8	116.8
	2	28.2	21.1	133.6
	3	16.5	21.4	77.1
	4	18.1	21.7	83.4
1986	1	25.2	22.0	114.5
	2	27.3	22.3	122.4
	3	18.5	22.6	81.9
	4	20.2	22.9	88.2

The problem at hand is to develop a 1987 quarterly sales forecast for the Dialnet Company. The quarterly trend estimates have been obtained from the regression model identified above. The quarterly sales forecast is obtained by multiplying the trend estimate by the seasonal index and dividing by 100. The results are shown in Figure 11.21. These quarterly forecasts are far more representative of future sales than the simple trend estimates. Notice, however, that the forecast does not contain either a cyclic or an irregular adjustment. Recall that it is usually not possible to mathematically develop a quantitative estimate of the irregular variation. Consequently, the normal practice is to set it equal to 1, unless something specific is known regarding the pending occurrence of an irregular event. With regard to the cyclic variation, the analysis presented in Figure 11.21 consists of only four years of quarterly data. Therefore, it is difficult to estimate a complete set of

FIGURE 11.21 1987 Quarterly Sales Forecast for Dialnet Company

Quarter	Trend Estimate	Seasonal Index	Sales Forecast
1	22.01	112.8	24.8
2	22.11	123.4	27.3
3	22.21	87.4	19.4
4	22.31	81.4	18.2

494 Chapter 11 *Time Series and Forecasting*

cyclic indexes over such a short time span. When a complete set of cyclic indexes are available, however, they can be included in the forecasting process. In these cases, the forecast would be modified by multiplying it by the appropriate cyclic index.

11.6 FORECAST VALIDATION

Validation is one of the most serious issues regarding the development and use of forecasting models. In this regard, it is important to keep in mind the actual purpose or objective of the forecast. The following list identifies several common objectives in the preparation of forecasts:

Key idea 4

- Very often a forecast is used as a motivator or as a strategic objective. For example, the manager may wish to get his salespeople to work harder by forecasting dire events.
- Frequently, the mere reporting of a forecast ensures that the forecast event will not occur. For example, when the government forecasts an expected shortage of 10,000 electrical engineers, market mechanisms are automatically activated to ensure that such will not be the case.
- Often, forecasts are prepared to justify an action with no necessary counterpart in reality. For example, Congress will be assailed for funding a new weapons system based upon a purported and forecast enemy "threat."

FIGURE 11.22 Definitions of MAD, MSE, and MPE Accuracy Measurements

Definition	Model	Remarks
Mean absolute deviation (MAD): The deviations are summed without algebraic sign. An average is computed by dividing the sum of the deviations by the number of data points.	$MAD = \dfrac{\Sigma|Y - y|}{N}$	This measure is easy to understand and can be used in most cases. However, a distorted picture of the forecasted results can occur when the variable values are large.
Mean square error (MSE): Each deviation is squared, and the results are summed and divided by the number of data points.	$MSE = \dfrac{\Sigma(Y - y)^2}{N}$	This measure is more difficult to understand and should not be used when there are significant deviations between the predicted and the actual values.
Mean percentage error (MPE): Each deviation is converted to a percentage of the actual value. These percentages are then summed and divided by the number of data points.	$MPE = \dfrac{\Sigma[(Y - y)/Y] \times 100}{N}$	This measure should be used when both the variable values and the deviations are large.

It would therefore appear that the validation criterion of a forecast sometimes is the event occurring and sometimes it is the event *not* occurring. How can this ambiguity be resolved? The normal use of the term *validation* may not apply here as it does in other branches of statistical measurement. Rather, the techniques introduced here should be "validated" by how well they predict the past. Thus, instead of using 1987 data to predict 1988, one might use 1986 data to predict 1987, for which data have already been collected. In this manner, appropriate techniques can be chosen for how well forecasts "fit" already existing data. The model that best fits the data can then be used as the basis for developing a forecast.

Typically, many indexes are available for measuring how well a forecast fits or matches the actual data. Several standard measures, such as the error of estimate for bivariate regression and residuals for multiple regression, have already been discussed. Figure 11.22 presents the three most frequently used error measures in time series analysis.

Figure 11.23 presents a comparison of MAD and MSE forecast error measurements for the Dialnet installation data for several different models. The fact that the regression model yields the smallest MAD and MSE values should not be too surprising since the actual time series data are nearly linear. Furthermore, none of the exponential smoothing models have been optimized with respect to their coefficients. For example, optimizing both the smoothing and trend coefficients for the two-parameter model results in MAD and MSE values of approximately 9 and 118, respectively.

FIGURE 11.23 Comparison of MAD and MSE Error Measurements for Dialnet Installation Data

Model	Error Measurements	
	MAD	MSE
Simple moving average		
3-period	24	672
5-period	40	1,680
Weighted moving average		
3-period	20	477
5-period	31	1,022
Exponential smoothing		
$\alpha = .2$	43	2,159
$\alpha = .5$	23	630
$\alpha = .5, \beta = .2$	9	106
Regression analysis	6	59

11.7 COMPUTER ANALYSIS

Key idea 5

In most cases, time series problems are solved using computer analysis. The use of computer models allows for the rapid solution of large problems and reduces the potential for numerical errors. Furthermore, computer analysis is a necessity when attempting to optimize the coefficients for the exponential smoothing model. The input requirements for most computer models used in analyzing time series are quite simple. Typically, the input consists of data for one or more time series along with the selection of a specific analytical model (e.g., moving averages). This section illustrates the use of CBS computer analysis for solving several of the problems that have been presented.

Figure 11.24 presents CBS-generated output for the Dialnet installation problem using a three-month moving average model. These results, which also include residuals, are identical to those presented in Section 11.4. Figure 11.25 shows a CBS analysis of the same problem using exponential smoothing with a smoothing coefficient of .5.

FIGURE 11.24 CBS Three-Month Moving Average Analysis of Dialnet Sales

Time Series Results

Model type:	Trend—Moving averages
Number of periods in average:	3
Number of periods:	11
Mean square error (MSE):	672.42
Mean absolute deviation (MAD):	24.32
Forecast for period 12:	168.00

Residual Analysis

Period	Y-Actual	Y-Pred	Residual
1	45.5000	45.5000	0
2	79.2000	79.2000	0
3	77.8000	77.8000	0
4	81.4000	67.5000	13.9000
5	102.2000	79.4667	22.7333
6	124.8000	87.1333	37.6667
7	132.6000	102.8000	29.8000
8	156.3000	119.8667	36.4333
9	161.2000	137.9000	23.3000
10	168.6000	150.0333	18.5667
11	174.2000	162.0333	12.1667

FIGURE 11.25 CBS Exponential Smoothing Analysis ($\alpha = .5$) of Dialnet Installation Data

Time Series Results

Model type:	Trend—Exponential smoothing
Smoothing coefficient	.5
Number of periods:	11
Mean square error (MSE):	600.02
Mean average deviation (MAD):	22.23
Forecast for period 12:	166.65

Residual Analysis

Period	Y-Actual	Y-Pred	Residual
1	45.5000	0	45.5000
2	79.2000	22.750	56.4500
3	77.8000	50.9750	26.8250
4	81.4000	64.3875	17.0125
5	102.2000	72.8938	29.3062
6	124.8000	87.5469	37.2531
7	132.6000	106.1734	26.4266
8	156.3000	119.3867	36.9133
9	161.2000	137.8434	23.3566
10	168.6000	149.5217	19.0783
11	174.2000	159.0608	15.1392

A CBS analysis of the same Dialnet data using regression modeling is shown in Figure 11.26. The printout also displays a variety of error measurements, including the standard error of the estimate and standard error of the slope (i.e., B_1). An examination of the residual analysis shows the presence of autocorrelation in that most of the residuals at the beginning are negative, followed by a series of positive residuals. The printout also features a forecast for period 14 using the model. (Recall that one of the advantages of the regression model over the other time series models is that it can be used to forecast more than one time period beyond the current data base.) Finally, Figure 11.27 presents a CBS seasonal analysis of Dialnet sales.

11.8 PRACTICAL APPLICATIONS

This section presents several specific business applications involving time series analysis and forecasting applications.

FIGURE 11.26 CBS Regression Analysis of Dialnet Installation Data

Time Series Results

Model type:	Trend—Least squares (linear)
Number of periods:	11
Mean square error:	58.7438
Mean average deviation:	6.4430
B_0 coefficient:	40.4455
B_1 coefficient:	13.0136
Coefficient of determination:	.9665
Standard error estimate:	8.4734
Standard error B_1:	0.8079
Forecast for period 12:	196.6091

Residual Analysis

Period	Y-Actual	Y-Pred	Residual
1	45.5000	53.4591	−7.9591
2	79.2000	66.4727	12.7273
3	77.8000	79.4864	−1.6864
4	81.4000	92.5000	−11.1000
5	102.2000	105.5136	−3.3136
6	124.8000	118.5273	6.2727
7	132.6000	131.5409	1.0591
8	156.3000	144.5545	11.7455
9	161.2000	157.5682	3.6318
10	168.6000	170.5818	−1.9818
11	174.2000	183.5955	−9.3955

Forecasting

phone = 40.445 + 13.014 (time)

Time Period ?14

phone = 222.64

Citrus Association Exchange (Cyclic Analysis)

The general manager of the Citrus Association is interested in analyzing trends in orange concentrate prices. U.S. orange production, which peaked in the late 1970s, has been falling throughout the 1980s. This decline can be attributed, in part, to freezes that occurred in four of the last five years. Additionally, aggressive marketing by Brazil further helped reduce U.S.

FIGURE 11.27 CBS Seasonal Analysis of Dialnet Sales

Time Series Results

Year	Quarter	Actual	Seasonal Index	Adjusted Sales
1983	1	22.2	112.8	19.7
	2	24.3	123.4	19.7
	3	16.5	87.4	18.9
	4	15.4	81.4	18.9
1984	1	21.1	112.8	18.7
	2	21.7	123.4	17.6
	3	17.2	87.4	19.7
	4	16.9	81.4	20.8
1985	1	24.3	112.8	21.5
	2	28.2	123.4	22.9
	3	16.5	87.4	18.9
	4	18.1	81.4	22.2
1986	1	25.2	112.8	22.3
	2	27.3	123.4	22.1
	3	18.5	87.4	21.2
	4	20.2	81.4	24.8

production. The decline in U.S. production initially brought about an increase in cash prices. However, over the last several years, prices have dropped. A summary of average concentrate prices and U.S. production is given in Figure 11.28.

The general manager has decided to estimate the trend in orange prices using a regression model. The developed model is:

$$\text{Orange price index} = 250.8 + 14.7 \times \text{Time}$$

The coefficient of determination is approximately .50, which means the model explains approximately 50% of the variance in orange prices. The large amount of unexplained variance is of concern to the general manager. One possible approach to improving the explanatory power of the model without resorting to more complex forecasting methods would be to measure the amount of change associated with cyclic activity. Recall that cyclic variations represent an upward or downward movement around the general direction of the trend line. A CBS analysis of the cyclic activity for this problem is given in Figure 11.29. The cyclic index was computed by dividing the actual price by the estimated trend price (i.e., Cyclic index (1979) = 260.2/265.5 = 1.020).

FIGURE 11.28 Orange Wholesale Index Prices and Production

Year	Average Wholesale Index (1967 = 100)	Production (000 metric tons)
1979	260.2	8,310
1980	250.7	10,734
1981	319.2	9,514
1982	315.5	7,025
1983	301.8	8,827
1984	385.2	6,573
1985	398.7	6,095
1986	304.7	6,815

A plot of the cyclic indexes reported in Figure 11.29 is shown in Figure 11.30. These data indicate the existence of a three-year cycle starting in 1979 and ending in 1982. The average indexes for the cycles are 1.0, 1.1, and 0.91.

The general manager can now use the computed cyclic indexes for adjusting the basic trend forecast. This computational process is presented in Figure 11.31. The trend estimates for the three-year forecast were developed using the regression model. The forecast values are determined by multiplying the trend estimates by the corresponding cyclic index. Notice how the forecasted results demonstrate a cyclic nature when compared to the simple trend line estimates. This adjusted forecast represents a more accurate estimate of future market prices.

U.S. Treasury Bonds (Autoregressive Forecasting)

The U.S. Secretary of the Treasury wishes to estimate treasury bond rates for the coming year as part of the departmental budgetary process. The actual bond rate will be determined by market conditions at the time of issuance.

FIGURE 11.29 CBS Trend and Cyclic Orange Price Estimates

Year	Actual	Trend Estimate	Cyclic Index
1979	260.2	265.5	1.020
1980	250.7	280.2	1.117
1981	319.2	294.9	0.924
1982	315.5	309.6	0.981
1983	301.8	324.4	1.075
1984	385.2	339.1	0.880
1985	398.7	353.8	0.887
1986	304.7	368.5	1.209

FIGURE 11.30 Plot of Orange Concentrate Cyclic Indexes

The department offers a variety of time-dependent investment instruments, but the secretary is particularly interested in the very popular three-year bonds. Figure 11.32 presents three-year treasury bond yields for the period 1975 to 1985.

The secretary's staff has decided to use an autoregressive forecasting approach in developing a forecast for 1986. More specifically, they plan to develop an autoregressive model based on the two previous time periods. The general form of this model is

$$\text{T-rate}(t) = b_0 + b_1 \times \text{T-rate}(t-1) + b_2 \times \text{T-rate}(t-2)$$

FIGURE 11.31 Orange Concentrate Price Forecast for 1986–1988

Year	Trend Estimate	Average Index	Forecast
1986 (8)	368.4	1.00	368.4
1987 (9)	383.1	1.10	421.5
1988 (10)	397.8	0.90	358.0

FIGURE 11.32 U.S. Treasury Three-Year Bond Yields, 1975–1985

Year	Bond Rate (%)
1975	7.49
1976	7.27
1977	6.69
1978	8.32
1979	9.52
1980	11.48
1981	14.24
1982	13.01
1983	10.80
1984	12.24
1985	10.13

A CBS analysis of the data using this model is presented in Figure 11.33. A preliminary examination of these results shows that the two-lagged-period autoregressive model explains approximately 72% of the variance in bond rates. This model requires modification because of the presence of significant collinearity between the independent variables and the fact that one of the variables is statistically insignificant at the 0.05 level. A stepwise analysis of the data base yields the revised model presented in Figure 11.34. This model can now be used to forecast three-year treasury bond rates for 1986. The developed forecast is also shown in Figure 11.34. The estimated bond rate of 10.35 is considerably higher than the actual rate for that year of 7.07. Part of the observed difference can be attributed to the fact that the revised model explains only 50% of the variance in the dependent variable. Obviously, the incorporation of additional predictor variables is needed to improve the predictive power of this model.

11.9 CASE STUDY: THERMHOUSE INSULATION CORPORATION

The Thermhouse Insulation Corporation manufactures and markets a complete line of insulation products for residential and commercial buildings. Although Thermhouse's sales and profits remained fairly stable during the 1960s and early 1970s, the company's financial situation changed dramatically as a result of the 1972–1973 worldwide energy crisis. The demand for insulation products increased due to new government regulations and industry conservation. More specifically, Thermhouse's sales increased from $11.8 million in 1970 to $44.9 million by 1980, while profits rose from $1.2 million to $5.4 million.

The company's rate of growth declined somewhat during the early 1980s, although sales for 1986 are estimated at a respectable $65 million. Several of

FIGURE 11.33 CBS Autoregressive Analysis of Treasury Bond Rates

Multiple Regression Results

Variable	B-*Coeff*	Std Err	Beta	t-value	p-*value*
tb-2	−0.571	0.330	−0.661	−1.730	.100
tb-1	1.147	0.348	1.258	3.293	.000

B_0 intercept: 4.393
Critical *t*-value: 2.571

Sum squares regression:	31.0167
Sum squares error:	12.2053
Sum squares total:	43.2220
Mean square regression:	15.5083
Mean square residual:	2.4411
C.O.D. (R^2):	.7176
Adjusted C.O.D. (R^2):	.6047
Multiple correlation coefficient (R):	.8471
df regression:	2
df error:	5
Standard error estimate:	1.5624
Computed F:	6.3531
Critical F:	5.79
Durbin-Watson statistic:	1.5

Correlation Matrix

	tb-2	tb-1	tb
tb-2	1	.7830	.3242
tb-1		1	.7407
tb			1

FIGURE 11.34 CBS Forecasting of Treasury Bond Rates

Forecasting

Model: tb = 3.631 + 0.664 × tb-1

Variable	b-*coefficient*	Significant	Value
tb-1	1.1468	Yes	10.13

T-bond est. (1986) = 10.357

the company's senior managers believe that this decline can be attributed almost entirely to the general drop in energy prices that occurred during this period. This growth trend coupled with declining profits has not escaped the attention of Thermhouse's CEO. At a recent management workshop, the CEO requested information on the current direction and characteristics of corporate sales. Mr. James Brook, marketing vice president, was assigned the task of formulating a response to the CEO's inquiry.

Mr. Brook decided to first analyze the historical sales data using a simple four-quarter moving average model. This would provide him with a baseline trend forecast. Next, he computed seasonal indexes in order to adjust the trend forecast for any seasonal effects. He then planned to evaluate whether or not the changes in energy prices were impacting corporate sales. Finally, he wished to develop a quarterly sales forecast for 1986.

A review of Thermhouse's marketing data base yielded the quarterly sales information presented in Figure 11.35. To obtain information on energy prices, Mr. Brook conducted a computer base data search. After reviewing a number of energy time series, he decided to use the producers price index for home heating fuel as the most appropriate indicator of energy prices. Quarterly time series information on the producers price index is given in Figure 11.36.

Mr. Brook's plot of both time series is shown in Figure 11.37. This data base was evaluated using seasonal time series analysis, and the results are given in Figure 11.38. Mr. Brook decided that the irregular component is insignificant for this situation. This could be a somewhat shaky assumption in light of the ability of OPEC (the international oil cartel) to alter oil prices and production levels at any time.

The computed average seasonal factors and seasonal indexes for the Thermhouse case are shown in Figure 11.39. Notice that the index ranges from 85 for the second quarter to 115 for the fourth quarter. This variance of

FIGURE 11.35 Thermhouse Sales, by Quarter

Year	Quarterly Sales ($000)			
	1	2	3	4
1977	3,833	3,857	5,821	7,575
1978	7,506	3,408	3,903	7,868
1979	7,340	5,939	7,983	8,852
1980	8,958	9,840	9,971	16,262
1981	12,371	9,332	12,412	14,105
1982	12,478	12,278	12,092	13,856
1983	15,060	13,193	13,605	15,900
1984	17,499	15,632	15,664	17,195
1985	17,544	13,934	15,685	17,923

11.9 Case Study: Thermhouse Insulation Corporation

FIGURE 11.36 Producers Price Index for Home Heating Oil

Producers Price Index
(1967 = 100)

	Quarter			
	1	2	3	4
1979	256.20	258.23	268.57	279.23
1980	287.20	301.83	308.80	311.07
1981	313.67	320.07	325.37	330.83
1982	343.83	377.60	433.13	477.77
1983	531.40	571.73	589.87	602.93
1984	666.20	707.93	704.23	699.57
1985	697.53	670.03	702.37	702.77

approximately 30% indicates the presence of a rather significant seasonal effect (not surprising, given the nature of demand for insulation). The fact that the indexes for the second and third quarters are both below 1 indicates that demand for insulation during this period runs below average.

Because of the relatively large variance in the seasonal indexes, Mr. Brook has been advised to use these factors to deseasonalize the time series data prior to preparing a forecast. This process is illustrated for the last two years of sales data in Figure 11.40. Adjusted or deseasonalized sales values, as illustrated earlier, are obtained by simply dividing the unadjusted sales value by the corresponding seasonal index. For example, 1st quarter 1984 = $(17,499/107.46)100 = 16,278$.

FIGURE 11.37 Plot of Thermhouse Sales and Producers Price Index

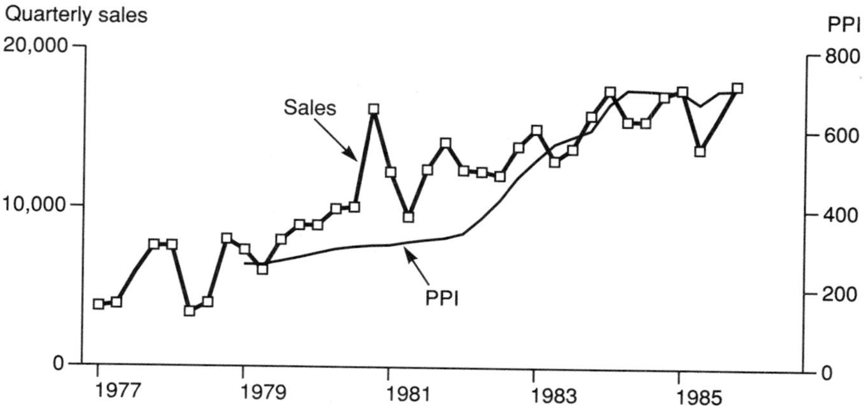

FIGURE 11.38 Seasonal Factors for Thermhouse Corporation

Time Series Results

Year	Quarter	Value	Centered Moving Average	Seasonal Component
1977	1	3,833	—	—
	2	3,857	—	—
	3	5,821	5,730.6250	101.58
	4	7,575	6,133.6250	123.50
1978	1	7,506	5,837.7500	128.58
	2	3,408	5,634.6250	60.48
	3	3,903	5,650.5000	69.07
	4	7,868	5,946.1250	132.32
1979	1	7,340	6,772.5000	8.38
	2	5,939	7,405.5000	80.20
	3	7,983	7,730.7500	103.26
	4	8,852	8,420.6250	105.12
1980	1	8,958	9,156.7500	97.83
	2	9,840	10,331.5000	95.24
	3	9,971	11,684.3750	85.34
	4	16,262	12,047.5000	134.98
1981	1	12,371	12,289.1250	100.67
	2	9,332	12,324.6250	75.72
	3	12,412	12,068.3750	102.85
	4	14,105	12,450	113.29
1982	1	12,478	12,778.2500	97.65
	2	12,278	12,707.1250	96.62
	3	12,092	12,998.7500	93.02
	4	13,856	13,435.8750	103.13
1983	1	15,060	13,739.3750	109.61
	2	13,193	14,184	93.01
	3	13,605	14,744.3750	92.27
	4	15,900	15,354.1250	103.56
1984	1	17,499	15,916.3750	109.94
	2	15,632	16,335.6250	95.69
	3	15,664	16,503.1250	94.92
	4	17,195	16,296.5000	105.51
1985	1	17,544	16,086.8750	109.06
	2	13,934	16,180.5000	86.12
	3	15,685	—	—
	4	17,923	—	—

11.9 Case Study: Thermhouse Insulation Corporation

FIGURE 11.39 CBS Generated Seasonal Indexes for Thermhouse Corporation

Seasonal Index, by Quarter

Quarter	Average Seasonal Component	Seasonal Index
1	107.71	107.46
2	85.39	85.18
3	92.79	92.57
4	115.18	114.89
	401.07	400.00

A graph of both the unadjusted and adjusted (i.e., deseasonalized) sales values for the total period is given in Figure 11.41. Mr. Brook plans to use the adjusted sales data to prepare a trend forecasting model.

After careful consideration, Mr. Brook has decided to use linear regression to formulate the trend model. The specifications for the developed regression model are presented in Figure 11.42. These results show that the time-based model explains approximately 89% of the variance in the adjusted sales values. The corresponding correlation coefficient is .94. These model results suggest that time is a very good predictor of adjusted company sales, although the existence of autocorrelation needs to be explored. A comparison of the adjusted data base and trend line is shown in Figure 11.43.

Mr. Brook also wishes to formulate a forecasting model that includes the energy price index as a second independent variable. As indicated earlier, he believes that incorporating energy prices will improve the sales forecast

FIGURE 11.40 Deseasonalized Quarterly Sales for Thermhouse Corporation (1984–1985)

	Quarter	Unadjusted Sales ($000)	Seasonal Index	Adjusted Sales ($000)
1984	1	17,499	107.46	16,278
	2	15,632	85.18	18,347
	3	15,664	92.57	16,916
	4	17,195	114.89	14,965
1985	1	17,544	107.46	16,320
	2	13,934	85.18	16,354
	3	15,685	92.57	16,938
	4	17,923	114.89	15,599

FIGURE 11.41 Unadjusted and Adjusted Quarterly Sales for Thermhouse Corporation

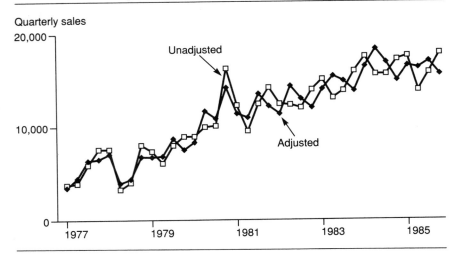

accuracy. The multiple regression model incorporating both time and energy prices is given in Figure 11.44. The data relating to energy prices were lagged by one period (i.e., one quarter) to account for the normal delay between changes in energy prices and changes in demand for insulation.

An inspection of the modeling results shown in Figure 11.44 reveals that the energy price variable is not statistically significant (i.e., the computed t is less than the critical t). Therefore, incorporating this variable into the model will not enhance forecasting accuracy. One might challenge this conclusion, since it seems somewhat inconsistent with the notion that rising energy prices should increase the demand for insulation. An examination of the

FIGURE 11.42 CBS Regression Forecasting Model for Thermhouse Corporation

Time Series Results

Number of periods:	36
B_0 coefficient:	4,142.1953
B_1 coefficient:	387.0615
Correlation coefficient:	.9441
Coefficient of determination:	.8914
Standard error estimate:	1,443.8200
Standard error B_1:	23.1642

Model: sale = 4,142.195 + 387.062 * (time)

11.9 Case Study: Thermhouse Insulation Corporation

FIGURE 11.43 Adjusted Sales and Trend Line for Thermhouse Corporation

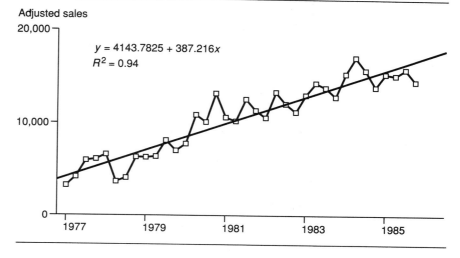

$y = 4143.7825 + 387.216x$
$R^2 = 0.94$

FIGURE 11.44 CBS Multiple Regression Analysis of Thermhouse Sales Data

Multiple Regression Results

Variable	B-Coeff	Std Err	Beta	t-value	p-value
time	441.775	74.024	1.078	5.968	.000
price	−3.271	4.198	−0.141	−0.779	.450

B_0 intercept: 4,524.448
Critical t-value: 1.96

Sum squares regression:	583,316,350
Sum squares error:	69,598,136
Sum squares total:	652,914,500
Mean square regression:	291,658,180
Mean square residual:	2,109,034
C.O.D. (R^2):	.8934
Adjusted C.O.D. (R^2):	.8869
Multiple correlation coefficient (R):	.9452
df regression:	2
df error:	33
Standard error estimate:	1,452
Computed F:	138
Critical F:	3.38
Durbin-Watson statistic:	2.4

FIGURE 11.44 (concluded)

Correlation Matrix

	Time	Price	Sales
time	1	.9492	.9442
price		1	.8822
sales			1

FIGURE 11.45 1986 Quarterly Sales Forecast for Thermhouse Corporation

Year	Quarter	Trend Estimate (000)	Seasonal Index	Forecast (000)
1986	37	$18,461	107.46	$19,838
	38	18,848	85.18	16,055
	39	19,235	92.57	17,806
	40	19,622	114.89	22,543

correlation matrix, however, reveals that a high correlation exists between time and energy prices, that is, .9492, which suggests the presence of collinearity. Therefore, incorporating both variables will only confound the forecasting process. In fact, if this problem had been analyzed using the stepwise regression model, the energy variable would not have appeared in the solution.

Based on the results from the multiple regression analysis, Mr. Brook has decided to use the bivariate model to prepare a quarterly sales forecast for 1986. A company sales forecast using the bivariate model is given in Figure 11.45. The trend estimate is based on the regression model presented in Figure 11.42. The quarterly forecasts were derived by multiplying the trend estimates by the seasonal indexes. Mr. Brook plans to present these results at the next management meeting.

11.10 SUMMARY

Time series modeling has emerged as one of the premier analytical tools for business forecasting. Developing accurate forecasts of future conditions is becoming increasingly important in all phases of business operations. For example, maintaining inventory levels based on the currently fashionable "just-in-time" strategy requires very precise forecasts of component and end-product demand.

This chapter introduced two basic forecasting techniques: time series models and multiple linear regression. Time series analysis utilizes past information as a basis for predicting future events. There are two main advantages in using time series models: First, these models do not require external variables to describe the behavior of a given dependent variable. Second, these models are easy to use. The primary disadvantages are that time series models do not identify which factors may be influencing the dependent variable, and they are usually limited to forecasting only one time period in the future. The normal approach in analyzing a time series is to decompose the data into four basic components: trend, cyclic, seasonal, and irregular. Specific time series trend models include moving averages, exponential smoothing, and bivariate regression.

Multiple regression, the second basic forecasting approach, attempts to explain the behavior of the dependent variable through the use of multiple "explanatory" or independent variables. The regression model assumes that a linear relationship exists between the dependent and independent variables. Multiple regression models may also contain dummy variables and lagged variables, including previous values of the dependent variable. The use of lagged values of the time series in the model is known as autoregression. This approach takes advantage of the existence of a trend in the time series in which past values serve as predictor variables for future values.

This chapter has also presented several measures for evaluating the accuracy of a particular forecast. Two of the more popular are mean square error (MSE) and mean absolute deviation (MAD). Both of these measures are designed to provide the user with an overall indication of model accuracy. Changes in model accuracy over time can be monitored using a tracking signal system based on acceptable tolerance limits.

11.11 GLOSSARY

autocorrelation The degree of time-lagged correlation in the time series.

autoregressive A method that uses regression analysis based on past time values to forecast future variable values.

cyclic component The long-term variations around the trend of a time series generally attributed to changing business and economic conditions.

exponential smoothing A forecasting method based on weighting the previous time series point using a smoothing coefficient.

forecasting A process for predicting the future.

irregular component The impact of random events on the time series.

mean absolute deviation (MAD) A method for measuring the accuracy of a forecast by summing the absolute differences between the actual and forecast values.

mean percentage error (MPE) A method for measuring the accuracy of a forecast by summing the percentage error.

mean square error (MSE) A method for measuring the accuracy of a forecast by summing the squared differences between the actual and forecast values.

nonstationary time series A time series containing a trend.

seasonal component Represents a pattern of change that is completed within one year and repeats itself regularly over the time series.

simple moving averages A forecasting method based on averaging two or more consecutive time series data points.

stationary time series A time series where the process average remains unchanged over time.

time series A set of data points recorded over successive time periods.

trend component The general long-term direction of the time series.

weighted moving averages A forecasting method based on weighting two or more consecutive time series data points.

11.12 BIBLIOGRAPHY

Abraham, B., and J. Ledolter. *Statistical Methods for Forecasting.* New York: John Wiley & Sons, 1983.

Gardner, E. S. "Exponential Smoothing: The State of the Art," *Journal of Forecasting* 4, no. 1 (March 1985).

Knowles, T. W. *Management Science: Building and Using Models.* Irwin, 1989.

Levenback, H., and J. P. Cleary. *The Modern Forecaster: The Forecasting Process Through Data Analysis.* Toronto: Wadsworth (LLP), 1984.

Thomopoulis, N. T. *Applied Forecasting Methods.* Englewood Cliffs, N.J.: Prentice Hall, 1980.

Whellwright, S. C., and S. Makridakis. *Forecasting Methods for Management,* 3rd ed. New York: John Wiley & Sons, 1984.

11.13 PROBLEMS

1. Describe the following forecasting techniques using specific business examples:
   a. Simple moving averages
   b. Weighted moving averages
   c. Exponential smoothing
2. Discuss the role of the smoothing coefficient in exponential smoothing.
3. Describe several methods for evaluating the effectiveness of a given forecast.
4. Develop time series graphics for the following data:

Year	U.S. Personal Income ($ billions)	U.S. Savings as % of Disposable Income
1978	1,812	7.1
1979	2,034	6.8
1980	2,258	7.1
1981	2,521	7.5
1982	2,671	6.8
1983	2,839	5.4
1984	3,110	6.3
1985	3,315	5.1
1986	3,487	3.9

5. Given a moving average model:
   a. Which is the first forecasted month based on a three-month average starting in January?
   b. Which is the first forecasted month based on a six-month average starting in May?
6. Describe how the value of the smoothing coefficient impacts the forecast.
7. Compare the strengths and weaknesses of moving averages and exponential smoothing.
8. The production manager at Baker Tool and Die is concerned about the apparent rise in manufacturing costs over the last year. The data processing department has assembled the following monthly unit cost data for review by the manager:

Month	Cost ($/unit)	Month	Cost ($/unit)
Jan	245.25	Jul	278.20
Feb	241.00	Aug	275.50
Mar	254.75	Sep	282.00
Apr	249.50	Oct	285.25
May	262.75	Nov	277.00
Jun	267.00	Dec	280.75

   a. Develop a scatter diagram for these data.
   b. Describe the general trend.
   c. Develop a three-period moving average analysis.
   d. Develop a simple exponential smoothing analysis with $\alpha = .2$.
   e. Determine the α value that minimizes the MAD error.
9. The vice president at Amar International wishes to analyze the general direction of the firm's sales over the past nine years. The company's controller has provided the vice president with the following annual sales data:

Year	Revenues (000)
1979	$490
1980	550
1981	575
1982	635
1983	595
1984	580
1985	600
1986	640
1987	655

    a. Develop a trend analysis using the following methods: regression, simple moving averages ($n = 3$), and simple exponential smoothing ($\alpha = .2$).
    b. Which method produces the smallest MAD error?
    c. Determine the α value that minimizes the MAD error.
    d. Develop a forecast for 1987 using each of the three methods.

10. The following data summarize the monthly cash flow requirements at Johnson Bakery for the past year:

Month	Cash ($)	Month	Cash ($)
Jan	2,800	Jul	1,600
Feb	2,500	Aug	1,500
Mar	2,200	Sep	1,800
Apr	2,000	Oct	2,000
May	1,500	Nov	2,300
Jun	1,300	Dec	3,100

    a. Develop a trend analysis using the following methods: regression, simple moving averages, and simple exponential smoothing ($\alpha = .2$, $\alpha = .4$).
    b. Which method produces the smallest MAD error?
    c. Develop a cash forecast for the following January.
    d. Why is the regression model particularly unsuited for this problem?

11. Given the following time series data on revenues for the Drexton Corporation:

Period	Revenues (000)
1	$254
2	237
3	264
4	275
5	284
6	305
7	293
8	307
9	315
10	323

    a. Develop an exponential smoothing analysis based on the following coefficient values: $\alpha = .2$, $\alpha = .4$, $\alpha = .6$.

    b. Which coefficient yields the minimum MAD error?

12. Given the following prime interest rates for the years 1975 to 1984:

Period	Rate
1975	7.9
1976	6.8
1977	6.8
1978	9.1
1979	12.7
1980	15.3
1981	18.9
1982	14.9
1983	10.8
1984	12.0

    a. Develop a linear regression model for characterizing the trend line.

    b. Estimate the cyclic component.

    c. Develop a forecast of interest rates for 1985 and 1986 using the trend model, adjusting for the cyclic variations.

13. Following are annual U.S. steel production figures (in thousands of short tons) for the period 1965 to 1984.

Year	Total	Year	Total
1965	92,666	1975	79,957
1966	89,995	1976	89,447
1967	83,897	1977	91,147
1968	91,856	1978	97,935
1969	93,877	1979	100,262
1970	90,798	1980	83,853
1971	87,038	1981	88,450
1972	91,805	1982	61,567
1973	111,430	1983	73,739
1974	109,472	1984	73,043

a. Develop a plot of these time series data.
b. Estimate the cyclic effect.
c. Formulate a regression model that incorporates the cyclic component.

14. Following are data on annual U.S. lumber production for the period 1965 to 1984. (Totals are in millions of board feet.)

Year	Total	Year	Total
1965	35,697	1975	31,583
1966	35,710	1976	36,120
1967	34,499	1977	38,438
1968	36,124	1978	39,088
1969	35,382	1979	38,746
1970	34,245	1980	33,361
1971	36,381	1981	30,928
1972	37,643	1982	28,849
1973	38,297	1983	35,370
1974	34,097	1984	37,436

a. Develop a plot of these time series data.
b. Estimate the cyclic effect.
c. Discuss the nature of the cyclic effect vis-à-vis the standard business cycle.

15. Meeker Electronics, a major supplier of automotive electronic devices, has experienced rapid growth over the past 10 years. The company's vice president for marketing is interested in analyzing the current direction of industrywide and company sales. A summary of industry and company sales for the reporting period 1975 to 1983 is given below:

Year	Industry Sales ($ millions)	Company Sales ($ millions)	Market Share (%)
1975	$27,076	$327.6	.0121
1976	30,637	1,072.3	.035
1977	35,476	1,383.5	.039
1978	41,719	1,793.9	.043
1979	45,963	2,160.3	.047
1980	46,083	2,534.5	.055
1981	48,001	3,072.1	.064
1982	44,005	3,608.4	.082
1983	49,058	3,581.2	.0730

a. Develop a trend forecast of industry sales for 1984 using several different forecasting methods.
b. Compare each forecast with the actual industry sales of $54,993 for 1984.
c. Develop a forecast of market share for 1984 using several different forecasting methods.
d. Estimate Meeker sales for 1984 using the market share estimate from part c.
e. Develop a regression model where industry sales is the independent variable and Meeker sales is the dependent variable.
f. Estimate Meeker sales for 1984 using the regression model developed in part c.
g. Develop a trend forecast of Meeker sales for 1984 using several different forecasting methods.
h. Compare the forecasts from parts d, f, and g.

16. Following are quarterly revenue data (in $000) for the Builtright Bicycle Company for the past five years.

	Quarter			
	1	2	3	4
1980	1,200	950	800	2,700
1981	1,400	800	775	2,500
1982	1,375	560	725	2,300
1983	1,250	700	650	2,250
1984	1,100	650	745	2,000
1985	1,200	825	915	2,250

a. Develop a trend line using both regression and moving averages.
b. Isolate the seasonal component and calculate the seasonal index.
c. Forecast sales for the first quarter of 1986.

17. The following table presents sales data ($000) for the Johnson Solar Corporation during the period 1977 to 1988.

Year	Sales	Year	Sales
1977	2,245	1983	2,970
1978	2,450	1984	2,560
1979	3,245	1985	2,225
1980	3,750	1986	2,100
1981	3,450	1987	2,350
1982	3,175	1988	2,275

Analyze these data using the two-parameter exponential smoothing model for the following cases:

a. $\alpha = .1, \beta = .4$
b. $\alpha = .4, \beta = .1$
c. The values of α and β that minimize MSE

18. The following table presents sales data ($000) for the Health Spa Corporation during the period 1977 to 1988.

Year	Sales	Year	Sales
1977	150	1983	345
1978	220	1984	650
1979	250	1985	725
1980	175	1986	535
1981	165	1987	475
1982	225	1988	425

Analyze these data using the two-parameter exponential smoothing model for the following cases:

a. $\alpha = .1, \beta = .2$
b. $\alpha = .4, \beta = .3$
c. The values of α and β that minimize MSE

19. Analyze the sales data given in Figure 11.35 using the three-parameter exponential smoothing model with optimized values for the coefficients.

20. Analyze the producers price index data given in Figure 11.36 using the three-parameter exponential smoothing model with optimized values for the coefficients.

21. Quarterly sales data (in units) for the best-selling undergraduate statistics book for the past four years are as follows:

Quarter	1983	1984	1985	1986
1	18,500	19,500	21,000	20,000
2	12,000	12,500	12,000	9,000
3	4,000	4,500	5,500	4,500
4	25,000	28,000	32,000	30,000

a. Develop a four-quarter moving average.
b. Compute seasonal factors and analyze their effects.

22. Quarterly manganese ore imports (in thousands of long tons) into the United States for the period 1977 to 1985 are as follows:

Year	Quarter			
	1	2	3	4
1977	153	288	259	134
1978	113	176	221	189
1979	172	264	85	234
1980	219	231	181	164
1981	153	259	195	167
1982	179	135	82	154
1983	127	103	124	130
1984	135	165	129	105
1985	190	228	174	208

a. Develop a four-quarter moving average.
b. Compute seasonal factors and analyze their effects.
c. Analyze these data using the three-parameter exponential smoothing model.

23. The Tunnel Diode Co. sells personal computers in 10 major Western cities. Following are monthly sales data (in thousands) for the past two years, along with the number of operating days:

	1986			1987	
Month	Sales	Days	Month	Sales	Days
Jan	1,201	25	Jan	1,679	25
Feb	1,040	26	Feb	1.442	25
Mar	1,323	25	Mar	1,622	24
Apr	1,295	25	Apr	1,709	25
May	1,507	26	May	1,828	24
Jun	3,493	28	Jun	1,614	26
Jul	1,338	25	Jul	1,910	23
Aug	1,832	27	Aug	1,688	22
Sep	1,758	26	Sep	2,214	29
Oct	1,449	25	Oct	1,544	24
Nov	2,073	27	Nov	1,365	25
Dec	1,911	23	Dec	2,409	30

a. Determine the extent of any seasonal effects.
b. Develop a first-quarter forecast for 1988 using several different forecasting methods.
c. Analyze the data using the two-parameter exponential smoothing model.

24. The following data base summarizes the average hours worked, weekly compensation, and the general business index (GBI) in the United States for the period 1970 to 1984.

Year	Weekly Hours	Weekly Compensation (1977 dollars)	GBI (1977 = 100)
1970	37.1	186.94	78.5
1971	36.9	190.58	79.6
1972	37.0	198.41	87.3
1973	36.9	198.35	94.4
1974	36.5	190.12	93.0
1975	36.1	184.16	84.8
1976	36.1	186.85	92.6
1977	36.0	189.00	100.0
1978	35.8	189.31	106.5
1979	35.7	183.41	110.7
1980	35.3	172.74	108.6
1981	35.2	170.13	111.0
1982	34.8	168.09	103.1
1983	35.0	171.37	109.2

a. Why is compensation reported in 1977 dollars?
b. Define GBI.
c. Is there a relationship between hours worked and compensation?
d. Is there a relationship between GBI and compensation?
e. Is there a relationship between GBI, hours worked, and compensation?
f. What other factors might help explain the changes in compensation?
g. Develop a compensation forecast for 1984.

25. Predicting change in the population (increase or decrease) represents a significant factor in the design of public policies. Birth rates still represent a major factor determining population change in most regions of the United States. The following table highlights birth rate data for the United States from the period 1978 to 1984.

Year	Live Births (millions)	Birth Rate (per 1,000)	Infant Mortality Rate (per 1,000)
1978	3,333	15.0	13.8
1979	3,494	15.6	13.1
1980	3,612	15.9	12.6
1981	3,629	15.8	11.9
1982	3,681	15.9	11.5
1983	3,639	15.5	11.2
1984	3,669	15.5	10.8

a. Has the birth rate changed significantly over the reporting period?
b. Have infant mortality rates changed significantly over this period?
c. Forecast the infant mortality rate for 1985 using several different methods (actual = 10.6).
d. What factors might be contributing to the steady decline in infant mortality?

26. Revise the analysis for the Thermhouse case (Section 11.9) by incorporating a cyclic component.
27. Determine the extent of the relationship between orange prices and U.S. production using the data given in Figure 11.28.
28. Determine average cyclic indexes for the following time series on sack paper index prices (1973 = 100). Develop a forecast for 1987 that includes an adjustment for cyclic variations.

Year	Index Price
1979	178.3
1980	191.3
1981	220.4
1982	214.2
1983	226.8
1984	240.3
1985	219.3
1986	216.8

29. Discuss the role dummy variables play in characterizing seasonal effects. How many dummy variables are required for quarterly analysis?
30. The following data base summarizes monthly absentee rates (excluding vacations and holidays) for a two-year period. Perform a time series analysis to determine the extent of seasonal effects.

	Jan.	Feb.	Mar.	Apr.	May	June	July	Aug.	Sept.	Oct.	Nov.	Dec.
1988	4.8	4.1	3.8	3.6	3.4	3.7	4.2	4.5	4.4	4.9	5.2	6.1
1989	4.7	3.9	3.3	3.5	3.7	4.1	3.9	4.3	4.6	5.1	5.7	6.0

31. The earnings per share for the Allied Signal Corporation between 1980 and 1989 are reported below:

1980	1981	1982	1983	1984	1985	1986	1987	1988	1989
5.43	6.11	3.68	4.17	5.02	3.40	3.00	2.25	3.10	3.40

a. Comment on the general trend of these earnings per share.
b. Perform a time series analysis on these data.
32. Explain the effect of seasonality on the trend for both the multiplicative and additive models. Which model would experience the larger effect due to seasonality?
33. Develop a regression model for the data given in problem 23. Utilize dummy variables to present the seasonal effects. Discuss the effect of any seasonality.
34. Describe the basic difference between a stationary and nonstationary time series. Identify one or more methods for transforming a nonstationary process into a stationary one.
35. The following table presents a regression analysis of the relation of sales to marketing expenses over a three-year period:

Parameter	1987	1988	1989
r^2	.85	.90	.95
r	.92	.81	.97
p-value	.003	.001	.000

a. Discuss the significance of these results.
b. Does it appear that an increase in marketing expense is warranted?
36. A CBS simple moving averages analysis of operating profits for California's telecommunication (tcap) industry for the period 1975 to 1986 is presented below:

CBS-Time Series and Forecasting
Information Entered

Model: Moving Averages (simple)
Number of Periods in Average: 3
Number of Periods: 12

	tcap
1 =	4.2
2 =	3.2
3 =	3.8
4 =	4.2
5 =	4.2
6 =	5.4
7 =	5.3
8 =	5.9
9 =	7.3
10 =	15.2
11 =	10.1
12 =	8.4

Results

Model:	Moving Averages (simple)
Number of Periods in Average:	3
Number of Periods:	12
Average tcap:	6.4333
Mean Square Error:	10.5314
Mean Average Deviation:	1.9778
Forecast for Period 13:	

    a. Develop a time series plot of the raw data.
    b. Describe the general trend.
    c. Develop a forecast for period 13.
    d. Would a four-period average reduce the error terms?

37. A CBS simple exponential smoothing analysis of average automobile maintenance cost as a percentage of estimated replacement cost(rela) for the period 1981 to 1988 is presented below:

Information Entered

Model:	Simple Exponential Smoothing
Data Smoothing Coefficient:	.6
Initial Data Value:	7
Number of Periods:	8

	rela
1 =	6.4
2 =	6.8
3 =	7.3
4 =	7.5
5 =	6.8
6 =	8.1
7 =	8.2
8 =	8.8

Results

Model:	Simple Exponential Smoothing
Data Smoothing Coefficient:	.6
Initial Data Value:	7
Number of Periods:	8
Average rela:	7.4333
Mean Square Error:	0.4123
Mean Average Deviation:	0.5896
Forecast for Period 9:	

a. Develop a time series plot of the raw data.
b. Describe the general trend.
c. Develop a forecast for period 9.
d. Would a larger smoothing coefficient reduce the error terms?

38. A CBS exponential smoothing with trend factoring analysis of the percentage of small cars in the U.S. vehicle fleet for the period 1978 to 1986 is presented below:

Information Entered

Model:	Smoothing and Trend Factoring
Data Smoothing Coefficient:	.7
Initial Data Value:	40
Trend Smoothing Coefficient:	.3
Estimate of Trend	5
Number of Periods:	9

```
  small
1 =  41
2 =  46
3 =  45
4 =  41
5 =  41
6 =  43
7 =  50
8 =  60
9 =  65
```

Results

Model:	Smoothing and Trend Factoring
Data Smoothing Coefficient:	.7
Initial Data Value:	40
Trend Smoothing Coefficient:	.3
Estimate of Trend:	5
Number of Periods:	9
Average small:	47.2000
Mean Square Error:	32.5046
Mean Average Deviation:	4.7852
Forecast for Period 10:	

a. Develop a time series plot of the raw data.
b. Describe the general trend.
c. Develop a forecast for period 10.
d. Would a larger smoothing coefficient reduce the error terms?

39. A CBS time series regression analysis of percentage of foreign sales for California's commercial aircraft(cair) industry for the period 1975 to 1986 is presented below:

Information Entered

Model:		Least Squares (linear)
Number of Periods:		12

	cair
1 =	49.9
2 =	59
3 =	55.8
4 =	38.1
5 =	45.3
6 =	52.8
7 =	55.3
8 =	39.1
9 =	39.7
10 =	30.9
11 =	37.3
12 =	38.1

Results

Model: cair = 56.642 − 1.744 * (time)

Model:	Least Squares (linear)
Number of Periods:	12
Average cair:	45.1083
Mean Square Error:	39.1230
Mean Average Deviation:	5.1998
Forecast for Period 13:	33.5742
B_0 Coefficient:	56.6424
B_1 Coefficient:	−1.7745
Coefficient of Determination:	0.4896
Standard Error Estimate:	6.8518
Standard Error B1	0.5730

a. Develop a time series plot of the raw data.
b. Describe the general trend.
c. Develop a forecast for periods 13 and 14.
d. How much of the explained variance in foreign sales is explained by the model?
e. Discuss the problem in using this model for predicting the percentage of foreign sales in 1991.
f. What other factors might contribute to explaining foreign sales?

40. Presented below are gross national product (GNP − constant 1973 $), gross domestic investment (GDI − constant 1973 $), and consumer price index (CPI − 1975 = 100) time series data for Peru from 1970 to 1981:

Period	GNP	GDI	CPI
1970	346,760	44,704	55
1971	365,623	51,593	59
1972	371,745	35,564	63
1973	386,292	65,595	69
1974	417,350	92,319	81
1975	436,372	87,903	100
1976	441,902	76,690	134
1977	440,614	57,205	184
1978	435,975	49,130	291
1979	448,373	50,027	485
1980	471,176	72,353	772
1981	480,215	85,231	1,354

   a. Develop a simple moving average analysis of GNP using three periods.
   b. Develop a simple exponential smoothing analysis of GNP using $\alpha = 0.2$.
   c. Determine the optimal smoothing coefficient for part b.
   d. Develop a simple regression model of GNP using time as the independent variable.
   e. Compare the results obtained in parts a through d. Which model minimizes the mean absolute error?
   f. Prepare a forecast for 1982 using each method. Compare the results.
41. Referring to problem 40, perform a similar analysis using GDI.
42. Develop a regression model for the data given in problem 16. Utilize dummy variables to represent the seasonal effects. Describe the extent of any seasonality. Develop a forecast for 1986.
43. Presented below are gross national product (GNP — Constant 1968 $), gross domestic investment (GDI — Constant 1968 $), and consumer price index (CPI — 1975 = 100) time series data for Turkey for 1970 to 1981:

Period	GNP	GDI	CPI
1970	125,400	25,500	43
1971	138,200	25,700	51
1972	148,500	27,800	58
1973	156,400	31,400	67
1974	168,000	39,200	83
1975	181,400	47,600	100
1976	195,700	49,700	118
1977	203,400	54,800	148
1978	209,200	53,800	240
1979	208,300	56,000	392
1980	206,100	49,000	761
1981	214,700	47,600	1047

a. Develop a simple moving average analysis of GNP using three periods.
b. Develop an simple exponential smoothing analysis of GNP using $\alpha = 0.2$.
c. Determine the optimal smoothing coefficient for part b.
d. Develop a simple regression model of GNP using time as the independent variable.
e. Compare the results obtained in parts a through d. Which model minimizes the mean absolute error?
f. Prepare a forecast for 1982 using each method. Compare the results.

44. Referring to problem 43, perform a similar analysis using GDP.
45. Referring to problem 43, perform a similar analysis using CPI. Provide a general description of the trend in CPI. What economic forces underlie this trend?
46. U.S. federal government receipts and expenditures data ($ billions) for the period 1976 through 1988 are presented below.

Period	Total Receipts	Total Expenditures
1976	298	372
1977	356	409
1978	400	459
1979	463	504
1980	517	591
1981	599	678
1982	618	746
1983	601	808
1984	667	852
1985	734	946
1986	769	990
1987	842	1,015
1988	917	1,024

a. Develop a simple moving average analysis of total receipts using the optimal number of periods.
b. Develop a simple exponential smoothing analysis of total receipts using $\alpha = 0.4$ and $\alpha = 0.6$.
c. Determine the optimal smoothing coefficient for part b.
d. Develop a simple regression model of total receipts using time as the independent variable.
e. Compare the results obtained in parts a through d. Which model minimizes the mean absolute error?
f. Prepare a forecast for 1982 using each method. Compare the results.

47. Referring to problem 46, perform a similar analysis using total expenditures.
48. Referring to problem 46, perform a similar analysis using the difference between receipts and expenditures.
49. The following data present the relative demand for home heating oil for a typical single family dwelling in Cleveland, Ohio, for the period 1985 to 1989.

Month	1985	1986	1987	1988	1989
Jan	650	750	650	775	675
Feb	700	800	575	825	725
Mar	450	675	450	650	600
Apr	250	550	300	375	450
May	50	100	50	75	100
Jun	0	10	0	0	10
Jul	0	0	0	0	0
Aug	0	0	0	0	0
Sep	200	150	100	50	125
Oct	375	275	250	225	375
Nov	550	475	450	375	500
Dec	600	600	650	550	650

    a. Determine the quarterly seasonal indexes.
    b. Determine the trend.
    c. Determine the cyclic indexes using a three-period moving average.

50. Analyze the GNP data given in problem 43 using the two-parameter exponential smoothing model with $\alpha = 0.1$, and $\beta = 0.3$. Determine the optimal values for *a* and *b* using this data.

51. Analyze the total receipts data given in problem 46 using the two-parameter exponential smoothing model with $\alpha = 0.4$ and $\beta = 0.2$. Determine the optimal values for *a* and *b* using this data.

52. Analyze the data given in problem 49 using the three-parameter exponential smoothing model with $\alpha = 0.3$, $\beta = 0.1$, and $\gamma = 0.2$. Determine the optimal values for *a*, *b*, and *g* using this data.

53. The following data present the percentage change in average weekly earnings between 1985 and 1986 for the United States.

Month	Current Dollars	1977 Dollars
Jan	3.1	−0.6
Feb	2.5	−0.5
Mar	2.4	0.4
Apr	2.1	0.9
May	1.9	0.7
Jun	1.1	−0.2
Jul	1.7	0.5
Aug	1.8	0.6
Sep	1.0	−0.3
Oct	1.4	0.2
Nov	2.0	1.1
Dec	0.7	0.1

a. Determine the extent of the relationship between the two variables.
b. Is the relationship significant at the 0.05 level?
c. What factors might contribute to the differences observed in the monthly data?

54. The production manager at Waco Electronics has obtained the following sample data on the percentage defectives found in incoming lots of capacitors over a 20-day period:

Day	Defects (%)	Day	Defects (%)
1	4	11	2
2	5	12	2
3	3	13	1
4	2	14	3
5	2	15	0
6	2	16	0
7	0	17	4
8	5	18	2
9	3	19	1
10	4	20	1

a. Develop a control chart showing the percent defectives as a function of time.
b. Calculate the average % defective level.
c. Develop a simple exponential smoothing analysis of the data using $\alpha = 0.3$.
d. Determine the optimal value of α that minimizes the mean absolute error.

55. The production manager at Johnson Cereals is concerned about the actual weight of the company's 10-oz cereal box. The manager has obtained the following data from a random sample of cereal boxes over a 20-day period. Each sample contained two observations.

Day	Weight (oz)	Day	Weight (oz)
1	10.2, 10.3	11	10.0, 9.9
2	10.1, 10.3	12	10.1, 10.1
3	9.9, 10.0	13	9.8, 9.9
4	10.3, 10.5	14	10.1, 10.2
5	9.8, 10.2	15	10.2, 10.2
6	10.1, 10.5	16	10.0, 10.3
7	10.0, 10.0	17	10.3, 10.4
8	10.1, 10.2	18	10.2, 10.3
9	10.2, 10.3	19	10.1, 10.1
10	10.3, 10.3	20	9.9, 10.1

a. Develop a control chart showing the measurement values as a function of time.

b. Calculate the average product weight.
c. Develop a simple exponential smoothing analysis of the data using $\alpha = 0.3$.
d. Develop a simple moving average analysis with four periods.

56. The quality control manager at Blade Tool and Die has been receiving a large number of customer complaints regarding the poor tolerance on its 3-inch diameter ball bearing line. The manager has obtained the following data from a random sample of ball bearings over a 20-day period. Each sample contained three observations.

Day	Diameter (inches)	Day	Diameter (inches)
1	3.03, 3.04, 3.07	11	3.07, 3.08, 3.08
2	2.98, 2.99, 3.01	12	3.01, 3.02, 3.02
3	3.03, 3.05, 3.08	13	2.98, 2.99, 3.00
4	3.00, 3.02, 3.03	14	2.97, 3.01, 3.04
5	2.98, 2.99, 3.01	15	3.03, 3.05, 3.07
6	3.01, 3.03, 3.05	16	2.96, 3.01, 3.08
7	3.03, 3.05, 3.09	17	3.02, 3.04, 3.06
8	2.94, 2.98, 3.02	18	2.96, 2.98, 3.05
9	2.98, 2.99, 3.02	19	3.04, 3.05, 3.07
10	3.01, 3.02, 3.02	20	3.05, 3.08, 3.08

a. Develop a control chart showing the average ball bearing diameter as a function of time.
b. Calculate the average diameter for the sample.
c. Develop a simple exponential smoothing analysis of the data using $\alpha = 0.1$.
d. Develop a simple moving average analysis with three periods.
e. What conclusions can be made regarding the current production practices?

57. Presented below is key financial data on California's semiconductor industry between 1975 and 1985.

Year	Capital Expended (% sales)	Operating Profit (% sales)	Foreign Sales (% sales)
1975	6.0	−3.6	27.9
1976	9.0	10.0	29.0
1977	11.0	10.6	30.4
1978	14.0	12.0	32.2
1979	13.0	13.2	32.7
1980	15.0	15.5	34.9
1981	18.0	7.0	30.4
1982	15.0	3.1	30.7
1983	14.0	10.1	31.7
1984	21.0	16.1	37.4
1985	17.0	−4.2	37.5

a. Develop a simple exponential smoothing analysis of capital expended using $\alpha = 0.3$.
b. Determine the optimal value of a that minimizes the mean absolute error.
c. Develop a regression model of capital expenditives versus operating profits.
d. Is the model in part c statistically significant at the 0.05 level?
58. Referring to problem 57, perform a similar analysis using operating profits.
59. A simple exponential smoothing model with $\alpha = 0.5$ is related to a moving average model with how many periods?
60. Presented below is key financial data on California'a machine tools industry from 1975 to 1986.

Year	Capital Expended (% sales)	Operating Profit (% sales)	Foreign Sales (% sales)
1975	2.7	7.5	35.1
1976	3.8	9.3	30.6
1977	2.9	10.1	28.4
1978	3.8	14.5	29.1
1979	4.5	16.7	26.6
1980	5.3	17.1	28.6
1981	5.3	16.7	24.4
1982	4.7	11.9	21.2
1983	4.1	−1.7	26.8
1984	7.3	0.6	16.4
1985	3.3	−0.3	17.0
1986	3.0	−0.1	23.3

a. Develop a simple exponential smoothing analysis of capital expended using $\alpha = 0.3$.
b. Determine the optimal value of α that minimizes the mean absolute error.
c. Develop a time series regression model.
d. Compare the results from parts b and c.
61. Referring to problem 57, perform a similar analysis using operating profits.
62. Referring to problem 57, perform a similar analysis using total expenditures.
63. Develop an autoregressive model for the data given in problem 57 where operating profits is the dependent variable and operating profits lagged one period, capital expenditures lagged one period and foreign sales are the candidate independent variables. Which variables are statistically significant at the 0.05 level?
64. Develop an autoregressive model for the data given in problem 60 where operating profits is the dependent variable and operating profits lagged one period, capital expenditures lagged one period and foreign sales are the candidate independent variables. Which variables are statistically significant at the 0.05 level?
65. Presented below are American and Japanese medical trade data for the period 1970 to 1986:

	U.S. Global	U.S. Trade with Japan	
Year	Trade Balance ($ billion)	Exports ($ billion)	Imports ($ billion)
1970	0.4	37	46
1971	0.4	41	43
1972	0.4	59	62
1973	0.5	92	97
1974	0.7	126	92
1975	0.7	128	95
1976	0.7	146	124
1977	0.8	193	110
1978	1.0	228	133
1979	1.2	314	130
1980	1.3	279	131
1981	1.3	258	110
1982	1.1	231	149
1983	0.7	243	217
1984	0.4	260	282
1985	0.2	279	396
1986	−0.01	355	526

a. Develop an autoregressive model for predicting U.S. global trade balance based on two previous periods.
b. How much of the variability in the dependent variable is explained by the model?
c. Characterize the extent of autocorrelation from an examination of the Durbin-Watson statistic.
d. What additional variables might be used to improve the predictive power of the model?
66. The following data present phone installations for the Dialnet exchange and the gross regional product (constant 1980 dollars) for the period 1976 to 1986.

Year	Installations (000)	Gross Regional Product ($ billion)
1976	45.5	446
1977	79.2	466
1978	77.8	488
1979	81.4	472
1980	102.2	457
1981	124.8	432
1982	132.6	441
1983	156.3	475
1984	161.2	502
1985	168.6	531
1986	174.2	562

a. Develop an autoregression model along the lines outlined in Section 11.4 (page 485).
b. Characterize the accuracy of the model.
c. Discuss the existence of autocorrelation vis-à-vis the Durbin-Watson statistic.
d. Evaluate several other lagged models. Which one appears to yield the best results?

67. Referring to the data given in problem 40, develop an autoregression model for GNP in which GNP and GDI are lagged one period. Describe the extent of autocorrelation for this model.

68. Referring to the data given in problem 60, develop an autoregression model for operating profits as a function of capital expenditures, foreign sales, and lagged profits. Describe the extent of autocorrelation for this model.

11.14 CASES

11.14.1 California, Here We Come

The governor of California wishes to conduct a major advertising campaign to attract new visitors to the state. This campaign is to be undertaken in connection with the state's 150th anniversary. One of the main selling points the governor wishes to emphasize is the low crime rate found throughout the state. The state's attorney general, however, has indicated to the governor that he is unsure as to the current status and direction of overall crime. Furthermore, the attorney general indicated that the term *crime* is too general, and that statements on the status of crime should be based on a more detailed definition, (e.g., assaults). Accordingly, the governor has asked the attorney general to conduct a statistical analysis of existing crime data from the past six years. Armed with these results, the governor should be in a better position to inform future visitors about the status of crime in California.

TABLE 1 California Crime Statistics, by Category

	Homicide	Robbery	Assault	Burglary	General Population (000)
1978					22,839
Men	748	1,959	720	1,926	
Women	62	84	62	69	
1979					23,255
Men	782	2,115	830	2,004	
Women	87	93	38	74	
1980					23,771
Men	917	2,305	1,003	2,300	
Women	62	117	50	78	
1981					24,216
Men	1,002	2,712	1,263	3,260	
Women	69	117	58	96	
1982					24,698
Men	1,315	2,994	1,262	3,707	
Women	99	118	70	116	
1983					25,186
Men	1,298	3,198	1,420	4,181	
Women	93	168	71	153	
1984					25,622
Men	1,028	2,458	1,476	3,750	
Women	80	119	77	194	

Table 1 provides criminal statistics, by year, offense, and sex, for the years 1979 to 1984. The data are based on the felons entered in California prisons. The attorney general believes that these data reflect the general status of crime in the state.

Suggested Questions

1. Design a crime index that might be used by the governor.
2. Is criminal activity increasing or decreasing?
3. Does the general growth in population have an impact on criminal activity?
4. Should there be a lagged relationship between population growth and incarceration?
5. What conclusions can be drawn about crimes committed by women?
6. Identify several other factors (e.g., unemployment) that might explain the current trends.
7. What types of statements regarding crime should the governor include in the advertising campaign?
8. Is the attorney general's assumption regarding the use of convicted felons as a gauge of general criminal activity valid?

11.14.2 U.S. and Japanese Exchange Rates and Trade

There is increasing concern throughout industry and government regarding the growing U.S. trade imbalance. This is a particularly serious problem for U.S. trade with Japan, where the trade gap (i.e., the difference between imports and exports) has risen from a few billion dollars in the early 1970s to nearly $50 billion in the mid-1980s. Presently, the trade deficit with Japan accounts for approximately 50% of the total U.S. trade imbalance. Traditionally, the U.S. dollar has served as the standard currency unit used to compare rates in the international market. The inability to correct these trends, especially in relation to Japan, has weakened the historically dominant role of the U.S. dollar in international trade and finance. Many economists believe that this erosion will continue unless strategies can be implemented to solve the trade problem.

Dr. Milton Friedman has written that the manipulation of the exchange rate is a method of adjusting foreign prices and related economic factors without disrupting domestic employment and prices. Many economists regard the devaluation of currency as particularly significant because a strong currency generally results in low prices for imports and high prices for exports. The overriding conclusion is that a fluctuation in the foreign exchange rate will either reduce or amplify the effect of prices and thereby influence trade balances.

Table 1 presents trade balance estimates, exchange rates, and GNP figures for the United States and Japan for the period 1975 to 1985. The economic data are in current U.S. dollars.

TABLE 1 United States and Japanese Trade Balance, Exchange Rates, and GNP Levels

	Trade Balance ($000)	Yen/Dollar	GNP ($ billions)	
			U.S.	Japan
1975	−2,773	297	1,598	466
1976	−6,778	297	1,783	532
1977	−9,666	269	1,991	593
1978	−13,577	210	2,250	654
1979	−10,584	219	2,508	744
1980	−11,901	227	2,732	849
1981	−18,081	221	3,053	964
1982	−18,965	249	3,166	1,060
1983	−21,665	238	3,406	1,138
1984	−36,796	238	3,765	1,198
1985	−46,152	250	3,988	1,301

Suggested Questions

1. Is there a relationship between the yen-to-dollar exchange rate and trade balance?
2. Should the analysis consider use of a lagged relationship between trade and exchange rates? If so, by how many periods?
3. Prepare a 1986 trade balance forecast using several different methods.
4. Which economy has grown faster during the reporting period?
5. Is there a relationship between the ratio of U.S. GNP to Japan's GNP and the trade balance?
6. How would converting the trade balance estimates and GNP estimates to constant 1985 dollars affect the analysis?
7. What other factors might contribute to the growing trade imbalance between the United States and Japan?
8. What strategies might the United States employ to redress the trade imbalance with Japan?

Chapter 12

Chi-Square Analysis

The truth does not become less true for being endlessly repeated.
Konstantin Fedin

CHAPTER OUTLINE

12.1 Introduction
12.2 Example Management Problem: Kwan Bottling Company
12.3 How to Recognize a Chi-Square Problem
12.4 Model Formulation
12.5 Computer Analysis
12.6 Practical Applications
12.7 Case Study: Leaky Pen Company
12.8 Summary
12.9 Glossary
12.10 Bibliography
12.11 Problems
12.12 Cases

CHAPTER OBJECTIVES

The primary objectives of this chapter are to develop an understanding of

1. How to recognize chi-square type problems.
2. The principles of goodness of fit.
3. How to perform tests of independence.
4. How to make inferences about a population variance.
5. How to solve chi-square problems using computer analysis.

There has been a growing interest in better understanding the relationship between the securities option market and other trading commodities. Of specific interest is whether or not the expiration of options has a significant impact on stock prices. An extensive analysis was undertaken to determine the extent of the relationship between stock prices and option selling.*

---

* R. A. Strong and W. P. Andrew, "Further Evidence of the Influence of Optional Expiration on the Underlying Common Stock," *Journal of Business Research* 15, no. 4 (August 1987).

The null hypothesis adopted in this analysis is that the movement of a stock's price with respect to the nearest option strike price is random. The alternative hypothesis is that stock prices tend to move in the direction of the nearest option strike price during the option expiration week. The data set consists of 584 securities with stock prices under $50. The results of the analysis show that common stocks that are also traded as options do show a tendency to converge toward the option strike price at the time of expiration.

12.1 INTRODUCTION

The previous material on statistical inference has focused on problems involving means and proportions. In many business situations, however, the study of variances can contribute significantly to the decision-making process. For example, revenue data recorded on a month-by-month basis over the last two years could result in the same annual average. However, monthly cash flows between years may vary considerably, which could have significant financial implications. The chi-square statistic is often used to measure the variance of such relationships. In addition to its use in making inferences about variances, chi-square can also be used to test hypotheses when the data are classified into categories. Thus, chi-square analysis can be used to test whether or not observable frequencies approximate a specific curve form (e.g., normal, binomial, or Poisson) or whether or not two variables are independent.

The following list presents some potential applications of chi-square analysis:

- Analyzing the effect of changes in the advertising budget on revenue
- Determining the impact of a job safety program on the number and severity of subsequent injuries
- Evaluating the effect of worker participation in decision making on productivity
- Investigating the relationship between customer characteristics and product use

This chapter discusses how to identify the data needed in a chi-square analysis, how to transform the data for computer modeling, and how to interpret the results. The previous chapters focused on testing various hypotheses when the data are interval or ratio and the samples can be assumed to be drawn from normal populations. In some business situations, neither of these conditions may hold. The application of the nonparametric statistic called chi-square is an appropriate method in these cases. Because of different ways of classifying data and different hypotheses, the chi-square methodology is divided into two major test formats: *goodness of fit* and *test for independence*. Each of these methods will be addressed separately.

> **HISTORICAL NOTE**
>
> In 1900 Karl Pearson (1857–1936) proposed the chi-square statistic, which dealt with the squared deviations from an expected value in reference to multinomial applications. He not only developed the statistic but also derived the basic distribution. He also contributed to the development of partial or multiple correlation and defined such basic terms as *variation* and *standard deviation*. Pearson is often viewed as the founder of modern statistics.

12.2 EXAMPLE MANAGEMENT PROBLEM: KWAN BOTTLING COMPANY

The Kwan Bottling Company of Hong Kong is planning a conversion from glass to plastic containers. Prior to initiating this change, Kwan's management decided to determine the public's reaction to the new bottles. They are most concerned about the possible effect an unfavorable response may have on sales. Accordingly, Kwan's marketing department developed a 10-week survey to measure the potential impact of the proposed change. Ten stores of comparable sales levels were selected. Five of the stores continued to sell the water in the glass bottles, while the other five sold water in the new plastic bottles. The following data summarize the results of the survey:

	Bottles Sold	
Week	Glass	Plastic
1	520	470
2	490	500
3	510	505
4	515	515
5	495	505
6	525	510
7	500	505
8	515	510
9	495	515
10	505	500

The company's management wishes to know what conclusions can be drawn from these survey results.

12.3 HOW TO RECOGNIZE A CHI-SQUARE PROBLEM

Key idea 1

Generally speaking, chi-square problems have the following characteristics:

1. The original measurement data are in categories.
2. The nature of the population distribution is unknown.
3. The objective is to determine:
   a. If a particular theoretical distribution "matches" a sample frequency distribution *(goodness of fit)*.
   b. If the hypothesis of independence between two variable sets is tenable *(test of independence)*.

12.4 MODEL FORMULATION

Chi-square is one of the most frequently used statistics in analyzing business problems when the data is organized into categories (e.g., income levels). This statistic is used to determine whether or not a relationship exists in frequency-based data. It should be remembered that frequency in mathematics represents the number of occurrences within a class of events. Most market studies, segmentation studies, and questionnaire analyses use the

> **INTERNATIONAL VIGNETTE**
>
> "Improving productivity is our number one priority," says Mr. Alfred Eugster, manager of purchasing at Robert Neher Ltd., a Swiss company. Currently, the department has no formal productivity program, nor has a method been defined for even measuring productivity. Mr. Eugster is interested in developing a measuring system that would indicate the contribution of his department to overall corporate performance. This represents a significant challenge, since the department is constantly being squeezed between market prices and the cost of resources, neither of which they can directly control.
>
> The Neher Company, founded by Robert Neher in 1910, specializes in producing foil for a wide range of packaging and design applications. Approximately 75% of the products are sold outside Switzerland. Neher is one of five aluminum operations owned by the Alusuisse group, the largest producer of aluminum in Switzerland. The basic role of Neher is to convert aluminum stock into foil products.
>
> Typically, purchasing requests are placed with the department on a day-to-day basis. The department processes approximately 6,000 orders per year. Mr. Eugster feels that a statistical analysis of the order type and cost distributions might provide some insight into how productivity might be measured in the department. With this in mind, Mr. Eugster has set about preparing a plan to implement his ideas.

chi-square statistic to compare various parts of the response sets. These comparisons permit inferences to be made regarding the population.

Chi-Square Statistic

Chi-square is considered a nonparametric statistic. That is, the population shape is not assumed to be normally distributed. Thus, the analyst need not worry about the effect of normality when applying this statistic. On the other hand, chi-square and other nonparametric statistics do not have the power of the more elegant parametric methods. A chi-square analysis requires the use of frequency data. The basic chi-square model is as follows:

$$\chi^2 = \Sigma \frac{(o_i - e_i)^2}{e_i}$$

where:

o = Observed frequency
e = Expected frequency, which is either provided or computed

Chi-Square Distribution

The **chi-square distribution** is a continuous probability distribution that, like the t distribution, is a function of the degrees of freedom. For applications with small numbers of degrees of freedom, the distribution is positively skewed, as illustrated in Figure 12.1. However, the chi-square distribution can be approximated by the normal distribution if the degrees of freedom exceed 10.

As with all statistical procedures, care should be exercised in applying the chi-square methodology. In particular, the chi-square statistic should not be used when more than 20% of the cells contain **expected frequencies,** or hypothetical values, less than 5. For example, consider the data shown in Figure 12.2 on management positions at the Kwan Bottling Company. These data show expected frequencies based on the current management growth plan.

The data set on the left can be analyzed using chi-square analysis since only 20% (i.e., one out of five) of the cells contain expected frequencies less than 5. However, the data set on the right should not be analyzed using chi-square since three categories out of seven have frequencies less than 5 (i.e., 43%). In such cases, the data for some categories can be combined or pooled to meet the minimum frequency requirements, as was done here for the three vice president categories.

FIGURE 12.1 Chi-Square Distributions for 1, 5, and 10 Degrees of Freedom

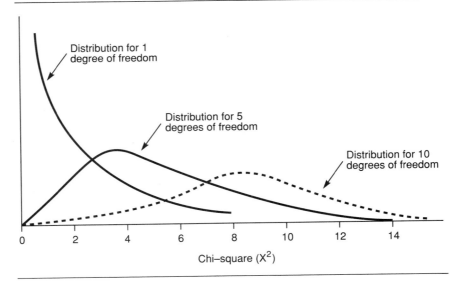

FIGURE 12.2 Distribution of Management Positions at Kwan Bottling Company

Position	Expected Frequency	Position	Expected Frequency
Foreman	35	Foreman	35
Supervisor	20	Supervisor	20
Director	15	Director	15
Scientist	2	Scientist	2
Vice president	10	Vice president	6
		Senior V.P.	3
		Executive V.P.	1

GOODNESS OF FIT

Key idea 2

The **goodness-of-fit test** is designed to determine whether or not sample data match a particular probability distribution. As can be seen from the preceding description, the computations required for the chi-square statistic are straightforward. The major difficulty in its use for goodness of fit lies in the development of an appropriate hypothesis. Some hypotheses are simple to generate; others are not. For example, suppose a coin is flipped 10 times, resulting in seven heads and three tails. The basic objective is to determine whether the coin is biased or whether the observed differences are merely a random variation (i.e., sampling error). The resultant chi-square analysis is shown in Figure 12.3.

For the goodness-of-fit test, the degrees of freedom equal the number of categories minus 1. Since this problem has two categories, the degrees of freedom are equal to 1 (i.e., df = 2 − 1 = 1). The chi-square table of critical values in Appendix A (Table A.3) shows that a value of 3.84 would be required for significance at the 5% level and df = 1. Since 1.6 is below the critical value, one can conclude that the observed departure from the expected value is a random variation. That is, there is insufficient evidence to reject the null hypothesis. It is important to note that the expected value of

FIGURE 12.3 Chi-Square Analysis for Coin Flip Experiment

	Heads	Tails	Total
Observed	7	3	10
Expected	5	5	10

$$\chi^2 = \frac{(7-5)^2}{5} + \frac{(3-5)^2}{5} = 1.6$$

FIGURE 12.4 Chi-Square Analysis for Extended Coin Flip Experiment

	Heads	Tails	Total
Observed	70	30	100
Expected	50	50	100

$$\chi^2 = \frac{(70-50)^2}{50} + \frac{(30-50)^2}{50} = 16$$

five heads and five tails represents the hypothesis to be tested. Note that the total expected frequencies must equal the total observed frequencies.

Consider the same coin flipped 100 times, resulting in the same ratio of heads and tails as for the previous experiment. The analysis is shown in Figure 12.4.

Comparing the computed chi-square of 16 with the critical value of 3.84 reveals that the departure from the expected frequency is significant. This suggests that the coin is biased, that this difference cannot be accounted for by sampling error. Note that even though the ratio of 7 to 3 and 70 to 30 are identical, one is significant and the other is not. The sensitivity of the chi-square analysis to frequency size in each cell is evident. This sensitivity is one of the main reasons that percentages or other data forms are not allowed for this type of analysis.

The goodness-of-fit test will now be used to analyze the Kwan Bottling data. Notice that in this problem the collected data from Section 12.2 are reported in raw form instead of in frequency form. A number of methods are available for making the transformation. Perhaps the simplest is to categorize the demand data for the plastic bottle with respect to the mean for the glass option (baseline case). The mean of the baseline is 510. Plastic demand values above this mean will be classified as "high," and data values below this mean will be classified as "low." Naturally the expected count for each of the two categories is 5. (If there were 11 observations, the expected count for each category would be 5.5). The resulting goodness-of-fit table and chi-square analysis are given in Figure 12.5.

FIGURE 12.5 Chi-Square Analysis of Kwan Bottling Problem

	High	Low	Total
Observed	4	6	10
Expected	5	5	10

$$\chi^2 = \frac{(4-5)^2}{5} + \frac{(6-5)^2}{5} = 0.4$$

Notice that four values were classified as "high" (above the mean of 510) and six values were classified as "low" (below the mean of 510). The null hypothesis established by the manager is that the demand for bottled water will not be influenced by the type of container. The resultant chi-square of 0.4 is well below the critical chi-square value (3.841) at the .05 level. Therefore, the manager can conclude that switching to plastic containers should not affect the demand for bottled water.

TEST OF INDEPENDENCE

The use of chi-square analysis for testing variable independence is somewhat different in format from that used for the goodness-of-fit test. The goodness-of-fit test required a hypothesis regarding the nature of an observed distribution. The **test of independence,** however, always uses the same null hypothesis, which is that there is no relationship between the two variables of interest (i.e., the variables are *independent*). It should be remembered that, for statistical applications, **dependence** means the same thing as *relationship* and *statistical significance.* Similarly, the term **independence** is synonymous with *no significant difference* and *no relationship.*

Key idea 3

The accounts receivable manager at Kwan Bottling is considering whether or not to initiate a discount plan for improving the collection of accounts on a companywide basis. The manager wishes to determine if a relationship exists between the age of an account (variable Y) and the level of discount the company offers its retail customers (variable X). The manager selected 200 accounts at random for evaluating the merits of a discount plan. Ninety of these accounts received no discount, 45 accounts received a 2% discount if paid within 30 days, and 65 accounts received a 5% discount if paid within 30 days. The test was run for 100 days on the 200 accounts. Figure 12.6, called a **contingency table,** shows the status of each account as a function of age (i.e., when paid) and discount level. For example, 40 accounts receiving no discount were paid within 30 days.

FIGURE 12.6 Contingency Table of Observed Frequencies for Kwan Accounts Receivable

Discount	Age of Account (days)				Total
	<30	30-60	60-90	>90	
0%	40	25	15	10	90
2%	20	10	10	5	45
5%	40	15	5	5	65
Total	100	50	30	20	200

Notice that the sum of the rows and columns is 200. The operational hypothesis, which, in general, is the same for a contingency table of any size, is that the Y variable (age of the account) is independent of the X variable (discount level). With this in mind, the following logic can be used to test the independence of the two variables.

- If Y and X are independent, then chi-square should be 0.
- For chi-square to be 0, the deviations of observed from expected frequencies should also be 0.
- The expected frequency for each cell should be in proportion to the totals.

To apply the chi-square statistic requires a set of expected frequencies for each cell. The following general model can be used for this purpose:

$$\text{Expected frequency (}i\text{th row, }j\text{th column)} = \frac{R_i \times C_j}{N}$$

where:

R_i = Total for row i
C_j = Total for column j
N = Total number of observations

Figure 12.7 presents a contingency table for the accounts receivable at Kwan Corporation.

For any contingency table, the degrees of freedom equal the number of rows minus 1 $(R - 1)$ times the number of columns minus 1 $(C - 1)$, or $(R - 1)(C - 1)$. Since this problem involves three rows and four columns, the number of degrees of freedom is 6. The critical chi-square statistic (from Table A.3) for df = 6 at the 5% level is 12.592. The computed chi-square of 7.64 is smaller than the critical chi-square, which supports the null hypothesis of independence between discount level and age of accounts. It would seem that offering a discount does not influence the time in which an account is paid. A more mathematical statement is that a departure from the expected value as large as the one obtained would occur by chance alone more than 5% of the time.

FIGURE 12.7 Contingency Table of Expected Frequencies for Accounts Receivable at Kwan Corporation

Discount	Age of Account (days)				Total
	<30	30–60	60–90	>90	
0%	45	22.5	13.5	9	90
2%	22.5	11.25	6.75	4.5	45
5%	32.5	16.25	9.75	6.5	65
Total	100	50	30	20	200

In conclusion, the following list summarizes the basic procedure for developing a test of independence:

1. The data for computation can be justified as frequency data.
2. Cells are mutually exclusive and independent. That is, no particular object of measurement can appear in more than one cell.
3. No cells should contain expected frequencies below 5. If the cell counts are too small, the table should be compressed (i.e., the number of rows and/or columns should be reduced).
4. The categories chosen for analysis should be logically related to the decision in question.

An important variation to the basic test of independence occurs when the number of total observations for the columns or rows is known in advance. In these cases, each column or row is treated as a separate population. The null hypothesis is that these populations are the same, i.e., homogenous. Consequently, the test procedure can be viewed as a test of independence as well as a **test of homogeneity.** The computation process for analyzing the contingency table in these cases is identical to the basic test of independence.

For example, suppose that the operations manager for Buyright Food Markets is interested in determining if the level of use of discount coupons is dependent on the location of the store. The manager has decided to take a random survey of 500 customers from the firm's three stores regarding the use of coupons. However, the sample size for the company's westside store should be three times the size of the other two stores because of the significantly larger sales volume. The results of the survey are:

Redemption Level	Store Location			Total
	Westside	Central	Eastside	
Low	100	50	25	175
Medium	150	45	40	235
High	50	5	35	90
Total	300	100	100	500

Notice that in this problem the column totals were specified in advance of the data collection phase. The corresponding null hypothesis can be stated as follows:

H_0: The three store populations exhibit the same level of coupon use,

or

H_0: Coupon use and store location are independent.

The degrees of freedom for this problem are 4 $(3-1)(3-1)$. The critical chi-square statistic at a 0.05 level of significance with four degrees of freedom is 9.487. The computed chi-square is 36.96. Based on these results, the manager can reject the null hypothesis of no difference and conclude that the level of coupon redemption varies between stores. That is, coupon redemption and store location are not independent nor are the stores homogenous with respect to coupon redemption.

INFERENCES ABOUT A POPULATION VARIANCE

Key idea 4

The primary emphasis of Chapter 8 (hypothesis testing) was on making inferences regarding means and proportions. In some testing situations, however, interest shifts to inferences about variances. Consider, for example, a manufacturing situation where a machine is used to fill boxes of cereal. The machine's filling mechanism is designed to provide 10 oz of cereal per box. It is unlikely, however, that every box will contain precisely 10 oz of cereal. Some boxes will contain more than 10 oz, and others will contain less than 10 oz. Nevertheless, it is possible that the average box contains 10 oz. Clearly, control of the variation in filling is important in this situation (imagine some boxes that are half-filled and some boxes that are overflowing with cereal). An estimate of the filling variance could be made by randomly sampling boxes from the production line. The sample variance will serve as an estimate of the true population variance. If the variance is within acceptable limits, then the production will continue. However, if the variance is outside these limits, the filling mechanism will need to be adjusted.

Interval Estimates

The chi-square statistic can be used to develop an interval estimate of the population variance (for applications in which the population is normally distributed). The ratio of the sample variance (s^2) to population variance (σ^2) has a chi-square distribution with $n - 1$ degrees of freedom. More specifically, the sampling distribution can be described as follows:

$$\chi^2 = \frac{(n-1)s^2}{\sigma^2}$$

where n represents the size of the sample.

Suppose that the production manager at Bland Foods randomly selected 25 cereal boxes from the production line, which yielded a sample variance of 0.1. The production manager knows that this point estimate is unlikely to equal the actual population variance. Therefore, the manager would like to develop an interval estimate that will provide an indication of the value of the population variance (σ^2). More specifically, the manager would like to be 95% confident that the developed interval contains the population variance.

The generalized model for developing an interval estimate for a population variance is

$$\frac{(n-1)s^2}{\chi^2_{(\alpha/2)}} \leq \sigma^2 \leq \frac{(n-1)s^2}{\chi^2_{(1-\alpha/2)}}$$

where $\chi^2_{(\alpha/2)}$ is the chi-square statistic that corresponds to a probability of $\alpha/2$ above the upper limit, and $\chi^2_{(1-\alpha/2)}$ is the chi-square statistic that corresponds to a probability of $\alpha/2$ below the lower limit. The critical chi-square values for a 95% confidence interval with 24 degrees of freedom are 12.40 (lower limit) and 39.36 (upper limit). The desired confidence interval can be specified as follows:

$$\frac{(24)(0.1)}{39.36} \leq \sigma^2 \leq \frac{(24)(0.1)}{12.40}$$

or

$$0.061 \leq \sigma^2 \leq 0.194$$

The corresponding interval estimate for the population standard deviation is simply obtained by taking the square roots of these interval limits:

$$0.247 \leq \sigma \leq 0.440$$

The chi-square distribution along with the developed variance interval estimate for this problem is shown in Figure 12.8. The manager can be 95% confident that this interval contains the population variance.

FIGURE 12.8 Chi-Square Distribution and 95% Confidence Interval for Bland Foods

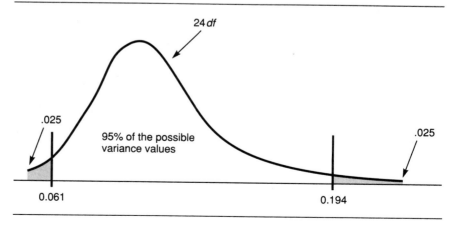

Hypothesis Testing

The hypothesis-testing process outlined in Chapter 8 can also be used for analyzing hypotheses about variances. Suppose the production manager at Bland Foods wishes to determine if the filling mechanism is operating within specifications at a 5% level of significance. The corresponding critical chi-square statistic for 24 degrees of freedom is 36.42 (from Table A.3). The manufacturer's tolerance (i.e., maximum variance) for the filling machine is 0.05. The null hypothesis is that the machine is within tolerance, and the alternative hypothesis is that the machine is operating outside tolerance. The decision rule for this problem is:

$$\text{Accept } H_0: \quad \chi^2 \leq 36.42$$
$$\text{Reject } H_0: \quad \chi^2 > 36.42$$

The computed chi-square with $s^2 = 0.1$, $n = 25$, and $\sigma^2 = 0.05$ is:

$$\chi^2 = \frac{24 \times 0.1}{0.05} = 48$$

Since the computed chi-square is greater than the critical chi-square, the null hypothesis can be rejected, which leads to the conclusion that the filling machine is operating outside manufacturer specifications. Figure 12.9 shows the chi-square distribution for this problem.

The chi-square statistic can also be used for drawing inferences regarding the variances of two populations. However, the more common approach in such cases is to use analysis of variance, which is the subject of the next chapter.

FIGURE 12.9 Chi-Square Distribution and Hypothesis Test for Bland Foods

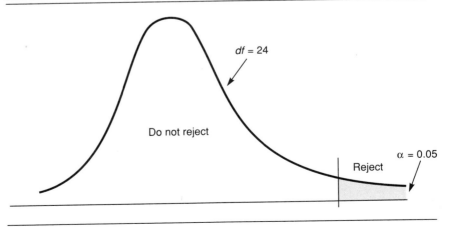

12.5 COMPUTER ANALYSIS

As illustrated here, the computations for a chi-square analysis are rather straightforward. Nevertheless, computer analysis does have a role, particularly in the test of independence with many categories. CBS can be used to solve problems involving both goodness of fit and test of independence. The CBS-generated solution includes both the computed chi-square and the corresponding p-value. (Recall that the p-value provides the user with the probability that the sample results occurred by chance only.) The following illustrates how CBS can be used to solve two problems introduced in the preceding section.

Extended Coin Flip Problem (Goodness of Fit)

Recall that the extended coin flip problem involved observing the outcome of 100 coin tosses. The null hypothesis was that a fair coin will produce 50 heads and 50 tails. The actual outcome was 70 heads and 30 tails. The input data and results for this goodness-of-fit test are given in Figure 12.10. These results confirm the previous analysis, in that the computed chi-square is greater than the critical chi-square value. Therefore, one can reject the notion that the coin is "fair."

FIGURE 12.10 CBS Chi-Square Analysis for Extended Coin Flip Problem

Information Entered

Number of rows:	2
Alpha error:	.05
Degrees of freedom:	1
Critical chi-square value:	3.841

	Observed	Expected
1 =	70	50
2 =	30	50

Results

Critical chi-square value:	3.8410
Computed chi-square value:	16
p-value:	.000

Conclusion: Reject null hypothesis

Kwan Accounts Receivable Analysis (Test of Independence)

CBS computer analysis for the Kwan accounts receivable problem is given in Figure 12.11. The printout includes both the contingency table and the computed expected frequency table. These data confirm the results developed earlier, namely, that the discount rate does not seem to encourage customers to pay on time. The computed chi-square of 7.64 is less than the critical chi-square value of 12.592, which supports the null hypothesis of independence between the two variables.

Figure 12.12 presents a CBS-generated chi-square distribution for this problem, including the critical and computed chi-square values.

FIGURE 12.11 CBS Chi-Square Analysis for Kwan Accounting

Information Entered

Number of columns:	4
Number of rows:	3
Alpha error:	.05
Degrees of freedom:	6
Critical chi-square value:	12.592

Observed

	<30	30-60	60-90	>90	Total
1 =	40	25	15	10	90
2 =	20	10	10	5	45
3 =	40	15	5	5	65
Total	100	50	30	20	200

Expected

	<30	30-60	60-90	>90	Total
1 =	45	22.50	13.50	9	90
2 =	22.50	11.25	6.75	4.50	45
3 =	32.50	16.25	9.75	6.50	65
Total	100	50	30	20	200

Results

Critical chi-square value:	12.5920
Computed chi-square value:	7.6353
p-value	.200

Conclusion: Do not reject null hypothesis

FIGURE 12.12 CBS Chi-Square Distribution for Kwan Accounting

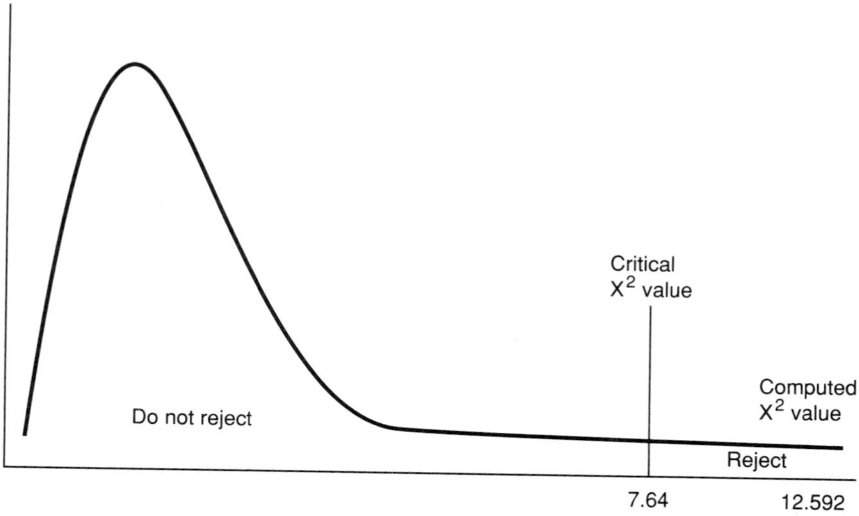

```
                                    Critical
                                    X² value
                                       |
                                       |              Computed
                                       |              X² value
         Do not reject                 |
                                       |      Reject
                                      7.64            12.592
```

12.6 PRACTICAL APPLICATIONS

The chi-square model is used extensively in the analysis of market studies, segmentation, and preference studies. Following are specific examples of these application areas.

Hot Stuff, Inc. (Test of Independence)

The director of marketing at Hot Stuff, Inc., has just completed a product preference study of 400 randomly selected individuals. For the study, the director asked a number of personal questions and sought scaled preference responses regarding Hot Stuff's Chili Delight. Two of the questions from the survey are listed below:

1. Age _____
2. Chili Delight has a better flavor than the other major brands.

   Strongly   Strongly
   agree   Agree   No opinion   Disagree   disagree

The manager would like to establish whether or not a relationship exists between age and preference for Chili Delight. By segmenting the responses, the problem can be solved using chi-square analysis. Since age is both a cardinal number and a magnitude measure, the issue becomes one of transforming the data for use in a contingency table. One approach would be to develop age categories, such as 21–30, 31–40, 41–50, and so forth. In general, however, most analysts prefer using the mean to classify the respondents. Those whose age is above the mean will be called the "high-age" group, and those who fall below the mean will be called the "low-age" group. Thus, for the age dimension, there will be two groups of questionnaires, one representing high age and the other low age. The resulting contingency table, including the data collected from the survey, is shown in Figure 12.13.

A CBS analysis of this problem is given in Figure 12.14. These results show that there appears to be a relationship between age and product preference. The computed chi-square of 10.13 is greater than the critical chi-square of 9.49 at a .05 level of significance. Therefore, the manager can reject the hypothesis of independence and conclude that the variables are dependent.

The analysis implies that each questionnaire would be tallied by age for each preference category. However, the manager may be uninterested in certain categories (e.g., no opinion) and the gradation of intensity. Further, more categories imply more degrees of freedom, which increases the size of the chi-square required for significance. Another approach is highlighted in Figure 12.15. High preference is defined as the "strongly agree" or "agree," and low preference is defined as "disagree" or "strongly disagree." The neutral point is then discarded.

A CBS analysis of this table also supports the premise that age influences preference for the product. The computed chi-square of 8.752 is greater than the critical chi-square of 3.841 at the 5% level for 1 degree of freedom.

It should be remembered that this suggested solution is only one of many ways of formulating a segmentation hypothesis. If the sample were large enough, one could, for example, use just the top 25% of the ages as "high" and the bottom 25% as "low." The rationale for this approach would be that if a relationship does in fact exist between age and preference, it would be most discernible at the extremes of the distribution. This type of analysis could appreciably aid the market researcher in understanding the character of the market.

Also, one should not automatically discard the "no opinion" midpoint in

FIGURE 12.13 Contingency Table for Hot Stuff Market Survey

	Preference					
Age	Strongly Agree	Agree	No Opinion	Disagree	Strongly Disagree	Total
High	30	60	100	20	15	225
Low	20	30	80	30	15	175
Total	50	90	180	50	30	400

FIGURE 12.14 CBS Analysis for Hot Stuff Marketing Study

Information Entered

Number of columns:	5
Number of rows:	2
Alpha error:	.05
Degrees of freedom:	4
Critical chi-square value:	9.488

Observed

	SA	A	N	D	SD	Total
1 =	30	60	100	20	15	225
2 =	20	30	80	30	15	175
Total	50	90	180	50	30	400

Expected

	SA	A	N	D	SD	Total
1 =	28.125	50.625	101.250	28.125	16.875	225
2 =	21.875	39.375	78.750	21.875	13.125	175
Total	50	90	180	50	30	400

Critical chi-square value:	9.4880
Computed chi-square value:	10.1305
p-value	.0370

Conclusion: Reject null hypothesis

---

the scale. In some cases, that category could be very meaningful. For example, many elections have been swung by the undecided voters going predominantly to one side. The following list presents some of the instances where one might choose to discard the center category:

- No interest in the category exists.
- The category is not a true center point.
- The total frequency in the category would not change the results even if it were to swing to one side or the other.

Albright Lighting Company (Goodness of Fit)

The chief engineer at Albright Lighting Company is concerned about the growing number of industrial accidents at the plant. Accidents have increased 20% from the preceding year. The chief is planning to initiate a new safety program and is interested in determining if a relationship exists be-

FIGURE 12.15 Reduced Contingency Table for Hot Stuff Market Survey

	Preference		
Age	High	Low	Total
High	90	35	125
Low	50	45	95
Total	140	80	220

tween level of industrial accidents and employee ethnic background. If there appears to be a relationship, the chief wishes to take this factor into account in designing the new safety program. Suppose the chief obtained company data on ethnic background and on history of reported accidents. Accordingly, the chief engineer could create a table of observed frequencies tallying accidents by ethnic group, as shown in Figure 12.16.

A total of 116 accidents occurred in the past year. A brief inspection of the table clearly shows that the white group had the greatest number of accidents. Could one conclude that whites have the largest accident rate? Such a conclusion would follow from a goodness-of-fit test, which would divide the total of 116 by the four categories, resulting in an expected frequency of 29 for each category. The computed chi-square in this case is 41.59. Comparing this result with the critical chi-square value at the .05 level for 3 degrees of freedom (i.e., 7.81) suggests that, as expected, ethnicity seems to contribute to the accident rate.

However, as shown in Figure 12.16, the distribution of ethnic groups employed at the plant is not uniform. Because whites constitute 50% of the work force, they have a greater probability of suffering accidents than any other group. Thus, an equal distribution of expected frequencies may turn out to be a meaningless hypothesis. What hypothesis would better reflect the objectives of this study? How would the null hypothesis translate into a computational methodology? To take into account the differing proportions of the various groups in question, one must assume that the accidents should

FIGURE 12.16 Accident Summary by Ethnic Group for Albright Company

Ethnic Group	Number of Accidents	Percent of Work Force
White	57	50
Black	15	15
Hispanic	30	25
Other	14	10
Total	116	100

FIGURE 12.17 Revised Accident Summary, Adjusted for Ethnic Distribution

Ethnic Group	Observed Accidents (number)	Expected Accidents (number)
White	57	58 (116 × .50)
Black	15	17 (116 × .15)
Hispanic	30	29 (116 × .25)
Other	14	12 (116 × .10)
Total	116	116

occur as a function of those proportions. More precisely stated, the number of accidents for each ethnic group should occur in proportion to their number in the work population. Using this logic, the analysis should be adjusted to incorporate the percentage of the total population that each group constitutes. The revised expected frequencies are shown in Figure 12.17.

The computed chi-square for the revised data is 0.62. Since this value is less than the critical value at 5% (i.e., 7.81), the chief engineer can conclude that ethnicity does not appear to influence accidents.

Hollywood Park (Goodness of Fit)

Suppose you are considering a wager at Hollywood Park. One factor you might consider in selecting a horse is that the gates closer to the railing should contain more winners than those farther away. The logic for this notion is simply that the starting gates closer to the rail imply a shorter running distance. Data could be obtained from the scratch sheets for all of last year's performances at the race track. A set of sample data taken for 122 races is shown in Figure 12.18.

If the gate assignment had no impact on winning, one would expect the

FIGURE 12.18 Sample Data for Horse-Racing Problem

Gate Number	Observed	Expected	Chi-Square
1	23	20.33	0.351
2	20	20.33	0.005
3	35	20.33	10.586
4	14	20.33	1.971
5	17	20.33	0.545
6	13	20.33	2.643
		Total	16.101

frequency of winning from any gate to be the same as for any other gate. Under this condition, it could be assumed that all gate assignments afford an equal opportunity of winning. This statement represents the null hypothesis, that is, the starting position has no impact on winning. The translation of this null hypothesis suggests that the expected value should equal 1/6 of the total, since there are six gates. Thus, each gate would have an expected frequency of 20.33 wins, (1/6)122 (122 ÷ 6 = 20.33).

Computing the chi-square statistic and comparing it to the critical value for 5 degrees of freedom would tell the bettor whether or not the distance run seems to relate to winning. Further, by comparing the observed and expected frequencies in each cell and the size of the contribution to chi-square (the last column of the computation), one could determine the significant contributors. Each could then affect the betting strategy. As in other cases, a significant chi-square indicates a departure from expectations that cannot be reasonably accounted for by chance.

A CBS analysis of this problem shows that at the 5% level one can reject the null hypothesis that gate assignment does not influence the outcome of the race. The computed chi-square of 16.101 exceeds the critical chi-square of 11.07. Therefore, you might be wise to take into account the gate position in placing a bet.

Blackman Tool and Die (Goodness of Fit—Poission Distribution)

The quality control manager at Blackman Tool and Die is interested in determining the nature of the company's defective rate. The manager has collected data from a sample of 100 production batches. Figure 12.19 shows the observed frequencies of production batches for given numbers of defec-

FIGURE 12.19 Observed Numbers of Defectives for a Sample of 100 Batches

Number Defective (per batch)	Observed Frequency (batches)
0	11
1	21
2	24
3	16
4	13
5	10
6	4
7	1
	100

tive parts. For example, 21 of the production batches were found to contain exactly one defective part.

The manager believes that the number of defectives observed can be characterized by a Poisson distribution. To test this hypothesis, the manager first estimates the mean by computing the weighted average of the number of defectives using the observed frequencies. The estimated mean of 2.5 will serve as the estimating parameter for the Poisson distribution. The manager then develops an expected frequency table based on the sample of 100 from the Poisson table with a mean of 2.5. The data are reported in Figure 12.20.

A CBS analysis of the observed and expected data using the goodness-of-fit test at the .05 level yields a computed chi-square of 4.088. Since this value is less than the critical value of 14.066, the manager can accept the null hypothesis and conclude that the failure rate is Poisson-distributed.

12.7 CASE STUDY: LEAKY PEN COMPANY

The agency manager at Van Gough Limited was very perplexed as he chaired the weekly artists meeting. He had been approached by the president of the Leaky Pen Company about placing a nationwide ad on television. The president wished to purchase a 30-second TV or radio spot promoting the top-of-the-line pen with very persuasive music in the background. He pounded the table with his gavel. "Will the meeting please come to order. Now, what the heck is persuasive music?" he asked in his benignly paternalistic manner.

The chief of the art department promptly presented her verdict: "There is no connection between music and advertising media. Why don't you suggest to the president a pen giveaway program at the next Beach Boys concert?" Almost everyone agreed that they would not compromise their art in response to an unreasonable demand for crass commercialism. The agency

FIGURE 12.20 Expected Frequencies of Defectives Based on a Poisson Distribution

Number Defective (per batch)	Poisson Probability ($\mu = 2.5$)	Expected Frequency (batches)
0	0.0821	8
1	0.2052	21
2	0.2565	26
3	0.2138	21
4	0.1336	13
5	0.0668	7
6	0.0278	3
7	0.0099	1

manager said, "All of you who don't think it worthwhile to work on this project may clear out your desks." Suddenly, they became interested in solving the problem confronting them.

The head of research, witnessing motivation in action, thought for 20 seconds—a record on Madison Avenue—and said, "I think I know how to approach this problem. First of all, it is quite obvious from the literature that people form pleasant as well as unpleasant associations with music. To study this problem, all we need to do is:

1. Form a focus group consisting of 200 members.
2. Expose the focus group to familiar music while airing the TV and radio ads.
3. Measure the group's preference for various combinations of music and media.

Presumably, the results from this experiment will indicate the best approach for advertising the pens."

The agency manager turned to the head of research and said, "You sound like you know what you're doing. Report back to the committee as soon as possible." As the meeting was breaking up, the head of finance said, "Your proposed design sounds to me like a simple chi-square test of independence." The research head smiled and quickly left the room.

The research head immediately set about designing an appropriate test instrument. She decided to expose 100 members of the focus group to the TV ad and the other 100 members to the radio ad. Both groups would be exposed to an ad three times: first with no music, second with classical music, and third with modern music. At the end of the session each member would be asked for his or her preference of music type. The survey results are presented in Figure 12.21. These data show, for example, that 33 members viewing the TV ad preferred it with no music.

At a subsequent meeting of the advertising committee, the research head made her report. She stated that the study revealed an apparent relationship between type of music and advertising medium. That is, the null hypothesis of independence of music and medium can be rejected at the .05 level. (The computed chi-square of 7.28 is greater than the critical chi-square of 5.99 for

FIGURE 12.21 Survey Results of Music Preference by Medium

Medium	Musical Preference			Total
	None	Classical	Modern	
TV	33	27	40	100
Radio	17	29	54	100
Total	50	56	94	200

2 degrees of freedom.) The agency manager responded by asking her to be more specific. The research head said, "Unfortunately, the test of independence provides only this conclusion. However, one could informally compare the observed and expected frequencies to obtain some indication of the level of the relationship between music and medium." The agency manager concluded the meeting by observing that this could be the subject of a subsequent session once the contract with Leaky Pen is signed.

12.8 SUMMARY

Chi-square is the most frequently used nonparametric method. The chi-square analysis is based on frequency measure instead of specific parametric estimates. Data from questionnaires and similar sources are typically analyzed by this method. Consequently, the business literature is replete with examples of hypotheses, data collection, and analyses using the chi-square test. The two most common uses of the chi-square statistic are:

1. *The goodness-of-fit test:* Used for testing the amount of departure of the actual data (observed frequencies) from hypothetical data (expected frequencies).
2. *Independence of variables:* Used to determine whether or not a relationship exists between two variables quantified in frequency form.

One of the major applications for the goodness-of-fit test is to determine whether or not one of the classic probability distributions (e.g., normal or Poisson) can be used to describe a specific business situation (e.g., the arrivals of customers at a bank). In these cases, the observed frequency data are compared with those estimated by the theoretical distribution. Typically, the null hypothesis is that the observed data come from a specific theoretical distribution. If the null hypothesis is rejected, one can conclude that they did not.

The test of independence is used to determine if two variables or factors are related. The null hypothesis is that the two variables are not related, that is, independent. If two variables are independent, then changes in one variable will not influence the state of the other variable. The test of independence usually involves the development of a contingency table, where observed and expected frequencies are compared. An important variation to the standard test of independence occurs when the number of total observations for the columns or rows is established in advance. In these cases, the test procedure can be viewed as a test of independence as well as a test of homogeneity.

The chi-square method can also be used to develop interval estimates and to test hypotheses regarding variances. The general procedure for developing interval estimates and testing hypotheses is very similar to the process outlined in Chapters 7 and 8 regarding means and proportions.

12.9 GLOSSARY

chi-square distribution A family of probability distributions based on the degrees of freedom and used for conducting tests on frequency-based data.

contingency table A matrix used in conjunction with the test of independence. The table contains two or more rows and two or more columns. Each row corresponds to a specific value for the first variable, and each column corresponds to a specific value for the second variable. The entries in the table correspond to the observed frequencies associated with the occurrence of both variable values.

dependence Variables are related.

expected frequencies A calculated or hypothetical set of variable values.

frequency The number of occurrences within a class of data.

goodness of fit A nonparametric test procedure for determining how much the observed data depart from expected values.

independence Variables are not related.

test of homogeneity A test procedure similar to the test of independence, in which the totals for the columns or rows are established in advance.

test of independence A nonparametric test procedure for determining whether two variables are independent or dependent.

12.10 BIBLIOGRAPHY

Aczel, A. D. *Complete Business Statistics.* Homewood, Ill.: Irwin, 1989.

Conover, W. J. *Practical Nonparametric Statistics.* New York: Wiley, 1980.

Gibbons, J. D. *Nonparametric Methods for Quantitative Analysis.* New York: Holt, Rinehart & Winston, 1976.

Guttman, I., and S. S. Wilks. *Introductory Engineering Studies.* New York: Wiley, 1986.

Mason, R. D. *Statistical Techniques in Business and Economics.* 7th ed. Homewood, Ill.: Irwin, 1990.

12.11 PROBLEMS

1. Discuss the differences between the goodness-of-fit test and the test for independence (use examples).
2. Using the goodness-of-fit test method, illustrate how the normal distribution can be used to estimate the binomial distribution under certain situations.
3. Using the goodness-of-fit test method, illustrate how the Poisson distribution can be used to estimate the binomial distribution under certain situations.
4. Can the chi-square test of independence be used to analyze problems involving three or more variables? If not, which procedure can be used?
5. Design a business problem consisting of three independent variables.

6. A specialist in heredity wishes to perform a recombinant DNA experiment on sweet corn. The literature clearly shows that the offspring should occur in an 8–2–5 ratio as follows:

Dark yellow corn	8
Light yellow corn	2
White corn	5

After cloning the corn, however, the specialist obtains the following results:

	Number
Dark yellow corn	45
Light yellow corn	7
White corn	38

   a. What statistical method should the specialist use for this problem?
   b. Is there a basis for concluding that the cloning has affected the outcome of the experiment.

7. A university sought to determine whether men or women are better in mathematics. Based on results from its entrance test, 48 females were divided into 52.1% high arithmetic scores and 47.9% low scores, and 66 males were divided into 60.6% high arithmetic scores and 39.4% low scores.
   a. Develop null and research hypotheses.
   b. Formulate a contingency table.
   c. Draw a conclusion regarding the test results at a 5% level of significance.

8. The Mike Row Chip Company has had a large turnover problem. The manufacturing vice president indicated that he thought the short-termers accounted for a disproportionate number of turnovers, as they represented only 20% of the total work force. In 1983, 31 short-termers quit, and 46 long-termers resigned. Did the short-termers contribute a disproportionate number of turnovers to the total in 1983?

9. The Setee Furniture Co. wishes to analyze employee turnover, since the costs of recruiting and training continue to rise dramatically, particularly for certain job classifications. The following data were collected for the 1985 period:

Job Classification	Turnovers (number)	Size (%)
Management	9	20
Craft	12	33
Clerical	6	18
Assemblers	123	34

   a. What is an appropriate research hypothesis for this problem?

b. What is the goodness-of-fit hypothesis?
c. What is the chi-square statistic, and which hypothesis does it support?

10. In certain manufacturing processes, it is customary to subject rejected metal to a high-temperature shock process to bring the stresses back into alignment. A normal scrap rate for forging is 5% or less. On one tested batch the rate was 35%. The rejected parts were subjected to the thermo shock, which resulted in the following data:

	Accepted	Rejected
Shocked	29	2
Not shocked	21	12

a. What is an appropriate null hypothesis?
b. What is an appropriate research hypothesis?
c. What conclusion should be drawn?

11. A large manufacturing firm is concerned about absenteeism. Recent literature seems to support the idea that a physical fitness program reduces absenteeism. Consequently, the firm has decided to allow one of its divisions to implement a physical fitness program while a different division serves as the control group. The pretest mean absence rate for both divisions was .65 per 1,000 hours. The results for the two divisions are stated in the following contingency table:

	Absenteeism	
	High Rate	Low Rate
Test division	2	10
Control division	10	4

a. What are the research and null hypotheses?
b. What is the computed chi-square statistic?
c. Which hypothesis is tenable at the 0.05 level?
d. Based on these results, what is your advice to management?

12. The Ajax Corporation, a market research firm, is concerned about increasing its response rate to mailed questionnaires. The general manager (GM) has stated that the problem can be studied during the course of normal business. Accordingly, the next set of mailings will be divided into three return postage types and one control group. The groups, of equal size, have been defined as follows:

Group	Definition
1	Standard first-class postage
2	Commemoration stamp
3	Business reply stamp
4	No return stamp

Each group contains 300 potential respondents. The following response rates were observed:

Group 1	35%
Group 2	37%
Group 3	38%
Group 4	20%

    a. What is the goodness-of-fit hypothesis?
    b. What is the chi-square value?
    c. How should the GM interpret these results?

13. The following data base presents two years of historical revenue data for the Kwan Bottling Company.

Month	Revenue Data ($000) 1984	1985
January	200	240
February	190	270
March	230	300
April	180	280
May	150	270
June	240	350
July	220	280
August	200	260
September	220	300
October	250	380
November	260	400
December	280	420

Management would like to know if sales have changed significantly over this two-year period. Comment on the computed p-value.

14. The manager at Sound City Stereo is interested in determining if a relationship exists between the type of TV purchased and the method of payment. The manager has compiled the following data on a sample of 150 recent purchases.

Payment Method	TV Model Type 19"	25"	48"	Total
Cash	75	20	5	100
Store credit	0	10	10	20
Credit card	5	20	5	30
Total	80	50	20	150

a. Is there evidence of a relationship between method of payment and TV model purchased at the .05 level?
b. Reduce the problem to a 2 × 2 contingency table, where the rows represent cash versus credit and the columns small versus large TV. Does this reduced problem change the results obtained in part *a*?
c. Interpret the *p*-values for parts *a* and *b*.

15. A survey was recently conducted by the American Management Institute on the importance of ethics in American business. The survey, consisting of 250 American managers selected at random, yielded the following data base:

| | Management Level | | | |
Concern	Supervisor	Middle	Top	Total
Very important	30	40	30	100
Important	50	30	20	100
Not important	35	15	0	50
Total	115	85	50	250

Is there a relationship between concern for ethics in business and level of management at the 5% level? What does the *p*-value mean in this case?

16. The audit manager at Hill Accounting Service discovered the following error pattern in 100 randomly selected accounts:

Errors per Account	Observed Frequency
0	18
1	20
2	18
3	17
4	15
5 or more	12

Can the manager conclude that the observed frequency can be described using the Poisson distribution? Use a 5% level of significance.

17. The quality control manager at Glen Electric observed the following defect pattern in 50 randomly selected disk drive assembly shipments.

Defectives per Shipment	Observed Frequency
0	15
1	18
2	9
3	5
4	2
5	1

Can the QC manager, at a 5% level of significance, conclude that the observed frequency comes from a population where the number of violations is binomially distributed? Comment on the computed p-value.

18. The chief fire inspector for the city of Los Angeles recently conducted an inspection of 150 pre-1950 buildings for possible fire code violations. Each building was inspected for four possible violations. The following table provides the results of the inspection:

Violations per Building	Observed Frequency
0	30
1	65
2	40
4	15
Total	150

Can the fire inspector, at a 5% level of significance, conclude that the observed frequency comes from a population where the number of violations is binomially distributed?

19. Conduct a goodness-of-fit test at a .05 level of significance to determine if the following sample test scores were selected from a normal population.

47	53	96	18	38
61	42	67	51	72
44	71	58	23	59
36	27	52	69	81
45	58	32	77	64

20. Prepare a histogram of the sample data given in problem 19. Does the histogram support the conclusions reached?
21. Conduct a goodness-of-fit test at a .05 level of significance to determine if the following sample data on car accidents at a certain intersection during a period of 100 days can be characterized by a Poisson distribution with a mean of 3.

Number of Accidents	Observed Frequency
0	3
1	9
2	16
3	28
4	22
5	13
6	7
7 or more	2

22. Prepare a histogram of the sample data given in problem 21. Does the histogram support the conclusions reached?
23. Conduct a goodness-of-fit test at a .05 level of significance to determine if the following sample data on the number of sales per day for an insurance salesman over a 30-day period can be characterized by a Poisson distribution with a mean of 2.

Number of Sales	Observed Frequency
0	3
1	5
2	9
3	8
4	4
5 or more	1

24. Prepare a histogram of the sample data given in problem 23. Does the histogram support the conclusions reached?
25. What informal conclusions can be made regarding the results observed in the contingency table for the Leaky Pen Case?
26. The following list presents the first and second numbers from a sample of 20 drawings for the California Lottery.

6–32	21–29	20–46	1–2	28–26
27–41	13–28	3–10	30–18	40–14
36–34	25–13	1–17	14–13	15–13
39–43	10–31	2–24	11–36	8–4

Can one conclude that the drawing for the first and second numbers is uniformly distributed at a 5% level of significance?

27. The U.S. secretary of education is interested in determining if a relationship exists between education level and income, at a 5% level of significance. The

department conducted a random survey of 200 individuals of age 40. The results of the survey are:

Education Level	Current Annual Income Level ($000)		
	Under 30	30–60	Over 60
High school	25	20	5
College	15	40	15
Graduate school	5	50	25

28. The editors of *Sports Illustrated* are interested in determining if sports interest is independent of sex, at a 1% level of significance. The magazine recently randomly surveyed 250 subscribers to ascertain their sports preference. The results of the survey are:

Sex	Baseball	Basketball	Football
Male	35	40	60
Female	30	55	30

Is a random sample of the subscribers to *Sports Illustrated* a representative sample of the general sporting population?

29. The quality control manager at Jensen Manufacturing is concerned about possible specifications violations in the company's rotor cup assembly. This product, which is used on the space shuttle, requires very narrow tolerances. The production standard calls for a maximum SD of 0.05 for the diameter of the rotor cup. A recent random sample of 20 assemblies showed a standard deviation of 0.06 centimeters.
    a. Develop a 95% confidence level for the SD of the rotor cup.
    b. Determine if the sample was taken from a population having a SD of 0.05 or less.

30. Develop a 95% confidence interval for a sample standard deviation of 10 and a sample size of 30.

31. Develop a 99% confidence interval for a sample standard deviation of 10 and a sample size of 100.

32. Develop a 95% confidence interval for a sample standard deviation of 5 and a sample size of 15.

33. The product manager at Compudesign, a manufacturer of numerical control systems, is interested in determining whether accident rates are related to shift operations. A random sample of 100 observations was made over a two-week period for the three shifts (columns). Accident data (rows) was recorded as either being above, within, or below the current company standard. A CBS analysis of the collected data is presented below:

570 Chapter 12 Chi-Square Analysis

Information Entered

Number of Columns:	3
Number of Rows:	3
Alpha Error:	0.05
Degrees of Freedom:	4
Critical chi-square:	9.4877

	S_1	S_2	S_3	Total
1	5	5	8	18
2	35	20	10	65
3	10	5	2	17
Total	50	30	20	100

Results — Expectations

	S_1	S_2	S_3	Total
1	9	5.4	3.6	18
2	32.5	19.5	13	65
3	8.5	5.1	3.4	17
Total	50	30	20	100

Critical chi-square:	9.4877
Computed chi-square:	8.9258
p-value:	0.0619
Conclusion: ?	

    a. What type of statistical test is most appropriate for this problem?
    b. Formulate the null and alternative hypotheses.
    c. What conclusions can the production manager make regarding a possible relationship between accidents and shift operations at the 0.05 level?
    d. Provide an interpretation of the reported p-value.

34. The vice president for audits at Union des Assurances, a major French financial services company, is interested in determining if a relationship exists between the number of loan defaults and loan size. A random sample of 150 loan records were selected for examination for the number of defaults (rows). The following four loan levels (columns) were used in recording the data (FR 000): 100<, 100–299, 300–499, 500+. A CBS analysis of the collected data is reported below:

Information Entered

Number of Columns:	4
Number of Rows:	2
Alpha Error:	0.05
Degrees of Freedom:	3
Critical chi-square:	7.81473

	L_1	L_2	L_3	L_4	Total
1	60	35	12	10	117
2	10	7	8	8	33
Total	70	42	20	18	150

Results — Expectations

	L_1	L_2	L_3	L_4	Total
1	54.6	32.76	15.60	14.04	117
2	15.4	9.24	4.40	3.96	33
Total	70	42	20	18	150

Critical chi-square:	7.8147
Computed chi-square:	12.1841
p-value:	0.0065
Conclusion: ?	

a. What type of statistical test is most appropriate for this problem?
b. Formulate the null and alternative hypotheses.
c. What conclusions can the production manager make regarding a possible relationship between loan defaults and loan levels at the 0.05 level?
d. Provide an interpretation of the reported p-value.

35. The production manager at Microteck Electronics is interested in determining if a relationship exists between absenteeism and plant location. A random sample of 280 employee records at the company's four plants (columns) were selected for evaluating the absentee rates. Absentee rates (rows) were recorded as either being above, within standard, or below company norms. A CBS analysis of the collected data is reported below:

Information Entered

Number of Columns:		4
Number of Rows:		3
Alpha Error:		0.025
Degrees of Freedom:		6
Critical chi-square:		14.4494

	P_1	P_2	P_3	P_4	Total
1	20	30	15	10	75
2	40	30	50	40	160
3	10	20	5	10	45
Total	70	80	70	60	280

Results — Expectations

	P_1	P_2	P_3	P_4	Total
1	18.750	21.492	18.750	16.071	75
2	40	45.714	40	34.286	160
3	11.250	12.857	11.250	9.643	45
Total	70	80	70	80	280

Critical chi-square:	14.4494
Computed chi-square:	23.0023
p-value:	0.0008
Conclusion: ?	

a. What type of statistical test is most appropriate for this problem?
b. Formulate the null and alternative hypotheses.
c. What conclusions can the production manager make regarding a possible relationship between employee absenteeism and plant location at the 0.05 level?
d. Provide an interpretation of the reported *p*-value.

36. The vice president of Greenwall Investments is interested in determining if a relationship exists between bank assets and earnings per share (EPS) for the European banking industry. The following data were obtained from a random sample of the top European banking institutions for 1990:

Assets ($ million)	Earnings per Share ($)	Assets ($ million)	Earnings per Share ($)
68,119	18.70	145,423	17.46
50,879	11.17	112,825	13.61
97,692	2.95	102,256	21.15
85,353	3.81	90,129	22.57
68,989	2.41	40,512	4.13
53,743	1.90	69,987	4.84
28,896	1.58	58,718	1.93
210,727	20.61	44,218	3.10
125,163	7.46	50,117	2.20
231,463	11.30	113,870	4.53
175,787	11.50	105,076	27.42
136,668	11.01	186,559	0.97
202,606	29.17	3,075	1.32
204,861	1.15	44,662	21.00

a. Formulate the null and alternative hypotheses.
b. Define a method for partitioning the data into a contingency table.
c. What conclusions can the vice president make regarding a possible relationship between EPS and bank assets at the 0.05 level?
d. If a relationship exists, how might it be defined?
e. Develop a simple linear regression using EPS as the dependent variable and assets as the independent variable.
f. Compare the results developed in parts c and e.

37. The maintenance manager at The Wilburn Group is interested in determining if a relationship exists between repair time and the types of computers manufactured by the company. The following sample data were collected on repair times for the three current computer models (L500, L600, and L750).

Computer Model	Time (minutes)	Computer Model	Time (minutes)	Computer Model	Time (minutes)
L500	24	L600	57	L750	59
L500	33	L600	71	L750	101
L500	27	L600	66	L750	87
L500	41	L600	45	L750	57
L500	45	L600	37	L750	43
L500	66	L600	61	L750	89
L500	31	L600	55	L750	84
L500	38	L600	51	L750	79
L500	35	L600	45	L750	106
L500	31	L600	39	L750	93
L500	29	L600	43	L750	80
		L600	49	L750	75

a. Formulate the null and alternative hypotheses.
b. Define an appropriate classification scheme for repair time.
c. What conclusions can the manager make regarding a possible relationship between repair time and computer model at the 0.05 level?

38. The operations manager at Big Time Pizza is interested in determining if a relationship exists between the size of the pizza ordered and whether it is delivered, picked up, or eaten on site. The following sample data were collected over a three-week period at Big Time's West Los Angeles restaurant:

Size	Method	Size	Method	Size	Method
Small	On site	Medium	On site	Large	On site
Small	On site	Medium	On site	Large	On site
Small	On site	Medium	On site	Large	On site
Small	On site	Medium	On site	Large	Pickup
Small	On site	Medium	On site	Large	Pickup
Small	On site	Medium	Pickup	Large	Pickup
Small	On site	Medium	Pickup	Large	Pickup
Small	On site	Medium	Pickup	Large	Delivery
Small	On site	Medium	Pickup	Large	Delivery
Small	Pickup	Medium	Pickup	Large	Delivery
Small	Pickup	Medium	Pickup	Large	Delivery
Small	Pickup	Medium	Delivery	Large	Delivery
Small	Pickup	Medium	Delivery	Large	Delivery
Small	Pickup	Medium	Delivery	Large	Delivery
Small	Pickup	Medium	Delivery	Large	Delivery
Small	Pickup	Medium	Delivery		
Small	Delivery				
Small	Delivery				
Small	Delivery				
Small	Delivery				

a. Formulate the null and alternative hypotheses.
b. What conclusions can the manager make regarding a possible relationship between pizza size and distribution method at the 0.05 level?
c. If a relationship exists, how might it be used in the planning process?

39. In a recent case in the Superior Court, County of Los Angeles, the following observations were made regarding the number of objections made by the attorneys for the plaintiff and for the defense.

Objections made during testimony of plaintiff's witnesses:

	Judge's Ruling		
Attorney	Sustained	Overruled	Total
Defense	51	10	61
Plaintiff	12	49	61
Total	63	59	122

Objections made during testimony of defendant's witnesses:

Attorney	Judge's Ruling		Total
	Sustained	Overruled	
Defense	74	23	97
Plaintiff	127	104	231
Total	201	127	328

a. Is there a difference in the judge's ruling on objections made by the plaintiff and defense attorneys during testimony of witnesses appearing for the *plaintiff* at the 0.05 level?
b. Is there a difference in the judge's ruling on objections made by the plaintiff and defense attorneys during testimony of witnesses for the *defense* at the 0.05 level?
c. Is there a difference in the judge's ruling on *total* objections made by the plaintiff and defense attorneys at the 0.05 level?
d. How would you explain these results in light of your understanding of a typical courtroom situation?

40. The research firm of Johnson and Baker has been commissioned by a local politician to determine the voting public's view on abortion. The director of the survey has decided to interview 1,000 voters selected at random to determine if income level and abortion position are related. However, the number of voters by income level has been preestablished to reflect the actual proportion in the voting population. The following contingency table reports the survey results:

Position	Income Level			Total
	Low	Medium	High	
For	325	150	40	515
Against	215	100	50	365
No opinion	60	50	10	120
Total	600	300	100	1,000

a. Formulate the null and alternative hypotheses.
b. In what way is this problem different from the standard test of independence?
c. Perform a test of homogeneity and independence at the 0.05 level.
d. Provide an interpretation of the results.

41. The marketing manager at Palos Verdes Community Bank is interested in determining the level of interest in home computer banking among its current customers. Furthermore, the manager would like to know whether the customer's level of education is related to interest in home banking. The manager has decided to conduct a random survey of 300 customers where the number selected per education level is pre-established according to the actual proportion in the customer population. The following contingency table reports the survey results:

Education	Interested	Not Interested	No Opinion	Total
Grade school	5	20	5	30
High school	30	90	30	150
Undergraduate	40	30	20	90
Graduate	20	5	5	30
Total	95	145	60	300

a. Formulate the null and alternative hypotheses.
b. In what way is this problem different from the standard test of independence?
c. Perform a test of homogeneity and independence at the 0.05 level.
d. Provide an interpretation of the results.

12.12 CASES

12.12.1 Audit Supply, Inc.

The president of Audit Supply, Inc., a mail-order financial house serving the Midwest, mused, "I wonder how we can increase the response rate to our monthly mailings. Next month is March, and that is an important time to sell our income tax preparation kit. I think I'll go consult a market research firm."

The market research firm head was a devout Rogerian and seemed committed to solving the problem. "I would like to increase my response rate from its average of 10%," said the mail-order executive. "Oh, you would like to increase your response rate," offered the researcher.

"My income tax kit inventory must be moved," the executive said. "Otherwise I will be stuck with its contents, which include:

1. An eraser
2. A foot-long pencil
3. A table of random digits
4. A bottle of disappearing ink
5. A record of the president laughing loudly."

"I hear you saying you would like to move your inventory," retorted the researcher. "I think the problem is that I have not been stamping the return envelopes, and people tend not to return those," proffered the executive.

"You think the problem is the lack of stamps on the return envelopes," the researcher responded, in sharp departure from his nondirective approach. "Yes," said the executive. "I think I should compare the return results using commemorative stamps, regular stamps, and business reply stamps. I will use the previous method of no stamps as a control measure."

"Why don't you study the problem by dividing the mailing list into four equal parts? For three groups use stamps of various kinds, and for the fourth use your old procedure," said the researcher in a forthright manner. "You are some kind of researcher," said the executive, "and worth every penny of the consulting fee."

The very next day the executive ordered the mailing procedure to be modified in accordance with the research design as he remembered it. He sent 250 mailings with commemorative stamps on the return envelopes, 250 with regular stamps, 250 with business reply stamps, and 250 with no stamps.

Ten days later the executive tallied the results, which were as follows:

1. Commemorative stamp returns equaled 18%, 12% from rural areas and 6% from urban areas.
2. Regular stamp returns were 16%, divided equally between rural and urban.
3. Business reply stamp returns were 15%, 10% urban and 5% rural.
4. No-stamp returns were 12%, 10% urban and 2% rural.

The executive looked at the data. "Commemorative stamps look most effective," he said happily to his market research consultant. "It seems possible that the commemorative stamps are the most effective," the researcher said, wondering how to get the executive to say that chi-square and further research is probably necessary.

Suggested Questions

1. Is the use of commemorative stamps an effective marketing strategy for the company?
2. What conclusions can be drawn regarding the other strategies?
3. How does the level of significance alter the general conclusions?

12.12.2 Horse Racing at San Ynez

"It's perfectly obvious," the physics professor said, "that horses with less distance to run can complete any race faster than horses with more distance to go." Smiley Beauregard countered, "Hey professor, there are many other variables entering into the equation—for example, the condition of the track and the physiological condition of the horse. The distance argument isn't even an example of Newton's first law." The brilliant empiricist I. Watson Skinner then broke in: "We could simply interview some of the horses and see if they thought the distance they ran in the race was less than for the others. After all, it's the perceptual distance that's important."

A masked statistician suddenly appeared. He suggested, "Why don't you follow the races at Santa Ynez for the 20-day season. You could record the amount of time by which each horse finished behind the winner. Of course, gate 1 is closest to the rail, and the distance that horse runs is less than for the horse at gate 2. If you plot and record the data properly, you should be able to determine whether the starting gate number (i.e., distance traveled) is a significant determinant of winning or losing."

"Who was that masked man?" said Beauregard. "I don't know who he was, but his approach is easier to implement than mine," said I. Watson Skinner. With that, the professor called in his graduate assistants and directed them to visit the track each day and record the data according to the procedure outlined by the masked man. At the end of the racing season the graduate students produced the data shown in Table 1 for 20 races. These data show the time, in seconds, that each horse finished behind the winner. (The winner is indicated by a 0.)

TABLE 1

Race	Gate Number							
	1	2	3	4	5	6	7	8
1	3.25	2.75	2.25	0.25	4.75	0	4.50	3.50
2	1.75	0	2.25	1.00	2.75	2.50	4.25	3.75
3	2.00	1.50	2.00	1.50	0	2.50	3.25	2.50
4	1.75	2.50	1.00	0	0.50	0.75	1.75	2.25
5	0.75	1.25	0	1.50	2.25	3.75	1.25	0.50
6	2.25	1.00	1.25	0	1.75	2.50	2.25	3.50
7	0	1.25	2.00	1.50	1.00	0.75	0.25	2.25
8	2.00	0	3.00	0.25	0.75	1.00	1.25	0.50
9	1.00	1.25	1.75	1.25	0	2.25	2.00	0.25
10	0.75	1.00	1.25	2.25	2.00	0.50	0	1.75
11	1.25	0	0.50	0.75	0.25	1.75	1.00	2.00
12	2.00	1.25	0	1.00	1.50	0.50	1.75	0.25
13	1.00	1.50	0.75	0.50	0	0.25	2.25	2.50
14	0	1.75	3.00	1.50	2.00	0.50	1.00	1.25
15	2.00	1.25	0	1.00	0.75	0.25	2.75	2.25
16	0	0.50	0.25	1.00	1.25	2.00	1.75	0.75
17	0	0.50	1.00	1.50	2.00	1.25	2.25	2.50
18	0.50	1.00	1.50	0.75	2.25	1.75	0.25	0
19	1.00	1.75	0.25	0.50	2.00	1.50	0	2.25
20	1.25	1.50	0	0.25	2.00	0.50	1.00	2.50

Note: 0 = Race won. Other numbers = Seconds behind winner.

Suggested Questions

1. Does gate assignment influence the outcome of the race?
2. Does the gate assignment have an impact on win, place, or show?
3. What is the relationship (if any) between gate assignment and winning?
4. What other factors might contribute to winning?

Chapter 13

Analysis of Variance

Truth is found more often through error than through confusion.
   Sir Francis Bacon

CHAPTER OUTLINE

13.1 Introduction
13.2 Example Management Problem: Mitterand Cable Company
13.3 How to Recognize an Analysis of a Variance Problem
13.4 Model Formulation
13.5 Computer Analysis
13.6 Practical Applications
13.7 Case Study: Microchip Electronics, Inc.
13.8 Summary
13.9 Glossary
13.10 Bibliography
13.11 Problems
13.12 Cases

CHAPTER OBJECTIVES

The primary objectives of this chapter are to develop an understanding of

1. The basic characteristics of analysis of variance problems.
2. The difference between fixed and random effects.
3. The standard one-factor and two-factor ANOVA models.
4. The relationship between analysis of variance and multiple regression.
5. How computer models can be used to solve analysis of variance problems.

Employee absenteeism and turnover represent major threats to productivity. The management at Struxs Corporation have noticed that absenteeism has been growing over the past several years. They decided to launch an incentive program to address this problem.*

The provisions of the program are as follows:

---

*D. Marth, "Making It," *Nation's Business* (June 1986), pp. 69–70.

1. Perfect attendance earns the employee a $5.00 monthly bonus plus credit for an annual bonus.
2. The annual bonus amounts to half of a day's pay for each month of perfect attendance for employees with less than five years of seniority, and a full day's pay for employees with five or more years of seniority.
3. Longer periods of perfect attendance are rewarded with additional bonuses and days off.

This prototype program reduced absenteeism from 23% to 4%. Furthermore, turnover declined by 50%. This program did not require additional administrative costs. Statistical analysis of the collected data formed the basis for evaluating the effectiveness of the program.

13.1 INTRODUCTION

Many business situations involve more than two populations. In these cases, the use of the classical hypothesis-testing model developed in Chapter 8 is not appropriate. Recall that the hypothesis tests in Chapter 8 analyzed a maximum of two populations. The **analysis of variance** (ANOVA) model provides a method for collecting, evaluating, and interpreting data from more than two populations. For example, a marketing manager is interested

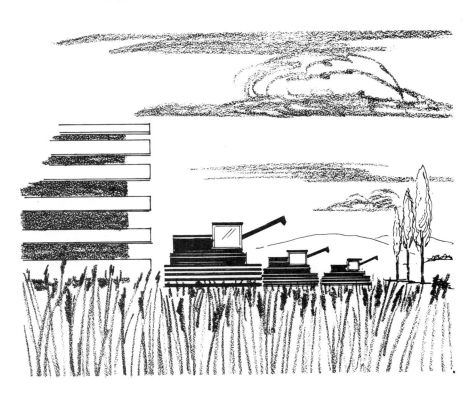

> **HISTORICAL NOTE**
>
> The analysis of variance model was introduced by R. A. Fisher (1890–1962) in 1924. Fisher began his statistical career at the Rothamstead Agricultural Experimental Station in Great Britain. There he worked out the famous F distribution, which was later named in his honor. To many, Fisher's contribution to statistics ranks with that of Karl Pearson, the father of modern statistical thought.

in testing the effects of three different advertising campaigns on product sales. The ANOVA model offers the manager a method of analyzing the effects of all three groups simultaneously. Using the standard hypothesis test procedure to compare the groups two at a time (i.e., multiple pairwise comparisons) would involve a total of three tests, and the potential for error increases as a direct function of the number of tests for a given level of significance.

Like hypothesis testing for one or two populations, the goal of the ANOVA model is to determine whether or not sets of data represent random samples from the same population (i.e., null hypothesis: $\mu_1 = \mu_2 = \mu_3 = \cdots = \mu_k$). Usually, each of the sample groups is exposed to a specific **treatment,** or possible cause of variation (e.g., a different advertising strategy). If the treatment produces significant results, the null hypothesis of no difference can be rejected. The analytical process involves decomposing the variance into its component parts and evaluating whether or not the treatment variance (between groups) is significantly larger than the corresponding sampling error variance (within a group).

Specific applications of the ANOVA model include:

- Comparing the mileage achieved with five different types of tires.
- Evaluating the impact of four different employee training programs on production.
- Assessing the impact of different CEO incentive compensation plans on the performance of small and large firms.

This chapter introduces the ANOVA model and presents applications involving one and two **factors** (i.e., variables). This chapter also discusses the relationship between the ANOVA model and multiple regression.

13.2 EXAMPLE MANAGEMENT PROBLEM: MITTERAND CABLE COMPANY

The production manager at the Mitterand Cable Company is interested in determining the impact of part-time employees on production. The Mitterand Company, located in southern France, produces high-quality electronic cable for telephone companies. Traditionally, the company has used part-timers during periods of peak production demand. However, the production manager is concerned that the actual level of production may drop if the mix of part-timers to full-time employees becomes too high. The production manager has decided to conduct a 10-day test using the following three levels of part-timers: low (less than 10%), medium (10–30%), and high (above 30%). Each team will be assigned to a specific production line during the experiment. The manager plans to measure production output, in thousands of feet of cable produced per day.

The following table presents production data and group means from the 10-day sample period:

	Part-Time Level		
Period	High	Medium	Low
1	0.659	0.911	1.230
2	0.330	1.010	1.490
3	0.680	1.000	0.998
4	0.820	0.898	1.010
5	0.626	0.992	1.230
6	0.831	1.110	1.110
7	0.595	0.900	1.440
8	0.439	0.998	1.000
9	0.649	1.010	1.150
10	0.800	0.944	1.960
Means	0.643	0.977	1.263

Grand mean = 0.961

The production manager wishes to draw conclusions from these data. What are the basic assumptions the manager needs to make to evaluate the impact of part-timers on production?

13.3 HOW TO RECOGNIZE AN ANALYSIS OF A VARIANCE PROBLEM

The following list highlights the general characteristics of the basic analysis of variance (ANOVA) model:

Key idea 1

- Each population can be described by a variable or factor (generally called the *main effect*).
- Each independent variable is subdivided into two or more groups.
- The variable being measured is called the dependent variable.
- The objective is to determine the impact of the main effects on the depen-

dent variable, as well as any interactions between the independent variables.

The basic assumptions associated with ANOVA are:

1. The error terms are normally distributed.
2. The sample variances are not significantly different.
3. The samples selected from each population are random and independent.
4. The measurement data are based on a ratio or interval scale.

13.4 MODEL FORMULATION

The concept of variance plays a key role in formulating the ANOVA model. Recall that variance is a measure of group variability. The basic approach is to determine the relative contributions to the total variance of the various factors. For example, the Mitterand Cable Company problem requires two variance estimates, one measuring the within-group variance and the other measuring the between-group variance.

INTERNATIONAL VIGNETTE

At a recent meeting of the Maillefer SA management team, product deliveries and prices were the primary topics on the agenda. Mr. Bonjour, who had recently accepted the position of marketing director, noted that he needed flexibility in establishing product prices and delivery schedules with potential customers. This was especially the case for the new pipe extruder equipment product line. He believed that, if given sufficient flexibility, he could significantly increase revenues for this product line over the next several years. The meeting concluded with the president requesting a detailed marketing plan from Mr. Bonjour.

Maillefer SA, established by Charles Maillefer in 1900 and located at Romainomoter near the French border, specialized in the manufacture of sophisticated cable and pipe extruder machines. By the mid-1980s, the company's total sales were approaching 100 million SFR, of which 90% involved exports. Historically, the cable extruder product line generated the bulk of the company's sales. However, the pipe extruder market was projected to grow twice as fast as the cable extruder market throughout the remainder of the century.

The basic options facing Mr. Bonjour included increasing sales, changing pricing, and improving market position. These issues needed to be resolved quickly due to growing pressure from Mr. Bonjour's management and from customers. With this in mind, Mr. Bonjour decided to analyze the impact of these basic decision options and to present his recommendations at the next management meeting.

Key idea 2

The actual model formulation depends directly on the number of problem factors (i.e., independent variables). In principle, the generalized ANOVA model can be used to analyze any number of factors and any number of groupings within each factor. In practice, however, the number of factors is often restricted to a maximum of two and the number of groups to a maximum of five.* Basically, there are two types of factors (i.e., effects) that can be analyzed using the ANOVA model structure: fixed effects and random effects. The **fixed-effects model** is one where only specific levels or treatments are of interest. The Mitterand Cable Company problem is an example of a fixed-effects problem. In the **random-effects model,** the factors represent only a sample of the population levels of interest. For example, a refinery manager may wish to determine the impact of temperature on product yield. In this situation, a random sample of control temperatures would be selected for the experiment. It is also possible to have a mixed-effects model, which includes both fixed effects and random effects. The primary presentation in this chapter involves the fixed-effects model. However, for one-factor applications, the computational process for both models is identical. A detailed discussion of the random-effects model can be found in the Neter and Wasserman text cited in the bibliography.

ONE-FACTOR ANOVA

Key idea 3

The computational procedure for one-factor ANOVA is to develop two separate estimates of the population variance. The first is calculated from the within-group variance, or that occurring from measure to measure. The second is calculated from the variance existing between the groups, or that occurring from condition to condition. The usual null hypothesis in ANOVA is that there are no differences among the groups as a result of a specific treatment effect. When the groups show greater differences than exist within each group, the null hypothesis of no difference can be rejected.

The general structure for the one-factor model is:

$$X_{ij} = \mu + a_j + e_{ij}$$

where:

X_{ij} = ith observed response from the jth population group
μ = Grand mean of the population
a_j = Additive effect for factor A (i.e., treatment effect)
e_{ij} = Random error term

---

*The general approach for solving problems with more than two factors is to use multiple regression analysis. The relationship between ANOVA and regression is presented later in this chapter.

For the Mitterand Cable Company problem, X_{ij} represents the ith production observation from part-time group j, μ_j is the true average production for part-time group j, μ is 0.961, and e_{ij} represents the random variation in actual production.

The general approach for determining the impact of each treatment effect involves computing the sum of the squares for the total (SST). This can be written as the sum of two separate terms: the sum of squares within groups (SSE) and sum of squares between groups (SSB). The symbolic expression for this relationship is

$$\text{SST} = \text{SSE} + \text{SSB}$$

SSE, also called SSW, indicates the extent of the within-group variability (i.e., from measure to measure). It is computed by summing the squared differences between each observation and the group mean. SSB represents the amount of variability between groups and is computed by summing the squared differences between each group mean and the grand mean. In general, a large SSB would indicate a large variation between the sample means, which in turn would suggest that the factor under investigation may be significant. However, both SSB and SSE are required in testing for statistical significance. The appropriate test procedure in this case utilizes the F statistic. As introduced in Chapter 10, F is defined as the ratio of the between-group variance (the numerator) to the within-group variance (the denominator). Recall that to conduct a hypothesis test, the computed F is compared with a critical F. The critical F value is obtained from either CBS or the F table (Appendix A, Table A.4), based on a given level of significance and the appropriate degrees of freedom. There are two degrees-of-freedom values used in determining the critical F value. These are:

1. The degrees of freedom for the numerator (between-group variance) equal the number of groups (K) minus 1, i.e., $K - 1$.
2. The degrees of freedom for the denominator (within-group variance) equal the number of individual measures (N) minus 1 for each condition or group, i.e., $N - K$.

Once the sum of the squares for each source has been obtained, the remaining computations are relatively straightforward. The next step is to determine the "mean" squares for each source (i.e., factor and error). This is accomplished by dividing the sum of the square for each source by the corresponding degree of freedom. Figure 13.1 presents the standard one-factor ANOVA model.

Generally speaking, the ANOVA model uses a 5% or a 1% level of significance as the rejection criterion. Again, the procedure is similar to the one used in hypothesis testing. That is, the null hypothesis can be rejected if the computed F is greater than the critical F. If the null hypothesis is not rejected, the observed results can be ascribed to chance factors. The F value is determined by dividing the **mean square for between-group variance (MSB)** by the **mean square of the error variance (MSE)** (see Figure 13.1). Once again, the term *significant* is used to denote rejection of the null hypothesis. In general, F will be significant if significance exists between any two of the groups. Thus, the F test is very sensitive to the presence of differences.

One-factor ANOVA represents the simultaneous testing of a number of values of a single variable. In the Mitterand Cable problem, the observed results will be significant if any of the three group means is statistically different. When significance does occur, there is usually interest in identifying which treatment caused the difference. This can be accomplished through a process called multiple pairwise comparison analysis. The idea is to test all possible pairs of means. The Mitterand Cable problem would require the following three null hypotheses for testing: $H_0: \mu_1 = \mu_2$, $H_0: \mu_1 = \mu_3$, and $H_0: \mu_2 = \mu_3$. Usually the process starts with the groups showing the largest difference in means and works downward until a comparison shows no significance. At that point it can be assumed that the smaller mean differences would likewise turn out nonsignificant. In conducting the pairwise analysis, the following adjustments to the standard t-test procedure should be implemented:

- The MSE variance is used.
- The t-value is based on $N - K$ degrees of freedom.

One additional issue needs to be considered concerning multiple pairwise tests. In principle, each test has the potential for a Type I error (i.e., alpha error). Therefore, multiple tests will increase the chances of a Type I error occurring for a given confidence level. For example, suppose the manager of Mitterand wishes to test the three pairwise null hypotheses at the .05 level. The probability that the conclusions on all three tests will be correct is not

FIGURE 13.1 ANOVA Structure for One-Factor Model

Source of Variance	Sum of Squares	Degrees of Freedom	Mean Squared	Computed F Value
Columns	SSB	$K - 1$	MSB = SSB/$(K - 1)$	MSB/MSE
Error	SSE	$N - K$	MSE = SSE/$(N - K)$	
Total	SST	$N - 1$		

SSB = Sum of squares between groups (treatment)
SSE = Sum of squares within groups (error)
K = Number of groups
N = Number of data points
MSB = Variance between groups
MSE = Variance within groups

95%, but (.95) (.95) (.95), or 86%. Therefore, the probability that one of the tests results in the unwarranted rejection of the null hypothesis is 14% and not 5%.

A simplified approach to this problem is to adjust α. This is accomplished by simply dividing the alpha used in ANOVA by the number of paired tests (i.e., α/m). In the present example, this would result in an alpha of approximately 1.66%. Generally speaking, a modified alpha of 1% can be used when the number of tests is relatively small or if only a subset of the total possible number of tests is performed. A more sophisticated, albeit complex, approach is to employ a statistical test procedure that compares all possible pairs of means in such a way that the chances of making one or more Type I errors is still a. The Tukey test is perhaps the most common technique for performing multiple comparisons in this way. The Tukey test statistic, which uses the smallest and largest sample means, is computed as follows:

$$D = Q_{\alpha,k,v} \sqrt{\frac{\text{MSE}}{N}}$$

where critical values of $Q_{\alpha,k,v}$ (found in Appendix A, Table A.8) are based on a given level of significance α, the number of groups k, and the degrees of freedom associated with the error term v. The Tukey test assumes that each sample contains the same number of observations.

The basic idea is to compare all possible combinations of sample means. If the difference between two sample means is greater than the parameter D, one can conclude that the corresponding population means are not equal, i.e., $\mu_i \neq \mu_j$. For example, a total of three such tests would be preformed for the Mitterand Cable problem. The specific comparisons would be $\overline{X}_1 - \overline{X}_2$, $\overline{X}_1 - \overline{X}_3$, and $\overline{X}_2 - \overline{X}_3$. In general, the number of comparisons will be equal to $k!/(2 \times (k-2)!)$, where again k represents the number of groups.

TWO-FACTOR ANOVA

Key idea 3

The ANOVA model can also be used to analyze problems containing two or more factors. The two-factor model introduces several additional sources of variance, which will be illustrated in the following example. Suppose the management at Johnson Carbide wish to study the productivity of their employees. Specifically, they are interested in determining if productivity is related to the management style of their supervisors and to plant location. Johnson manufactures and distributes industrial solvents from three plants located in different parts of the country. The personnel director has determined that plant supervisors can be characterized as either autocratic or democratic in leadership style. The chief engineer has collected weekly production data from the three plant locations. The basic design scheme and sample data for this problem are given in Figure 13.2.

FIGURE 13.2 Data Base for Johnson Carbide Company

Leadership Style	Plant			Mean
	A	B	C	
Autocratic	205	275	315	244.8
	225	195	254	
Democratic	390	350	313	344.5
	365	345	304	
Mean	296.3	291.3	296.5	294.7

The numerical values represent the production rates (in thousands of gallons of solvent), by leadership type and plant location. For example, the production rate under an autocratic leader in plant A was 205,000 gallons of solvent. The overall average production is 294,700 gallons. This term is often referred to as the grand mean. Using these data, it is possible to segregate the variance from the effects of plant location alone. This is accomplished by "pooling" the data for plant location to derive a single factor (i.e., leadership style). The same pooling process can be used for isolating the impact of plant location. Further, a third variance can be computed that is attributable to the interaction of leadership style and plant location. **Interactive effects** can be important since they suggest, when significant, a synergy between the two factors independent of their individual contributions. For example, autocratic leadership style in plant A may lead to a significantly different level of productivity from that related to autocratic leadership in plant C. The three effects are summarized as follows:

1. The effect on production of leadership style.
2. The effect on production of plant location.
3. The joint effect of leadership and plant location on production.

Accounting for a second factor and for potential interactive effects requires a more complex model than the one-factor ANOVA model. The general structure for the two-factor model is:

$$X_{ijk} = \mu + a_i + b_j + (ab)_{ij} + e_{ijk}$$

where:

i = Level of factor A
j = Level of factor B
k = Observation number per cell
X_{ijk} = Variable value for ith row, jth column, and kth point
μ = Grand mean of the population
a_j = Additive effect for factor A
b_i = Additive effect for factor B

$(ab)_{ij}$ = Interactive effect
e_{ijk} = Random error term

For example, X_{231} refers to row 2 (level 2 of factor A), column 3 (level 3 of factor B), and observation 1. This corresponds to the value 313 in Figure 13.2.

This model assumes that the error term e_{ijk} is an independent, normally distributed random variable with a mean of 0 and constant variance over all factors. Furthermore, this model requires more than one data point per cell (i.e., **replications**) in order to isolate the interactive and error terms. A special case of the two-factor model is the block design. Block designs are useful because they improve the experimental accuracy without increasing the size of the data base. The block design requires only one data value per cell; the interactive factor is eliminated. The block design model is presented in the next section.

The first step in developing the two-factor ANOVA model is to calculate the sum of squares for each effect. In this case, the model consists of two main effects, A and B, the interactive effect, AB, and an error term, E.

$$SST = SSA + SSB + SSAB + SSE$$

Figure 13.3 presents the ANOVA structure for the two-factor model. Basically, interaction can be viewed as the joint effect of two or more factors. In the two-factor ANOVA model, there will be only one interactive effect (i.e., AB). However, in the three-factor model, the number of interactions grows to four (i.e., AB, AC, BC, and ABC). Interactions can be positive or negative. That is, they can increase or decrease the cumulative effect of the factors.

Providing an adequate interpretation of interaction effects is often difficult, especially when more than two factors are involved. For example, the data shown in Figure 13.4 represent a plot of mean production versus leadership style for several of Johnson Carbide's plants. In Panel A, no interaction is present since the two linear functions are parallel. The data in Panel B,

FIGURE 13.3 ANOVA Structure for Two-Factor Model

Source of Variance	Sum of Squares	Degrees of Freedom	Mean Squared	Computed F Value
Rows	SSA	$a-1$	MSA = SSA/$(a-1)$	MSA/MSE
Columns	SSB	$b-1$	MSB = SSB/$(b-1)$	MSB/MSE
Interactive	SSAB	$(a-1)(b-1)$	MSAB = SSAB/c	MSAB/MSE
Error	SSE	$N-ab$	MSE = SSE/$(N-ab)$	—
Total	SST	$N-1$	—	—

a = Number of levels for factor A
b = Number of levels for factor B
N = Number of data points

FIGURE 13.4 Illustration of Potential Interactive Effects for Johnson Carbide

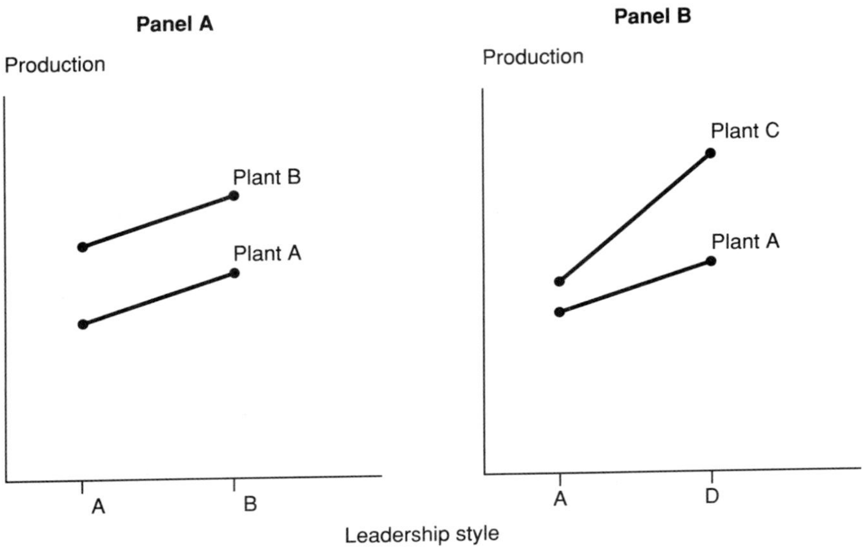

[Panel A: Production vs. Leadership style (A to B), showing Plant B and Plant A as nearly parallel lines, both sloping upward.]

[Panel B: Production vs. Leadership style (A to D), showing Plant C with a steep upward slope and Plant A with a gentler upward slope, the lines not parallel.]

however, indicate the presence of interaction because the two lines are not parallel. The plot indicates a positive interaction between the democratic leadership style and production in plant C. That is, a democratic leadership style at plant C has significantly increased production compared to plant A, with the same type of leadership. This example merely illustrates one of the many types of interactions that are possible for multifactor ANOVA applications.

BLOCK DESIGNS

Generally, experimental error can be reduced by increasing the size of the data base. However, there are important cost and time implications associated with this approach. Another strategy is to reduce the error variance by eliminating the variation associated with a particular dimension of measurement (e.g., sex or age). This approach is somewhat similar to adding a dummy variable to improve R^2 in regression analysis. A randomized block design can be used to control the variation of some secondary factor and thus remove its effect from the analysis. The two-factor ANOVA model is used to solve block design problems. This is because both a treatment variable and a blocking variable are included in the model. The primary interest, however, is in the treatment variable, and it is assumed that there is no interaction between the treatment and blocking factors.

The format for the block design is very similar to the two-factor model, the only difference being the absence of the interactive effect. In the blocking model the total sum of squares is partitioned as follows:

$$SST = SSA + SSB + SSE$$

where:

SST = Sum of squares total
SSA = Sum of squares for factor
SSB = Sum of squares for blocking factor
SSE = Sum of squares for error term

The determination of the degrees of freedom and the interpretation of the statistical results are performed in exactly the same way as for two-factor ANOVA. The following problem illustrates the general design of the blocking model.

The portfolio manager at Craig Investments International is interested in reviewing the performance of a new computer-based decision support system. The primary objective of the new system is to help reduce the number of "bad" loans made by the institution. More specifically, the manager wishes to know whether the new decision support system outperforms the more traditional methods of loan evaluation. Historically, loan decisions were based on the applicant's balance sheet or, in some cases, key financial ratios. A data base consisting of the lending performance (i.e., percentage of bad loans) of 12 loan officers was selected at random from the company's records. To isolate the effects of experience, the manager organized the officers into four age blocks (0–4 years, 5–9 years, 10–14 years, and 15–19 years). The resultant data (single observation per cell) are presented in Figure 13.5. This table presents the standard block design, where the columns represent the treatment factor, in this case the different methods used in loan evaluation, and the rows represent the blocking factor. The null hypothesis for the primary factor, i.e., method of evaluation, is that each method yields the same percentage of "bad" loans. The alternative hypothesis is that they do not. The null hypothesis for the blocking factor is that there is no block effect, that is, experience level does not inpact the proportion of "bad" loans.

FIGURE 13.5 Random Block Design for Craig International

	Loan Evaluation Method		
Block	Balance Sheet	Key Ratios	Decision System
0–4	7	6	2.5
5–9	5	5	2
10–14	4	3	1
15–19	3	2	0.5

The alternative hypothesis is that there is a blocking effect. The test procedure used for evaluating the statistical significance of these two null hypotheses is the same as in the two-factor model. That is, if the computed F is greater than the critical F, the null hypothesis can be rejected.

REGRESSION ANALYSIS

Key idea 4

Generally speaking, the ANOVA model represents a special case of multiple regression analysis. Recall that the fixed-effects ANOVA model involves specific variable groupings, as compared with the continuous or random variables associated with the regression model. In some situations, the use of regression modeling yields clear advantages over the ANOVA model. Figure 13.6 presents some of the advantages and disadvantages of regression analysis compared with ANOVA.

The Mitterand Cable Company case will be used to illustrate the application of regression analysis to a single-factor fixed effects ANOVA. Recall that the manager was interested in determining whether or not the proportion of part-timers influenced productivity. This case involves three different levels of part-time participation: high, medium, and low. The three levels correspond to three independent dummy variables in the regression model. These variables are assigned a value of either 0 or 1 as follows:

$X_1 = 1$ if production data are from high group, otherwise 0
$X_2 = 1$ if production data are from medium group, otherwise 0
$X_3 = 1$ if production data are from low group, otherwise 0

In practice, one of the dummy variables (e.g., X_3) can be dropped since it is already represented when the other two variables are set to 0. In general, the number of independent variables is equal to the number of groups minus 1. The dependent variable in this case is production level. The regression model can be described as follows:

$$Y = b_0 + b_1 X_1 + b_2 X_2$$

The same approach is taken as for applying regression to two-factor problems. Provision must be made to handle both the main effects and the

FIGURE 13.6 Advantages and Disadvantages of Regression versus ANOVA

Advantages	Disadvantages
• Forecasting	• Handling discrete variables
• Handling multiple factors	• Interpreting interactive effects
• Handling missing data	• Handling dependent data

interactive effects. The number of independent variables for a two-factor problem is determined as follows:

$$N_v = (r-1) + (c-1) + (r-1)(c-1)$$

where:

N_v = Number of independent variables
r = Number of row groups
c = Number of column groups

For example, the total number of independent variables for the Johnson Carbide problem is five.

13.5 COMPUTER ANALYSIS

Key idea 5

Most ANOVA problems are solved using computer analysis. Computer analysis allows for the rapid solution of complex problems while reducing the potential for numerical errors. The computer input process for most ANOVA problems is relatively straightforward. CBS uses a spreadsheet format, which greatly simplifies both data input and editing. The CBS input prompts for the Mitterand Cable Company problem are highlighted in Figure 13.7.

FIGURE 13.7 CBS Input Data for Mitterand Cable Company

```
Table Commands Enter Raw Data                    FILE:none       MODULE:DSTA

Quick Mvmt              gold

HOME    Row 1   1       371.300
END     Row n   2       386.400
PgUp    Up 16   3       394.700
PgDn    Dn 16   4       382
L       Col 1   5       377.700
R       Col m   6       378.100
                7       346.800
←- -→ Edit      8       348.100
                9       341.300
ESC restore    10       340.600
     value     11       341.500
               12 |     319.500
INS add
    value
DEL delete
    value

 ↑   Mvmt   Press F    Enter Command OR Position Cursor & enter value for point
←- -→       when       press ←┘ to complete entry of value for data point
 ↓   ←┘     finished
```

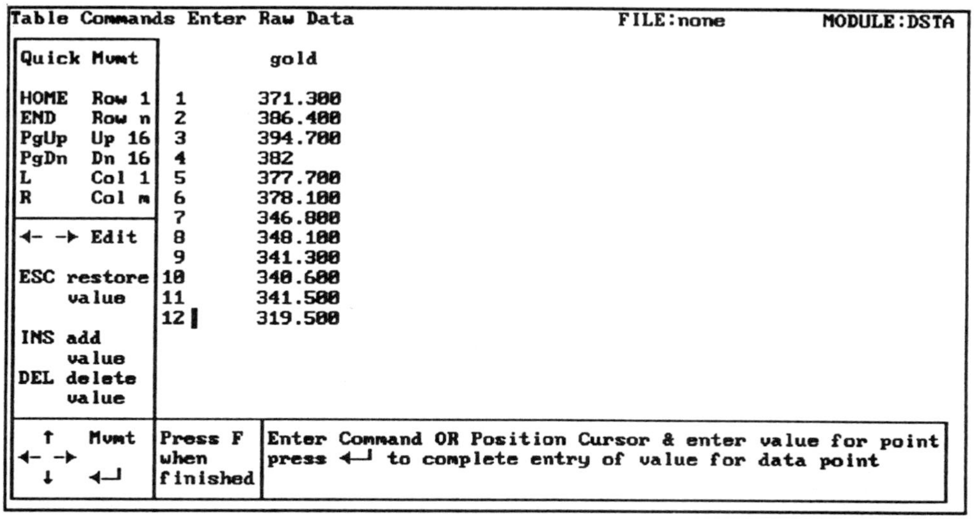

Mitterand Cable Company (One-Way ANOVA)

Figure 13.8 presents a CBS analysis of the Mitterand Cable Company problem. The numbers in parentheses refer to a specific section in the following discussion. Recall that this problem involves one factor, i.e., level of part-time employees. The null hypothesis is that there is no difference in productivity as a function of the ratio of part-time employees. These results show that the impact of part-time employees on production is significant. The computed F statistic is greater than the critical statistic (i.e., $23.898 > 3.35$), which leads to rejection of the null hypothesis; that is, the effect of part-timers is real. Therefore, the production manager may wish to consider reducing the number of part-time employees. However, shifting to a lower percentage of part-timers may have adverse financial implications.

1. *Source of variance:* The sources of variance include the effect (i.e., the column), the error, and the total.
2. *Sum of squares:* The sum of the squares is the difference between the actual and the average values for each variance source, squared and summed.
3. *Degrees of freedom:* The degrees of freedom represent the number of information items, for each variance source, that are independent of each other. For the column effect, the degrees of freedom are equal to the number of groups minus one (i.e., 2). For the error effect, the degrees of freedom are equal to the number of data points minus the number of groups (i.e., 27).
4. *Mean square:* The mean of the sum of the squares, or **mean square,** is the variance. It is computed for each variance source by dividing the sum of the squares by the corresponding number of degrees of freedom.
5. *Computed F:* The computed F is determined by dividing the mean squared value for the effect (i.e., column) by the mean squared value for the error. The critical F of 3.35 was obtained from the F table for df = 2 (numerator) and df = 27 (denominator) for a level of significance of .05.

FIGURE 13.8 CBS One-Factor ANOVA Output for Mitterand Cable Company

(1) Source of Variance	(2) Sum of Squares	(3) Degrees of Freedom	(4) Mean Squared	(5) Computed F
Columns	1.919	2	0.960	23.898
Error	1.084	27	0.040	
Totals	3.003	29		

Critical F(col): 3.35   Reject null hypothesis

The null hypothesis can be rejected since the computed F is greater than the critical F.

Since the null hypothesis was rejected, the manager now wishes to determine which of the effects are significant. As outlined earlier, this can be accomplished through the use of multiple pairwise comparisons. Figure 13.9 presents a multiple pairwise comparison analysis based on the Tukey's test procedure. The critical Q value from Appendix A (Table A.8) based on $\alpha = 0.05$, k = 3 and v = 27 is approximately 3.5. The corresponding Tukey statistic (D) is:

$$D = 3.5 * \sqrt{\frac{0.04}{10}} = 0.063$$

These results show that there is a statistically significant difference between each pair of means, (i.e., the null hypothesis of no difference can be rejected). Therefore, one can conclude that production varies significantly between high, medium, and low levels of part-timers. This conclusion should not be surprising since the mean for the low group is twice the size of the mean for the high group.

As outlined in the previous section, regression analysis can frequently be used to solve ANOVA problems. Figure 13.10 presents the regression analysis for the Mitterand Cable case. These results show that both independent variables, X_1 = High and X_2 = Medium, are significant, as indicated by the computed t values. Furthermore, the overall model is significant, based on comparison of the computed F value with the critical F value. Although these results appear encouraging, the relatively low R^2 (.47) suggests that other factors are needed to help explain the variability in the observed production levels. The fact that both X_1 and X_2 are significant implies that the effects they represent are real. Thus, the regression approach avoids the need to undertake pairwise comparisons to determine which effects are significant. The developed regression model can be used to estimate group means, as indicated in Figure 13.11.

FIGURE 13.9 Tukey's Pairwise Analysis of Mitterand Cable Company

$Q = 3.5$ $\qquad\qquad\qquad\qquad$ $D = 0.063$

$|x_1 - x_2| = |0.643 - 0.977| = 0.334 > 0.063$

$|x_1 - x_3| = |0.643 - 1.263| = 0.620 > 0.063$

$|x_2 - x_3| = |0.977 - 1.263| = 0.286 > 0.063$

Conclusion: $\mu_1 \neq \mu_2, \mu_1 \neq \mu_3, \mu_2 \neq \mu_3$

FIGURE 13.10 CBS Regression Analysis of Mitterand Cable Company

		Results		
Variable	B-Coeff	Beta	t-value	p-value
X_1	−0.480	−0.606	−4.201	.000
X_2	−0.403	−0.510	−3.532	.000

B_0 intercept: 1.137
Critical t-value: 2.571

Sum of squares regression:	1.4202
Sum of squares error:	1.5834
Sum of squares total:	3.0036
Mean square regression:	0.7101
Mean square residual:	0.0586
Coefficient of determination (R^2):	.4728
Adjusted C.O.D. (R^2):	.4338
Multiple correlation coefficient (R):	.6876
df regression:	2
df error:	27
Standard error estimate:	0.2422
Computed F:	12.1089
Critical F:	5.49
$F(p$-value):	.0000

FIGURE 13.11 Estimation of Group Means for Mitterand Cable Company

Variable	Actual	Predicted	Residual
$X_1 = 1$, $X_2 = 0$	0.643	0.657	−0.014
$X_1 = 0$, $X_2 = 1$	0.977	0.734	0.243
$X_1 = 0$, $X_2 = 0$	1.263	1.146	0.117

Johnson Carbide (Two-Way ANOVA)

Figure 13.12 presents CBS output for the Johnson Carbide problem introduced in Section 13.4. Recall that the company's management was interested in determining if leadership style or plant location had an impact on production. The impact of leadership style (i.e., row effect) is shown to be

FIGURE 13.12 CBS Two-Factor ANOVA Output for Johnson Carbide Company

Source of Variance	Sum of Squares	Degrees of Freedom	Mean Squared	Computed F Value
Rows	29,800.334	1	29,800.334	31.781
Columns	70.167	2	35.083	0.037
Inter	9,838.167	2	4,919.084	5.246
Error	5,626	6	937.667	
Totals	45,334.668	11		

Critical F value (row): 5.99 — Reject null hypothesis
Critical F value (int): 5.14 — Reject null hypothesis
Critical F value (col): 5.14 — Do not reject null hypothesis

significant since the computed F value is greater than the critical F value at a .05 level of significance. On the other hand, the impact of plant location on production does not seem significant, as indicated by the very small F value (i.e., 0.037). The interactive effect, however, is significant. This suggests that the impact of leadership style on production varies according to plant location.

Craig Investments International (Block Design)

Figure 13.13 presents a CBS analysis for the Craig Investments problem. Recall that the manager wished to know if the new decision support system outperformed either of the two more traditional methods used in loan evaluation, i.e., balance sheet or financial ratios. These results show that both the primary factor (col) and the blocking factor (blk) are statistically significant at the 0.05 level. That is, the null hypothesis that the method of evalua-

FIGURE 13.13 CBS Block Design ANOVA Analysis for Craig International

Source of Variance	Sum of Squares	Degrees of Freedom	Mean Squared	Computed F Value
Columns	21.9	2	10.95	21.06
Blocks	19.4	3	6.47	12.44
Error	3.1	6	0.52	
Totals	44.4	11		

Critical F value (col): 3.98 — Reject null hypothesis
Critical F value (blk): 3.59 — Reject null hypothesis

tion has no effect on the percentage of "bad" loans can be rejected (12.06 > 3.98). Furthermore, the null hypothesis for the blocking variable can also be rejected (12.44 > 3.59). This result suggests that the experience of the loan manager *does* impact the loan performance. At this point, the portfolio manager may wish to identify which of the evaluation methods are different. Again, this can be accomplished using the Tukey test procedure.

13.6 PRACTICAL APPLICATIONS

Analysis of variance is used extensively in applications where the effects of specific treatments are of interest. Several examples on the use of the ANOVA methodology are presented in the following sections.

Product Placement at Allnight Drug Store (Two-Way ANOVA)

The display manager for Allnight Drug Stores is interested in determining if shelf location or store location will significantly affect the sales of a new health kit. The manager plans to implement a factorial design that will involve nine of the company's stores. The manager plans to collect weekly sales data from each store over a three-week period (a total of 27 data points). Each store will be randomly assigned a unique combination of in-store location and shelf location. For example, store 1 will locate the new product at the front of the store on a middle shelf. Unfortunately, during the experiment the product is removed from three stores because of the need for shelf spacing to accommodate a manager's special. Therefore, only 24 data points were actually collected. The data base is reported in Figure 13.14.

This data base was evaluated using CBS's two-factor ANOVA model. The results are given in Figure 13.15. These results show that both in-store location and shelf position affect product sales. That is, the null hypothesis of no effect for both variables can be rejected, as indicated by the large computed F values. However, there appear to be no interactive effects, since the computed F value for this effect is well below the critical statistic. The manager's next step is to determine which in-store location and which shelf position yield the most sales. This can be accomplished through a multiple pairwise comparison of means.

Affirmative Action in Government Contracting (Two-Way ANOVA)

The head of the congressional watchdog committee on affirmative action is interested in evaluating the status of minority and female participation in government contracts. Congress has been receiving increased pressure from a number of action groups to expand contract opportunities for firms owned by minorities and women. The committee was interested in both defense

FIGURE 13.14 Results from Allnight Product Placement Survey

Shelf Location	In-Store Location		
	Front	Middle	Back
Low	$1,500	$1,200	$900
	1,800	800	700
	1,200	950	675
Middle	950	750	250
	800	600	400
		875	550
High	1,100	850	500
	925	750	450
	850		

and nondefense contracts. The committee's staff surveyed four major defense contractors and six major nondefense contractors to measure the percentage of subcontracts let to minority- and female-owned firms (based on dollar value). Summarized in Figure 13.16 are the results of the survey in three major subcontracting areas (service, software, and hardware).

This data base was evaluated using CBS's two-factor ANOVA model. The results are given in Figure 13.17. These results show a statistically significant difference in subcontracting both for type of contract (defense versus nondefense) and contract area (service, software, and hardware). This conclusion is drawn from the fact that both of the computed F values are larger than the critical F values. On the other hand, the interactive effects between contract type and contract area are not significant. The head of the watchdog committee can conclude that the issuance of government-based subcontracts is not uniform throughout industry. The next step is to identify which specific contract areas are different by contract type.

FIGURE 13.15 CBS Two-Factor ANOVA of Allnight Drug Store

Source of Variance	Sum of Squares (000)	Degrees of Freedom	Mean Squared (000)	Computed F Value
Rows:	847.622	2	423.811	14.943
Columns:	1,380.625	2	690.313	24.340
Inter:	82.977	4	20.744	0.731
Error:	425.417	15	28.361	
Totals:	2,736.641	23		

Critical F value (row): 3.68   Reject null hypothesis
Critical F value (int): 3.06   Accept null hypothesis
Critical F value (col): 3.68   Reject null hypothesis

FIGURE 13.16 Results from Subcontractor Survey

Contract Area	Contract Type	
	Defense	Nondefense
Service	15	20
	12	22
	8	16
	11	18
		19
		8
Software	5	11
	2	6
	4	7
	1	3
		9
		0
Hardware	0	3
	2	5
	1	6
	3	7
		2
		5

FIGURE 13.17 CBS Two-Factor ANOVA for Government Contracting Problem

Source of Variance	Sum of Squares	Degrees of Freedom	Mean Squared	Computed F Value
Rows:	787.400	2	393.700	36.600
Columns:	112.022	1	112.022	10.414
Inter:	10.711	2	5.356	0.498
Error:	258.167	24	10.757	
Totals:	1,168.300	29		

Critical F value (row): 3.4 — Reject null hypothesis
Critical F value (int): 3.4 — Do not reject null hypothesis
Critical F value (col): 4.26 — Reject null hypothesis

13.7 CASE STUDY: MICROCHIP ELECTRONICS, INC.

A number of customer complaints have recently hit the desk of Mr. Ed Barns, vice president of operations for Microchip Electronics, Inc. (MEI). MEI is a major producer of power transformers for the computer industry. Because of the growth in product demand, Mr. Barns has had to increase production by adding second and third shifts. This has resulted in a product defect rate that is unacceptable to Microchip's customers.

Mr. Barns decided to call a meeting of the production and quality control staff to address this growing problem. Basically, he was interested in determining the impact of both production volume and shift on product quality. Mr. Barns opened the meeting with an overview of the problem and stressed the need to develop an effective solution. He indicated that the transformer business is extremely competitive. He further stated that unless the company solved this problem, the business might move to other firms. At that point, Ms. Johnson, the quality assurance manager, suggested that a better understanding of the problem could be obtained through a statistical analysis of recent production data. Mr. Rines, the production supervisor, broke in and stated that this problem could be easily solved simply by scaling back production. He observed that a lower production rate would allow more time to be spent on product quality. Although this was an appealing argument, Mr. Barns rejected it out of hand, stating that one of the company's primary goals for the year was to increase market share. Clearly, the proposal of Mr. Rines was inconsistent with this objective.

At that point, Mr. Barns turned to Ms. Johnson and told her to "fix the problem." As the meeting was breaking up, Ms. Johnson asked Mr. Fred Yellin, a recent MBA graduate, if he was interested in helping solve this problem. Mr. Yellin responded affirmatively, so Ms. Johnson asked him to collect recent data on product defects as a function of production level and operating shift. After a few days, Mr. Yellin presented Ms. Johnson with the data shown in Figure 13.18.

Mr. Yellin explained that these data showed the defect rate (measured as percentage of production) by production level (Low = 500/shift, Medium = 750/shift, High = 1,000/shift) and shift (1st, 2nd, or 3rd). Mr. Yellin observed that a preliminary inspection of the data indicated that the defect rate seemed to increase with production level. However, he noted that a more analytical approach was needed to determine the actual effect, since the data represented only a sample of the actual production data. Ms. Johnson, recalling her graduate days, remembered a statistical procedure called analysis of variance. This methodology, she stated, was useful in analyzing problems with two or more variables where each variable is divided into two or more groups. This description seemed to aptly characterize the current problem. Accordingly, Mr. Yellin processed the data sample he had developed using the CBS computer model. The results of this analysis are shown in Figure 13.19.

At the next meeting of the combined staffs, Mr. Barns asked Ms. Johnson for a report on the progress made to date on the problem. She presented the sample data followed by the computer results presented in Figure 13.18. Mr. Barns, who had a limited background in computer analysis, asked Ms. Johnson to help him read these results and to translate them into plain English. Responding to Mr. Barns's request, Ms. Johnson stated that the analysis

FIGURE 13.18 Product Defective Levels at Microchip Electronics

Shift	Production Level		
	Low	Medium	High
1 (8:00–4:00)	2	2	3
	3	4	7
	2	4	5
	4	5	6
	5	6	7
	3	5	5
	2	6	6
	3	6	4
2 (4:00–12:00)	5	5	6
	3	6	7
	4	4	8
	3	5	6
	2	6	5
	2	7	7
3 (12:00–8:00)	3	5	6
	5	6	7
	4	3	8
	6	4	7
	5	5	5

showed that the level of production did influence the defect rate. That is, the computed F exceeded the critical F statistic (22.2 versus 3.2). On the other hand, shift operations did not, statistically speaking, seem to affect the defect rate: the computed F statistic was less than the critical F. Finally, Ms. Johnson stated that there do not appear to be any interactive effects between

FIGURE 13.19 CBS Two-Factor ANOVA for Microchip Electronics

Source of Variance	Sum of Squares	Degrees of Freedom	Mean Squared	Computed F Value	Computed P-value
Rows:	8.743	2	4.371	3.049	.300
Columns:	63.614	2	31.807	22.188	.000
Inter:	9.080	4	2.270	1.584	.600
Error:	68.808	48	1.434		
Totals:	150.245	56			

Critical F value (row): 3.2 — Do not reject null hypothesis
Critical F value (int): 3.2 — Do not reject null hypothesis
Critical F value (col): 3.2 — Reject null hypothesis

production level and shift operations. Mr. Barns thanked Ms. Johnson for her presentation and then turned to the smiling Mr. Rines. "Well," Mr. Barns said, "I guess you were right after all." He then instructed Mr. Rines to develop a new production schedule that would reflect the need to improve product quality.

13.8 SUMMARY

This chapter introduced the basic structure of the analysis of variance (ANOVA) model. The ANOVA model represents a logical extension to the two-population tests presented in Chapter 8, in the sense that it can be used for testing the means for more than two populations. ANOVA is often used to analyze business problems that contain one or more variables where each variable is partitioned into two or more distinct groups. The standard problem involves analysis of the effects of specific *treatments* (fixed-effects model). The typical null hypothesis is that there is no difference among the means of the groups. When the null hypothesis is rejected, additional analysis can be carried out to identify which effects are statistically different, i.e., Tukey's multiple pairwise comparison test. The ANOVA model can be used, with appropriate modifications, to study both fixed effects as well as random effects.

In the single-factor ANOVA, two estimates of the variance are made using the between-group variance and the within-group variance. An F value is computed by dividing the between-group variance by the error variance. This statistic is used to determine the statistical significance of the test effect. The ANOVA model can also be used to determine the effect of two or more independent variables operating together or of one independent and a second blocking variable. Problems involving two or more variables generate so-called interaction effects. The analysis of these effects is based on the same logic as one-factor analysis, by isolating their contribution to the total variance. The block design model represents an important variation to the basic two-factor ANOVA model. Multiple regression can be used to solve ANOVA-type problems, particularly ones involving more than two variables.

13.9 GLOSSARY

analysis of variance (ANOVA) A statistical technique used to draw inference regarding whether more than two population means are equal.

factor A variable of interest.

fixed-effects model An ANOVA procedure in which the levels being evaluated are the only ones of interest.

interactive effect The joint impact of two or more factors.

mean square between-group variance (MSB) The sum of the squares of the between-group variance divided by the corresponding degree of freedom.

mean square of the error variance (MSE) The sum of the squares divided by the corresponding number of degrees of freedom.

random-effects model An ANOVA procedure in which the levels being evaluated are only a sample of the levels of interest.

replications The number of observations per group.

treatment A probable cause of variation in the dependent variable.

13.10 BIBLIOGRAPHY

Luck, D. J., H. G. Wales, D. A. Taylor, and R. S. Rubin. *Marketing Research.* Englewood Cliffs, N.J.: Prentice Hall, 1978.

Miller, R. D., and D. W. Wichem. *Intermediate Business Statistics.* New York: Holt, Rinehart & Winston, 1977.

Neter, J., W. Wasserman, and M. H. Kutner. *Applied Linear Regression Models.* 2nd ed. Homewood, Ill.: Irwin, 1989.

———. *Applied Linear Statistical Models,* 3rd ed. Homewood, Ill.: Irwin, 1990.

13.11 PROBLEMS

1. Identify whether each of the following business situations involves fixed or random effects.
   a. The effect of milling machine operators on product output.
   b. The effect of various fertilizers on crop yield.
   c. The effect of three proposed marketing plans on sales revenues.
   d. The effect of four new engine options on market share.
2. Identify whether each of the following public policy situations involves fixed or random effects.
   a. The effect of the new tax policy on income.
   b. The effect of new automobile fuel economy standards on oil imports.
   c. The effect of new standardized tests on improving educational quality.
   d. The effect of multisource government contracts on reducing waste.
3. Design a business-oriented experiment that involves fixed effects.
4. Design a business-oriented experiment that involves random effects.
5. Identify several business situations involving two factors where the interactive effects might be significant.
6. Formulate a business application involving three factors where the three-factor interaction might be significant.
7. The credit manager at the Jiffy Loan Company wishes to determine if there is a relationship between the amount of delinquent payments and the annual income of its clientele. The manager has selected 36 accounts at random and has arranged the delinquent payment data according to three income groups (low, medium, and high). The resulting data base is:

Low (<$15,000)	Medium ($15,000–$40,000)	High (>$40,000)
$123	$476	$325
98	195	624
145	321	786
101	98	143
66	224	315
127	356	557
23	378	478
45	621	331
77	287	845
167	333	715
204	197	434
54	324	614

Does there appear to be a relationship between income and delinquent payments at the .05 level?

8. The personnel director at Jade Electronics, an integrated circuit board manufacturer, is interested in determining if level of education has an impact on employee turnover. The director has selected, at random, the files of 25 ex-employees. The director found that 8 employees have only a high school degree, 7 only a bachelors degree, 6 only a masters degree, and 4 a Ph.D. The director arranged the turnover data (measured in months) as follows:

Level of Education			
High School	Bachelors	Masters	Ph.D.
48	32	40	24
12	18	24	18
8	22	18	22
6	19	15	19
10	15	11	
17	14	28	
24	26		
10			

What conclusions can the director draw regarding these data at a .05 level of significance?

9. The director of research and development at the Eternal Candle Company surveyed 25 companies to determine the impact of R&D expenditures on corporate profits. The director believes there is a lagged relationship between earnings (net profits on net sales) and R&D expenditures. Therefore, the survey instrument requested current earnings data along with R&D expenditures from two years before. Further, R&D expenditures were categorized into the three groups: high (above 20% of sales), medium (between 10% and 20% of sales) and low (below 10% of sales). The data generated from the survey are as follows:

Level of R&D Expenditures		
High	Medium	Low
10.4	10.3	6.8
14.3	8.7	9.7
19.8	5.4	−1.2
15.8	13.9	4.2
20.9	7.6	3.8
11.4	8.9	4.4
14.3	10.4	−5.5
8.7	9.3	6.1
13.4		

a. What conclusions can the director draw regarding the impact of R&D expenditures on earnings at a .01 level of significance?

b. Comment on the director's assumption of a lagged relationship between R&D and profits.

10. The marketing director at Drexway Pharmaceutical is interested in determining whether age and level of education have an effect on income. The director undertook a survey, which resulted in the following data:

	Educational Level		
Age	High School	Undergraduate	Graduate
<40	$15,500	$22,000	$27,000
	12,000	20,000	30,000
	13,000	19,000	25,000
	17,000	24,000	24,000
>40	21,000	32,000	41,000
	23,500	30,000	45,000
	24,000	35,000	48,000

a. Formulate null and alternative hypotheses.
b. Test the impact of each factor for $\alpha = .01$.
c. What is the significance of the interactive effect?

11. The sales manager at Whirlright City, a retail outlet specializing in home appliances, is interested in determining if the model type (deluxe, standard, or economy) of the appliance and the skills of a particular salesperson have an impact on the number of units sold. The manager has collected the following data on sales performance (measured in units) as a function of both model type and individual salesperson over the last three months.

	Salesperson		
Model Type	Mr. Smith	Ms. Jones	Mr. Faith
Deluxe	6	3	2
	5	1	2
	5	0	1
Standard	4	3	3
	4	4	2
	2	3	5
Economy	1	7	6
	0	8	5
	2	7	8

  a. Formulate null and alternative hypotheses.
  b. Test the hypothesis at a .01 level of significance.
  c. What conclusions can be drawn from the analysis?

12. Following is a CBS analysis of a single-factor ANOVA problem. Fill in the missing numerical values and draw a conclusion at the .05 level of significance.

Source	SS	df	MS	F
Columns	392	2		
Error				
Total	788	11		

13. Following is a CBS analysis of a two-factor ANOVA problem. Fill in the missing numerical values and draw a conclusion at the .01 level of significance.

Source	SS	df	MS	F
Rows	276,125	1		
Columns	385,583	2		
Inter				
Error	25,166	12		
Total	694,426	17		

14. Solve the Johnson Carbide problem presented in Section 13.4 using regression analysis.
15. Perform a pairwise hypothesis test for the remaining effects in the Mitterand Cable problem.
16. Perform a pairwise hypothesis test for the Johnson Carbide problem.
17. The Los Angeles chapter of the Citizens for Accuracy in Advertising (LACAA) wishes to test the accuracy of several recent ads run by the three leading supermarket chains. Each chain is claiming to offer the lowest overall prices. The LACAA recently conducted a random survey in seven local stores of each chain. The survey consisted of purchasing the exact same basket of goods in each store

and recording the total price. The results from the survey are reported in the following table.

Store	Supermarket Chain		
	Ralph's	Von's	Alpha Beta
1	$89.43	$92.13	$94.56
2	84.32	90.41	95.06
3	88.76	89.42	92.67
4	90.45	89.01	93.90
5	85.82	91.87	92.39
6	87.96	92.62	95.11
7	89.05	91.11	94.76

    a. Formulate the null and alternative hypotheses.
    b. Is there a significant difference in prices among the three supermarkets?
    c. If differences exist, which supermarket offers the lowest prices?

18. The commercial loan officer at the Palos Verdes Community Bank is interested in analyzing the performance characteristics of selected industry groups. More specifically, she wishes to determine if there are differences in the ratio of current assets to current debt among four food group industries. The results of her survey are:

Industry			
Bakery Products	Canned Foods	Dairy Products	Confectionery Products
1.72	1.79	1.49	2.35
1.58	1.83	1.32	2.64
2.02	1.54	1.68	2.25
2.25	2.02	1.81	1.98
1.95	1.92	1.21	2.08
1.79	1.48	1.34	2.71
1.82	1.69		

    a. Are there significant differences in the assets-to-debt ratio among the four groups at the .05 level?
    b. If differences exist, which industry group has the highest assets-to-debt ratio?

19. The general manager for Brooks Emporium wishes to determine whether or not sales at three of the company's stores are the same. A random sample of monthly store sales over the last two years is presented below:

Los Angeles	Chicago	Atlanta
$256,354	$212,789	$167,871
187,555	200,445	176,121
183,111	187,314	189,317
145,849	156,367	195,387
192,133	188,743	202,551
225,337	208,663	218,391

a. Develop an appropriate ANOVA model, assuming no interactions.
b. Formulate the null and alternative hypotheses.
c. Are sales for the three stores the same or different at the .05 level?
d. If sales are different, which stores produce different sales?

20. The marketing manager at Wade Home Electronics is interested in determining if the cost of radio advertising is significantly different throughout the state according to the type of transmission (i.e., AM or FM). The manager conducted a phone survey of several radio stations located around the state regarding the cost of running a 30-second advertisement. The following data summarize the results of the phone survey:

Radio Type	Region		
	Southern	Central	Northern
AM	$425	$250	$475
	445	235	400
FM	220	175	240
	200	150	275

a. Develop an appropriate ANOVA model, assuming no interactive effects.
b. Determine if the cost of advertising is significantly different throughout the state at a .05 level.

21. Analyze problem 20 assuming the existence of interactive effects.

22. The general manager of Healthnet Corporation is interested in determining the impact of exercise and diet on weight loss at a 1% level of significance. Currently, Healthnet offers three different diet plans and four exercise plans. The manager selected a random sample of 36 individuals that participated in both the diet and the exercise plans. The weight loss (in pounds) for each participant is as follows:

	Diet Plan		
Exercise Plan	A	B	C
I	11	8	13
	9	7	15
	18	5	16
II	15	7	20
	17	6	18
	12	4	17
III	13	9	12
	15	4	16
	22	5	20
IV	16	2	12
	19	4	15
	25	2	17

a. Do the different diet plans yield different results?
b. If so, which diet plans are different?
c. Do the different exercise plans yield different results?
d. If so, which exercise plans are different?
e. Do diet and exercise both contribute to weight loss?

23. The Consumers Media Guide of Southern California (CMGSC) plans to report in next month's publication the prices on 25-inch color television sets. CMGSC's research department recently conducted a survey of the three leading brands of 25-inch sets at four of the area's discount chains. Prices for the 25-inch set were obtained from two stores of each retail chain. The CMGSC likes to report all survey findings at a 95% level of confidence. The survey data are:

	Brand		
Retail Chain	Toshiba	Magnavox	Sony
Federated	$425	$375	$495
	405	350	510
Circuit City	390	345	485
	400	330	500
Target	415	350	510
	395	340	515
Big Jack's TV	410	365	495
	395	360	505

a. Do prices differ significantly among the four stores?
b. Is there a significant price difference among the three brands?
c. Comment on the meaning and significance of any interactive effects.
d. Perform a multiple pairwise comparison using Tukey's method on the significant effects.

24. The fair pricing committee of the American Bar Association (ABA) is charged

with monitoring legal fees throughout the country. The committee recently completed a random sample of the legal fees for forming either a corporation, a partnership, or a proprietorship in California, New York, and Florida. The results from the survey are:

	Corporation	Partnership	Proprietorship
California	$1,500	$1,250	$1,300
	1,250	1,400	1,750
	1,650	1,650	1,200
New York	2,100	1,800	1,550
	2,400	1,750	1,700
	2,000	1,950	1,850
Florida	1,300	1,100	1,200
	1,450	1,000	1,450
	1,500	1,250	1,350

a. Perform a two-way ANOVA using a .01 level of significance.
b. Do legal fees differ by state?
c. Do legal fees differ by type of business?
d. Provide an interpretation of the interactive effects.

25. Solve problem 24 using multiple regression analysis with legal fees as the dependent variable. How are the dummy variables defined? Which of the explanatory variables are significant at the .05 level?

26. The general manager of the Realtors® Association of Southern California recently conducted a survey on housing prices in Southern California. Among the data recorded were the general location (urban or rural), the view (Exceptional = Yes, Typical = No), the type of the house (modern or Spanish), and, of course, the price. The results of the survey are:

Location	View	Style	Price
Urban	Yes	Modern	$875,000
Urban	Yes	Modern	915,000
Urban	Yes	Spanish	950,500
Urban	Yes	Spanish	895,000
Urban	No	Modern	475,000
Urban	No	Modern	425,500
Urban	No	Spanish	515,000
Urban	No	Spanish	605,000
Rural	Yes	Modern	700,000
Rural	Yes	Modern	750,000
Rural	Yes	Spanish	765,500
Rural	Yes	Spanish	725,000
Rural	No	Modern	375,000
Rural	No	Modern	425,000
Rural	No	Spanish	450,000
Rural	No	Spanish	510,500

a. What type of ANOVA model is appropriate for solving this problem?
b. Why is ANOVA analysis somewhat more effective than regression analysis for solving this problem?
c. Provide an interpretation for possible three-way interactive effects.
d. How could the data base be reconstructed for solution using a two-way ANOVA model?
e. Develop a two-way model.
f. Which factors are statistically significant at the .05 level?
g. Perform a multiple pairwise comparison analysis using Tukey's method on the significant effects.

27. Solve problem 26 using multiple regression analysis with price as the dependent variable. How are the dummy variables defined? Which of the explanatory variables are significant at the .05 level?

28. The manager of *Consumer Goods* magazine recently conducted a survey on refrigerator prices for 10-cubic-foot models throughout the country. Among the data recorded were the location, the brand name, the store chain, and the price. The results of the survey are:

Location	Brand	Store Chain	Price
Los Angeles	Maytag	Fedmart	$498
Los Angeles	Maytag	Fedmart	525
Los Angeles	Maytag	Sears	455
Los Angeles	Maytag	Sears	479
Los Angeles	G.E.	Fedmart	545
Los Angeles	G.E.	Fedmart	565
Los Angeles	G.E.	Sears	498
Los Angeles	G.E.	Sears	515
Chicago	Maytag	Fedmart	525
Chicago	Maytag	Fedmart	545
Chicago	Maytag	Sears	515
Chicago	Maytag	Sears	520
Chicago	G.E.	Fedmart	575
Chicago	G.E.	Fedmart	590
Chicago	G.E.	Sears	535
Chicago	G.E.	Sears	550
New York	Maytag	Fedmart	565
New York	Maytag	Fedmart	545
New York	Maytag	Sears	525
New York	Maytag	Sears	539
New York	G.E.	Fedmart	569
New York	G.E.	Fedmart	575
New York	G.E.	Sears	569
New York	G.E.	Sears	555

a. What type of ANOVA model is appropriate for solving this problem?
b. Why is ANOVA analysis somewhat more effective than regression analysis for solving this problem.

c. Provide an interpretation for possible three-way interactive effects.
d. How could the data base be reconstructed for solution using a two-way ANOVA model.
e. Develop a two-way model.
f. Which factors are statistically significant at the .05 level?

29. Solve problem 28 using multiple regression analysis with price as the dependent variable. How are the dummy variables defined? Which of the explanatory variables are significant at the .05 level?

30. The Terrestrial Cell Company is planning to introduce a new-generation solar cell. The company's chief engineer is interested in evaluating three different manufacturing procedures. Each procedure will involve "baking" the solar cell at specific oven pressures (1 atmosphere and 4 atmospheres) and temperatures (600°F, 700°F, and 800°F). Additionally, each cell will be subject to outside climatic conditions (50% and 75% humidity). The chief engineer wants to identify the combination of oven pressure and temperature that maximizes cell life. The chief plans to evaluate 100 cells for each of the specified conditions.
a. Identify the major factors and interactions.
b. Formulate a set of hypotheses for this problem.
c. How many degrees of freedom are associated with the main and interactive effects?
d. Formulate an appropriate data collection table.

31. The marketing director at Yang Consulting Services is interested in conducting a market survey on the vehicle emissions test industry. More specifically, the director would like to determine the potential impact of geographic location (East or West), site location (urban or rural), type of enterprise (chain or independent), and equipment used (manual or automatic) on prevailing prices.
a. Identify the major factors and interactions.
b. Formulate a set of hypotheses for this problem.
c. How many degrees of freedom are associated with the main and interactive effects?
d. Formulate an appropriate data collection table.

32. The general manager at Sanford International wishes to determine whether an increase in production rate and training will have a significant impact on the number of defective parts. The following data were collected on a two-week experimental program that measured the number of defectives produced under various production rates and training conditions.

	Production Rate (units per hour)		
	300	350	400
No training	45	46	49
	37	41	53
	42	38	46
	38	42	52
	44	48	54
Training	39	42	38
	32	36	41
	27	32	37
	40	44	46
	29	33	32

   a. What type of ANOVA model is appropriate for solving this problem?
   b. Provide an interpretation for possible two-way interactive effects.
   c. Which factors are significant at the .05 level?
   d. How could the data base be reconstructed for solution using a one-way ANOVA model?
   e. Develop a one-way model.
   f. Which factors are statistically significant at the .05 level?
   g. Perform a multiple pairwise comparison analysis using Tukey's method on the significant effects.
33. Solve problem 32 using multiple regression analysis with number of defectives as the dependent variable. How are the dummy variables defined? Which of the explanatory variables are significant at the .05 level?
34. The general manager at Wong Computer Repair is interested in determining if the time required to repair computer disk drives is dependent on the type of drive or the experience of the repairperson. This information will be helpful in providing cost estimates to Wong's customers. The general manager has collected the following information on repair times (in minutes) from a sampling of the company's historical records.

Experience	Computer Type		
	IBM	Apple	DEC
Less than one year	45	51	62
	53	56	53
	49	49	57
	62	47	58
	49	54	61
One to three years	40	44	45
	38	45	48
	41	39	41
	43	49	52
	37	41	45
More than three years	36	38	40
	40	42	45
	37	35	36
	44	38	33

- a. What type of ANOVA model is appropriate for solving this problem?
- b. Provide an interpretation for possible two-way interactive effects.
- c. Which factors are significant at the .05 level?
- d. How could the data base be reconstructed for solution using a one-way ANOVA model?
- e. Develop a one-way model.
- f. Which factors are statistically significant at the .05 level?
- g. Perform a multiple pairwise comparison analysis using Tukey's method on the significant effects.

35. Solve problem 34 using multiple regression analysis with repair time as the dependent variable. How are the dummy variables defined? Which of the explanatory variables are significant at the .05 level?

36. The U.S. secretary of agriculture is interested in determining if the method of irrigation or type of corn influences corn yield. The secretary has collected the following information on corn yield (bushels/acre) from a random sample of historical data.

Corn Type	Irrigation Method					
	Manual		Semi-Automatic		Automatic	
No. 1 yellow	85	87	92	89	114	99
	91	81	94	97	101	104
No. 2 yellow	78	89	88	84	94	97
	82	84	91	93	90	101
No. 3 yellow	92	87	101	105	123	116
	97	94	111	104	118	131

a. What type of ANOVA model is appropriate for solving this problem?
b. Provide an interpretation for possible two-way interactive effects.
c. Which factors are significant at the .05 level?
d. How could the data base be reconstructed for solution using a one-way ANOVA model?
e. Develop a one-way model.
f. Which factors are statistically significant at the .05 level?
g. Perform a multiple pairwise comparison analysis using Tukey's method on the significant effects.

37. Solve problem 36 using multiple regression analysis with yield as the dependent variable. How are the dummy variables defined? Which of the explanatory variables are significant at the .05 level?

38. The production manager at Honda Motors is interested in determining the potential impact of employee training and gender on production. The following production data (units/day) were collected from a random sample of the company's historical records.

	Level of Training		
Sex	Low	Medium	High
Male	312	454	565
	343	423	578
	389	398	545
	356	413	523
	313	445	515
Female	387	420	497
	404	435	488
	413	456	505
	429	476	489
	408	438	515

a. What type of ANOVA model is appropriate for solving this problem?
b. Provide an interpretation for possible two-way interactive effects.
c. Which factors are significant at the .05 level?
d. How could the data base be reconstructed for solution using a one-way ANOVA model?
e. Develop a one-way model.
f. Which factors are statistically significant at the .05 level?
g. Perform a multiple pairwise comparison analysis using Tukey's method on the significant effects.

39. Solve problem 38 using multiple regression analysis with output as the dependent variable. How are the dummy variables defined? Which of the explanatory variables are significant at the .05 level?

40. The marketing manager for Union Oil Company is interested in determining the potential impact of location and octane level on retail gasoline prices. The following price data ($/gal) were collected from a random sample of retail outlets in the Southern California area.

	Octane Level		
Location	89	91	93
Urban	$0.95	$1.05	$1.21
	0.97	0.99	1.19
	0.94	1.11	1.18
	1.02	1.19	1.27
	0.92	1.14	1.21
Rural	0.99	1.18	1.29
	1.06	1.17	1.33
	1.11	1.21	1.31
	1.08	1.11	1.33
	1.09	1.18	1.39

   a. What type of ANOVA model is appropriate for solving this problem?
   b. Provide an interpretation for possible two-way interactive effects.
   c. Which factors are significant at the .05 level?
   d. How could the data base be reconstructed for solution using a one-way ANOVA model?
   e. Develop a one-way model.
   f. Which factors are statistically significant at the .05 level?
   g. Perform a multiple pairwise comparison analysis using Tukey's method on the significant effects.

41. Solve problem 40 using multiple regression analysis with price as the dependent variable. How are the dummy variables defined? Which of the explanatory variables are significant at the .05 level?

42. Describe the differences between a one-factor randomized block design and a two-factor completely randomized design. Provide several examples.

43. Describe several applications where analysis of variance is more effective than regression techniques.

44. What role does Tukey's test play in analysis of variance?

45. The chief designer at Spectrum Software has obtained the following information on the execution speed (seconds) for a standard software package for three different compilers and three different IBM series computers:

	Compiler		
Computer	1	2	3
283	10.8	11.4	9.7
383	7.6	8.9	7.1
483	6.8	7.1	6.3

a. Identify the control and block variables for this problem.
b. Is there a significance difference in compiler performance at the 0.05 level?
c. Is there a significance difference due to the block variable at the 0.05 level?

46. The vice president of Agricorp Inc. has obtained the following information on the corn yield (bushels per acre) for three different corn types and four fertilizer types:

Corn Type	Fertilizer			
	A	B	C	D
X–12	62.3	68.1	59.4	66.1
Y–31	58.3	61.2	66.3	58.7
Z–44	55.5	60.8	61.2	69.3

a. Identify the control and block variables for this problem.
b. Is there a significance difference in fertilizer performance at the 0.05 level?
c. Is there a significance difference due to the block variable at the 0.05 level?

47. A completely randomized one-factor ANOVA with three groups yielded the following results:

$$SST = 150 \quad SSC = 110 \quad N = 15$$

a. Develop the standard ANOVA table using these data.
b. What conclusions can be drawn at the 0.05 level regarding the significance of the variable effect?

48. A completely randomized one-factor ANOVA with three groups yielded the following results:

$$SST = 1162 \quad SSC = 809 \quad N = 23$$

a. Develop the standard ANOVA table using these data.
b. What conclusions can be drawn at the 0.05 level regarding the significance of the variable effect?

49. A completely randomized two-factor ANOVA with three column groups and four row groups yielded the following results:

$$SST = 3000 \quad SSR = 1100 \quad SSC = 900$$
$$SSE = 250 \quad N = 25$$

a. Develop the standard ANOVA table using these data.
b. What conclusions can be drawn at the 0.05 level regarding the significance of the two factors?
c. Is the interactive effect significant?

50. A completely randomized two-factor ANOVA with three groups yielded the following results:

$$SST = 110 \quad SSR = 55 \quad SSC = 21$$
$$SSE = 5 \quad N = 14$$

a. Develop the standard ANOVA table using these data.

b. What conclusions can be drawn at the 0.05 level regarding the significance of the two factors?
c. Is the interactive effect significant?

51. A CBS analysis of the results from three different employee productivity training programs is presented below. The data reports the post-training production levels, in units per hour, obtained by each employee involved in the program.

Information Entered

Number of Variables:	1
Number of Columns:	3
Alpha Error:	.05

	T_1	T_2	T_3
1 =	72	81	79
2 =	77	83	84
3 =	79	80	87
4 =	68	78	91
5 =	66	77	89
6 =	71	83	93
7 =	74	85	
8 =	72	88	
9 =		83	

Results

Source of Variance	Sum of Squares	Degrees of Freedom	Mean Squared	Computed F Value
Columns:	809.292	2		
Error:	352.708			
Totals:	1,162	22		

Critical F (Col): 3.49

a. Determine the mean square values.
b. Compute the F statistic.
c. Formulate the null and alternative hypotheses.
d. What conclusions can be made regarding the statistical significance of the training program?

52. A CBS analysis of the results of a quality control survey at four different electronic assembly plants is presented below. The data report the percent defectives found in a random sample of 11 outgoing shipments.

Information Entered

Number of Variables: 1
Number of Columns: 4
Alpha Error: .05

	P1	P2	P3	P4
1 =	5	3	2	6
2 =	4	2	5	3
3 =	2	1	6	2
4 =	1	2	7	4
5 =	0	4	4	6
6 =	2	0	5	5
7 =	3	3	6	3
8 =	4	1	4	2
9 =	2	2	7	4
10 =	6	0	4	3
11 =	4	3	7	5

Results

Source of Variance	Sum of Squares	Degrees of Freedom	Mean Squared	Computed F Value
Columns:	63.545	3		
Error:	95.455			
Totals:	159	43		

Critical F (Col): 2.84

a. Determine the mean square values.
b. Compute the F statistic.
c. Formulate the null and alternative hypotheses.
d. Is there a statistical difference in quality, as measured in percent defectives, between the four plants?

53. A CBS analysis of the results from an industrywide survey on leadership ability by sex is reported below. The data report the scores obtained on a leadership test given to three different levels of management (FL = first line supervisor, MM = middle management, UM = upper management). The data have been organized into two groups (GP1 = male, GP2 = female). In general, higher test scores tend to indicate greater leadership ability.

Information Entered

		Number of Variables:			2		
		Number of Rows:			2		
		Number of Columns:			3		
		Alpha Error:			.05		
GP1	FL	MM	UM	GP2	FL	MM	UM
1 =	36	42	55	1 =	37	51	39
2 =	38	40	37	2 =	35	54	44
3 =	44	38	34	3 =	41	50	57
4 =	37	51	44	4 =	44	47	51
5 =	34	45	47	5 =	39	44	49
6 =	29	56	51	6 =	40	37	50
7 =	31	37	39	7 =	44	46	48
8 =	36	32	48	8 =	49	48	53
9 =		41		9 =	51	51	49
10 =		37		10 =		55	57
				11 =		59	

Results

Source of Variance	Sum of Squares	Degrees of Freedom	Mean Squared	Computed F Value
Rows:	601.085	1		
Columns:	666.926	2		
Inter:	5.998	2		
Error:	1,818.487	50		
Totals:	3,080.500	55		

Critical F (Row): 4.04
Critical F (Int): 3.19
Critical F (Col): 3.19

a. Determine the mean square values.
b. Compute the F statistics.
c. Formulate the null and alternative hypotheses.
d. What conclusions can be made regarding the impact of management level on leadership ability?
e. What conclusions can be made regarding the impact of sex on leadership ability?
f. How could the interactive effect between management position and sex be interpreted?

54. A CBS analysis of the results from a survey of absenteeism in the electronics manufacturing industry is presented below. The data report the level of absenteeism, hours per month, by shift (S1, S2, S3) for small (GP1), medium (GP2), and large (GP3) businesses.

Information Entered

Number of Variables:						2				
Number of Rows:						3				
Number of Columns:						3				
Alpha Error:						.05				

GP1	S1	S2	S3	GP2	S1	S2	S3	GP3	S1	S2	S3
1 =	8	2	11	1 =	14	15	12	1 =	14	10	5
2 =	12	14	14	2 =	10	12	16	2 =	18	2	0
3 =	4	6	0	3 =	0	7	0	3 =	21	0	9
4 =	0	8	6	4 =	7		7	4 =	15	14	11
5 =		0		5 =	0		2	5 =	14	18	14
				6 =	14			6 =		22	4

Results

Source of Variance	Sum of Squares	Degrees of Freedom	Mean Squared	Computed F Value
Rows:	170.490	2		
Columns:	56.246	2		
Inter:	220.751	4		
Error:	1,278.150	35		
Totals:	1,725.637	43		

Critical F (Row): 3.275
Critical F (Int): 2.65
Critical F (Col): 3.275

a. Determine the mean square values.
b. Compute the F statistics.
c. Formulate the null and alternative hypotheses.
d. What conclusions can be made regarding the impact of shift on abseentism?
e. What conclusions can be made regarding the impact of business size on absenteeism?
f. How could the interactive effect between shift and business size be interpreted?

55. A random survey of basic homeowner insurance rates for 13 communities in the Southern California area for selected insurance companies is reported below:

		Annual Premium		
Area	State Farm	Allstate	Farmers	20th Century
1	416	494	456	329
2	397	553	420	342
3	668	553	696	450
4	397	494	421	342
5	432	404	426	466
6	419	367	426	324
7	398	367	426	324
8	406	390	412	306
9	364	390	377	280
10	384	418	501	325
11	484	418	501	325
12	472	425	484	368
13	378	306	416	279

a. Which ANOVA model is most appropriate for solving this problem?
b. Formulate a null and alternative hypothesis.
c. Is there a statistical difference of insurance rates at the 0.05 level?
d. If the answer to part c is yes, which company appears to have the lowest rates?

56. Solve problem 46 using multiple regression analysis. What role do dummy variables play in formulating the model?

57. Presented below are earnings per share data for 1989 and 1990 for the top three Japanese trading, banking and automotive companies:

		Industry	
	Trading	Banking	Automotive
1989	$0.36	$0.46	$0.87
	0.17	0.35	0.32
	0.17	0.30	0.58
1990	0.31	0.35	0.85
	0.19	0.35	0.38
	0.23	0.26	0.63

a. Identify the problem variables.
b. Which ANOVA model is most appropriate for solving this problem?
c. Formulate the null and alternative hypotheses.
d. Which of the effects are statistically significant at the 0.05 level?
e. Provide an overall interpretation of the results.

58. The data presented below, obtained from a random survey of the top 1,000 U.S. firms, represent the changes in stock value over a two-year period. The data are reported as a function of company size and the characteristics of the CEO incentive plan.

	Incentive Plan		
	Poor	Average	Good
Small	$0.45	$7.25	$35.5
	0.11	8.13	27.34
	0.31	5.25	29.31
Medium	0.25	6.23	47.65
	0.41	4.37	51.25
	1.21	3.25	29.45
Large	0.36	2.34	25.67
	0.89	3.45	17.56
	1.16	4.44	22.45

    a. Identify the problem variables.
    b. Which ANOVA model is most appropriate for solving this problem?
    c. Formulate the null and alternative hypotheses.
    d. Which of the main effects are statistically significant at the 0.05 level?
    e. Provide an overall interpretation of the results.

59. The data presented below, obtained from a random survey of the top 1,000 U.S. firms, represent corporate performance based on a 5-point scale (1 = poor). The data are reported as a function of planning level (low or high) and the planning horizon (short, medium, and long).

	Planning Level	
	Low	High
Short	3.57	3.81
	3.36	3.73
Medium	3.08	3.59
	3.32	3.47
Long	2.76	3.13
	3.02	3.14

    a. Identify the problem variables.
    b. Which ANOVA model is most appropriate for solving this problem?
    c. Formulate the null and alternative hypotheses.
    d. Which of the effects are statistically significant at the 0.05 level?
    e. Provide an overall interpretation of the results.
    f. Perform a multiple pairwise comparison for each group.

60. The data presented below, obtained from a random survey of the top 100 U.S. venture capital firms, represent the average expected rate of return on funded projects. The data are reported as a function of product status and the status of management (I = individual founder, II = partnership, III = partial management team, IV = professional management team) at the time of funding.

| | Management Status | | | |
Product Status	I	II	III	IV
Concept	131	99	80	62%
	125	103	75	58
Prototype	82	77	61	38
	78	71	55	44
Fully developed	45	36	28	25
	51	44	25	31

a. Identify the problem variables.
b. Which ANOVA model is most appropriate for solving this problem?
c. Formulate the null and alternative hypotheses.
d. Which of the main effects are statistically significant at the 0.05 level?
e. Provide an overall interpretation of the results.
f. Perform a multiple pairwise comparison for each group.

61. Perform a multiple pairwise comparison using Tukey's method on the Johnson Carbide problem given in Section 13.5.
62. Perform a multiple pairwise comparison using Tukey's method on the Craig Investments problem given in Section 13.5.
63. Perform a multiple pairwise comparison using Tukey's method on the Allnight Drug Store problem given in Section 13.6.
64. Perform a multiple pairwise comparison using Tukey's method on the government contractor's problem given in Section 13.6.
65. Consider a random block design where the total sum of squares is 490, the mean square for the blocks is 48, and the error mean square is 28. Determine the F value if the design consists of four blocks and three factor groups. Are any of the effects significant at the 0.05 level?
66. Consider a two-factor ANOVA design where the total sum of squares is 139.5, the mean square for Factor 1 is 13.25, the mean square for Factor 2 is 18.5, and the error mean squares is 8.5. Determine the F value if the design consists of four groups for Factor 1 and three groups for Factor 2. Are any of the effects significant at the 0.05 level?
67. Given the following data:

$$SST = 2250 \quad SSA = 1125 \quad SSB = 450 \quad SSAB = 37$$
$$n = 30 \quad a = 5 \quad b = 3$$

a. Construct an appropriate ANOVA table.
b. Identify which effects are statistically significant at the 0.01 level.

68. Given the following data:

$$SST = 483 \quad SSB = 33 \quad n = 15 \quad a = 3$$

a. Construct an appropriate ANOVA table.
b. Identify which effects are statistically significant at the 0.01 level.

69. Perform a multiple pairwise comparison using Tukey's method for problem 57.
70. Perform a multiple pairwise comparison using Tukey's method for problem 58.

13.12 CASES

13.12.1 Mehta Corporation

The home products division of Mehta Corporation manufactures and markets two kinds of baseboard molding: a plastic molding and a wood molding. At a recent meeting of the advanced planning group, the group vice president, Ms. Alice Cook, expressed an interest in how the two products were selling throughout the company's western region. Mr. Herb Smith, the vice president for marketing, indicated, "Both products seem to be moving well." Herb's response seemed to leave everyone at the meeting a little cold because of the lack of specificity. At that point, Ms. Cook turned to Mr. Smith and said, "I would like you to work up some numbers on product sales by market area for the next planning meeting." Armed with this assignment, Herb set off to find the company's statistics department to obtain the necessary information to complete his task.

At a subsequent meeting with the director of the statistics department, Herb discovered that the home products division's primary markets are Los Angeles, San Francisco, and San Diego. During the meeting, the director also provided Herb with sales data for the last quarter for the two products by marketing area. This information is reported in Table 1. As the meeting was breaking up, Herb asked the director if

TABLE 1 Sales Data

Plastic Molding (units)	Wooden Molding (units)	Location*
9,050	8,000	1
7,200	7,000	2
9,600	6,500	3
10,000	7,200	1
8,200	8,500	2
6,300	9,600	3
8,600	7,800	1
7,700	7,600	2
9,000	8,050	3
8,700	8,500	1
8,500	8,800	2
7,000	8,600	3
10,350	6,300	1
9,320	8,010	2
8,700	7,490	3
9,337	8,630	1
8,645	6,900	2
7,397	7,501	3
7,510	8,230	1
9,900	6,205	2
8,700	5,400	3

*1 = Los Angeles; 2 = San Francisco; 3 = San Diego.

he had any suggestions on how the data should be analyzed. The director responded with a sly grin, "I think a 95% level of confidence should do nicely."

Suggested Questions

1. What ANOVA model best describes this application, assuming there are no interactive effects?
2. Are sales for plastic and wood molding the same for all three regions?
3. How would the model change if interactive effects were included?
4. Which marketing region produces significantly different sales?
5. Comment on the possibility of any interactive effects.

13.12.2 Rocket Chemical Company

Mr. Ed Bates, senior project manager for mechanical systems, called a meeting of his design team. He started the meeting by observing: "Some of our test results show that the propellants used by the launch vehicle are eating away at the propellant tank weld

TABLE 1 Results of Propellant Test Program

Test Medium	Test Duration (months)	Yield Strength (KSI)	Percent Elongation (1/2")	Test Medium	Test Duration (months)	Yield Strength (KSI)	Percent Elongation (1/2")
MMH:				NTO:			
MMH	1	86.88	21	Control	0	87.4	22
		89.5	23			92.3	22
		87.8	17			98.8	20
		84.0	19			90.0	23
		86.1	20			86.3	19
		88.8	18			90.8	19.5
MMH	3	87.2	16	NTO	1	91.2	20
		88.4	27			87.9	23
		91.6	22			96.3	18
		83.5	20			89.5	20
		83.5	20			88.9	21
		94.9	22			90.8	18
MMH	6	89.4	20	NTO	3	88.7	24
		89.5	20			90.6	19.5
		98.6	18			97.0	18
		84.7	22			84.3	22
		87.5	21			94.5	17.5
		96.8	18			93.1	18
				NTO	6	86.3	20
						89.3	19
						98.6	18
						85.7	22
						94.1	16

joints." He went on to add: "Eroding of the joints could lead to premature engine shutdown or worse."

"OK," said Bill Morton, assistant project manager. "Maybe we should conduct a series of tests to measure the effects of the two propellants, monomethyline hydrazine (MMH) and nitrogen tetroxide (NTO), on the tank weld joints over time.

"My proposed test plan," said Jim Carlson, associate engineer, "is to test our titanium weld joints. For each of the two propellants, we need to test the effect for a period of one month, three months, and six months. During those periods we should measure (1) yield strength and (2) percentage of elongation. All these results should be compared against two control groups, one for MMH and the other for NTO."

The meeting concluded with Bates instructing Mr. Carlson to conduct a six-month experimental test along the lines outlined in the meeting. The results from the test program are given in Table 1.

Suggested Questions

1. What is the effect of either MMH or NTO on the titanium joints?
2. What is the effect of either propellant in various time intervals?
3. Do you think the weld joints will hold for use in future launches?
4. Comment on the possibility of any interactive effects.

Chapter 14

Nonparametric Statistics

Round numbers are always false.
Samuel Johnson

CHAPTER OUTLINE

- **14.1** Introduction
- **14.2** Example Management Problem: Perpetual Savings
- **14.3** How to Recognize a Nonparametric Problem
- **14.4** Basic Nonparametric Models
- **14.5** Computer Analysis
- **14.6** Practical Applications
- **14.7** Case Study: Fujti Motors
- **14.8** Summary
- **14.9** Glossary
- **14.10** Bibliography
- **14.11** Problems
- **14.12** Cases

CHAPTER OBJECTIVES

The primary objectives of this chapter are to develop an understanding of

1. The advantages and disadvantages of nonparametric statistics.
2. How to recognize nonparametric problems.
3. How to select a nonparametric model for a particular application.
4. How to solve nonparametric problems using computer analysis.

Management information systems (MISs) have become an essential element for effective business operations in today's increasingly competitive environment. Historically, most MIS applications were developed using a life-cycle approach, based on a set of requirements and specifications. More recently, a second approach, based on developing a prototype system at the very beginning, has emerged. This method is more advantageous in that the system designer and system user can work together in completing the project.

> **HISTORICAL NOTE**
>
> The father of modern nonparametric statistical methods was Dr. Frank Wilcoxon (1892–1965). In 1945, he delivered his now-famous paper introducing the rank-sum test and the signed ranked test. Both of these nonparametric techniques have been named in his honor. Dr. Wilcoxon's basic idea of replacing actual sample data, although conceptually straightforward, represents the cornerstone for the development of nonparametric statistics. Subsequent to the pioneering work by Dr. Wilcoxon, nonparametric statistics have found widespread application in both business and economics.

A recent study was undertaken to test for differences between the two design approaches.* In this study, two sets of students (MBAers) and (MISers) were paired into nine teams for the purpose of designing an MIS for a new chemical plant. The nine teams were randomly assigned one of the two design approaches. The MBAers took the role of the users and the MISers acted as system designers. Several weeks into the design process, each team was surveyed regarding their response to the design experience. The survey consisted of a series of statements about system performance, which could be answered using a 1-to-5 scale. The generated data were evaluated using nonparametric analysis. The results showed that the users (i.e., MBAers) were generally more pleased with the prototype design approach.

14.1 INTRODUCTION

The discussion so far has focused on the application of **parametric tests** as a basis for inferring the characteristics of a population (e.g., the t-test). In general, these test procedures use sample data for determining the presence or absence of specific relationships. Typically, the use of parametric methods is based on the following set of assumptions:

1. Samples are drawn from a "normal" population.
2. Data are measured on interval or ratio scales.
3. Variances are homogeneous.
4. Data are independent.

In many business situations, however, uncertainty exists about the shape of the population distribution. Furthermore, the data may be ranked or or-

---

* M. Alavi, "An Assessment of the Proto-typing Approach to Information Systems Development," *Communications of the ACM* 27 (1984), pp. 556–63.

dered into categories and may not fulfill the provisions of the second assumption. To avoid the risk of wrong answers from the application of parametric tests, a number of "distribution-free" methods called **nonparametric tests** have been developed. They are logically similar to the parametric tests but are not based on any of the preceding assumptions. The chi-square model, an example of a nonparametric test, has already been introduced in Chapter 12. Figure 14.1 lists the major advantages and disadvantages of nonparametric methods compared with parametric test methods.

14.2 EXAMPLE MANAGEMENT PROBLEM: PERPETUAL SAVINGS

Perpetual Savings & Loan (S&L) is a relatively new lending institution specializing in first trust deeds for midrange housing projects. Historically, S&Ls have been the primary market for these products. However, recent federal legislation has opened the door for full-service banks to enter the real estate loan market. The vice president for mortgage loans, Mr. William Peterson, is interested in comparing interest rates on first trust deeds between the local S&Ls and banks. His concern is that banks will gain market share by offering additional services normally not available through the S&Ls (e.g., free checking accounts). Mr. Peterson believes that the best way to combat this new threat is by maintaining significantly lower lending rates. To better understand the dynamics of the present situation, he has compiled the following information on a sample of current S&L and bank mortgage rates within the Southern California area.

Savings & Loans	Banks
9.50	10.50
9.25	11.00
9.00	9.50
10.25	10.00
10.00	11.00
9.75	10.50
9.50	10.25
10.00	9.75
10.50	
9.75	
10.25	

Mr. Peterson plans to use nonparametric methods for analyzing these data because of the relatively small sample size and the uncertainty regarding the shape of the population distributions.

632 Chapter 14 Nonparametric Statistics

FIGURE 14.1 Advantages and Disadvantages of Nonparametric Methods

Advantages	Disadvantages
• Applicable to problems where insufficient information is available about the population from which the sample is drawn. • Applicable to problems where measurements are based on nominal or ordinal data.	• Less powerful than parametric tests under similar conditions (i.e., sample size and level of confidence). • Less efficient than parametric tests because of the tendency to ignore some of the information contained in the sample.

Key idea 1

14.3 HOW TO RECOGNIZE A NONPARAMETRIC PROBLEM

Key idea 2

The following list highlights the primary characteristics of nonparametric problems:

- The data are measured on or transformed to a nominal or ordinal scale.
- The data base is relatively small (e.g., less than 30) and the shape of the population distribution is unknown.

> **INTERNATIONAL VIGNETTE**
>
> "I know the direction this company should take, and I feel we should start now! My intuition will lead the way!" said Fred Welschen, commercial director of Inalfa BV, a manufacturer of fabricated metal parts in Venray, a town in southern Holland near Eindhoven. He was considering a proposal from Xerox Corporation that would require considerable Inalfa investment. A yes would bring Inalfa an increasing share of Xerox's business in exchange for two things: lower prices and much higher quality. A no would mean that Inalfa would no longer be a supplier to Xerox. Fred was strongly in favor of accepting the Xerox proposal in order to boost historically flat sales. However, many questions remained. Even if he could persuade the managing director, how would the new demands be implemented?
>
> Of Xerox's requirements, the introduction of statistical process control would be the most difficult for Inalfa to implement. A series of statistical techniques would be applied to the production process to control quality and reduce costs. If Inalfa agreed to cooperate with Xerox, the initial impact would be on the company's quality control department and its relationship with production. Mr. Welschen had made contact with several other vendors that had received Xerox's proposal. The response from the vendors was consistent: "We are going to sign. We have an opportunity we shouldn't pass up; we can help Xerox, they can teach us, and everyone will be better off."

14.4 BASIC NONPARAMETRIC MODELS

Key idea 3   A large number of nonparametric models have been developed for use in business. This section presents some of the more frequently used models.

RANK-SUM TESTS

Several **rank-sum tests** have been developed to test for differences between two populations. In general, they are used to determine whether or not two independently drawn samples can be considered to have come from the same population (i.e., $H_0: \mu_1 = \mu_2$). As such, these tests are similar to the hypothesis test statistics presented in Chapter 8. However, these tests are nonparametric. Accordingly, no assumptions are needed about the populations. The two most popular rank-sum tests are the **Wilcoxon** and the **Mann-Whitney**. Both of these tests use ordinal data. The Perpetual Savings problem will be used to illustrate the application of these two models.

Figure 14.2 presents the combined rank ordering of the two samples in the Perpetual Savings problem. Notice that some samples receive the same

FIGURE 14.2 Data Ranking for Perpetual Savings

Savings & Loans			Banks		
Sample	Rate	Rank	Sample	Rate	Rank
1	9.50	4	1	10.50	16
2	9.25	2	2	11.00	18.5
3	9.00	1	3	9.50	4
4	10.25	13	4	10.00	10
5	10.00	10	5	11.00	18.5
6	9.75	7	6	10.50	16
7	9.50	4	7	10.25	13
8	10.00	10	8	9.75	7
9	10.50	16		Sum of ranks	103
10	9.75	7			
11	10.25	13			
	Sum of ranks	87			

ranking (e.g., samples 1 and 7 for the first population and sample 3 for the second population are all ranked fourth). This is due to the fact that each of these samples has the same interest rate value (i.e., 9.5%). The normal practice when a tie occurs is to assign each item the average of the ranks associated with the tied values. A cursory examination of the results shows that there is a 15-point difference between the rank sum of the two samples. Is this difference large enough to reject the null hypothesis of no difference?

The basic statistical parameters (i.e., mean, standard deviation, and test statistic) for the Wilcoxon and Mann-Whitney tests are shown in Figure 14.3, along with numerical estimates for Perpetual Savings. It is assumed that the sampling distributions for the two tests can be approximated by a normal curve with mean μ and standard deviation σ. For this application, the vice president has selected a 5% level of significance (i.e., $\alpha = .05$).

The sampling distribution and critical limits for the Wilcoxon test are illustrated in Figure 14.4 for the Perpetual Savings case. (The Mann-Whitney test is very similar.) Again, the hypothesis test procedure is the same as that outlined in Chapter 8. Namely, the null hypothesis will be rejected only if the test statistic falls outside the critical limits. The critical limits are determined as follows:

Upper limit $= \mu + Z \times \sigma = 110 + 1.96(12.11) = 133.74$

Lower limit $= \mu - Z \times \sigma = 110 - 1.96(12.11) = 86.26$

Comparing the computed rank-sum test statistic (87) with the computed limits reveals that the test statistic falls within the limits. Therefore, the vice president can conclude that there is no significant difference between the loan rates offered by the S&Ls and the banks. The Mann-Whitney test would have produced exactly the same conclusion.

FIGURE 14.3 Wilcoxon and Mann-Whitney Analysis of Perpetual Savings

Parameter	Wilcoxon	Mann-Whitney
Mean (μ)	$0.5 \times N_1(N_1 + N_2 + 1)$ (110)	$0.5 \times N_1 N_2$ (44)
Standard deviation (σ)	$\sqrt{N_1 N_2 (N_1 + N_2 + 1)/12}$ (12.11)	$\sqrt{N_1 N_2 (N_1 + N_2 + 1)/12}$ (12.11)
Test statistic	Sum of ranks (87)	$N_1 N_2 + 0.5[N_1(N_1 + 1)]$ − Sum of ranks (57)

THE WILCOXON SIGNED-RANK TEST

The **Wilcoxon signed-rank test** is designed to test a hypothesis of no change (i.e., $\mu_1 = \mu_2$) when the measures are paired or matched. Thus, the test is ideal for a pre/post, or "before and after," design to determine if the measured changes are significant. Suppose, for example, the vice president of human resources at Allied-Signal needs to measure the impact of sending the company's managers to a human relations course. The vice president wishes to test the hypothesis that the course makes an impact on the manager and consequently on the morale of the employees. Each participating manager is interviewed by a trained psychologist before and after the course. The psychologist evaluated each manager on his or her ability to provide effective leadership on a 1–100 scale. The resultant pre- and post-test data along with the ranking are given in Figure 14.5. Notice that the second and fifth items

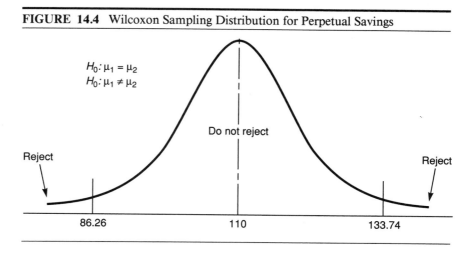

FIGURE 14.4 Wilcoxon Sampling Distribution for Perpetual Savings

FIGURE 14.5 Manager Evaluation Scores for Allied-Signal Corporation

Manager	Pretest Score	Posttest Score	Difference	Signed Rank
1	29	45	16	9
2	35	37	2	1.5
3	28	28	0	—
4	38	49	11	7
5	36	38	2	1.5
6	42	38	−4	−4
7	38	41	3	3
8	37	45	8	6
9	52	57	5	5
10	48	60	12	8
			Sum of signed ranks	37

are given the same signed rank. This is because the test score differences for these two managers are the same. As indicated above, when ties occur, the normal practice is to allocate the average rank to the tied values. The final step is to assign the sign of the original difference to each rank. Thus, some ranked items have a negative value (e.g., item 6).

The null hypothesis for this problem is that the test scores are not significantly affected by the training program (i.e., $H_0: \mu_1 - \mu_2 = 0$). The procedure for the Wilcoxon signed-rank test is very similar to the rank-sum tests outlined previously. Figure 14.6 shows the sampling distribution for the

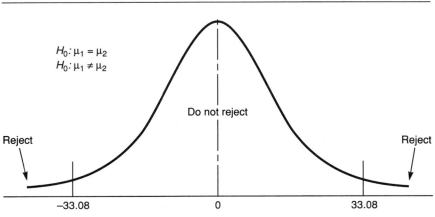

FIGURE 14.6 Sampling Distribution for Allied-Signal Problem

Allied-Signal problem. The expected mean and standard deviation for the population are as follows:

$$\mu = 0, \qquad \sigma = \sqrt{N(N+1)(2N+1)/6} = \sqrt{(10)(11)(21)/6} = 16.88$$

The critical limits for $\alpha = .05$ are determined as follows:

Upper limit = $0 + 1.96(16.88) = 33.08$

Lower limit = $0 - 1.96(16.88) = -33.08$

Comparing the computed Wilcoxon test statistic, 37, with these limits reveals that the vice president can reject the null hypothesis of no difference and conclude that the observed differences are real. That is, the training program significantly changes the leadership characteristics of the participating managers.

SIGN TEST

In some applications, quantitative measures of differences are not available, e.g., customer preference. Therefore, it is not possible to rank-order the absolute differences. In these cases the standard Wilcoxon signed-rank test is not applicable. Instead, a less effective nonparametric method called the **sign test** can be used. Perhaps the most common use of the sign test involves the analysis of customer preferences. Frequently, firms are interested in determining customer preferences for certain product brands (e.g., Coke versus Pepsi). Typically, a survey is undertaken in which customers are asked to express their preference for brand A or brand B. The preference survey data are usually recorded in the form of a + (prefer product A), a − (prefer product B), or a 0 (no preference). In using the sign test, one merely counts the numbers of plus, minus, and zero observations.

The null hypothesis for these applications is that no differences in preference exist between the two brands. This suggests that the numbers of observed pluses and minuses should be the same (i.e., $H_0: p = .5$). The alternative hypothesis is that there is a significant difference in the numbers of pluses and minuses (i.e., $H_1: p \neq .5$). This suggests that a difference in preference does exist. The null hypothesis for sign test problems is similar to the hypothesis associated with flipping a coin. Recall that the general expectation for a fair coin is that equal numbers of heads and tails will occur over a large number of flips.

Consider that the Los Angeles chapter of the Citizens for Accuracy in Advertising is concerned about conflicting claims made by the two leading soft drink companies. In several recent commercials, both Coke and Pepsi have claimed that their brand is preferred by more buyers than the competing brand. The committee has decided to conduct an independent survey to determine the accuracy of these claims. A blind taste test was given to 14 customers selected at random at a local supermarket. The results of this survey are presented in Figure 14.7. For the purposes of this test, a + indicates that the customer preferred Pepsi over Coke, and a − indicates that the customer selected Coke over Pepsi. A 0 recording indicates that the customer was unable to rank the two brands.

The analysis procedure for the sign test is exactly the same as that for the test for proportions introduced in Chapter 8. In this case, the estimated proportion is the number of observed pluses divided by the total of pluses and minuses (i.e., $\frac{6}{12}$, or .5). Notice that customers 7 and 13 are eliminated since they were unable to rank the two brands. Based on the null hypothesis of $p = .5$, $\sigma = 0.14$, and $\alpha = .05$, the upper and lower test limits can be determined as follows:

$$\text{Upper limit} = p + Z\sigma = .5 + 1.96(0.14) = 0.77$$

$$\text{Lower limit} = p - Z\sigma = .5 - 1.96(0.14) = 0.23$$

Figure 14.8 shows the sampling distribution for the number of positive ratings. Comparing the observed proportion of .5 with these limits reveals that the committee can accept the null hypothesis of no difference in brand preference. That is, it appears that the preferences for Pepsi and for Coke in Southern California are about the same.

The major weakness in the sign test is that it does not take into account the magnitude of the measures. Consequently, it is not as effective as the Wilcoxon signed-rank test. However, the test is so easy to use that it should be considered whenever conditions warrant.

FIGURE 14.7 Soft Drink Preference Survey Data

Customer	Brand Preference	Sign
1	Coke	−
2	Coke	−
3	Pepsi	+
4	Pepsi	+
5	Pepsi	+
6	Coke	−
7	Neither	0
8	Coke	−
9	Coke	−
10	Pepsi	+
11	Coke	−
12	Pepsi	+
13	Neither	0
14	Pepsi	+

FIGURE 14.8 Sampling Distribution for the Soft Drink Preference Survey

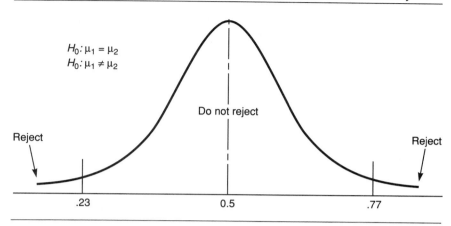

An important variation to the basic sign test is the so-called runs test. This test procedure is used extensively in statistical quality control applications. Simply stated, the runs test is used to determine if an observed sequence containing two outcomes is occurring randomly or not. Recall that one of the assumptions associated with linear regression is that the residuals are independent. This means that the signs of these errors should be random. One way of measuring if a sequence of observations or residuals is occurring randomly is to measure the number of runs occurring in the sequence. A run, in these cases, is defined as the number of identical values occurring in a row. For example, suppose a coin is flipped twelve times, which results in the following observed sequence: T T T T T T H H H H H H.

This sequence contains only two runs, one consisting of six tails followed by one consisting of six heads. Clearly, this sequence was not generated in a random manner. Similarly, the sequence T H T H T H T H T H T H, which contains 12 runs, a large number, most likely did not occur at random. While these limit examples are fairly easy to assess, the process for detecting more complex sequences requires a more analytical approach.

While there exists a number of runs tests, one technique that is used frequently in quality control applications is to count the largest number of runs occurring in the sequence. This measurement, along with the number of values above and below the center line, is used in conjunction with the statistical table reported in Appendix A (Table A.9) to determine whether the observed sequence occurred in a random manner. The table provides the critical number of runs that need to be observed to reject the null hypothesis for various levels of significance.

Suppose the Q.C. manager at Allied Electronics is concerned about the quality level on the soldering line. A recent survey of 20 random samples, with each sample containing four observations, yielded the control chart (i.e., p-chart) shown in Figure 14.9. It shows the average proportion of defective solders for each of the 20 samples. This chart can be used to measure the number of runs based on the average defective level. An inspection of this chart reveals that the largest number of runs is 6, with samples 11,12,13,14,15,16 below the mean. Furthermore, 10 samples were above the mean, and 10 were below the mean. The manager's hypotheses for this problem are as follows:

H_0: The soldering process is producing defectives in a random manner

H_1: The soldering process is not producing defectives in a random manner

The acceptance of the alternative hypothesis would suggest that one or more nonrandom factors are impacting the soldering process. Consulting Appendix Table A.9 based on 10 observations above the average and 10 below the average reveals a critical value of seven runs for a 0.05 level of significance. Since the critical value is greater than the largest run observed from the survey, i.e., six, there is insufficient evidence to reject the null hypothesis.

FIGURE 14.9 Control Chart for Allied Electronics Showing the Proportion of Defective Solders for 20 Random Samples.

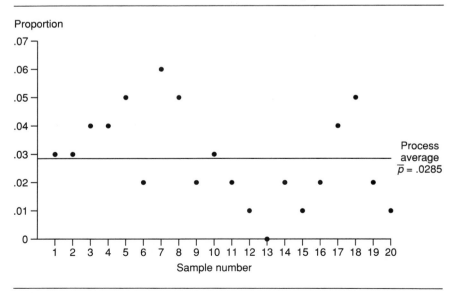

Therefore, the manager can conclude that the soldering process is currently producing defectives in a random manner.*

SPEARMAN RANK CORRELATION

Decision makers often need to find a relationship between data sets that are quantified by some ranking system. Although there are many methods, the Spearman test (named in honor of C. E. Spearman) seems to be used most frequently. The **Spearman rank correlation** coefficient is similar to the Pearson correlation coefficient introduced in Chapter 9. Both coefficients indicate the extent of the relationship between two variables. Like the correlation coefficient, the Spearman statistic can take on values between -1 and $+1$. The Spearman measure, however, is based on the relative ranking of the two variables and not on their actual values. Therefore, one basic difference between these two measures is that the Spearman coefficient will equal positive one (negative one) when the dependent variable increases (decreases) as the independent variable increases (decreases). Thus, while the Pearson coefficient measures the degree of linear association between the two variables, the Spearman simply measures whether the two variables are positively or negatively related. Specific applications for the Spearman include:

- Determining the level of relationship between sales ranking and years of seniority of the sales forces.
- Evaluating if a relationship exists between employee age and absenteeism rate.
- Analyzing if the level of worker training has an impact on productivity.

The chief operating officer (COO) at Rector Transportation is in the process of evaluating the performance of the company's 10 regional marketing managers. The data reported in Figure 14.10 show the leadership rankings (based on a series of psychological tests) and sales rankings (based on sales revenues for the past year). For example, manager 8 has the lowest leadership ranking and the second lowest sales ranking. The COO is interested in knowing the extent of the relationship between leadership and sales rankings. These findings would be helpful in the selection of new managers.

The procedure for computing the Spearman rank correlation coefficient (r_s) is very straightforward. The primary step is to compute the squared differences between the two rankings. This is shown in Figure 14.10. The

---

*For processes involving large sample sizes, the runs distribution can be approximated by the normal curve with:

$$\mu_r = 2n_1 n_2/(n_1 + n_2) + 1 \text{ and } \sigma_r = \sqrt{\frac{2n_1 n_2 (2n_1 n_2 - n_1 - n_2)}{(n_1 + n_2)^2 (n_1 + n_2 - 1)}}.$$

FIGURE 14.10 Ranking Data for Rector Transportation Company

Manager	Leadership Ranking	Sales Ranking	Difference	Squared Difference
1	2	3	−1	1
2	4	7	−3	9
3	7	5	2	4
4	9	9	0	0
5	6	4	2	4
6	3	1	2	4
7	8	10	−2	4
8	1	2	1	1
9	5	8	−3	9
10	3	6	−3	9
			Total	45

Spearman model is:

$$r_s = 1 - \frac{6\Sigma d_i^2}{n(n^2 - 1)}$$

The rank correlation coefficient using these data is computed as follows:

$$r_s = 1 - \frac{6 \times 45}{10 \times 99} = 1 - \frac{270}{990} = .7273$$

This result indicates a large positive relationship between leadership rank and sales rank. However, is this estimate statistically significant? To answer this question requires the formulation of the usual hypothesis-testing procedure and decision rule. The null hypothesis is that no relationship exists between the two ranks (i.e., H_0: $p_s = 0$). The alternative hypothesis is that a positive relationship exists.

The sampling distribution for r_s can be approximated by the normal curve with the following mean and standard deviation:

$$\mu_r = 0 \qquad \sigma = \sqrt{\frac{1}{n-1}}$$

The corresponding decision limits when $\alpha = .05$ are:

Upper limit = $\mu + Z\sigma = 0 + 1.96 \times 0.33 = 0.65$
Lower limit = $\mu - Z\sigma = 0 - 1.96 \times 0.33 = -0.65$

Since the computed r_s falls outside these limits, the COO can reject the null hypothesis of no relationship and can conclude that a positive relationship exists between leadership rank and sales rank. That is, there is a relationship between the ranks of the two variables.

In summary, the major advantages of the Spearman test are as follows:

- Ranking reduces extreme measures.
- It is easy to understand and compute.
- Naturally ranked data approximate the Pearson correlation coefficient.

THE KRUSKAL-WALLIS TEST

The **Kruskal-Wallis (K-W) test** expands the range of the classic Wilcoxon rank-sum test to more than two statistical groups. This method is similar in approach to the one-way analysis of variance introduced in Chapter 13. The basic difference is that rankings are used instead of the actual numerical values in computing the test statistic. The sampling distribution for the Kruskal-Wallis test statistic (named after W. H. Kruskal and W. A. Wallis) can be approximated by a chi-square distribution with $N-1$ degrees of freedom.

Suppose that the manager at Briggs Department Store is interested in determining if there is a relationship between monthly sales and the method of payment used by the store's customers. Typically, customers pay for merchandise by one of three methods: cash, store credit, or bank credit. The manager has collected sales data by method of payment for the past six months. This information is reported in Figure 14.11. The manager has decided to use the K-W test because of the relatively small sample and the general uncertainty regarding the shape of the population distribution. The data show actual sales in $000 and corresponding rank. For example, cash sales in March amounted to $44,500, which was the fifth lowest total.

The null hypothesis for the K-W test is that there are no differences among the defined groups (i.e., $H_0: \mu_1 = \mu_2 = \mu_3$). For the Briggs sales data, this translates into a null hypothesis that monthly sales are not influenced by the method of payment. Computation of the K-W test statistic is based on a rather complicated formula using the rank totals. A computer analysis of this problem is presented in the next section.

FIGURE 14.11 Data for Briggs Department Store

Month	Cash Sales	Rank	Store Card Sales	Rank	Bank Card Sales	Rank
January	58.5	9.5	61.5	17	64.5	18
February	52.0	3.5	57.5	15	60.0	16
March	44.5	5	42.0	3.5	48.0	7.5
April	48.0	7.5	38.0	1	41.5	2
May	50.5	9.5	51.0	11	45.5	6
June	56.0	14	55.0	13	52.5	12
	Total	49.0	Total	60.5	Total	61.5

14.5 COMPUTER ANALYSIS

key idea 4

This section illustrates the use of computer analysis in solving several of the basic nonparametric models introduced in this chapter. Computer analysis eliminates the potential for numerical errors and vastly reduces the time required to solve most nonparametric problems.

Wilcoxon Rank-Sum (Perpetual Savings)

Figure 14.12 presents a CBS-generated Wilcoxon rank-sum analysis for the Perpetual Savings problem. These results show that there is insufficient evidence to reject the null hypothesis of no difference between S&L and bank lending rates. The sums for group 1 (S&Ls) and group 2 (banks) fall within the critical limits at a 5% level of significance. The standard error term is simply the sample standard deviation multiplied by the Z statistic at the .05 level (exactly the same as in Chapter 8).

Mann-Whitney Rank-Sum (Perpetual Savings)

Figure 14.13 presents a CBS-generated Mann-Whitney analysis of the Perpetual Savings problem. These results support the conclusions from Figure 14.12, namely, there is insufficient evidence to reject the null hypothesis. The computed Mann-Whitney test statistic falls within the critical limits at the 5% level.

FIGURE 14.12 CBS Wilcoxon Analysis of Perpetual Savings Problem

Test procedure:	Wilcoxon rank-sum test
Alpha error:	.05
Population mean:	110
Standard deviation:	12.1106
Sum of group 1:	87.0000
Sum of group 2:	103.0000
Standard error:	23.7368
Critical upper limit:	133.7368
Critical lower limit:	86.2632

Conclusion: Do not reject null hypothesis

FIGURE 14.13 CBS Mann-Whitney Analysis of Perpetual Savings Problem

Test procedure:	Mann-Whitney test
Alpha error:	.05
Population mean:	44
Standard deviation:	12.1106
Test statistic:	67.0000
Standard error:	23.7368
Critical upper limit:	67.7368
Critical lower limit:	20.2632

Conclusion: Do not reject null hypothesis

Wilcoxon Signed-Rank Test (Allied-Signal Corporation)

Figure 14.14 presents a CBS analysis of the Allied-Signal problem using the Wilcoxon signed-rank test procedure. These results indicate that the null hypothesis can be rejected. That is, the computed Wilcoxon test statistic falls outside the critical limits at the 5% level. Therefore, the vice president can conclude that the training program does improve leadership.

Rank Correlation (Rector Transportation)

Figure 14.15 presents a CBS analysis of the Rector Transportation problem using rank correlation. The computed correlation coefficient falls outside the critical limits. Therefore, the COO can reject the null hypothesis of no rank relationship and conclude that leadership and sales are related. This information could be used to help recruit future regional sales managers.

FIGURE 14.14 CBS Wilcoxon Signed-Rank Analysis of Allied-Signal Problem

Test procedure:	Wilcoxon signed-rank test
Alpha error:	.05
Population mean:	0
Standard deviation:	16.8819
Test statistic:	37.0000
Critical upper limit:	33.0886
Critical lower limit:	−33.0886

Conclusion: Reject null hypothesis

FIGURE 14.15 CBS Rank Correlation Analysis of Rector Transportation Problem

Test procedure:	Spearman rank correlation
Alpha error:	.05
Population mean:	0
Standard deviation:	0.3333
Spearman coefficient:	.7273
Critical upper limit:	0.6500
Critical lower limit:	−0.6500

Conclusion: Reject null hypothesis

Kruskal-Wallis Test (Briggs Department Store)

Figure 14.16 presents a CBS analysis of the Briggs Department Store problem using the Kruskal-Wallis test procedure. An inspection of the computer printout reveals that there is insufficient evidence to reject the null hypothesis of no difference in the amount of sales generated by the different payment methods. The critical chi-square value is significantly larger than the computed Kruskal-Wallis value at the .05 level. The printout also shows the

FIGURE 14.16 CBS Kruskal-Wallis Analysis of Briggs Department Store Problem

	Data Rankings		
	r1	r2	r3
1 =	9.5	17	18
2 =	3.5	15	16
3 =	3	3.5	7.5
4 =	7.5	1	2
5 =	9.5	11	6
6 =	14	13	12

Test procedure:	Kruskal-Wallis test
Alpha error:	.05
Sum of group 1:	49
Sum of group 2:	60.5000
Sum of group 3:	61.5000
Critical chi-square:	5.9910
Kruskal-Wallis test statistic:	0.5643

Conclusion: Do not reject null hypothesis

overall ranking of the survey data. For example, the fourth observation in the second group (i.e., store credit) represented the lowest total and was therefore ranked first.

14.6 PRACTICAL APPLICATIONS

This section presents several business applications that illustrate the use of nonparametric statistics.

National Football Conference (Rank Correlation Analysis)

The competition committee of the National Football League (NFL) is interested in determining the extent of the relationship between winning and defense. The committee has decided to use the National Conference final standings for the 1986 season as a basis for the analysis. These data are shown in Figure 14.17.

The 14 teams are then rank-ordered based on winning percentage. For example, since both the Chicago Bears and N.Y. Giants ended up with the same winning percentage, they are both ranked 1.5. Similarly, the 14 teams

FIGURE 14.17 National Conference Final Standings for 1986

	Wins	Losses	Percent	Points For	Points Opponents
Eastern Division					
N. Y. Giants	14	2	.875	371	236
Washington	12	4	.750	368	296
Dallas	7	9	.438	346	337
Philadelphia	5	10	.344	256	312
St. Louis	4	11	.281	218	351
Central Division					
Chicago	14	2	.875	352	187
Minnesota	9	7	.563	398	273
Detroit	5	11	.313	277	326
Green Bay	4	12	.250	254	418
Tampa Bay	2	14	.125	239	473
Western Division					
San Francisco	10	5	.656	374	247
L. A. Rams	10	6	.625	309	267
Atlanta	7	8	.469	280	280
New Orleans	7	9	.438	288	287

FIGURE 14.18 Ranked Data for National Conference Standings

Final Standing	Defense Ranking
1.5	2
3	8
8.5	11
10	9
12	12
1.5	1
6	5
11	10
13	13
14	14
4	3
5	4
7	6
8.5	7

are then ranked based on the number of points scored by their opponents (i.e., a measure of defensive capability). For example, the Chicago Bears are ranked first since they allowed the fewest points to be scored against them. The total rankings based on winning percentage and defense are given in Figure 14.18. The null hypothesis for this problem is that final standing and defense are not related; that is, teams ranking high in the standings do not necessarily rank high on defense.

A CBS computer analysis of these data is given in Figure 14.19. These results show that the null hypothesis of no difference can be rejected at the .01 level of significance. The computed Spearman rank correlation coefficient lies outside the critical range (i.e., the correlation value of .91 is greater than the upper critical limit). Accordingly, one can conclude that a significant rank correlation exists between standing and defense. The result suggests that concentrating on defense may lead to an improvement in final conference standings.

Merken Software Co. (Rank Correlation Analysis)

Merken Software is a relatively young company specializing in the development and marketing of business software products. Historically, the company has had a very liberal policy with regard to employee absenteeism. However, recent competitive pressures in the marketplace have caused Merken's management to reexamine a number of personnel policies. As a result of this new emphasis, the personnel director at Merken is interested in

FIGURE 14.19 CBS Rank Correlation Analysis of National Football Conference Standings

Test procedure:	Spearman rank correlation
Alpha error:	.01
Population mean:	0
Standard deviation:	0.2774
Spearman coefficient:	.9121
Critical upper limit:	0.5436
Critical lower limit:	−0.5436

Conclusion: Reject null hypothesis

determining the extent of relationship between absenteeism and several employee characteristics (age, salary, time employed, and sex). Presented in Figure 14.20 are the data collected from a random sample of 20 employee records. Salary figures are in thousands of dollars, and time employed (tenure) is in months. Absenteeism is measured in number of hours per year.

FIGURE 14.20 Data Base for Merken Software Problem

Employee	Age	Salary ($000)	Tenure (months)	Sex*	Absenteeism (hrs/yr)
1	23	16	1	0	16
2	64	47	10	1	36
3	29	33	2	1	94
4	44	22	1	0	50
5	27	47	10	1	72
6	33	32	3	0	42
7	36	20	7	0	40
8	44	21	2	0	4
9	43	62	4	1	10
10	50	55	6	1	6
11	31	22	8	1	0
12	43	56	21	1	3
13	53	72	14	1	0
14	29	35	5	1	24
15	37	42	10	1	45
16	28	20	12	1	38
17	25	26	5	1	69
18	30	17	3	0	64
19	31	30	3	1	55
20	47	65	15	1	0

* Male = 1; female = 0.

The personnel director has decided to examine the impact of several of these factors on absenteeism. More specifically, the director wishes to determine if absenteeism is influenced by either annual salary or age. A CBS rank correlation analysis of salary versus absentee rate is given in Figure 14.21. The input data show the rankings of salary and absentee rate. For example, the first employee surveyed had the lowest salary (i.e., ranked first) and had the eighth lowest absentee rate. Fractional rankings indicate the presence of ties (e.g., the second and the fifth employees receive the same pay). The null hypothesis for this problem is that salary and absenteeism are not related. An analysis of these results reveals that there is insufficient evidence to reject the null hypothesis. The computed Spearman coefficient falls within the critical limits. Therefore, the director can conclude that salary level does not appear to relate to absenteeism.

A similar analysis of the data base is presented in Figure 14.22. In this case, the director wishes to determine the extent of the relationship between age

FIGURE 14.21 CBS Rank Correlation Analysis of Salary versus Absenteeism

Information Entered

Test procedure: Spearman rank correlation
Alpha error: .05

	Salary	Absent		Salary	Absent
1 =	1	8	11 =	6.5	2
2 =	14.5	10	12 =	17	4
3 =	11	20	13 =	20	2
4 =	6.5	15	14 =	12	9
5 =	14.5	19	15 =	13	14
6 =	10	13	16 =	3.5	11
7 =	3.5	12	17 =	8	18
8 =	5	5	18 =	2	17
9 =	18	7	19 =	9	16
10 =	16	6	20 =	19	2

Results

Test procedure: Spearman rank correlation
Alpha error: .05

Population mean: 0
Standard deviation: 0.2294
Spearman coefficient: −.3440
Critical upper limit: 0.4497
Critical lower limit: −0.4497

Conclusion: Do not reject null hypothesis

FIGURE 14.22 CBS Rank Correlation Analysis of Age versus Absenteeism

Information Entered

Test procedure: Spearman rank correlation
Alpha error: .05

	Age	Absent		Age	Absent
1 =	1	8	11 =	8.5	2
2 =	20	10	12 =	13.5	4
3 =	5.5	20	13 =	19	2
4 =	15.5	15	14 =	5.5	9
5 =	3	19	15 =	12	14
6 =	10	13	16 =	4	11
7 =	11	12	17 =	2	18
8 =	15.5	5	18 =	7	17
9 =	13.5	7	19 =	8.5	16
10 =	18	6	20 =	17	2

Results

Test procedure: Spearman rank correlation
Alpha error: .05

Population mean: 0
Standard deviation: 0.2294
Spearman coefficient: −.5383
Critical upper limit: 0.4497
Critical lower limit: −0.4497

Conclusion: Reject null hypothesis

and absenteeism. Again, the input data show the relative rankings of the sampled employees by age and absentee rate. These results indicate the presence of a relationship between the two variables. That is, older employees tend to have a lower level of absenteeism than younger employees. This conclusion is based on the fact that the Spearman coefficient both is negative (reflecting an inverse relationship) and lies outside the critical limits. The director can, with a 95% level of confidence, reject the null hypothesis of no relationship and conclude that age and absenteeism are related.

14.7 CASE STUDY: FUJTI MOTORS

Fujti Motors Limited manufactures power-steering units for General Motors and Ford. The company produces about 25,000 units per month. The company's production supervisor, Mr. Kurita, is concerned about falling product quality. Mr. Kurita believes that production quality is dropping on second and third shifts. Accordingly, Mr. Kurita has decided to quantitatively estimate the impact of shift operations on product quality by taking a sample over a two-week period. The sampling process measures the percentage of defective units produced on a daily basis. The results of the survey can be found in Figure 14.23. For example, 1% of the power-steering units produced on the first shift on April 4 were found defective.

Mr. Kurita has decided to use nonparametric modeling to analyze the developed data base. A nonparametric approach is judged to be most appropriate in this case because of the relatively small sample and the uncertainty associated with the population distribution. More specifically, the Kruskal-Wallis test procedure is selected due to the fact that there are more than two populations (i.e., three populations, corresponding to the three shifts). If the problem situation had consisted of data from only two shifts, then either the Wilcoxon or Mann-Whitney test would have sufficed. Mr. Kurita's null hypothesis for this situation is that there is no difference in product quality due to shift. This is in contrast with his hunch that product quality is better on the first shift. The alternative hypothesis is that there is a difference in product quality by shift. This hypothesis can be accepted if the null hypothesis is rejected.

A CBS analysis of the data using the Kruskal-Wallis test procedure for a .05 level of significance is given in Figure 14.24. The output reports the overall ranking by group (e.g., the defective level observed on the first day and on the first shift had the fourth lowest value). These computer results show that there is sufficient evidence to reject the null hypothesis of no difference. This conclusion is based on the fact that the computed test statistic is larger than the critical chi-square statistic. Therefore, Mr. Kurita can conclude that product quality is influenced by production shift. His next step

FIGURE 14.23 Product Quality Data Base for Fujti Motors

	Product Quality by Shift (% defective)		
Date	1	2	3
4/2	0.75	4.25	9
4/3	2.5	3	4.5
4/4	1	1.75	5
4/5	3.5	2	8.5
4/6	0.5	0	1.5
4/9	4.25	6.25	4.5
4/10	2	3	6.5
4/11	5.5	5	2.5
4/12	0	1.5	4
4/13	3.5	6	9.5

FIGURE 14.24 CBS Kruskal-Wallis Analysis of Fujti Motors Problem

	Data Ranking		
	First	Second	Third
1 =	4	18.5	29
2 =	11.5	13.5	20.5
3 =	5	8	22.5
4 =	15.5	9.5	28
5 =	3	1.5	6.5
6 =	18.5	26	20.5
7 =	9.5	13.5	27
8 =	24	22.5	11.5
9 =	1.5	6.5	17
10 =	15.5	25	30

Results

Test procedure:	Kruskal-Wallis test
Alpha error:	.01
Sum of group 1:	108.0000
Sum of group 2:	144.5000
Sum of group 3:	212.5000
Critical chi-square:	5.9900
Kruskal-Wallis test statistic:	7.2587

Conclusion: Reject null hypothesis

would be to isolate the shifts that have different product quality (e.g., shift 1 versus shift 3). This could be accomplished by conducting pairwise Wilcoxon tests. However, recall that the probability of committing a Type I error increases as a function of the number of tests (as in multiple pairwise *t*-tests). Finally, Mr. Kurita should identify those factors that are causing the variability in product quality, for example, lack of appropriate training and lack of production discipline.

14.8 SUMMARY

This chapter introduced a number of the so-called distribution-free or nonparametric methods. The primary advantage of these methods is that they are not limited by the major assumptions underlying the application of parametric statistics. A secondary advantage is that they are relatively easy to use. The main disadvantage with nonparametric tests is that they are less powerful than parametric analyses under similar conditions. Nonparametric methods are used most frequently when the data are measured on either a nominal or an ordinal scale or when the sample size is small and the shape of the population distribution is unknown. Nevertheless, the hypothesis-testing procedure for nonparametric methods is the same as for parametric methods.

Specific models introduced in this chapter include the Wilcoxon rank-sum test, the Wilcoxon signed-rank test, the sign test, the Spearman rank correlation, and the Kruskal-Wallis test. Selection of a specific nonparametric model depends, for the most part, on the characteristics of the application. The Wilcoxon rank-sum test is most appropriate when the problem consists of two independent random samples and the objective is to determine whether the two populations are the same or different. The Wilcoxon signed-rank test is used when the data consist of matched pairs. The sign test is employed when the data are measured on a binomial scale (e.g., preference date). An important variation to the sign test is the runs test, which is used extensively in statistical quality control. The Spearman model is similar to the Pearson correlation model, which was introduced in Chapter 9. Both models provide a quantitative indication of the extent of the relationship between two variables. The Kruskal-Wallis model extends the Wilcoxon rank-sum test to more than two variables.

14.9 GLOSSARY

Kruskal-Wallis test An extension of the Wilcoxon rank-sum test to include more than two populations.

Mann-Whitney rank-sum test A nonparametric test procedure that is equivalent to the Wilcoxon rank-sum test.

nonparametric tests A group of statistical test procedures that do not require any assumptions regarding the population distribution. Nonparametric tests can be applied when the problem consists of only rank-ordered data or when the sample size is small and the population distribution is unknown.

parametric tests A group of statistical test procedures that require assumptions regarding the population distribution.

runs test A nonparametric test procedure used to determine if an observed sequence is occurring randomly. This test is used extensively in quality control applications.

sign test A nonparametric test used for analyzing data measured on a binomial scale.

Spearman rank correlation A nonparametric test used to determine the extent of correlation between two variables involving ranged data.

Wilcoxon rank-sum test A nonparametric test used for evaluating two independent samples.

Wilcoxon signed-rank test A nonparametric test used for evaluating matched or dependent samples.

14.10 BIBLIOGRAPHY

Conover, W. J. *Practical Nonparametric Statistics.* 2nd ed. New York: John Wiley & Sons, 1980.

Daniel, W. W. *Essentials of Business Statistics.* 2nd ed. Boston: Houghton Mifflin, 1987.

14.11 PROBLEMS

1. Define and provide examples for the following nonparametric tests:
   a. Wilcoxon rank sum
   b. Wilcoxon signed-rank
   c. Spearman rank correlation
2. Discuss the differences between parametric and nonparametric methods. Give examples for both.
3. Why are nonparametric methods less frequently employed than parametric tests, particularly in light of their ease of use?
4. Formulate the null hypothesis and select the most appropriate nonparametric method for each of the following applications:
   a. A new diet that results in an eight-pound weight loss in the first month
   b. A higher number of banking transaction errors at branch 1 compared to branch 2
   c. Customer preference between domestic and foreign sports cars
5. The Automobile Club of Southern California recently conducted a fuel economy test for both domestic and foreign cars. The sample consisted of 10 domestic and 12 foreign 1988 compact automobiles selected at random. Each vehicle was driven over the standard urban cycle. The resultant fuel economy data (measured in mpg) are as follows:

Domestic	Foreign	Domestic	Foreign
25.4	32.1	28.1	31.2
23.6	29.6	27.2	30.3
22.5	27.4	29.4	27.7
25.5	24.8	26.2	28.2
26.1	29.8		30.7
27.2	25.3		28.1

   a. Formulate an appropriate hypothesis for this problem.
   b. Test the hypothesis at the .05 level.
6. The manager at Eastern Electric is interested in determining the impact of marital status on employee absenteeism. The following data were collected from a random sample of current employees. The data report the number of hours that the employee was absent during the previous year.

Single	Married
28	45
77	121
187	76
38	42
36	37
11	16
40	48
56	0
64	36
34	32
28	7
	12
	28
	55
	41
	32
	72

    a. Select the appropriate nonparametric test.
    b. Does marital status significantly affect absenteeism?
    c. Assuming that the sample came from a normal population, analyze the data using a parametric *t*-test.
    d. Compare the results obtained in parts *b* and *c*.

7. The U.S. Weather Service reported the following normal minimum daily temperatures and precipitation levels for the month of March for selected cities throughout the United States. These data represent 30-year averages for the period 1950 through 1980.

City	Temperature (°F)	Precipitation (inches)
Mobile	49.8	6.48
Little Rock	41.2	4.69
Hartford	28.0	4.15
Boise	30.9	1.03
Chicago	27.6	2.59
New Orleans	51.6	4.73
Portland	23.7	3.98
St. Louis	33.0	3.28
Reno	25.4	0.74
Cleveland	28.4	2.99
Los Angeles	49.7	1.76
Memphis	41.9	5.44
Houston	49.8	2.68

The chief weather forecaster is interested in knowing the extent of relationship between normal daily minimum temperature and precipitation level.

a. Select a nonparametric model for determining the extent of the relationship at a .95 level of confidence.
b. Develop a parametric test using the same confidence level.
c. Compare the results obtained in parts *a* and *b*.

8. Palos Verdes Community Bank is interested in improving customer service. The bank's management has decided to initiate a training program for its tellers. The primary objective of the program is to reduce customer waiting time. Each of the bank's eight tellers was timed before and after the training program over a period of two weeks. The following data base presents the average customer processing time for each teller:

	Minutes	
Teller	Before	After
1	7.8	6.9
2	5.9	5.5
3	4.1	4.3
4	5.2	4.4
5	7.8	7.6
6	9.4	8.2
7	5.5	5.1
8	6.2	6.0

a. Use a nonparametric test to determine if the training program significantly reduces customer transaction time at the .05 level.
b. Use a parametric test (assume that the population is normally distributed) to determine whether the training program significantly reduces customer transaction time at the .05 level.
c. Compare the results obtained in parts *a* and *b*.
d. Discuss the impact of the normality assumption in part *b*.

9. Following is a 30-day sample of the day-to-day percentage change in the average bond and stock prices. Determine if the two variables are related, using rank-order correlation analysis.

Day	Bond	Stock	Day	Bond	Stock	Day	Bond	Stock
1	.008	.012	11	−.004	.010	21	0.003	0
2	−.008	−.065	12	.015	−.004	22	−.015	0
3	−.002	.012	13	−.008	−.010	23	−.004	−.004
4	0	−.008	14	.004	.010	24	0	.007
5	.007	.004	15	.004	.020	25	0	0
6	.004	.013	16	.009	.008	26	.002	−.015
7	.004	.013	17	0	.015	27	.004	.004
8	.002	.020	18	.002	.019	28	0	.007
9	−.008	−.010	19	.006	−.004	29	0	−.004
10	.020	−.004	20	−.002	−.006	30	−.008	.013

10. Following is a group of general economic indicators for the United States. Use a nonparametric sign test to determine if there are any differences in the indicators between a year ago and today.

Indicator	Today	One Year Ago
Index of leading indicators	193.3	187.5
Consumer price index (1982 = 100)	117.1	112.7
Producer price index (1982 = 100)	106.9	105.1
Retail sales (billions)	131.5	124.7
Auto sales (millions)	10.1	12.2
Housing Starts (000)	1,561	1,699
Unemployment rate (%)	5.4	6.3

11. The chief surgeon at the Royal Hospital in London wishes to determine if there is a relationship between quitting smoking and weight gain. The following sample data were collected on a group of 10 volunteers undergoing a smoking withdrawal program at the hospital. Each volunteer was weighed before and after the 60-day program. Use a sign test to determine if there is a significant weight gain.

	Weight (lbs)	
Candidate	Before	After
Smith	150	150
Jones	153	155
Barnes	160	162
Williams	175	178
Johnson	168	173
Keck	170	169
Brown	175	180
Wellman	158	166
Drake	161	167
White	171	175

12. First-year graduate students at the University of Manchester are required to take an introductory course in business statistics. Last trimester a choice of two options was offered. One was based on the traditional lecture method, and the second featured extensive computer-based instruction. A comprehensive examination was administered to both sections at the end of the trimester. The exam results are reported in the following table. The dean wishes to know whether or not the computer-based approach resulted in higher test scores.

Principal Instruction Mode	
Lecture	Computer
75	81
79	80
91	95
88	93
76	78
83	72
81	83
67	81
85	80
78	
91	

13. Refer to the NFL problem data given in Section 14.6. Determine if a significant rank correlation exists between final conference standing and offense (as measured by points scored against opponents) at the .01 level.
14. Following are international exchange rates for 13 countries for 1985 and 1986 (Bank of England Index: 1975 = 100).

Country	1985	1986
Australia	55.7	57.8
Belgium	100.1	100.3
Canada	76.9	80.8
West Germany	147.2	148.5
Holland	135.4	137.5
Italy	47.4	46.3
Japan	226.9	246.1
Spain	48.4	52.0
Switzerland	175.2	174.8
Great Britain	73.2	78.5
United States	100.1	92.6

   a. Use the sign test to determine if exchange rates have changed significantly over the one-year period (use $\alpha = .05$).
   b. Could a parametric test be used to answer the question posed in part a?
   c. If the answer to part b is yes, then perform the appropriate test and compare the results with those obtained in part a.
15. The following data show the number of new plants and expansions for a random sample of states for 1986 and 1988.

State	1986	1988
FL	235	335
CA	242	313
NC	174	282
IL	196	193
TX	189	180
GA	151	204
KY	135	214

a. Use the sign test to determine if the rankings have changed significantly over the two-year period (use $\alpha = .05$).
b. Under what conditions could a parametric test be used to answer the question posed in part *a*?
c. Perform the appropriate test and compare the results with those obtained in part *a*.

16. Following are numbers showing the percentage of worldwide automobile production for selected countries:

Country	1977	1986
United States	29.8	23.4
France	10.0	8.3
Japan	17.6	23.4
South Korea	0.1	1.4
Great Britain	4.3	3.1
West Germany	12.3	12.9

a. Use the sign test to determine if the ranking by production percentage has changed significantly over the nine-year period (use $\alpha = .05$).
b. Could a parametric test be used to answer the question posed in part *a*?
c. If the answer to part *b* is yes, then perform the appropriate test and compare the results with those obtained in part *a*.

17. Following are average annual percentage growth rates for manufacturing labor productivity for selected countries:

Country	1973–1985	1982–1985
United States	2.2%	4.8%
Canada	1.9	4.4
France	4.5	4.3
Italy	3.6	4.2
Japan	5.6	5.8
Great Britain	2.8	4.8
West Germany	3.8	6.1

a. Use the sign test to determine if the ranking by labor productivity has changed significantly at the 0.05 level.

b. Under what conditions could a parametric test be used to answer the question posed in part *a*?

c. Perform the appropriate test and compare the results with those obtained in part *a*.

18. Following are total productivity growth rates for selected U.S. industries:

Industry	1970–1980	1980–1983
Food and tobacco	2.3	0.3
Textile	2.9	4.5
Pulp and paper	1.9	4.3
Chemicals	2.5	1.7
Primary metals	0.2	−2.9
Fabricated metal	0.9	1.8
Nonelectrical machinery	1.2	1.0
Electrical machinery	3.7	1.1

a. Use the sign test to determine if the ranking by total productivity has changed significantly (use $\alpha = .05$)

b. Could a parametric test be used to answer the question posed in part *a*?

c. If the answer to part *b* is yes, then perform the appropriate test and compare the results with those obtained in part *a*.

19. Many economists believe that there is a strong negative relationship between industrial production and unemployment; that is, unemployment should fall as industrial output rises. The following production and unemployment data were collected for 13 countries for 1985.

Country	Industrial Production (% change)	Unemployment Rate
Australia	6.2	7.4
Belgium	0.7	10.9
Canada	5.4	7.7
West Germany	2.3	8.9
Holland	−2.8	14.0
Italy	7.8	15.7
Japan	12.8	2.6
Spain	4.0	20.1
Switzerland	5.6	0.8
Great Britain	3.3	9.0
United States	5.7	5.4

a. Use rank-order correlation analysis to determine the extent of the relationship between changes in industrial production and unemployment.

b. Could regression analysis be used to answer the question posed in part *a*?

c. If the answer to part *b* is yes, then develop a regression model and compare the results with those obtained in part *a*.

20. The following employee absenteeism data (measured in hours per year) were

recently collected from a random sample of two departments at the Aerospace Corporation. Use a rank-sum test to determine if absenteeism within the two departments is the same.

Manufacturing	Administration
28	42
121	48
187	32
65	32
32	58
39	36
28	38
40	16
32	12
64	56
34	
46	
162	
7	

21. The Strecker Co. uses a chemical etching process to prepare aluminum parts for adhesive binding and painting. The process generates a hazardous liquid waste. Company management is interested in predicting the amount of waste that will be generated during the next quarter. The following data base reports plant production and the amount of waste generated over the past 12 months.

Month	Waste (gal)	Production (lb)
1	3,067	7,107
2	2,828	6,373
3	2,891	6,796
4	2,994	9,208
5	3,082	14,792
6	3,898	14,564
7	3,502	11,964
8	3,060	13,526
9	3,211	12,656
10	3,286	14,119
11	3,542	16,691
12	3,125	14,571

a. Determine the extent of the relationship between production and waste using rank correlation analysis.
b. Determine the extent of the relationship between production and waste using correlation analysis.
c. Compare the results from parts *a* and *b*.

22. The Thompson Corporation is considering the installation of a new automated painting machine to replace the current manual operation. Recently, a test was conducted to compare the reliability of the new device with the manual process. A random sample of 30 matched observations was taken measuring the number of defects produced by the two processes per batch. Use a Wilcoxon signed-rank test to determine if the accuracy of the automated device is comparable to that of the manual procedure.

Sample	Manual	Auto	Sample	Manual	Auto
1	2	1	16	0	0
2	0	0	17	0	0
3	2	0	18	0	1
4	0	1	19	0	1
5	0	0	20	2	1
6	0	1	21	2	0
7	0	0	22	2	0
8	1	0	23	1	1
9	0	0	24	1	2
10	0	1	25	0	2
11	1	2	26	1	1
12	2	1	27	0	0
13	0	1	28	0	0
14	1	0	29	0	1
15	0	0	30	1	2

23. The Committee for Community Fairness (CCF) believes that the business sector receives better telephone service than the local residents. The committee has collected the following data from telephone company records made available through the Public Utilities Commission (PUC). The data show the percentage of residential phone traffic in a given area and the corresponding service ranking made by the PUC. It is the CCF's view that those communities with high levels of residential phone traffic have the lowest service rating.

Residential Phone Traffic	Service Ranking	Residential Phone Traffic	Service Ranking
74.3	7	75.4	9
72.6	8	78.9	21
72.7	10	78.4	20
87.5	22	73.5	16
79.1	18	77.7	17
74.7	11	16.2	1
75.3	19	66.9	6
83.7	23	74.3	12
88.0	24	65.3	2
81.2	13	73.1	5
79.0	14	68.7	3
70.9	4	81.7	15

    a. Formulate an appropriate hypothesis.
    b. What conclusions can be drawn from these data at the 0.05 level?
    c. Why is nonparametric analysis more useful in this problem?

24. The foreman at Altrop Manufacturing is concerned about the use of the same stamp press to manufacture both English and metric parts. Company records show that the press was originally designed in English units. The foreman recorded the number of defects observed over an 11-week period when the press was used for processing both types of parts. What conclusions can be drawn from the following data?

Week	Metric	English
1	5	4
2	9	2
3	6	8
4	4	3
5	3	2
6	8	7
7	10	6
8	15	5
9	11	5
10	6	4
11	4	1

25. Referring to the data base given in Figure 14.20 (for Merken Software), determine the extent of the relationship between tenure and company absenteeism.
26. The following table presents sample air quality data from four U.S. cities. The data were collected randomly over a period of one year and are measured on a scale of 1 to 100 (readings above 80 are considered a serious health problem). Use the Kruskal-Wallis test to determine if air quality is the same in all four cities.

Los Angeles	Houston	Seattle	Atlanta
85	76	49	52
75	45	38	48
77	53	52	57
81	59	55	51
66	63	46	63
62	71	51	50
56		55	
59			

27. The investment manager at Wedd Financial is interested in determining if firms in the electronics sector outperformed those in either the automotive or the energy sector. The manager has decided to use price-to-earnings ratio as the measure of corporate performance. A random sample of 18 Fortune 500 firms yielded the following data on price/earnings ratio for the three sectors of interest. Use the Kruskal-Wallis test to determine if the electronics sector outperformed either of the other two groups.

Automotive	Electronics	Energy
4.5	12.1	8.9
5.2	15.5	7.7
7.2	13.2	8.3
5.8	10.2	5.5
3.9	14.4	6.2
4.8		7.3
		10.2

28. Analyze the case study given in Section 14.7 using one-way analysis of variance at a .01 level of significance. Compare these results with those developed using the Kruskal-Wallis model for the same alpha level.
29. Following are sample data taken from a survey of delinquent patient accounts from five Southern California hospitals:

		Hospital		
A	B	C	D	E
$275	$501	$985	$475	$505
515	405	225	378	155
228	374	175	567	245
175	657	89	774	315
198	545	137	634	175
89	487	167	538	239

Analyze these data using a one-way analysis of variance at a .01 level of significance. Compare these results with those developed using the Kruskal-Wallis model for the same alpha level.

30. Preference ranks of eight radio shows by sex are presented below.

	Program							
	1	2	3	4	5	6	7	8
Men	6	4	1	5	3	8	7	2
Women	2	7	3	8	4	6	5	1

Determine if a significant degree of rank correlation exists at the .05 level.

31. Following are the preference ranks for eight radio shows by income level:

Income	Program							
	1	2	3	4	5	6	7	8
Below $25,000	6	4	1	5	3	8	7	2
$25,000 and above	5	4	1	8	7	2	3	6

Determine if a significant degree of rank correlation exists at the .05 level.

32. Evaluation ranks for seven professors by type of student are as follows:

Student Type	Professor						
	A	B	C	D	E	F	G
Graduate	4	5	1	3	6	7	2
Undergraduate	5	4	2	3	7	6	1

Determine if a significant degree of rank correlation exists at the .01 level.

33. Following are the evaluation ranks of six automotive manufacturers by age group:

Age	Manufacturer					
	GM	Ford	Chrysler	Toyota	Volvo	V.W.
Below 35	1	2	6	3	5	4
35 and above	1	3	5	2	6	4

Determine if a significant degree of rank correlation exists at the .05 level.

34. Identify an appropriate nonparametric test for each of the following situations:

a. Evaluating which of two drugs more effectively reduces high blood pressure
b. Determining consumer preference between two new soft drinks
c. Measuring the extent of the relationship between sales and market share standing
d. Analyzing the effectiveness of five ad campaigns on sales

35. Identify an appropriate nonparametric test for each of the following situations:
    a. Evaluating the impact of three different training techniques on productivity
    b. Determining whether the goals of a CEO are the same as those of the board of directors
    c. Analyzing the number of defective items after the introduction of a new assembly procedure
    d. Comparing the differences in miles per gallon between foreign and domestic automobiles

36. The director of research at Bjorn Corporation wishes to determine if differences exist in machine downtime among three types of computers. The following downtime data, in minutes, were collected over a six-week period.

Computer Model		
A	B	C
67	86	104
54	102	77
70	75	92
81	119	105
77	98	111
34	58	82

37. The manager of production at Lieuw International wishes to determine if differences exist in daily productivity among the three shifts. The following production data (units/hr) were collected over a two-week period (use $\alpha = 0.05$).

		Shift	
	1	2	3
Week 1	315	274	239
	333	303	312
	298	298	345
	378	298	279
	338	312	267
Week 2	311	344	277
	355	278	278
	361	349	333
	381	328	328
	364	288	301

38. Conduct a weekly analysis of the data in problem 37 to determine if production is improving.

39. A prospective buyer is interested in the depreciation of three luxury cars over a six-year period. The following data show the percentage of depreciation by year.

Year	Automobile		
	A	B	C
1	23	22	19
2	17	15	16
3	14	12	13
4	9	7	8
5	4	4	3
6	1	1	1

Which car is the best value?

40. The production manager at Allied Technologies wishes to determine the impact of a new training course on productivity. Two groups of employees were selected at random to participate in the study. The first group, consisting of 11 employees, has undergone a three-week training course on productivity. The second group, consisting of 12 employees, has received no special training in productivity. If the sum of the ranks for the first group is 70, what conclusions can the production manager make regarding the effectiveness of the training program at the 0.05 level?

41. The vice president at Walker Financial Service is concerned about the growing number of customer complaints. The vice president believes that the qualifications of the financial consultant may be related to this problem. The vice president has randomly selected two groups of tax consultants from the firm's data base. The first group consists of 9 senior tax consultants while the second group consists of 11 junior tax consultants. If the sum of the ranks for the first group is 50, what conclusions can the vice president make regarding the impact of consultant qualifications on the level of customer complaints at the 0.05 level.

42. A Spearman rank correlation coefficient of 0.45 was obtained from 18 paired observations. Test the null hypothesis of no positive relationship between the two sampled variables at the 0.05 level.

43. A Spearman rank correlation coefficient of -0.72 was obtained from 25 paired observations. Test the null hypothesis of no negative relationship between the two sampled variables at the 0.05 level.

44. Conduct a runs test, at the 0.05 level, on the following process data (A = above process average, B = below process average) to determine if the production sequence has occurred randomly.

    A A B B A A A B A A A B B B B B A A A B B B A A B

45. Conduct a runs test, at the 0.05 level, on the following process data (A = above process average, B = below process average) to determine if the production sequence has occurred randomly.

    A A A A B B A A A B B B B B B A A B B A B A A B B

46. The following data were obtained from a random sample of 20 production lots of

sports radios. Each sample, which measured the proportion of defectives in each lot, contained three observations.

Sample	Average Proportion of Defectives	Sample	Average Proportion of Defectives
1	0.02	11	0.04
2	0.03	12	0.05
3	0.01	13	0.05
4	0.00	14	0.04
5	0.01	15	0.06
6	0.02	16	0.05
7	0.03	17	0.03
8	0.04	18	0.04
9	0.04	19	0.05
10	0.03	20	0.06

    a. Determine the average defective level for the process.
    b. Conduct a runs test, at the 0.05 level, to determine if the process is occurring randomly.

47. The following data were obtained from random samples of 20 computer diskettes. Each sample, which measured the length of each diskette, contained four observations.

Sample	Average Length (inches)	Sample	Average Length (inches)
1	5.18	11	5.22
2	5.22	12	5.24
3	5.28	13	5.29
4	5.26	14	5.25
5	5.29	15	5.27
6	5.25	16	5.23
7	5.19	17	5.18
8	5.27	18	5.29
9	5.24	19	5.27
10	5.26	20	5.24

    a. Determine the average diskette length for the process.
    b. Conduct a runs test, at the 0.05 level, to determine if the process is occurring randomly.

48. The following data present a comparison of the 1988 and 1989 profit rankings taken from a random sample of the top 75 U.S. corporations:

	Rank	
Company	1989	1988
General Electric	2	5
Texaco	10	18
Merck	17	23
Pacific Telesis	23	24
American Express	24	34
Johnson & Johnson	29	36
Westinghouse	36	45
PepsiCo	39	52
BankAmerica	45	84
Nynex	48	17
J.C. Penney	50	48
Security Pacific	55	72
Pfizer	62	50
American Electric	68	74

a. Identify the most appropriate nonparametric test for determining if the rankings are statistically different.
b. Formulate a null and alternative hypothesis.
c. Conduct the test at the 0.05 level.
d. What general conclusions can be drawn about corporate profits between 1988 and 1989?

49. The following data present a comparison of the 1988 and 1989 market value rankings taken from a random sample of the top 75 U.S. corporations:

	Rank	
Company	1989	1988
Chevron	15	16
GTE	18	20
Eli Lilly	25	34
Pacific Telesis	21	27
Sears	38	17
Eastman Kodak	39	19
Pfizer	45	41
USX	48	49
Berkshire	58	84
Lows	64	73
Toys "R" Us	66	101
Kellogg	70	48

a. Identify the most appropriate nonparametric test for determining if the rankings are statistically different.
b. Formulate a null and alternative hypothesis.
c. Conduct the test at the 0.05 level.
d. What general conclusions can be drawn about corporate profits between 1988 and 1989?

50. The following data present a comparison of the 1988 and 1989 sales rankings for the top 15 U.S. corporations:

	Rank	
Company	1989	1988
General Motors	1	1
Ford	2	2
Exxon	3	3
IBM	4	4
General Electric	5	6
Sears	6	5
Mobil	7	7
Philip Morris	8	14
Citicorp	9	12
ATT	10	9
Du Pont	11	11
Chrysler	12	8
Texaco	13	10
K mart	14	13
Chevron	15	15

   a. Identify the most appropriate nonparametric test for determining if the rankings are statistically different.
   b. Formulate a null and alternative hypothesis.
   c. Conduct the test at the 0.05 level.
   d. What general conclusions can be drawn about corporate profits between 1988 and 1989?

51. The results of a CBS analysis of the amount spent by men and women at a local food store chain, in dollars per purchase, is reported below:

Alpha error:	.05
Population mean:	138
Standard deviation:	15.1658
Sum of group 1:	179
Sum of group 2:	74
Standard error:	29.7249
Critical upper limit:	167.7249
Critical lower limit:	108.2751

   a. Identify the test procedure used in this analysis.
   b. Specify the null and alternative hypotheses.
   c. What conclusions can be drawn regarding the differences in the amount spent between men and women?

52. A CBS analysis of the outcome of a weight reduction program for 12 senior citizens is reported below. The data represent each participant's weight, in pounds, before and after the program.

Alpha error:	.05
Population mean:	0
Standard deviation:	22.4944
Sum of signed ranks:	−59
Critical lower limit:	−37.0034

    a. Identify the test procedure used in this analysis.
    b. Specify the null and alternative hypotheses.
    c. What conclusions can be drawn regarding the results of the weight reduction program?

53. A CBS analysis of a random sample of 10 residential homes is reported below. The data represent the relative ranking of the selling price and view. The data are ranked in ascending order where a 1 represents the lowest price and the poorest view.

Alpha error:	.05
Population mean:	0
Standard deviation:	0.3333
Spearman coefficient:	0.8667
Critical upper limit:	0.6533
Critical lower limit:	−0.6533

    a. Identify the test procedure used in this analysis.
    b. Specify the null and alternative hypotheses.
    c. What conclusions can be drawn regarding the extent of the relationship between selling price and view?

54. A CBS analysis of the results of an experiment which evaluated the impact of three different types of fertilizers on wheat production, measured in bushels per acre, is reported below:

Alpha error:	.05
Sum of group 1:	53
Sum of group 2:	131
Sum of group 3:	116
Critical chi-square:	5.9915
Test statistic:	8.6038

    a. Identify the test procedure used in this analysis.
    b. Specify the null and alternative hypotheses.
    c. What conclusions can be drawn regarding the impact of the different types of fertilizers on wheat production?

55. The following data represents 16 sample diameter measurements (in inches) taken from a ball-bearing production line:

1.05	1.04	1.03	1.02	1.03	1.02	1.01	0.99
1.00	1.02	0.98	0.97	0.96	0.97	1.01	0.97

a. Determine the number of measurements above and below and mean.
b. Conduct a runs test at the 0.05 level to ascertain if the production sequence has occurred randomly.

56. The following data represents 14 sample weight measurements (in ounces) taken from a cereal box filling line:

9.8	10.2	10.1	10.0	9.9	9.8	9.9
10.1	10.0	10.3	9.8	9.9	10.1	10.2

a. Determine the number of measurements above and below the mean.
b. Conduct a runs test at the 0.05 level to ascertain if the production sequence has occurred randomly.

57. The home loan officer at Palos Verdes Community Bank believes there is a negative relationship between the lending rate and the number of new loans. A random survey of eight months, taken from the bank's historical files, yields the following data:

Lending Rate (%)	Number of Loans	Lending Rate (%)	Number of Loans
10.4	47	11.3	38
10.2	51	10.5	45
9.7	67	10.7	41
11.1	44	9.9	53

a. Which nonparametric model is most appropriate for analyzing this problem?
b. Do the data indicate a relationship at the 0.05 level?

58. The results of a preference survey between American-made and Japanese cars are presented below. This survey was based on a random sample of 12 potential buyers.

Buyer	Car Preference	Buyer	Car Preference
1	Japan	7	U.S.
2	Japan	8	Japan
3	U.S.	9	Japan
4	U.S.	10	—
5	Japan	11	U.S.
6	Japan	12	Japan

Conduct a sign test to determine is there is a difference is preference at the 0.05 level.

59. The results of a preference survey between American and European airlines are presented below. This survey was based on a random sample of 14 recent passengers.

Passenger	Preference	Passenger	Preference
1	American	8	—
2	European	9	American
3	American	10	European
4	American	11	American
5	American	12	American
6	European	13	European
7	European	14	American

Conduct a sign test to determine if there is a difference in preference at the 0.05 level.

60. The QC manager at Windel Metal Products is concerned about the number of defectives being produced on the stamping press. The manager has collected 50 random samples over the past week with 20 samples above the mean and 30 samples below the mean. The largest number of runs observed in the data was eight. What conclusions should the manager draw at the 0.05 level? (Hint: Use the normal approximation to the runs distribution.)

14.12 CASES

14.12.1 Women's International Golf Association

The vice president and general manager of the Women's International Golf Association (WIGA), Ms. Sally Wright, is interested in evaluating trends in tournament prize monies. She wishes to determine if a relationship exists between player income and the player ranking in the WIGA. Ms. Wright has collected a sample of the salary and ranking history of 10 players from a random computer search of the association's data base. The results of the survey are reported in Table 1.

Suggested Questions

1. Do higher-ranked players make more money on the tournament circuit than lower-ranked players?
2. Have prize winnings grown over the five-year period?
3. Have the relative rankings changed over the five-year period?

14.12.2 Analyzing Personality Differences

It is commonly accepted that there are certain personality differences among various ethnic groups. Numerous studies have investigated cross-cultural traditions, values, and habits as a means of understanding these differences. Another approach to studying these differences is to investigate the major components of personality as manifested in one's behavior and in the functioning of the brain.

Two major personality constructs are *type A/type B* behavior patterns and *hemisphericity*, or cerebral dominance. After studying a number of coronary artery and heart disease patients, Friedman and Rosenman developed the following definition

TABLE 1 WIGA Ranking and Earnings Data for 1981 to 1985

Player	1981	1982	1983	1984	1985
1	$111,093 31	$88,118 47	$61,066 85	$207,543 28	$109,096 74
2	$46,214 90	$12,474 166	$52,800 96	$113,336 62	$126,177 65
3	$105,755 35	$208,627 15	$181,246 24	$422,995 3	$190,871 33
4	$30,034 110	$57,608 76	$149,909 33	$177,289 34	$76,038 97
5	$24,167 118	$54,165 84	$18,105 157	$29,094 141	$139,257 57
6	$105,395 37	$181,864 18	$62,371 83	$167,848 37	$233,352 21
7	$33,945 102	$64,622 68	$284,434 6	$126,400 53	$58,689 114
8	$134,710 14	$184,600 17	$85,575 64	$41,837 117	$81,121 94
9	$7,766 178	$44,796 97	$44,455 110	$25,712 146	$48,383 120
10	$100,847 38	$80,804 50	$223,810 15	$110,875 64	$86,214 88

of the type A behavior pattern: an action-emotion complex that can be observed in any person who is aggressively involved in a chronic, incessant struggle to achieve more and more in less and less time and, if required to do so, against the opposing efforts of other things or other persons. It is important to note that for a type A behavior pattern to explode into being, the environmental challenge always serves as the fuse. The person with the type B behavior pattern is the exact opposite of the type A subject. The type B individual is rarely harried by desires to attain a wildly increasing number of objectives or participate in an endlessly growing series of events in an ever-decreasing amount of time.

Another major personality construct is hemisphericity, or cerebral dominance. Segalowitz (1983) defines hemisphericity as the tendency for one cerebral hemisphere to be generally dominant and show greater control over behavior than the other hemisphere, no matter what the task. This does not mean that the other hemisphere is generally inactive but, rather, that the dominant one leads in decision making. For example, some people may tend to look at the world in a holistic fashion, others more analytically. Each style has its own merits and deficiencies, but it is a rare person who has equal command of both styles and can alternate comfortably. Rather, most people tend to adopt one cognitive style or the other.

Western societies have developed what is considered a cultural gap between the two styles of thinking. Those with left-brain dominance are characterized by an

TABLE 1 Results of Cerebral Dominance Test by Behavior Type and Age

Behavior Type		Age Group		
A	B	25–40	40–55	55–70
84	79	73	70	65
98	84	85	87	81
90	83	88	84	84
89	81	89	79	92
85	88	98	84	95
93	76	90	74	93
92	78	84	92	78
92	87	76	81	92
78	84	78	76	90
89	73	83	89	
92	74	82		
95	65			
84	90			
81	82			
	70			
	76			

orderly mentality and are epitomized by such professionals as mathematicians, accountants, scientists, and lawyers. Those with right-brain dominance are characterized by an attempt to avoid an emphasis on strict order and logic and are epitomized by artists and musicians. Left-brain-dominant individuals are seen as having a cold, analytical, verbal approach to life. Right-brain-dominant individuals are seen as having emotional, holistic, and imagistic attitudes. Such personality styles taken to the extreme may border on psychopathology. However, as with type A/type B behaviors, it is important to recognize that hemisphericity exists on a spectrum. Individuals can range anywhere along the left-brain–right-brain continuum.

A survey was recently conducted to measure the extent of the relationship between behavior type and cerebral dominance. A second objective of the study was to determine the impact of age on cerebral dominance. The survey consisted of administering a test to measure cerebral dominance to 30 randomly selected business professionals (higher test scores indicated increased right-brain dominance). A pretest interview was used to determine the individual's behavior type (A or B). The results of the survey are presented in Table 1.

Suggested Questions

1. Are age and cerebral dominance related?
2. Are behavior type and cerebral dominance related?
3. What analytical approach could be used for analyzing the impact of both age and behavior type on cerebral dominance?
4. What other factors might be considered for explaining the variance observed in the test scores?

Chapter 15

Managerial Decision Analysis

Decide not rashly; the decisions made can never be recalled.
H. W. Longfellow

CHAPTER OUTLINE

15.1 Introduction
15.2 Example Management Problem: Perpetual Investments Corporation
15.3 How to Recognize Decision Analysis Problems
15.4 Formulating Decision Analysis Models
15.5 Computer Analysis
15.6 Practical Applications
15.7 Case Study: Cleanall Corporation
15.8 Summary
15.9 Glossary
15.10 Bibliography
15.11 Problems
15.12 Cases

CHAPTER OBJECTIVES

The primary objectives of this chapter are to develop an understanding of

1. The basic decision analysis models.
2. How to analyze decision problems using payoff tables and decision trees.
3. The principles of decision making with additional information.
4. How utility can be used as a decision criterion.
5. How to solve decision problems using computer analysis.

Selecting the best type of mortgage to finance a specific real estate purchase usually requires considering a number of factors. A primary consideration is expected future interest rates. Basically, there are two types of mortgages: fixed and adjustable. The initial rate for an adjustable mortgage is generally lower than that for a fixed mortgage. However, the adjustable rate can change depending on general trends in the economy, whereas the fixed rate will remain unchanged over time. It is quite possible in an inflationary economy for an adjustable mortgage rate to grow significantly over the term

> **HISTORICAL NOTE**
>
> The analytical approach for decision making under risk with additional information is credited to the Reverend Thomas Bayes (1702–1761), an 18th-century British clergyman. Bayes, an expert mathematician, reversed the traditional approach of reasoning from the population to the sample to one of inferring from the sample to the population. His most famous contribution, involving the process of revising probability data using additional information, was published after his death. Fortunately, his work lives on, and modern decision theory is often called Bayesian decision analysis in his honor.

of the loan (e.g., 2–3 points). Many adjustable mortgages contain a rate cap to guard against excessive growth.

Classical decision theory can be used to help select the optimal mortgage type.* In one study, a decision tree was used to evaluate a wide range of mortgage alternatives. This study utilized four decision criteria to assess the cost differences between specific mortgage rate options over a five- to seven-year planning horizon. The decision criteria ran the gamut from pessimistic to optimistic. The results showed that an adjustable mortgage is most attractive if the property is held for a relatively short period (i.e., less than six years). The option of refinancing the property in the event that mortgage rates decreased also supported selection of the adjustable rate.

15.1 INTRODUCTION

Chance and uncertainty are key factors in decision making because the results of a decision can be significantly affected by events beyond the control of the decision maker. For example, an investor considering a stock purchase must account for the subsequent behavior of the stock market in making a final decision. The investor does not control the actions of the stock market; therefore, the outcome of the investment is uncertain. Nevertheless, a decision must be made based on an assessment of the probable influence of these external forces.

Capital investment decisions provide another example of situations where the decision maker does not have total control over the outcome. In such cases, the decision maker must choose among a number of alternatives (e.g., building size, different models of computers, or which new products to

---

* R. E. Luna and R. A. Reid, "Mortgage Selection Using a Decision Tree Approach," *Interfaces* 16, no. 3 (May–June 1986), pp. 73–81.

introduce). The outcome will be influenced by a number of factors or **states of nature** (e.g., future sales level, weather, future material costs, or competitors' prices), none of which are controllable by the decision maker.

In general, problems involving alternative decisions and states of nature fall into one of the following categories:

Key idea 1

- Decision making under certainty: The future state of nature is known.
- Decision making under risk: More than one state of nature can occur, and probability values can be assigned to each state of nature.
- Decision making under uncertainty: More than one state of nature can occur; however, no data are available regarding the probabilities of these future events.

Decision making under certainty is the simplest of these categories, but decision models of this type can be extremely complex. The payoff associated with each decision alternative is known with certainty (i.e., there is only one state of nature, with a probability value of 1). In analyzing problems under certainty, the decision maker merely selects the alternative that optimizes attainment of the specified objective.

Both **decision making under risk** and **decision making under uncertainty** involve selecting the best alternative when faced with more than one state of nature. In decision making under risk, the decision maker knows the probability of occurrence of each state. For example, a farmer knows that weather (either rain or sunshine) can affect the decision to harvest now or wait for two weeks. The crop will be worth more if the weather is sunny and it is harvested later, but rain will reduce the crop value. By consulting the weather service, the farmer can determine the probability of rain or sunshine over the next two weeks. Decision making under uncertainty also involves multiple states of nature; however, the probability of occurrence of each state of nature is unknown. Figure 15.1 summarizes the basic characteristics of the three decision-making environments.

This chapter focuses primarily on decision making under risk; this decision process is also known as **objective analysis.** Normally, decision makers can develop probability estimates for each state of nature from historical

FIGURE 15.1 Decision Analysis Classifications

Characteristic	Decision Environment		
	Certainty	Risk	Uncertainty
States of nature	One	Many	Many
Decision methodology	Objective	Objective	Subjective
Decision criteria	Deterministic	Expected value	Multiple
Probability of occurrence	One	Known	Unknown

patterns or from surveys. If probability data are not available, however, decision making under uncertainty must be used to analyze the problem. This chapter also presents the methodology for addressing problems of this type.

15.2 EXAMPLE MANAGEMENT PROBLEM: PERPETUAL INVESTMENTS CORPORATION

The general partner of Perpetual Investments, a small but rapidly growing financial house, is interested in selecting the optimal investment instrument for the next six months. The company has raised $500,000 from approximately 20 investors. Typically, Perpetual charges a 5% fee for managing the investment. The partner has identified the following three promising investment alternatives: construction company, certificate of deposit (CD), and precious metals. The partner knows that the rates of return for these investments can be influenced by a number of external factors, such as interest rates. For example, investing in a construction company when interest rates are increasing can lead to a low if not negative rate of return. Unfortunately, the future state of interest rates cannot be estimated with certainty. In this regard, the general partner will need to estimate the potential return for each

INTERNATIONAL VIGNETTE

Giorgio Pierotti, president and CEO of Frisbee Frozen Foods, explained to the group in his office located in Como, Italy, "Your study is very timely, especially in light of our plans to build a new central warehouse. We have the potential for adding a great deal of value in the distribution, even more than in manufacturing. The first task is to find the most competitive use of this potential." He was talking to a special ad hoc team that had been assembled to do a thorough review of the company. Frisbee's product range focuses primarily on frozen foods and ice cream. The wholesale market for these products in northern Italy is about 900 billion lire annually. Frisbee's distribution system consists of 11 depots and a semiautomated central warehouse located next to the factory in Como.

Historically, sales growth had come from new product introductions. The ad hoc team found that a large number of products were nearing the maturity stage in their respective life cycles. Clearly, there was a need to identify new candidate products and to estimate their impact on corporate performance. In this regard, the team visited a number of independent distributors. Finally, the team estimated that a new facility could operate with 13 fewer employees (at an annual savings of about 650 million lire) and with 2–3% lower operating expenses. Armed with these data, the ad hoc team was prepared to analyze the alternatives and make a final recommendation.

---

investment based on different interest rates. To simplify this situation, the partner has characterized the external environment under the conditions that interest rates will either increase, remain the same, or decrease over the investment period. Based on this simplification, the partner estimates the following gross returns for each investment:

Condition of Interest Rates	Investment Alternative		
	Construction Company	Certificate of Deposit	Precious Metals
Increasing	3	11	16
Unchanged	8	11	13
Decreasing	19	11	7

Clearly, the incorporation of alternative future interest rates into the problem has made the task of selecting the best investment more difficult. Nevertheless, the general partner must decide on a specific investment.

15.3 HOW TO RECOGNIZE DECISION ANALYSIS PROBLEMS

The classic decision analysis problem has the following characteristics:

- Two or more decision alternatives
- Two or more states of nature
- Prior probabilities for each state of nature are known (objective analysis) or unknown (subjective analysis)
- Economic payoffs specified for each combination of decision alternative and state of nature
- The objective to identify the alternative that yields the optimal payoff based on a given decision criterion

15.4 FORMULATING DECISION ANALYSIS MODELS

A decision analysis problem is usually described with the use of a **payoff table.*** The table specifies the payoff (e.g., profits) for every combination of decision alternative and state of nature. The use of a payoff table is appropriate for single-stage decision problems (i.e., a single decision) but usually not for multistage decision problems (i.e., a sequence of interrelated decisions). The use of a decision tree (discussed later in this chapter) is more effective for the latter type of problem. This section first presents how payoff tables can be used in analyzing problems involving decision making under uncertainty and risk.

DECISION MAKING UNDER UNCERTAINTY

In some decision analysis situations, data on the probability of occurrence for the various states of nature may have a low level of reliability or may not be available (e.g., demand for a new product). In these cases, **subjective analysis** is used, in which the decision criterion is dependent on the philosophy of the decision maker or institution. For example, if the decision maker is a risk taker, the optimal action may be to select the alternative with the highest gain regardless of the potential loss that might occur. It will be demonstrated that the use of different decision criteria will lead to different gains and losses and, often, to different decisions. Figure 15.2 presents a schematic of the model for decision making under uncertainty. The discussion that follows will show how this model can be applied to problems where probability data on the states of nature are not available.

---

* This is the standard approach when the probability distribution of events is discrete. The analysis becomes considerably more complex when the event distribution is continuous. The Bierman text listed in this chapter's bibliography provides an in-depth treatment of this approach.

15.4 Formulating Decision Analysis Models   683

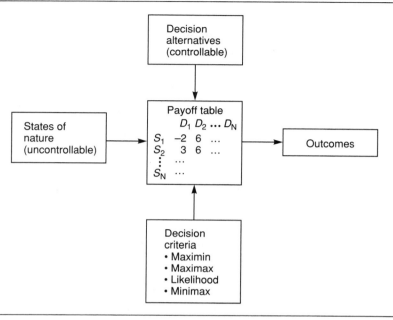

FIGURE 15.2 Subjective Decision Analysis Model

Recall that the general partner at Perpetual Investments (see Section 15.2) is interested in selecting among three investment alternatives: construction company (D_1), certificate of deposit (D_2), and precious metals (D_3). A revised payoff table that reflects Perpetual's management fee of 5% is presented in Figure 15.3. Notice that the columns reflect the decision alternatives and the rows the various interest rates (states of nature).

Again, selecting the optimal decision under conditions of uncertainty can be based on a wide variety of decision criteria. The following discussion introduces four of the most widely used subjective decision criteria and explains their impact on the decision-making process. These are by no means the only criteria that may be applied to a particular problem. Actually, there are as many subjective decision criteria as there are philosophical views of risk.

FIGURE 15.3 Net Payoff Table for Perpetual Investments Corporation

Condition of Interest Rates	Decision Alternatives		
	D_1	D_2	D_3
Increasing (I)	−2	6	11
Unchanged (U)	3	6	8
Decreasing (D)	14	6	2

Maximax Decision Criterion

The **maximax decision criterion,** as the name suggests, identifies the decision that yields the largest positive return. For Perpetual Investments Corporation, this corresponds to decision D_1, since the largest return in the Figure 15.3 is 14%. A list of the maximum return for each decision is given in row A of Figure 15.4. Notice that the maximax strategy does not consider the large potential loss associated with D_1, if interest rates increase rather than decrease. The maximax decision criterion would be used by those with a risk-oriented philosophy.

Maximin Decision Criterion

The **maximin decision criterion** is appropriate when the decision maker wishes to ensure a minimum return against all possible outcomes. Again referring to Figure 15.3, the approach is to identify the minimum return for each of the three decision alternatives. These values are shown in the row labeled B in Figure 15.4. The maximum value in this row is 6, which is associated with decision D_2. Thus, investing in a CD account yields the largest minimum return of all alternative investments. In general, the approach is to identify the minimum payoff for each alternative and then select the decision that yields the largest return from these payoffs.

Equal-Likelihood Decision Criterion

Another approach to decision making when probability data are not available is to assume that each state of nature has an equal likelihood of occurring; thus, this approach is known as the **equal-likelihood decision criterion.** For example, there are three states in the Perpetual Investments problem, so the equal-likelihood criterion assumes a 33% chance of each state occurring.

FIGURE 15.4 Decision Summary for Perpetual Investments

Decision Criterion	Decision Alternative			Optimal Decision
	D_1	D_2	D_3	
A. Maximax	(14)	6	11	D_1
B. Maximin	−2	(6)	2	D_2
C. Equal likelihood	5	6	(7)	D_3
D. Minimax	13	(8)	12	D_2

With this method, if there are n states of nature, the probability of a specific state occurring is $1/n$. Application of this criterion tends to "even out" the values in the payoff table. That is, since each payoff receives the same weight, large positive and large negative payoffs tend to cancel each other out numerically. Once probability values based on the equal-likelihood principle have been assigned, the next step is to weight the payoffs in the table by these values. This procedure is surprisingly similar to expected value analysis (the basic difference is that these probability values are subjective). For the investment problem, the resultant calculations are shown in row C of Figure 15.4. An inspection of these results indicates that decision D_3 yields the largest return.

Minimax Decision Criterion

A central theme throughout this text is the view that most managers, particularly in large institutions, are risk avoiders. This is not surprising given the structural incentives and disincentives associated with most large organizations. It is within this context that a fourth strategy, the **minimax decision criterion,** will be explored. This decision criterion is designed to minimize potential losses associated with a given problem.

Applying the minimax criterion requires the development of a relative loss table, as presented in Figure 15.5. The maximum potential losses for each decision alternative are recorded at the bottom of Figure 15.4, in row D. The optimal decision using the minimax criterion is D_2. Minimax decision analysis, because of its focus on minimizing loss, has found widespread acceptance in the business community.

The previous discussion has shown how the use of alternative decision criteria can result in different decisions. This result should not be surprising. The selection of the appropriate criterion is left to the decision maker. There is no single correct criterion when one is dealing with subjective analysis. The best advice is to select a criterion that fits the decision maker's philosophy, the general circumstances of the problem environment, and the experience gained from previous decisions.

FIGURE 15.5 Loss Table for Perpetual Investments

Condition of Interest Rates	*Decision Alternative*		
	D_1	D_2	D_3
Increasing (I)	13	5	0
Unchanged (U)	5	2	0
Decreasing (D)	0	8	12

DECISION MAKING UNDER RISK

In some instances, decision makers can make reasonably accurate estimates of the chances of alternative future events occurring. One approach to estimating the chances of various future events is to use historical or **prior probability** data. In this case, the decision maker is assuming that the future will mimic the past. An important variation on this theme involves the use of additional information regarding the future. Typically, this takes the form of revising the prior probability estimates using data collected from a survey or other forecasting instrument. As in the situation of decision making under uncertainty, a number of different criteria can be used for choosing among various alternatives when probability data regarding future events are available. The following discussion outlines several of the most popular criteria for decision making under risk.

Maximum Likelihood

The **maximum-likelihood criterion,** as the name implies, focuses exclusively on the event with the highest probability of occurrence. For example, suppose that the manager at Perpetual Investments estimates, from historical experience, the following probabilities for the direction of interest rates: increase = 20%, unchanged = 50%, decrease = 30%. In this case, only the payoffs associated with unchanged interest rates will be considered by the decision maker. More specifically, the manager at Perpetual Investments should invest in precious metals (D_3) since this alternative will yield the largest return if interest rates remain unchanged. This criterion has its share of critics as well as proponents. Critics argue that this criterion ignores too much of the problem information. In the Perpetual problem, for example, two thirds of the payoffs are not considered. On the other hand, proponents suggest that many decision makers do plan based on the most likely outcome.

Expected Value Analysis

Recall that expected value is the weighted average of the various outcomes associated with a given decision. Many problems involving decision making under risk are based on maximizing **expected monetary value (EMV).** That is, the objective is to identify the decision that, on the average, yields the largest expected monetary payoff. The basic principle in expected value analysis is to multiply each payoff by its probability of occurrence, which results in a weighted payoff value.

Expected value analysis is most applicable to decision problems of a repetitive nature. For example, decisions regarding stock market investments are made repeatedly over the life of the portfolio. Expected value (EV) analysis identifies the decision that, on the average, maximizes the payoff. In the short run, however, there is no assurance that the decision selected using this method will yield the largest return. Therefore, care must be exercised in applying this analysis technique to short-run problems or one-time decisions.

The following illustrates the method of computing EMVs for the Perpetual Investments problem using the probability data specified interest rate for movement:

$$\text{EMV}(D_1) = .2(-2) + .5(3) + .3(14) = 5.3$$
$$\text{EMV}(D_2) = .2(6) + .5(6) + .3(6) = 6$$
$$\text{EMV}(D_3) = .2(11) + .5(8) + .3(2) = 6.8$$

Applying the decision rule of maximizing expected monetary value results in choosing D_3 (i.e., invest in precious metals).

EMVs can be computed for problems with any number of alternatives and states of nature by simply summing the products of the payoff values and probabilities over the total number of states. The same procedure can be used to solve cost minimization problems. In these cases, the payoffs (i.e., costs) are multiplied by the probability estimates, resulting in an expected cost value for each decision alternative. The alternative with the minimum expected cost is the optimal decision.

Expected Opportunity Loss

Often, the decision maker is interested in minimizing loss rather than maximizing gain. Problems of this type can also be solved using expected value analysis by modifying the payoff table. Basically, two types of losses may be encountered: absolute and relative. The former involves a net reduction of the decision maker's assets, and the latter indicates the incremental gain that would have resulted if a better decision had been made. In these cases, one must determine the **expected opportunity loss (EOL)** for each decision and select the alternative with the smallest EOL. The opportunity losses are multiplied by the corresponding probabilities of occurrence for each state of nature and summed. This results in an expected or average loss for each decision alternative. The appropriate decision rule is to select the decision that minimizes the expected loss. The expected opportunity losses for Perpetual Investments using the loss table reported in Figure 15.5 are computed as follows:

$$\text{EOL}(D_1) = .2(13) + .5(5) + .3(0) = 5.1$$
$$\text{EOL}(D_2) = .2(5) + .5(2) + .3(8) = 4.4$$
$$\text{EOL}(D_3) = .2(0) + .5(0) + .3(12) = 3.6$$

Notice that the decision that minimizes the opportunity loss for Perpetual Investments is also the one that maximizes expected monetary value, that is, D_3. This result is not a coincidence! The logic behind this relationship is discussed in the following section.

Expected Payoff with Perfect Information

The focus of the discussion so far has been in identifying the optimal decision when the future state of affairs is uncertain. Suppose the decision maker has access to a "device" that forecasts the future with 100% accuracy. That is, it provides the decision maker with perfect information. This device will allow the decision maker to make better decisions than those made using only prior probability data. Typically, these "devices" take the form of business surveys, computer model forecasts, and intelligence gathering instruments.

Let's suppose that the manager at Perpetual Savings can obtain 100% accurate forecasts of future interest rates (a highly unlikely possibility). Based on this information, how much could the manager increase the return on the portfolio? The objective here is to determine the **expected payoff with perfect information** (EPPI). On the one hand, if the manager receives a forecast that interest rates will increase or remain the same, the optimal decision is to invest in precisions metals (D_3), with returns of 11% and 8%, respectively. On the other hand, if the forecast indicates that interest rates will decrease, the optimal choice is to invest in the construction company (D_1), with a return of 14%. The EPPI can then be computed by simply multiplying the returns by the probabilities of interest rates increasing, remaining unchanged, or decreasing, respectively. This yields an EPPI of 10.4%. The difference between expected payoff with perfect information (EPPI) and EMV is called the **expected value of perfect information (EVPI).** It is defined as the maximum amount the decision maker would be willing to pay for perfect information.

Notice that the EOL for the optimal decision is equal to the maximum the decision maker would pay for perfect information (EOL = EPPI − EMV). Thus, EOL analysis provides a direct method for determining the expected value of perfect information. The relationship between EMV, EOL, and EPPI is further illustrated in Figure 15.6. The sum of EMV and EOL equals EPPI for all decision alternatives. This approach provides another method for computing EPPI. The relationship between EMV, EOL, and EPPI holds regardless of the number of decision alternatives, states of nature, and payoff values. Remember, however, that the maximum amount the decision maker is willing to pay for perfect information is equal to the optimal EOL payoff.

FIGURE 15.6 Relationship between EPPI, EMV, and EOL for Perpetual Investments

	Decision Alternative		
Criterion	D_1	D_2	D_3
EMV	5.3	6.0	6.8
EOL	5.1	4.4	3.6
EPPI	10.4	10.4	10.4

In summary, the first step in the solution process is to prepare a payoff table indicating the various outcomes as a function of each decision alternative and state of nature. Once the payoffs have been formulated, the next step is to compute an expected value for each alternative by multiplying the payoff value by the corresponding state of nature probability. The alternative with the largest (in the case of a maximization problem) or smallest (in the case of a minimization problem) expected value is selected as optimal. Sometimes several alternatives yield the same optimal solution. In these cases, each alternative should be identified.

DECISION TREES

Key idea 2

A **decision tree** provides a graphical representation of the problem structure and is a particularly effective approach for evaluating multisequenced decision problems. Multisequenced decision making is characteristic of many problems occurring in today's business environment. For example, a marketing manager may wish to introduce a new product immediately or wait for the results of a market survey. If the latter course is taken, then, based on the survey results, the manager may elect to introduce the new product or continue to withhold it from the marketplace.

Basically, a decision tree consists of **branches** and **nodes.** There are three types of nodes in a decision tree: decision, chance, and terminal. Decision nodes, symbolized by a box, indicate the number of decision alternatives at a specific point in the tree. Chance nodes, symbolized by a circle, indicate points involving two or more states of nature with known probabilities. Terminal nodes, symbolized by a triangle, indicate the end of a specific branch and the associated payoff. To avoid confusion, a numbering system should be used to assign a unique number to each node. Branches then are used to interconnect the various nodes.

The following three steps outline the analytical procedure for solving problems using the decision tree methodology.

1. Formulate the decision tree structure.
2. Assign numerical values to the various branches.
3. Analyze the decision tree by working backward from the terminal nodes.

FIGURE 15.7 Decision Tree for Perpetual Investments

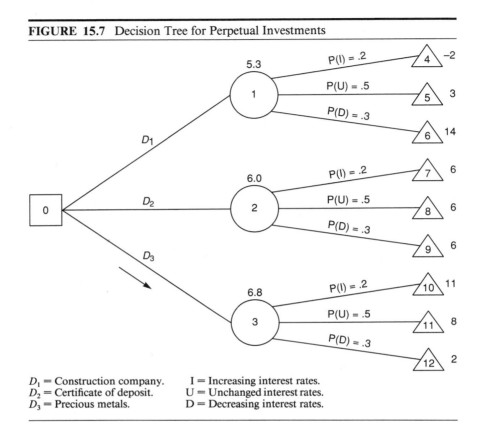

D_1 = Construction company.
D_2 = Certificate of deposit.
D_3 = Precious metals.

I = Increasing interest rates.
U = Unchanged interest rates.
D = Decreasing interest rates.

Figure 15.7 illustrates the use of a decision tree in describing the Perpetual Investments problem. The tree is composed of a decision node, three chance nodes, and nine terminal nodes. Each decision alternative is assigned a unique identifying symbol to the right of the decision node. The chance nodes are characterized in terms of the number of branches to the right and the probability value for each branch. The terminal nodes indicate the specific payoff value for each alternative and state-of-nature combination.

The first two steps in solving the Perpetual Investments problem using the decision tree method have now been completed—that is, formulation of the tree structure and assignment of numerical values. The final step is to determine which of the decision alternatives yields the maximum payoff. A solution to the problem can be developed by working backward from the terminal nodes on the right to the starting node on the left. The following two rules provided a guide to this process:

- At a chance node, the values of the connected branches to the right are weighted by the probabilities.

- At a decision node, an alternative is selected based on the preestablished decision criterion (e.g., maximization of profit).

Applying these rules to the decision tree in Figure 15.7 results in the values shown in parentheses above each node. Each of these values indicates the payoff at that point. To illustrate this process, consider node 1. Since node 1 is a chance node, the rule is to weight each branch value to the immediate right by the corresponding probability value. This procedure yields the following result:

$$\text{Node } 1 = .2(-2) + .5(3) + .3(14) = 5.3$$

The same procedure is used to determine the expected payoffs for the remaining chance nodes.

The backward process continues until the starting node is reached (i.e., node 0). This node is a decision node. The expected values for the three alternatives at this node are 5.3, 6.0, and 6.8. Since Perpetual's management is interested in maximizing return, the optimal decision is to invest in D_3, or precious metals. The optimal decision yields a 6.8% return, which is precisely the same result obtained using the payoff table.

The decision tree method offers two major benefits over the payoff table approach. First, the tree provides a graphic overview of the entire problem (i.e., a picture is worth a thousand words!). Second, for multisequenced decision problems, a decision tree provides a more efficient solution approach.

DECISION MAKING WITH ADDITIONAL INFORMATION

Key idea 3

The previous discussion has shown how uncontrollable events can affect the decision process in a dramatic way. The impact of changes in model data (in particular, the prior probabilities) on the decision process can be significant. In some cases modest changes in the data can result in a different optimal solution. Accordingly, the decision maker may be uncomfortable selecting from a set of alternatives for an analysis based solely on prior probabilities.

One way to reduce the level of uncertainty is to acquire additional information on the probabilities of future events. The acquisition of new facts can be used to update or revise the prior data. This updating will normally lead to an improved forecast of future events. Firms traditionally undertake market surveys on future product demand and conduct tests on product reliability prior to releasing a product into the marketplace. These activities are designed to provide the decision maker with a more precise projection of the future and thus improve the decision process.

Since the collection of additional data is not free, the decision maker must weigh the potential benefits of this new information with its costs. In general, if the benefits derived from the additional data exceed the costs of obtaining the data, the decision maker should acquire the data. The maximum one should be willing to pay for perfect information has already been discussed.

In most instances, however, perfect information is not available. Therefore, the decision maker must settle for an imperfect but, hopefully, an improved estimate of future events. The following discussion provides an approach for analyzing the benefits and costs of additional information that is not perfect.

Incorporating Additional Data into the Model

Estimates regarding the future can come from many sources (e.g., market surveys and economic forecasts). However, the decision maker must have confidence in a source's overall accuracy. In the case of a market survey, the decision maker should be concerned about the methods used in collecting the sample and in processing the data. Suppose that a market research firm has approached Perpetual's management with a proposal to conduct a survey on the future direction of interest rates for a fee of $15,000. The survey would provide Perpetual with a revised and presumably improved estimate of the direction of interest rates. To evaluate this proposal, Perpetual's management have requested from the research firm a summary of the firm's past performance. Management wish to know whether they should spend the $15,000 to obtain the forecast. Based on previous analysis, they already know that the company would be willing to pay up to 3.6% of the total investment portfolio for perfect information. Recall that Perpetual's management has raised $500,000. This suggests that they would be willing to pay up to $18,000 for perfect information (i.e., $500,000 × .036 = $18,000). Therefore, the management should further evaluate the proposal since the fee proposed by the consulting firm is less than this maximum. On the other hand, if the proposed fee were greater than $18,000, the proposal could be dismissed without further consideration.

The second part of this evaluation involves the consulting firm's past record in predicting interest rates. The accuracy or track record of the **indicator** (the terms *estimator* and *predictor* are also used in this regard) is often presented in the form of a conditional probability table. This format is used because it provides a direct comparison between the prediction and the state of nature (i.e., the actual outcome). Figure 15.8 presents such a table showing

FIGURE 15.8 Conditional Probability Table for Perpetual Investments

	Prediction		
Actual	Increase (i)	Unchanged (u)	Decrease (d)
Increase (I)	.8	.2	0
Unchanged (U)	.2	.7	.1
Decrease (D)	.2	.2	.6

FIGURE 15.9 Perfect Information Conditional Probability Table for Perpetual Investments

	Prediction		
Actual	Increase (i)	Unchanged (u)	Decrease (d)
Increase (I)	1.0	0	0
Unchanged (U)	0	1.0	0
Decrease (D)	0	0	1.0

the historical performance of the consulting firm in predicting interest rates. For example, the data show that the firm predicted that interest rates would increase 8 times out of 10 when they actually did increase. Two times out of 10 the firm predicted that interest rates would remain unchanged when they actually increased. Similarly, the firm predicted that interest rates would decrease 6 times out of 10 when they actually did decrease. In general, this table reports conditional probability values in terms of the prediction given the actual state, for example, $P(\text{increase}|\text{Increase}) = .8$.

Notice that the firm does not have a perfect record in predicting trends in interest rates. A perfect record would have been indicated by the presence of 1's along the main diagonal and 0's for the remaining values in the conditional table. Figure 15.9 indicates the case of perfect information for the current problem. The worst case (i.e., where the data provide no additional information) is illustrated in Figure 15.10. Notice that the probability of predicting a particular state given an actual state is one third, which is equivalent to the equal-likelihood subjective criterion, where the probability of occurrence is equal to 1 divided by the number of states. This suggests that the data contain no real information.

Unfortunately, the conditional probabilities given in Figure 15.8 cannot be used directly. The model requires that the conditional probability statements appear in the form $P(\text{state}|\text{prediction})$, whereas the data supplied by

FIGURE 15.10 Worst-Case Conditional Probability Table for Perpetual Investments

	Prediction		
Actual	Increase (i)	Unchanged (u)	Decrease (d)
Increase (I)	.33	.33	.33
Unchanged (U)	.33	.33	.33
Decrease (D)	.33	.33	.33

the consulting firm is in the form $P(\text{prediction}|\text{state})$; this is, by the way, how such data are usually presented. Therefore, the data need to be converted to a form that can be incorporated into the decision model.

Bayesian Analysis

Bayesian analysis is the method used to compute revised or **posterior probabilities** given prior and conditional probabilities. Figure 15.11 provides an overview of the Bayesian methodology. This diagram shows how the data are combined to generate the revised conditional probabilities, of the form $P(\text{state}|\text{prediction})$.

Figure 15.12 presents a probability tree (probability trees, introduced in Chapter 4, are somewhat similar to decision trees) showing the nine basic combinations of states and indicators and the associated probabilities. The diagram also shows the marginal probabilities for each of the indicators. Each marginal probability, which indicates the chances of a particular prediction by the consulting firm, was obtained simply by summing the products of the prior probabilities and conditional probabilities for a given indicator. For example, the probability that the consulting firm will indicate that interest rates will increase is 32%. Similarly, the chances that the consulting firm will predict that interest rates will remain unchanged are 45%, and the probability of a predicted decrease is 23%. Notice that the sum of the predictions must equal 1.

The probability data presented in Figure 15.12 can now be used to determine revised conditional probabilities. This is done by dividing the product of the prior and conditional probabilities by the marginal probability. For example, the conditional probability $P(\text{Increase}|\text{increase})$ is computed as:

$$P(\text{Increase}|\text{increase}) = P(\text{I}|\text{i}) = \frac{P(\text{I})\ P(\text{i}|\text{I})}{P(\text{i})} = \frac{.16}{.32} = 0.5$$

FIGURE 15.11 Bayesian Methodology for Revising Prior Probabilities

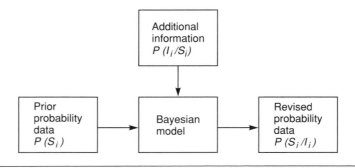

FIGURE 15.12 Probability Tree for Determining Marginal and Joint Probabilities for Perpetual Investments

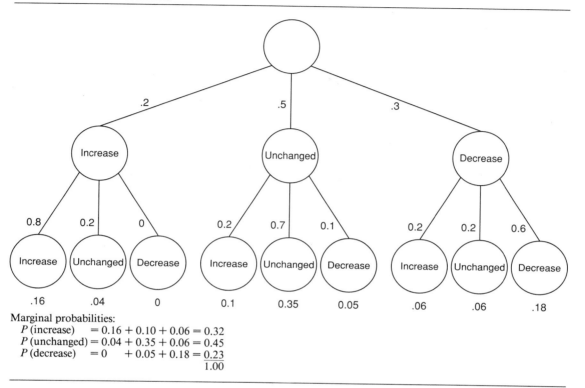

The revised probability values for the entire problem are reported in Figure 15.13.

Figure 15.14 shows a decision tree incorporating the revised probability data. In this case, Perpetual's management must first decide whether to purchase the survey. This decision is reflected at node 0 (i.e., purchase/do not purchase). If the management decide not to buy, the problem becomes the same as the original situation. If Perpetual does purchase the survey, the next step is to use the survey results to help identify the optimal investment decision. This relationship is shown at node 1. The marginal probability data show that there is a 32% chance that the survey will predict increasing interest rates, a 45% chance of a prediction of unchanged interest rates, and a 23% chance of a predicted decrease in interest rates. Based on these results, management can decide on the optimal investment strategy given an interest rate prediction from the survey. For example, the decision to select one of the

FIGURE 15.13 Revised Probability Values for Perpetual Investments

	Prediction		
Actual	Increase (i)	Unchanged (u)	Decrease (d)
Increase (I)	.50	.09	0
Unchanged (U)	.31	.78	.22
Decrease (D)	.19	.13	.78

three alternatives given a forecast of increasing interest rates is shown at node 2. Notice that alternative D_1 appears even though the survey results suggest otherwise. Clearly, management can decide to invest in D_1 even though the survey results do not support this decision. In general, a decision tree should include all decision alternatives, even those that appear illogical.

This decision tree can now be evaluated using the procedures outlined earlier. The numerical values above the nodes indicate the payoff at that point (e.g., the payoff at node 5 is 2.6). The arrows at the decision nodes indicate the optimal decision path. For example, the optimal decision at node 3 is to select decision D_3 based on maximizing expected value. The cost-effectiveness of the consulting firm's bid can now be evaluated using the expected value reported at node 1. Recall that the consulting firm's proposed fee for the survey was $15,000, or 3% of the investment budget. The **expected payoff of the additional information (EPAI)** is 8.7%. This is in contrast to the expected value of not purchasing the survey, 6.8%. Therefore, Perpetual's management should be willing to pay up to a maximum of 1.9% of the total investment, or $9,500, for the survey. This quantity is often known as the **expected value of additional information (EVAI).** Since the proposed price is greater than this amount, the optimal decision sequence is for Perpetual's management not to purchase the survey at the proposed price and to select decision alternative D_3 based on the prior analysis. The preceding presentation illustrates the power of the decision tree approach in analyzing problems with multiple-sequenced decisions.

UTILITY AS A DECISION CRITERION

Key idea 4

So far the selection criteria used in the decision-making process have been based on expected monetary value. In many business situations, however, this approach can lead to suboptimal decisions because of the different emphasis individual decision makers place on the value of money. To illustrate this issue, suppose a contestant on a TV game show has been given the choice between a cash prize of $1,000 or a box that has a 50-50 chance of containing either $10,000 or $0. What decision should the contestant make?

FIGURE 15.14 Decision Tree for Revised Perpetual Investments Problem

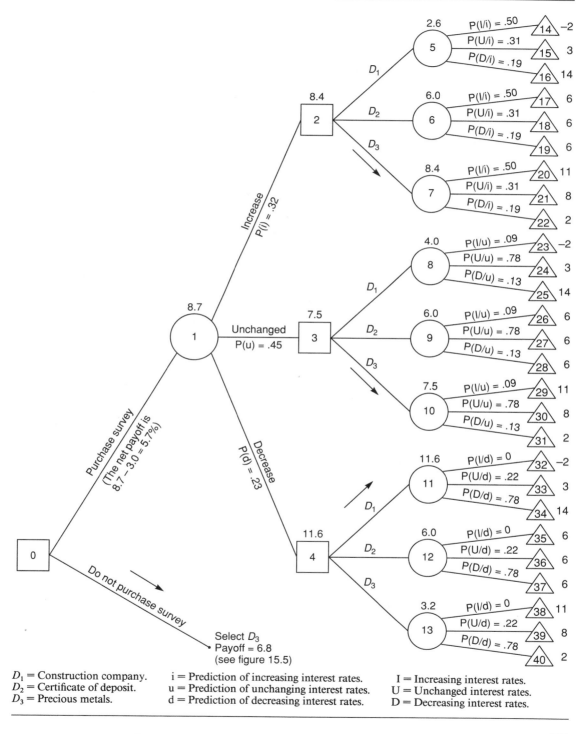

D_1 = Construction company.
D_2 = Certificate of deposit.
D_3 = Precious metals.

i = Prediction of increasing interest rates.
u = Prediction of unchanging interest rates.
d = Prediction of decreasing interest rates.

I = Increasing interest rates.
U = Unchanged interest rates.
D = Decreasing interest rates.

If the contestant makes the decision based on expected value analysis, the optimal choice is to select the box. The expected value of the box is $5,000 (.5 × 5,000 + .5 × 0). Most decision makers, however, would probably chose the cash prize of $1,000 because of the inherent risk in the alternative. Most individuals are basically risk avoiders (in both their personal and their business behavior). Since this is the case, the individual's attitude toward risk needs to be incorporated into the decision analysis model for certain types of problems. **Utility** theory provides a method for introducing risk preferences into the model. The objective is to select the decision that maximizes expected utility rather than expected monetary outcome. Utility analysis should be considered one's preference for money varies considerably from a risk-neutral position. The most difficult problem in applying utility analysis is developing an appropriate utility function.

Utility can be defined in several ways, including the notion of cardinal utility. The general procedure behind cardinal theory is to assign numerical values (utiles) to various monetary outcomes. Utility is measured on a scale between 0 and 1. Typically, the worst monetary outcome is assigned a value of 0 and the best monetary outcome a value of 1. The remaining monetary outcomes are assigned utility values between 0 and 1 depending on one's preference for risk. The process for developing a utility function is illustrated in the following section.

Developing a Utility Function

The utility functions for three different classes of decision makers are depicted in Figure 15.15. The risk preferences for the three decision makers are quite apparent. The risk avoider is conservative and tends to avoid situations where larger losses might occur. As the monetary outcomes decrease, the risk-avoider's utility decreases at an ever-increasing rate.

The risk-neutral decision maker's incremental value of utility is constant over the full range of monetary outcomes. This individual receives about the same increase in utility for a given increase in monetary income as the decrease in utility suffered for the same given decrease in monetary income. For example, if a $50,000 increase in the risk-neutral individual's income yields a .2 increase in utility, then a $50,000 decrease in income would result in a .2 decrease in utility. When the decision maker is risk-neutral, either utility considerations or expected value analysis yield the same results. Both analytical approaches, in this case, yield the same result.

The risk seeker obtains considerably more utility for a given monetary increase than the decrease in utility suffered for the same monetary loss. As monetary income increases, the risk seeker's utility increases at an ever-increasing rate. Generally speaking, one's utility function may be a composite of all three of these basic cases. Utility functions of this type are often constructed by interviewing the individual using a series of monetary outcomes for the specific problem.

FIGURE 15.15 Basic Risk Preference Models

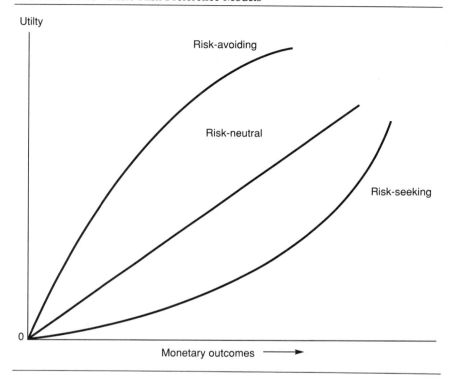

As indicated earlier, the construction of a utility function is based on assigning a utility value of 0 for the worst outcome and a value of 1 for the best outcome. For Perpetual these limits are:

$$\text{Utility}(-2) = U(-2) = 0$$
$$\text{Utility}(14) = U(14) = 1$$

Utility values for the remaining payoffs can be obtained using the following lottery process.

Suppose Perpetual's management is asked whether they prefer a 6% return with certainty (i.e., alternative D_2) or would be willing to engage in a lottery where:

- Perpetual obtains a payoff of 14% with probability P.
- Perpetual obtains a payoff of -2% with probability $1 - P$.

Generally, if P is near 1, Perpetual's management would select the lottery over the certain payoff. Conversely, if P is near 0, then the company would select the certain payoff. Obviously, as P moves from 0 to 1, management's preference for certainty will shift to preference for the lottery. Suppose management has indicated indifference between the certain payoff and the lottery when $P = .8$. Then the utility for a payoff of 6% can be determined as follows:

$$U(6) = P \times U(14) + (1 - P) \times U(-2)$$
$$U(6) = .8(1) + .2(0) = .8$$

Notice that the probability is exactly equal to the utility value. This result is always true; therefore, utility values can be assigned directly based on the estimates of P obtained from the lottery process.

The same procedure would be repeated until the utility values for all of the intermediate payoffs are defined. Figure 15.16 presents the resultant utility function reflecting the company's overall risk preference.

Expected Utility Analysis

The utility function given in Figure 15.16 can be used to convert the payoffs given in Figure 15.3 into utility values. The approach is to substitute the utility estimates for each return outcome. Figure 15.17 presents the payoff table showing both return values and corresponding utility, in parentheses. The expected utility value for the three alternatives is computed in exactly the same way as for the expected monetary value. For example, $\text{EUV}(D_1) = (.2)0 + (.5).6 + (.3)1 = .6$. From these results, the optimal decision, based on maximizing expected utility value, is alternative D_2. Observe that the use of utility instead of monetary outcomes yields a different optimal decision.

SENSITIVITY ANALYSIS

Sensitivity analysis can be used for studying the impact of changes in model parameters (e.g., probability values) on the optimal solution. This capability is important in light of the fact that most of the data used in business decisions involve some degree of uncertainty. The primary focus of sensitivity analysis in this chapter is to identify the range of values for the model coefficients within which the current solution remains optimal. In some cases, analysis will reveal that a slight change in one or more model parameters will change the decision. Accordingly, this reassessment may indicate the need to collect additional data prior to making a final decision.

The primary task of sensitivity analysis is to determine the range of feasibility for each model parameter (i.e., probability and payoff coefficients). The range of feasibility indicates the range over which the model coefficients can vary without resulting in a change from the current optimal decision.

15.4 Formulating Decision Analysis Models

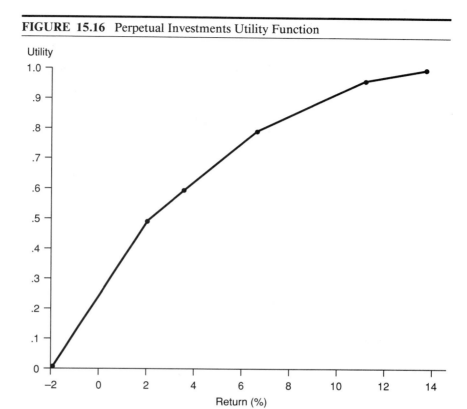

FIGURE 15.16 Perpetual Investments Utility Function

FIGURE 15.17 Perpetual Investments Utility Payoff Analysis

State of Nature	Decision Alternatives			Probability
	D_1	D_2	D_3	
Increasing (S_1)	−2	6	11	.2
	(0)	(.8)	(.95)	
Unchanged (S_2)	3	6	8	.5
	(.6)	(.8)	(.8)	
Decreasing (S_3)	14	6	2	.3
	(1)	(.8)	(.5)	
Expected utility	.6	.8	.74	

Recall that the optimal decision for Perpetual Investments, based on EV analysis, is to invest in precious metals (alternative D_3). One task is to determine the range of probability values over which this decision remains optimal. Clearly, increasing the probability that interest rates will rise or remain unchanged improves the desirability of this decision. However, the other two alternatives become more attractive as the probability increases that interest rates will decline.

To determine the critical probability values, a variable is assigned to each state of nature (e.g., P_1 for the first state, P_2 for the second state, and P_3 for the third state). Next the expected values for *two* of the decision alternatives are analytically equated. This equivalence relationship embodies the *principle of indifference*. At this point, the decision maker is indifferent between the two decision options. That is, either decision alternative yields exactly the same expected value. The computational process for determining the critical probabilities for alternatives D_2 and D_3 is presented in Figure 15.18. The fact that P_1, P_2, and P_3 sum to one is an important consideration in the analysis. The calculations in Figure 15.18 show that as long as the probability that interest rates will increase remains above approximately 11%, alternative D_3 remains the optimal decision with respect to decision D_2. On the other hand, if the probability of interest rates increasing drops below 11%, then alternative D_2 is the right choice. Note that both of these conclusions are based on P_2 remaining at 50%. The same computational process would be repeated for comparing alternatives D_1 and D_3. In general, a total of $n - 1$ comparisons are required, where n is the number of decision alternatives.

FIGURE 15.18 Probability Sensitivity Analysis for Perpetual Investments

State of Nature	Decision Alternative			Probability
	D_1	D_2	D_3	
Increasing (I)	-2	6	11	P_1
Unchanged (U)	3	6	8	P_2
Decreasing (D)	14	6	2	P_3

$$EV(D_2) = EV(D_3)$$
$$6P_1 + 6P_2 + 6P_3 = 11P_1 + 8P_2 + 2P_3$$
$$P_1 + P_2 + P_3 = 1$$

Substituting $(1 - P_1 - P_2)$ for P_3 yields the following:

$$P_1 = 4/9 - (2/3)P_2$$

If P_2 remains at .5, then:

$$P_1 = .11$$

Conclusion: Invest in precious metals (D_3) as long as the probability that interest rates will increase is greater than or equal to .11, given that $P_2 = 0.5$.

This same general approach can be used to analyze the impact of changes in the payoff values on decision selection. Figure 15.19 presents the range of feasibility for the first payoff value, -2%. In this case, a variable value (e.g., X) is assigned in place of the original payoff value. For example, equating the expected values for alternatives D_1 and D_3 and solving for X results in a critical value of 5.5%. As long as the first payoff remains equal to or below this value, the optimal decision is to select alternative D_3. The same technique can be used to analyze the impact of changes in the remaining payoff values on the optimal decision. Typically, these computations are best performed by computer analysis, as will be demonstrated in the next section.

The results from sensitivity analysis provide the decision maker with a guide to the impact of changes in model data on the optimal solution. Often, the decision maker will collect additional information to improve the accuracy of the data base prior to making a decision. It is important to remember that the preceding discussion is based on changing only one of the payoffs at a time. The simplest approach for analyzing the impact of multiple data changes is to use computer analysis. Sensitivity analysis can also be employed for problems involving utility considerations or subjective criteria. The procedure outlined here provides the general framework for analyzing the impact of changes in each of these cases on the optimal decision.

15.5 COMPUTER ANALYSIS

Key idea 5

Business problems involving large numbers of decisions are usually solved using computer analysis. CBS-generated results for Perpetual Investments are presented in Figure 15.20. The output identifies the expected monetary value (EMV) for each alternative, the expected opportunity loss (EOL), the

FIGURE 15.19 Payoff Value Sensitivity Analysis for Perpetual Investments

State of Nature	Decision Alternative			Probability
	D_1	D_2	D_3	
Increasing (S_1)	X	6	11	.2
Unchanged (S_2)	3	6	8	.5
Decreasing (S_3)	14	6	2	.3

$$EV(D_1) = EV(D_3)$$
$$.2X + .5 \times 3 + .3 \times 14 = .2 \times 11 + .5 \times 8 + .3 \times 2$$
$$.2X = 1.1$$
$$X = 5.5$$

Conclusion: Invest in precious metals (D_3) as long as the payoff for state S_1 and decision D_1 remains less than or equal to 5.5%.

FIGURE 15.20 CBS Payoff Analysis for Perpetual Investments

```
CBS-Decision Analysis                              03-15-1990 - 10:17:20
                           Results

                    Decision Alternatives

        Criteria    D1          D2          D3

        EMV         5.30        6           6.80
        EOL         5.10        4.40        3.60
        EPPI       10.40       10.40       10.40


               Optimal Decision:   D3 (# 3)




                           press ←┘
```

FIGURE 15.21 CBS Payoff Sensitivity Analysis for Perpetual Investments

```
CBS-Decision Analysis                              05-16-1990 - 14:15:58
                       Sensitivity Analysis

                    Lower       Current     Upper
        Decision    Limit       Value       Limit

        D1          NO LIMIT    -2           5.500
                    NO LIMIT     3           6.000
                    NO LIMIT    14          19.000

        D2          NO LIMIT     6          10.000
                    NO LIMIT     6           7.600
                    NO LIMIT     6           8.667

        D3           7          11          NO LIMIT
                     6.400       8          NO LIMIT
                    -0.667       2          NO LIMIT




                           press ←┘
```

expected payoff with perfect information (EPPI), and the optimal decision. Notice that these results are consistent with those developed earlier.

Computer modeling is also employed for performing sensitivity analysis. Figure 15.21 shows a CBS-generated sensitivity analysis of the payoff values for the Perpetual Investments problem. This output indicates the range (i.e., lower and upper limits) over which the current decision remains optimal. These results show that the payoff values associated with the current optimal decision, D_3, can be increased without limit and not change the current solution. On the other hand, the optimal decision will change if either payoff associated with D_3 drops below the values reported in the lower-limit column. The results generated for alternative D_2 show that the payoff values associated with this decision can be decreased (without limit) and not change the current optimal solution. However, if one of the payoffs associated with D_2 is increased above the reported upper limits, the optimal decision would shift from D_3 to D_2 (e.g., increasing the payoff from 6 to above 10). Typically, the most efficient approach for identifying the optimal solution when the revised data values fall outside the reported limits is to rerun the model with the new data.

15.6 PRACTICAL APPLICATIONS

This section outlines several application areas where the techniques introduced in this chapter can be effectively applied.

Peacock Bakery (Marginal Analysis Using Discrete Distributions)

The Peacock Bakery sells birthday cakes. A birthday cake costs $2 per unit and sells for $8 per unit. Cakes left over at the end of the day have no value. The daily demand distribution for cakes is shown in Figure 15.22. The owner wishes to know the number of cakes that maximizes profit. This problem

FIGURE 15.22 Demand Distribution for Peacock Bakery

Daily Demand	Demand Probability	Cumulative Demand Probability
1	.10	1.00
2	.20	.90
3	.30	.70
4	.20	.40
5	.10	.20
6	.10	.10

could be solved using a standard payoff table. However, incremental analysis tends to be a more effective approach for problems with a large number of discrete demand states ($n > 5$).

The basic decision rule is to increase the number of units to be stocked (in this case, cakes) until the expected marginal loss exceeds the expected marginal profit. Mathematically, this decision rule can be expressed using expected value analysis:

$$EV(MP) \geq EV(ML)$$

where:

MP = Marginal profits
ML = Marginal losses

Letting P stand for the probability that demand will be greater or equal to supply, the above relationship can be rewritten as follows:

$$P(MP) \geq (1 - P)ML$$

Solving for P yields the critical probability value in terms of marginal profits and marginal losses:

$$P \geq \frac{ML}{ML + MP}$$

This relationship indicates that as long as the probability of selling one more cake is greater than or equal to the ratio of ML to the sum of ML plus MP, then the owner should stock an additional cake.

The marginal profit for the Peacock Bakery is $6 (i.e., $8 − $2), and the marginal loss is $2. Therefore, the critical probability value is .25 (2/8). Referring to Figure 15.22, the optimal policy is to stock four cakes since the cumulative probability of selling four or more cakes is greater than the critical probability of .25. If the owner bakes five instead of four cakes, the actual profit level will be reduced since marginal losses will exceed marginal profits.

Demptol Company (Marginal Analysis with Continuous Distributions)

The Demptol Company is considering the introduction of a new compact hair drier. The production setup cost is $100,000, and the variable cost is $2 per drier. The projected demand is 30,000 units with a standard deviation of 10,000 units. The wholesale price of the drier is $7 per unit. Management wishes to know the break-even point and the probability of making a profit.

The break-even point is defined as the minimum number of units that need to be sold to cover the costs of production (i.e., for total revenues to equal total costs). This relationship can be expressed mathematically as:

$$\text{Break-even point (units)} = \frac{\text{Fixed costs}}{\text{Price/unit} - \text{Variable cost/unit}}$$

For this problem:

$$\text{Break-even point} = \frac{\$100,000}{\$7/\text{unit} - \$2/\text{unit}} = 20,000 \text{ units}$$

The probability of exceeding the break-even point can be determined from the normal distribution. Figure 15.23 shows the normal distribution for the Demptol problem. The shaded area to the right of the 20,000-unit line represents the required probability.

The probability table given in Appendix A (Table A.5) or CBS (Appendix B) along with the basic Z-score relationship will be used to determine the probability of making a profit. The Z-score relationship in terms of the current problem is:

$$Z = \frac{\text{Break-even point} - \text{Mean demand}}{\text{Standard deviation}}$$

Substituting the appropriate values yields:

$$Z = \frac{20,000 - 30,000}{10,000} = -1$$

Referring to Table A.5 or CBS, the probability value corresponding to a Z score of -1 is .8413. Management therefore knows that there is approximately an 84% chance of making a profit.

FIGURE 15.23 Demand Distribution for Demptol Driers

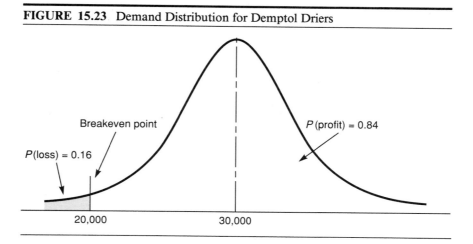

FIGURE 15.24 Payoff Table for Graybuilt Plant Expansion

State of Economy	Decision Alternative		
	D_1	D_2	D_3
Expanding	1,000	2,000	3,000
Declining	1,000	800	−1,000

Graybuilt Corporation (Subjective Decision Analysis Using Weighted Averages)

The general manager at Graybuilt Corporation is evaluating three different plant expansion options: add on to the existing facility (D_1), erect a new building on the existing site (D_2), or erect a new building at a new site (D_3). These options along with their respective net payoffs (in $000) are given in Figure 15.24. These payoffs are influenced by the future state of the economy (S_1 = Improving economy, S_2 = Declining economy). Unfortunately, the general manager has been unable to estimate the probability of either state of nature and therefore has decided to use subjective analysis. The general manager is a prudent decision maker and wishes to select a decision rule that is a compromise between the standard optimistic (maximax) and pessimistic (maximin) subjective decision criteria. This criterion is often called the *weighted average* or *Hurwicz rule*.

In applying this rule, the general manager must select the value of the weighting coefficient a. This coefficient ranges in value between 0 and 1. A value near 0 indicates that the decision maker is pessimistic, and a value near 1 indicates that the decision maker is optimistic. The relationship for computing a weighted average based on a given a is as follows:

Weighted average = $a \times$ (Maximum in column) + $(1 - a) \times$ (Minimum in column)

Notice that when $a = 0$ the optimal decision is based on the minimum value in the column and when $a = 1$ the decision is based on the maximum in the column. Figure 15.25 presents an analysis of the Graybuilt problem for various weighting coefficient values. Notice how the optimal decision shifts from D_1 to D_2 to D_3 as the weighting coefficient increases.

FIGURE 15.25. Optimal Payoff Analysis for Graybuilt Expansion

Weighting Coefficient	Payoff ($000)	Optimal Decision
0	1,000	D_1
.3	1,160	D_2
.5	1,500	D_3
.7	1,800	D_3
1.0	3,000	D_3

15.7 CASE STUDY: CLEANALL CORPORATION

Cleanall Corporation, a market leader in the production of specialized detergents, is considering the introduction of a new spot remover product. Cleanall's manufacturing facility can produce 50,000 cases of the new product per year with a setup cost of $100,000. The average variable cost per case of product is $5. Orders above 50,000 cases will be subcontracted overseas to a Korean firm at a variable cost per case, of $9. Whether the product is manufactured by Cleanall or by the Korean vendor, there is a $12 cost per case for marketing and overhead.

Cleanall's marketing manager is considering placing an order for an extra 50,000 cases, 75,000 cases, and 100,000 cases. The actual level of sales will depend on the state of the economy. Sales estimates prepared by the marketing manager indicate that if the economy is strong, Cleanall will sell 100,000 cases; if the economy is moderate, they will sell 70,00 cases; and if the economy is weak, they will sell only 40,000 cases. The marketing department forecasts a 45% chance of a strong market and a 20% chance of a weak market.

A case of the new product is priced to sell at $39.95. Unsold units will be sold to an overseas distributor at a 55% discount. Before deciding on a method and level of production, the production manager wishes to evaluate the possibility of hiring a local market research firm to conduct an economic survey. The manager has received a proposal from a local marketing survey firm. This firm has offered to conduct the survey for $20,000. The historical performance of the survey firm in predicting economic activity is summarized as follows:

- Correctly predicted a strong economy 12 out of 16 times (4 times they predicted a moderate economy and 2 times a weak economy).
- Correctly predicted a moderate economy 7 out of 10 times (2 times they predicted a strong economy and 1 time a weak economy).
- Correctly predicted a weak economy 9 out of 10 times (1 time they predicted a moderate economy).

The marketing manager wishes to know:

1. The optimal quantity of cases to produce.
2. The number of cases to be produced by the overseas vendor.
3. The expected value of the optimal decision.
4. Whether or not the company should purchase the economic survey.
5. The maximum the company should pay for the survey.
6. The maximum the company should pay for perfect information.
7. What production policies should Cleanall employ based on the results of the market survey?

This is a problem involving decision making under risk with additional information where the objective is maximize profit. The first step is to compute the profit level for each pairing of production level and state of economy. These estimates will be used to develop a payoff table. It is assumed that there are no goodwill costs associated with failure to meet demand.

The following highlights the development of the profit function and payoff table:

- The profit function when $P \leq D$:

$$\text{Revenue} = 39.95P$$

$$\text{Cost of goods sold} = 21P - 100,000$$

$$\text{Profit} = 18.95P + 100,000$$

For example, when $P = 50,000$ units and $D = 70,000$ units, then Profit = $1,047,500.

- The profit function when $P > D$ is:

$$\text{Revenue} = 21.97D + 17.98P$$

$$\text{Cost of goods sold} = 21P - 100,000$$

$$\text{Profit} = 21.97D - 3.02P + 100,000$$

For example, when $P = 75,000$ units and $D = 70,000$ units, then Profit = $1,411,400. Figure 15.26 presents the developed payoff table.

Presented in Figure 15.27 are the conditional and prior probabilities derived from the problem statement. A CBS analysis of this problem is given in Figure 15.28.

The following list highlights the results from the CBS analysis:

1. The optimal production level is 100,000 units.
2. Produce 50,000 units in-house and contract 50,000 units with the Korean vendor.
3. The expected value of the optimal decision is $1,500,675.
4. The expected value of the additional information is less than the proposed cost ($17,086 versus $20,000). Therefore, Cleanall should *not* purchase the service.

FIGURE 15.26 Payoff Table for Cleanall Corporation

Economy	Production Alternative (units)		
	50,000	75,000	100,000
Weak	827,800	752,300	676,800
Moderate	1,047,500	1,411,400	1,335,900
Strong	1,047,500	1,521,250	1,995,000

5. A maximum of $17,086 should be paid for the survey.
6. The maximum that should be paid for perfect information is $56,625.
7. The company should keep the same production policy (i.e., 100,000 units).

FIGURE 15.27 Conditional and Prior Probabilities for Cleanall Corporation

	Prediction			
Actual	Weak	Moderate	Strong	Prior
Weak	.90	.10	0	.20
Moderate	.10	.70	.20	.35
Strong	0	.25	.75	.45

FIGURE 15.28 CBS Analysis of Cleanall Data

Model: Payoff Table Analysis — Output

	Decision Alternative		
Criterion	P_1	P_2	P_3
EMV	1,003,560	1,329,012	1,500,675
EOL	553,740	228,288	56,625
EPPI	1,557,300	1,557,300	1,557,300

Bayesian Analysis

Marginal Probabilities

Prediction	Probability
Weak	0.215
Moderate	0.378
Strong	0.407

Posterior Probabilities (state)

Prediction	Weak	Moderate	Strong
Weak	0.837	0.163	0
Moderate	0.053	0.649	0.298
Strong	0	0.172	0.828

Worth of Additional Information

EPAI = 1,517,761
EVAI =   17,086

15.8 SUMMARY

This chapter has introduced several models for analyzing decision making under risk and uncertainty. The chapter also introduced the payoff table and the decision tree as two alternative methods for solving decision problems. The primary objective in decision analysis problems is to find the optimal decision(s) from among a set of alternatives based on maximizing the expected value of the resultant payoff (decision making under risk) or on a specific subjective decision criterion (decision making under uncertainty). The basic data required monetary values for each decision alternative and state-of-nature combination, and the prior probabilities for each state of nature. For a payoff table analysis, the data are presented in the form of a payoff table, with the rows describing the states of nature and the columns the decision alternatives. The incorporation of additional information and utility represent two common extensions to the basic decision analysis model.

15.9 GLOSSARY

Bayesian analysis A method used to compute posterior conditional probabilities from prior probabilities.

branches The connecting lines between nodes in a decision tree.

decision making under certainty The process of selecting from a set of alternatives when the outcome of all future events is known.

decision making under risk The process of selecting from a set of alternatives when the outcome of all future events is unknown but can be characterized probabilistically.

decision making under uncertainty The process of selecting from a set of alternatives when the outcome of all future events is unknown.

decision tree A graphical tool used to solve multisequenced decision models.

equal-likelihood decision criterion A decision strategy that yields the largest (or smallest) payoff from a set of alternatives based on weighting each state of nature equally.

expected monetary value (EMV) The weighted monetary payoff (i.e., average) of each decision alternative over the states of nature.

expected opportunity loss (EOL) The weighted loss (i.e., average loss) of each decision alternative over the states of nature.

expected payoff with additional information (EPAI) The largest expected payoff based on the use of additional information. This value is always less than or equal to EPPI depending on the accuracy of the information.

expected payoff with perfect information (EPPI) The maximum expected payoff (i.e., average) when the future states of nature can be predicted with certainty.

expected value of additional information (EVAI) The difference between the expected payoff with additional information and the expected payoff without the use of the information.

expected value of perfect information (EVPI) The difference between the expected payoff with perfect information and the expected payoff without the use of perfect information. This is the maximum the decision maker should pay for the information.

indicator An estimator of future events.

maximax decision criterion A decision strategy that maximizes the maximum payoff from a set of alternatives.

maximin decision criterion A decision strategy that maximizes the minimum payoff from a set of alternatives.

maximum likelihood criterion A decision strategy that maximizes the payoff based on the highest probability of occurrence for a set of alternatives.

minimax decision criterion A decision strategy that minimizes the maximum potential loss from a set of alternatives.

nodes The connecting points on a decision tree. There are three types of nodes: decision, chance, and terminal.

objective analysis The evaluation of alternative decisions when the probabilities of future states of nature are known.

payoff table A matrix that specifies the payoffs associated with each combination of decision alternative and state of nature.

posterior probabilities Probability values developed from revising prior probabilities with additional data.

prior probabilities Probability values based on historical data that are used to describe the chances of occurrence of the states of nature.

states of nature The uncontrollable future events influencing the outcome of a decision.

subjective analysis The evaluation of alternative decisions when data on the future states of nature are unavailable.

utility A decision maker's preference for money.

15.10 BIBLIOGRAPHY

Bierman, H., Jr., C. P. Bonini, and W. H. Hausman. *Quantitative Analysis for Business Decisions.* 2nd ed. Homewood, Ill.: Irwin, 1986.

Bunn, D. W. *Applied Decision Analysis.* New York: McGraw-Hill, 1984.

Jones, J. *Introduction to Decision Theory.* Homewood, Ill.: Irwin, 1977.

Sampson, D. A. "Corporate Risk Philosophy for Improving Risk Management." *Journal of Business Research* (April 1987), pp. 107–21.

Samson, D. *Managerial Decision Analysis.* Homewood, Ill.: Irwin, 1988.

Winkler, R. L. *Introduction to Bayesian Inference and Decision.* New York: Holt, Rinehart & Winston, 1972.

15.11 PROBLEMS

1. Discuss the differences between objective and subjective decision making (give examples).
2. Identify which of the following situations involve decision making under certainty (C), risk (R), and uncertainty (U).
   a. Choosing a factory location based on local labor rates
   b. Purchasing an automobile based on price
   c. Deciding on the quality of a production batch
   d. Choosing from several different types of mortgages
3. Identify which of the following situations involve decision making under certainty (C), risk (R), and uncertainty (U).
   a. Investing in the stock market
   b. Selecting from three investment alternatives with known returns
   c. Purchasing an automobile based on fuel economy
   d. Playing the California lottery
4. Describe the process of decision making with additional information using several examples. How does Bayesian analysis play a role in this process?
5. Why is the option that maximizes expected value the same as the one that minimizes expected loss?
6. Consider the following payoff table:

	Alternative				
State	A_1	A_2	A_3	A_4	A_5
S_1	10	20	30	40	50
S_2	60	40	20	0	−20
S_3	40	0	80	20	60

Determine the optimal decision based on:
a. The maximax criterion.
b. The maximin criterion.
c. The equal-likelihood criterion.
d. The minimax criterion.

7. Consider the following payoff table:

	Alternative					
State	A_1	A_2	A_3	A_4	A_5	A_6
S_1	25	50	75	25	100	50
S_2	50	35	0	75	50	125
S_3	100	25	50	100	75	75
S_4	0	75	25	50	50	0

Determine the optimal decision based on:
a. The maximax criterion.
b. The maximin criterion.

c. The equal-likelihood criterion.
d. The minimax criterion.
8. Consider the following payoff table:

| | Alternative | | | | | | |
State	A_1	A_2	A_3	A_4	A_5	A_6	A_7
S_1	25	50	75	25	100	50	-25
S_2	50	-35	0	75	50	-125	100
S_3	-100	25	50	100	-75	75	50

Determine the optimal decision based on:
a. The maximax criterion.
b. The maximin criterion.
c. The equal-likelihood criterion.
d. The minimax criterion.

9. Consider the following payoff table:

| | Alternative | | | | | |
State	A_1	A_2	A_3	A_4	A_5	A_6
S_1	275	375	125	425	550	200
S_2	275	-50	350	100	225	-75
S_3	275	100	200	150	375	100
S_4	275	200	-50	300	-75	50

Determine the optimal decision based on:
a. The maximax criterion.
b. The maximin criterion.
c. The equal-likelihood criterion.
d. The minimax criterion.

10. Evaluate problem 6 using the following weights: $a = 0$, $a = .2$, $a = .4$, $a = .6$, $a = .8$, $a = 1$.
11. Evaluate problem 7 using the following weights: $a = 0$, $a = .2$, $a = .4$, $a = .6$, $a = .8$, $a = 1$.
12. Select the optimal decision for problem 8 based on maximizing expected value when $P(S_1) = .3$, $P(S_2) = .3$, $P(S_3) = .4$.
13. Select the optimal decision for problem 9 based on minimizing expected loss when $P(S_1) = .2$, $P(S_2) = .2$, $P(S_3) = .2$, $P(S_4) = .4$.
14. Consider the following payoff table:

| | Alternative | | | | |
State	A_1	A_2	A_3	A_4	A_5
S_1	100	200	300	150	250
S_2	150	100	75	150	200

If $P(S_1) = .3$ and $P(S_2) = .7$, determine:
a. The alternative that maximizes the expected value.
b. The alternative that minimizes the expected loss.
c. The expected payoff with perfect information.
15. Perform a complete sensitivity analysis of problem 14.
16. Consider the following cost payoff table:

	Alternative					
State	A_1	A_2	A_3	A_4	A_5	A_6
S_1	-100	200	0	75	250	-75
S_2	150	-100	75	75	-200	300
S_3	200	50	100	75	100	0

If $P(S_1) = .3$, $P(S_2) = .3$, and $P(S_3) = .4$, determine:
a. The alternative that minimizes the expected value.
b. The alternative that minimizes the expected loss.
c. The expected payoff with perfect information.
17. Perform a complete sensitivity analysis of problem 16.
18. Consider the following cost payoff table:

	Alternative		
State	A_1	A_2	A_3
S_1	35	50	25
S_2	75	50	35
S_3	100	50	100
S_4	50	50	125
S_5	25	50	0

If $P(S_1) = .2$, $P(S_2) = .2$, $P(S_3) = .2$, $P(S_4) = .2$, and $P(S_5) = .2$, determine:
a. The alternative that minimizes the expected value.
b. The alternative that minimizes the expected loss.
c. The expected payoff with perfect information.
19. Perform a complete sensitivity analysis of problem 18.
20. Consider the following payoff table:

	Alternative				
State	A_1	A_2	A_3	A_4	A_5
S_1	100	200	300	150	250
S_2	150	100	75	150	200
S_3	200	50	100	150	100
S_4	175	0	200	150	300

If $P(S_1) = .2$, $P(S_2) = .2$, $P(S_3) = .2$, and $P(S_4) = .4$, determine:
a. The alternative that maximizes the expected value.
b. The alternative that minimizes the expected loss.
c. The expected payoff with perfect information.
d. How much does the payoff for alternative A_1 and state S_1 have to change for A_1 to be selected?

21. Perform a complete sensitivity analysis for problem 20. Identify the range of probability and payoff values over which the current solution is optimal.
22. Solve problem 18 using a decision tree and compare the results.
23. Solve problem 20 using a decision tree and compare the results.
24. Stuttgart Works is considering building an assembly plant in southern Germany. Three different size plants are under consideration. The following payoff table relates the discounted cash flow for each market condition and plant size. Also shown are the estimated initial construction costs. (All figures are in millions of deutsche marks.)

	Plant Size		
Market	Small	Medium	Large
Moderate	30	85	100
Strong	65	135	350
Initial cost	50	100	150

The chances for a strong market are estimated by the management to be 50%.
a. What size plant should be built based on maximizing expected value?
b. What size plant should be built based on minimizing expected loss?
c. What is the impact on the solution if the probability of a strong market is increased to 65%?

25. Referring to problem 24, management has the option of initially building the smaller plant and, if the market is strong, expanding the plant to a large facility. The cost of the expansion is estimated at 125 million deutsche marks. Use a decision tree to determine the optimal decision based on maximizing expected value.
26. Solve problem 24 using a decision tree and compare the results with those developed in problem 24.
27. The management of Banzai Limited wishes to determine the number of Sushi cafes that should be operated in the Sendai area of northern Japan. The estimated monthly cost per outlet is 5,000,000 yen. The following table provides gross monthly revenue estimates (millions of yen) as a function of potential demand.

	Number of Cafes				
Demand	1	2	3	4	5
Good	10	22	35	47	60
Moderate	8	17	26	36	45
Poor	4	8	13	19	26

Banzai's management estimates that there is a 30% chance that demand will be good and a 50% chance that demand will be moderate. How many Sushi cafes should the company open?

28. Perform a complete sensitivity analysis for problem 27.
29. Kay Johnson is interested in starting a boutique dress shop that will cater to upper-middle-income women. She is currently evaluating two possible sites. The first one is a 1,500-square-foot facility located in a shopping mall, and the second is a 3,000-square-foot stand-alone building. Kay knows that customer demand represents the largest unknown factor in terms of bottom-line profits. She has prepared the following payoff table, which provides net profit estimates ($000) as a function of demand and facility size:

	Facility	
Demand	1,500	3,000
High	100	180
Low	50	30

Kay estimates that there is a 50–50 chance that customer demand will be high.
   a. Which of the two stores should Kay rent?
   b. At what probability level for high demand is Kay indifferent between the two store alternatives?
   c. How much would the $100,000 payoff have to increase before she would consider switching to the smaller store?

30. Zwolle Chemicals Ltd. of Holland produces a chemical solvent at a cost of $10 per gallon. The product sells for $20 per gallon. For planning purposes the company is considering possible demands of 500, 1,500, and 2,000 gallons per week. If demand is greater than production, the firm will satisfy excess demand with a special production run at a cost of $30 per gallon. The product, however, always sells for $20 per gallon. The marketing department has prepared the following probability estimates for product demand: $P(500) = .2$, $P(1,000) = .5$, and $P(1,500) = .3$. Determine the optimal production run based on maximizing expected profit.

31. Perform a complete sensitivity analysis for problem 30.
32. Each year Juan Estrada operates a pumpkin stand during the month of October. Juan rents the stand from the county for $300 for the month. Juan's supplier will provide, in a single shipment, pumpkins in truck lots at $200 per lot. A lot contains approximately 50 pumpkins. Juan estimates that the average selling price will be $10 per pumpkin. After Halloween, it will cost Juan $0.50 per pumpkin to remove any that are unsold from the lot. To maintain his reputation

from year to year, Juan has arranged with another supplier to purchase additional pumpkins for $12 apiece. Juan estimates that the demand for pumpkins can be described by the following probability distribution:

Demand	Probability
50	.15
100	.15
150	.15
200	.25
250	.20
300	.10

a. How many truck lots of pumpkins should Juan stock?
b. If the cost of a pumpkin from the supplier is increased from $4 to $6, how many lots should Juan stock?
c. If Juan is not interested in always meeting demand, how many pumpkin lots should he stock?

33. Use discrete marginal analyses to solve problem 32.
34. The production manager at Highteck Plastics, Inc., wishes to determine the optimal inventory policy for a new composite material for the next six months. The manager can purchase the basic material from a supplier in quantities of 10,000, 20,000, or 30,000 pounds at $20 per pound. The manager estimates that there is a 60% chance that demand will exceed 10,000 pounds and a 20% chance that demand will exceed 20,000 pounds over the next six months. In the event demand exceeds the available supply, additional material can be purchased on the spot market at $28 per pound. Additionally, all inventory remaining at the end of six months must be "reprocessed" at a cost of $2 per pound. Identify the inventory policy that minimizes total costs.
35. Perform a complete sensitivity analysis for problem 34.
36. Zelco Corporation is considering the introduction of a new laptop computer. Historically, products of this type have recorded a 60% chance of being successful in the marketplace. The following payoff table presents discounted cash flows ($000) for each decision alternative.

	Action	
	Introduce	Wait
Successful	500	−100
Unsuccessful	−200	100

Before making a marketing decision on the new product, Zelco has the opportunity of purchasing a market survey for $60,000. The track record of the marketing firm is detailed in the following conditional probability table:

	Indicator	
State	Successful	Unsuccessful
Successful	.8	.2
Unsuccessful	.3	.7

a. What is Zelco's optimal decision and payoff if they do not buy the market survey?
b. How much should Zelco pay for perfect information?
c. Should Zelco buy the survey?
d. What is the maximum Zelco should be willing to pay for the market survey?

37. Zwolle's management (see problem 30) wish to retain a market survey group to estimate the demand for chemical solvent. The following conditional probability table indicates the historical performance of the survey firm in predicting market conditions for similar products.

	Indicator		
Demand	Good	Moderate	Poor
Good	.7	.2	.1
Moderate	.2	.6	.2
Poor	0	.2	.8

Here, "good" represents a demand of 1,500 pounds per week, "moderate" a demand of 1,000 pounds per week, and "poor" a demand of 500 pounds per week.
a. What is the maximum Zwolle should pay for the survey?
b. What is the maximum Zwolle should pay for perfect information?

38. The management at Quark Limited are considering the installation of a new testing system. The system is designed to identify out-of-specification components prior to final assembly. Currently, 10% of the final assemblies contain out-of-specification components that result in rejection by the quality control department. A recent test of the new system on 100 components selected at random revealed the following performance data:

	Indicator	
Actual	In Spec	Out of Spec
In Spec	60	2
Out of Spec	4	34

Quark manufacturers 50,000 final assembly units per year at an average cost of $2,000. A failed unit costs the company $400 to rebuild. How much should the company be willing to pay, on an annual basis, for the new device?

39. The product quality manager at Sunbeam Electric has the following two quality control procedures under investigation for its 1/2 horsepower motor:

   I. A 100% inspection, costing $2 per unit.

   II. A no-inspection plan, with the following estimated defective rate:

Percentage Defective			
0	5	10	15
.30	.40	.20	.10

   The company plans to produce 100,000 units next month. The manager estimates that the total cost to Sunbeam of shipping a defective motor is $40 per unit. Which of the two plans minimizes the quality control costs for this product?

40. The production manager at Edwin Optical purchases lens frames from two vendors. The manager plans to place an order for 1,000 lenses. The quality and price of the frames from each vendor can be described with a Poisson distribution as follows:

Vendor	Average Defective Rate	Price ($ per lens)
A	1%	22.50
B	2%	18.75

   a. If a defective lens can be repaired for $5, which of the vendors should be selected based on minimizing costs?

   b. How much would vendor B's price have to change for the production manager to be indifferent between the two vendors?

41. Consider the following utility function for the payoff table given in problem 14.

Payoff	Utility
75	0
100	.2
150	.4
200	.8
250	.9
300	1.0

   a. Determine the optimal decision based on maximizing expected utility.

   b. Compare the results with those developed in problem 14.

42. Consider the following utility function for the payoff table given in problem 27.

$$U = \log(\$)/2$$

a. Determine the optimal decision based on maximizing expected utility.
b. Compare the results with those developed in problem 27.

43. Consider the following utility function for the payoff table given in problem 18.

$$U = \sqrt{\$/2}$$

a. Determine the optimal decision based on maximizing expected utility.
b. Compare the results with those developed in problem 18.

44. Consider the following utility function for the payoff table given in problem 18.

Payoff	Utility
0	0
25	.3
35	.5
50	.7
75	.9
100	1.0

a. Determine the optimal decision based on maximizing expected utility.
b. Compare the results with those developed in problem 18.

45. The owner of the 1-Hour Print Shop, located in downtown Fresno, is considering the purchase of a new offset printer. The owner has two systems under construction: the Mark 7, made by Xerox, and the System 2000, made by Rico. The Mark 7 is a low-cost, low-volume machine that should handle most orders. Then System 2000 is a high-speed machine that is twice as expensive as the Mark 7. The following table presents the discounted net payoffs for each alternative.

	Alternative	
Demand	Mark 7	System 2000
High	$40,000	$90,000
Moderate	40,000	40,000
Low	20,000	−10,000

The owner estimates that there is 40% chance that demand will be high, a 40% chance that demand will be moderate, and a 20% chance that demand will be low. The owner's utility function is:

Level	Utility
$90,000	1
40,000	.6
20,000	.4
−10,000	0

Determine:
a. The owner's optimal decision based on EMV considerations.
b. The owner's optimal decision based on utility considerations.

46. Suppose that the owner of the 1-Hour Print Shop (see problem 45) has the opportunity to purchase a market survey on the future of demand for printing services in the general area for $500. The firm providing the service has a good track record, as it correctly predicted:

    - High demand 7 times out of 10. (For the remaining 3 times of high demand, the firm predicted moderate demand.)
    - Moderate demand 6 times out of 8. (For the other 2 times, the firm predicted high demand.)
    - Low demand 5 times out of 7. (For the remaining 2 times, the firm predicted moderate demand.)

    a. Should the owner purchase the service based on expected monetary considerations?
    b. Should the owner purchase the service based on expected utility considerations?
    c. What is the maximum the owner should pay for the service?

47. Wade Industries has a new optical scanner under development. The R&D manager, who is responsible for this project, is considering a crash effort to ensure that the new scanner is available for the upcoming international trade show. The crash effort will require considerable additional resources. The R&D manager estimates that there is an 80% chance that the scanner will prove successful under the crash program and a 95% chance that it will prove successful under the current program. Another factor influencing the financial viability of this program is market demand. Wade's marketing department has estimated that there is a 70% chance that demand will be high and a 30% chance that demand will be low if the product is introduced at the trade show. If the product is not introduced at the trade show, the probability of high demand is 50%. If the project proves unsuccessful, then the company will lose $1,000,000 and $2,500,000 for the current and crash efforts, respectively. If the project proves successful and the demand is high, the company will realize a future net gain of $8 million for the current program and $6.5 million for the crash program. If demand is low, the company will generate gains of $5 million and $3.5 million, respectively.

    Wade's management team is very conservative and wishes to minimize the possibility of a loss. A composite corporate utility table follows:

Level ($ millions)	Utility
8	1
6	.9
5	.8
3.5	.6
−1	.4
−2.5	0

Determine:
  a. The optimal development program based on maximizing EMV.
  b. The optimal development program based on maximizing utility.
  c. What is the maximum Wade should pay for perfect information on market conditions?
48. Perform a complete sensitivity analysis on problem 47.
49. Transco Corporation is currently working on a new keel design for the U.S. entry in the *America's* Cup sailing race. Transco's chief engineer estimates that there is a 40% chance that the new design will prove successful, a 30% chance that the design will prove partially successful, and a 30% chance that the design will be unsuccessful. The committee funding the U.S. entry can either sponsor the new keel design or use their limited resources to support additional crew training. The committee's chairperson estimates that there is a 70% chance of winning if the new design proves successful, a 60% chance if the design is partially successful, and a 40% chance if the design is unsuccessful (In this case, the old design would be used.) The chairperson also estimates that there is a 50–50 chance of winning with the old design and additional crew training.

Formulate this problem using a decision tree, and determine the optimal strategy.

50. Thompson Corporation is considering bidding on one phase of the Space Defense Initiative program. The government contract will be competitive; therefore, determining the optimal bid price is important. The general manager has identified the following three bid levels for the project and the corresponding chances of winning.

Bid Level ($000)	Probability of Winning
100,000	.3
80,000	.5
60,000	.6

The general manager estimates that the proposal will cost $2 million and that the project will cost the company $45 million using standard design technology. The company's R&D department has recently announced the development of a new revolutionary design system that holds great promise. The estimated cost for completing the proposed government contract using the new technology is $36 million. Unfortunately, the new technology has not been adequately tested, and management therefore is reluctant to use it on this important project. The R&D manager has proposed a pilot test to determine the performance of the new technology, which will cost the company $4 million. The manager has developed the following estimates regarding the ability to predict the performance of the technology in the field from the test results.

| | Prediction | |
Actual	Success	Failure
Success	.9	.1
Failure	.3	.7

The general manager wishes to know:
a. What are the basic alternatives available to the company?
b. Should the test be conducted and, if so, when?
c. What bid decision should be made, and what is the expected outcome?

51. Tidro Industries is considering the introduction of a new tape-duplicating machine. The management has identified three alternative marketing strategies: low effort (D_1), medium effort (D_2), and aggressive effort (D_3). The success of the new product will depend on actual market demand. The company's president has prepared the following payoff table showing the estimated net profits ($000) for the next quarter as a function of marketing strategy and product demand.

Product Demand	Decision Alternatives		
	D_1	D_2	D_3
Low	20	10	0
Average	40	45	50
Good	40	60	80
Excellent	40	60	120

Management wishes to know the impact of the following weighting coefficients on the optimal decision: $\infty = 0, \infty = .5, \infty = 1$.

52. Keystone Corporation operates three stamping presses at its West Virginia facility. The probability of a single press malfunction during an eight-hour production run is 10%. The probability that two presses will malfunction simultaneously is 5%, and the probability of three simultaneous failures is 2%. It costs the company $300 per hour in lost production whenever a press line is not operating. The average time for repairing a press with the existing maintenance team (consisting of one repairperson) is four hours. Management is considering hiring an additional repairperson to reduce total costs (i.e., production losses plus repair costs). The use of a second person would reduce repair time by 50%. Each repairperson costs the company $15 per hour. Should the company hire a second person?

53. Forcheck Publishing Company is planning to produce a microcomputer-based quantitative analysis textbook. The company estimates that total sales will average 50,000 copies with a standard deviation of 5,000. The textbook will cost $15 to produce and will retail for $40. The fixed production costs are estimated at $200,000.
a. Determine the break-even production quantity.
b. Determine the expected value for this project.
c. Determine the amount of volume required to produce a profit of $500,000.

54. Maple manufacturing is planning to introduce a new portable printer. The company estimates that total sales will average 200,000 printers with a standard deviation of 20,000 printers. The printer will cost $125 to produce and will wholesale for $195. The fixed production costs are estimated at $3,000,000.
    a. Determine the break-even production quantity.
    b. Determine the expected value for this project.
    c. Determine the amount of volume required to produce a profit of $4,000,000.
55. Perform a complete sensitivity analysis for problem 54.
56. Adelman Enterprises is planning to produce a new film on freshman statistics. The production manager estimates a filming time of 100 days with a standard deviation of 20 days. Furthermore, the manager estimates the cost of production at $500 per day.
    a. What is the probability that the project will take over 120 days?
    b. What is the estimated cost if the project takes over 120 days?
    c. What is the estimated cost if the project takes less than 100 days?
57. Chandler Construction Company is preparing a bid for a new office complex. The developer wishes that the project be completed within 200 days from the start of construction. The developer has assigned a penalty of $10,000 for every day the project extends beyond the target versus a bonus of $5,000 for every day under the target. Chandler's management estimates that the project should take 190 days to complete with a standard deviation of 20 days.
    a. What is the probability that the job will be completed within 200 days?
    b. What is the expected penalty cost?
    c. What is the expected amount of the bonus?
    d. What is the net payoff?
58. Discuss the rationale behind why most car owners purchase accident insurance.
59. Identify several business applications where utility considerations would play a role in the decision-making process.
60. The managing partner at Santos Foundry, located in Sao Paulo, Brazil, is considering the possibility of constructing a new production facility near Brasilia. The partner estimates that the market for the new plant will be either $50 or $100 million annually. Furthermore, he estimates that the share of the market will be either 30% or 60%, depending on the capabilities of the local competition. Finally, he estimates that the annual operating costs will be either $20 or $40 million.
    a. Identify the eight basic possibilities.
    b. Based on the equal likelihood assumption, what is the expected value of their new product?
61. A CBS analysis of a decision-making-under-risk problem involving four alternatives and two states is presented below. The payoffs are given in thousands of dollars.

Results

Decision Alternatives

Criteria	D_1	D_2	D_3	D_4
EMV	240	220	90	200
EOL				
EPPI	340	340	340	340

Sensitivity Analysis

Decision	Lower Limit	Current Value	Upper Limit
D_1	166.667	200	No limit
	250	300	No limit
D_2	No limit	100	133.333
	No limit	400	450
D_3	No limit	50	300
	No limit	150	525
D_4	No limit	300	366.667
	No limit	50	150

a. Identify the optimal decision that maximizes the expected value.
b. Determine the expected opportunity loss values.
c. Provide a complete discussion of the sensitivity analysis.

62. A CBS analysis of a decision-making-under-risk problem with additional information involving three alternatives, four states, and two indicators is given below. The payoffs are reported in thousands of dollars. The estimated cost for the additional information is $25,000.

Results

Decision Alternatives

Criteria	D_1	D_2	D_3
EMV	205	105	150
EPPI	255	255	255

Additional Information Analysis

EPAI:	205
EVAI:	0
COST:	25

Revised Conditional Table

	I_1	I_2
1 =	0.36	0.04
2 =	0.36	0.24
3 =	0.24	0.56
4 =	0.04	0.16

Marginal Probabilities

	I_1	I_2
1 =	0.5	0.5

a. Identify the optimal decision that maximizes the expected value.
b. Determine the expected opportunity loss values.
c. Is the additional information worth $25,000?
d. How much is the additional information worth?
e. Provide an interpretation of the conditional and marginal probabilities. How would you characterize the information content of this data?

63. A CBS analysis of a decision-making-under-risk problem with additional information involving four alternatives, three states, and three indicators is given below. The payoffs are reported in thousands of dollars. The estimated cost for the additional information is $10,000.

Results

Criteria	Decision Alternatives			
	D_1	D_2	D_3	D_4
EMV	1,020	760	900	720
EOL	300	560	420	600
EPPI	1,320	1,320	1,320	1,320

Additional Information Analysis

EPAI:	1,211
EVAI:	191
COST:	10

Revised Conditional Table

	I_1	I_2	I_3
1 =	0.938	0.152	0
2 =	0.063	0.727	0.158
3 =	0	0.121	0.842

Marginal Probabilities

	I_1	I_2	I_3
1 =	0.48	0.33	0.19

  a. Identify the optimal decision that maximizes the expected value.
  b. Determine the expected opportunity loss values.
  c. Is the additional information worth $10,000?
  d. How much is the additional information worth?
  e. Provide an interpretation of the conditional and marginal probabilities. How would you characterize the information content of this data?

64. A CBS analysis of a decision-making-under-uncertainty problem involving three alternatives and four states is given below. The payoffs are given in percent.

Results

	Decision Alternatives		
Criteria	D_1	D_2	D_3
Maximax	18	8	15
Maximin	5	8	3
Equal likelihood	11.50	8	9.50
Minimax regret	10	10	15

  a. Select the alternative based on the maximax criterion.
  b. Select the alternative based on the maximin criterion.
  c. What are the complete set of payoff values for the second alternative?
  d. Which option is optimal based on the minimax regret criterion?

15.12 CASES

15.12.1 Ogden Aircraft Company

The management at Ogden Aircraft is considering bidding on a three-phase, multi-year strategic stealth bomber program for the U.S. Department of Defense (DOD). The total contract value for the three phases is $4.5 billion. The first phase is a study and design contract worth approximately $100 million. Five aerospace firms with equal chances of winning the study contract have been asked by DOD to bid on the first phase. The cost of preparing a proposal for the first phase is $5 million. DOD will select the top three firms, based on a review of their proposed designs, for a phase 2 "fly off" competition. The phase 1 report contains the proposal for phase 2 (the costs for preparing the phase 2 proposal are covered in the phase 1 contract).

The second phase is worth $600 million. The cost for developing the prototype is uncertain and is characterized by the data given in Table 1.

TABLE 1 Cost of Developing the Prototype

Cost ($ millions)	Probability
450	.2
500	.3
550	.2
600	.2
650	.1

Ogden Aircraft estimates that, bidding alone, they have a 40% chance of winning one of the three phase 2 contracts. On the other hand, if they team with either of the other two finalists, the probability of winning increases to 70%. However, they will have to share the phase 2 award on a 50–50 basis.

Based on the "fly off" results, DOD will ask two firms to submit a proposal for the production contract. If Ogden Aircraft bids alone and wins a phase 2 contract, then the probability of winning the phase 3 contract is 50%. If Ogden teams on phase 2 and wins, the probability of winning phase 3 is 75%. In this case, Ogden will obtain 40% of the phase 3 contract.

Suggested Questions

1. What is Ogden's optimal decision path?
2. What is the impact on the optimal bidding strategy if Ogden has only a 30% chance of winning a phase 2 contract?
3. What is the impact on the optimal bidding strategy if the total contract value is reduced to $3 billion?

15.12.2 Zerg Communications

Ms. Titus Zerg, founder and owner of Zerg Communications, located in Warsaw, Poland, decided to explore the possibility of manufacturing and selling a new touch-tone telephone. The production team at Zerg was given the task of developing the new system. The basic idea was a low-cost system that would retail at $25. The introduction of this new product would be timed with the general upgrading of Poland's telephone system.

The research team estimated it would cost approximately $1,000,000 to develop the equipment and procedures needed to manufacture the new system. Titus estimates that there is a 70% chance that approximately 200,000 of the new units could be sold over the next three years. Furthermore, she estimates that there is a 30% chance that 50,000 units could be sold over the same period. The estimated cost of producing the new telephone is $14.00. To better understand the marketplace, Titus decided to obtain an independent market appraisal. Accordingly, she sent proposal requests to several well-known market research firms. In addition to their proposed price, Titus also requested information on each firm's record in predicting the market response to new electronic products.

Two firms responded to Titus's request. Dudley, one of the largest marketing research firms in Western Europe, proposed a comprehensive survey costing $200,000. Dudley's track record is impressive. Their chance of gaining a favorable survey result given a successful venture is 90%. On the other hand, the chance of getting an unfavorable survey result given an unsuccessful venture is 85%.

J&J, a Warsaw firm specializing in communication systems, offered a more modest effort; their proposed price was $95,000. J&J's track record is less impressive than Dudley Forecasting's. J&J's record showed that when the market demand was favorable, they forecasted a favorable market demand 70% of the time. When the market demand was unfavorable, they predicted an unfavorable market demand 65% of the time.

Suggested Questions

1. Which market research team, if any, should be hired?
2. Should the new project be implemented?
3. What is the impact on the decision if the chance of selling 200,000 units is only 60%? 50%?
4. What is the impact on the decision if the cost of producing the new phone is increased by $3?
5. Suppose Titus Zerg has the following utility for money:

$(000)	-1,000	-500	0	500	1,000	1,500
Utility	0	.3	.5	.7	.9	1

Does the decision sequence change, based on utility considerations?

Appendix A

Statistical Tables

- **A.1** Areas under the Normal Curve
- **A.2** Student t Distribution
- **A.3** Critical Values of Chi-Squared
- **A.4** Critical Values of the F Distribution
- **A.5** Binomial Probability Distribution
- **A.6** Poisson Distribution: Probability of Exactly x Occurrences
- **A.7** Critical Values of the Durbin-Watson Test Statistic
- **A.8** Critical Values of the Studentized Range Distribution (Tukey Test)
- **A.9** Cumulative Distribution Function (Runs Test)

TABLE A.1 Areas under the Normal Curve

Example:
If $z = 1.96$, then
$P(0 \text{ to } z) = 0.4750$

z	0.00	0.01	0.02	0.03	0.04	0.05	0.06	0.07	0.08	0.09
0.0	0.0000	0.0040	0.0080	0.0120	0.0160	0.0199	0.0239	0.0279	0.0319	0.0359
0.1	0.0398	0.0438	0.0478	0.0517	0.0557	0.0596	0.0636	0.0675	0.0714	0.0753
0.2	0.0793	0.0832	0.0871	0.0910	0.0948	0.0987	0.1026	0.1064	0.1103	0.1141
0.3	0.1179	0.1217	0.1255	0.1293	0.1331	0.1368	0.1406	0.1443	0.1480	0.1517
0.4	0.1554	0.1591	0.1628	0.1664	0.1700	0.1736	0.1772	0.1808	0.1844	0.1879
0.5	0.1915	0.1950	0.1985	0.2019	0.2054	0.2088	0.2123	0.2157	0.2190	0.2224
0.6	0.2257	0.2291	0.2324	0.2357	0.2389	0.2422	0.2454	0.2486	0.2517	0.2549
0.7	0.2580	0.2611	0.2642	0.2673	0.2704	0.2734	0.2764	0.2794	0.2823	0.2852
0.8	0.2881	0.2910	0.2939	0.2967	0.2995	0.3023	0.3051	0.3078	0.3106	0.3133
0.9	0.3159	0.3186	0.3212	0.3238	0.3264	0.3289	0.3315	0.3340	0.3365	0.3389
1.0	0.3413	0.3438	0.3461	0.3485	0.3508	0.3531	0.3554	0.3577	0.3599	0.3621
1.1	0.3643	0.3665	0.3686	0.3708	0.3729	0.3749	0.3770	0.3790	0.3810	0.3830
1.2	0.3849	0.3869	0.3888	0.3907	0.3925	0.3944	0.3962	0.3980	0.3997	0.4015
1.3	0.4032	0.4049	0.4066	0.4082	0.4099	0.4115	0.4131	0.4147	0.4162	0.4177
1.4	0.4192	0.4207	0.4222	0.4236	0.4251	0.4265	0.4279	0.4292	0.4306	0.4319
1.5	0.4332	0.4345	0.4357	0.4370	0.4382	0.4394	0.4406	0.4418	0.4429	0.4441
1.6	0.4452	0.4463	0.4474	0.4484	0.4495	0.4505	0.4515	0.4525	0.4535	0.4545
1.7	0.4554	0.4564	0.4573	0.4582	0.4591	0.4599	0.4608	0.4616	0.4625	0.4633
1.8	0.4641	0.4649	0.4656	0.4664	0.4671	0.4678	0.4686	0.4693	0.4699	0.4706
1.9	0.4713	0.4719	0.4726	0.4732	0.4738	0.4744	0.4750	0.4756	0.4761	0.4767
2.0	0.4772	0.4778	0.4783	0.4788	0.4793	0.4798	0.4803	0.4808	0.4812	0.4817
2.1	0.4821	0.4826	0.4830	0.4834	0.4838	0.4842	0.4846	0.4850	0.4854	0.4857
2.2	0.4861	0.4864	0.4868	0.4871	0.4875	0.4878	0.4881	0.4884	0.4887	0.4890
2.3	0.4893	0.4896	0.4898	0.4901	0.4904	0.4906	0.4909	0.4911	0.4913	0.4916
2.4	0.4918	0.4920	0.4922	0.4925	0.4927	0.4929	0.4931	0.4932	0.4934	0.4936
2.5	0.4938	0.4940	0.4941	0.4943	0.4945	0.4946	0.4948	0.4949	0.4951	0.4952
2.6	0.4953	0.4955	0.4956	0.4957	0.4959	0.4960	0.4961	0.4962	0.4963	0.4964
2.7	0.4965	0.4966	0.4967	0.4968	0.4969	0.4970	0.4971	0.4972	0.4973	0.4974
2.8	0.4974	0.4975	0.4976	0.4977	0.4977	0.4978	0.4979	0.4979	0.4980	0.4981
2.9	0.4981	0.4982	0.4982	0.4983	0.4984	0.4984	0.4985	0.4985	0.4986	0.4986
3.0	0.4987	0.4987	0.4987	0.4988	0.4988	0.4989	0.4989	0.4989	0.4990	0.4990

TABLE A.2 Student t Distribution

df	Level of significance for one-tailed test					
($1-\alpha$ or $1-\alpha/2$)	.10	.05	.025	.01	.005	.0005
(α)	Level of significance for two-tailed test					
	.20	.10	.05	.02	.01	.001
1	3.078	6.314	12.706	31.821	63.657	636.619
2	1.886	2.920	4.303	6.965	9.925	31.598
3	1.638	2.353	3.182	4.541	5.841	12.941
4	1.533	2.132	2.776	3.747	4.604	8.610
5	1.476	2.015	2.571	3.365	4.032	6.859
6	1.440	1.943	2.447	3.143	3.707	5.959
7	1.415	1.895	2.365	2.998	3.499	5.405
8	1.397	1.860	2.306	2.896	3.355	5.041
9	1.383	1.833	2.262	2.821	3.250	4.781
10	1.372	1.812	2.228	2.764	3.169	4.587
11	1.363	1.796	2.201	2.718	3.106	4.437
12	1.356	1.782	2.179	2.681	3.055	4.318
13	1.350	1.771	2.160	2.650	3.012	4.221
14	1.345	1.761	2.145	2.624	2.977	4.140
15	1.341	1.753	2.131	2.602	2.947	4.073
16	1.337	1.746	2.120	2.583	2.921	4.015
17	1.333	1.740	2.110	2.567	2.898	3.965
18	1.330	1.734	2.101	2.552	2.878	3.922
19	1.328	1.729	2.093	2.539	2.861	3.883
20	1.325	1.725	2.086	2.528	2.845	3.850
21	1.323	1.721	2.080	2.518	2.831	3.819
22	1.321	1.717	2.074	2.508	2.819	3.792
23	1.319	1.714	2.069	2.500	2.807	3.767
24	1.318	1.711	2.064	2.492	2.797	3.745
25	1.316	1.708	2.060	2.485	2.787	3.725
26	1.315	1.706	2.056	2.479	2.779	3.707
27	1.314	1.703	2.052	2.473	2.771	3.690
28	1.313	1.701	2.048	2.467	2.763	3.674
29	1.311	1.699	2.045	2.462	2.756	3.659
30	1.310	1.697	2.042	2.457	2.750	3.646
40	1.303	1.684	2.021	2.423	2.704	3.551
60	1.296	1.671	2.000	2.390	2.660	3.460
120	1.289	1.658	1.980	2.358	2.617	3.373
∞	1.282	1.645	1.960	2.326	2.576	3.291

TABLE A.3 Critical Values of Chi-Squared

This table contains the values of χ^2 that correspond to a specific right-tail area and specific numbers of degrees of freedom df.

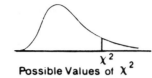

Possible Values of χ^2

DEGREES OF FREEDOM df	RIGHT-TAIL AREA			
	0.10	0.05	0.02	0.01
1	2.706	3.841	5.412	6.635
2	4.605	5.991	7.824	9.210
3	6.251	7.815	9.837	11.345
4	7.779	9.488	11.668	13.277
5	9.236	11.070	13.388	15.086
6	10.645	12.592	15.033	16.812
7	12.017	14.067	16.622	18.475
8	13.362	15.507	18.168	20.090
9	14.684	16.919	19.679	21.666
10	15.987	18.307	21.161	23.209
11	17.275	19.675	22.618	24.725
12	18.549	21.026	24.054	26.217
13	19.812	22.362	25.472	27.688
14	21.064	23.685	26.873	29.141
15	22.307	24.996	28.259	30.578
16	23.542	26.296	29.633	32.000
17	24.769	27.587	30.995	33.409
18	25.989	28.869	32.346	34.805
19	27.204	30.144	33.687	36.191
20	28.412	31.410	35.020	37.566
21	29.615	32.671	36.343	38.932
22	30.813	33.924	37.659	40.289
23	32.007	35.172	38.968	41.638
24	33.196	36.415	40.270	42.980
25	34.382	37.652	41.566	44.314
26	35.563	38.885	42.856	45.642
27	36.741	40.113	44.140	46.963
28	37.916	41.337	45.419	48.278
29	39.087	42.557	46.693	49.588
30	40.256	43.773	47.962	50.892

TABLE A.4 Critical Values of the F Distribution

1 percent level of significance, $\alpha = 0.01$

Degrees of freedom for numerator

	1	2	3	4	5	6	7	8	9	10	12	15	20	24	30	40	60	120	∞
1	4,052	5,000	5,403	5,625	5,764	5,859	5,928	5,982	6,023	6,056	6,106	6,157	6,209	6,235	6,261	6,287	6,313	6,339	6,366
2	98.5	99.0	99.2	99.2	99.3	99.3	99.4	99.4	99.4	99.4	99.4	99.4	99.4	99.5	99.5	99.5	99.5	99.5	99.5
3	34.1	30.8	29.5	28.7	28.2	27.9	27.7	27.5	27.3	27.2	27.1	26.9	26.7	26.6	26.5	26.4	26.3	26.2	26.1
4	21.2	18.0	16.7	16.0	15.5	15.2	15.0	14.8	14.7	14.5	14.4	14.2	14.0	13.9	13.8	13.7	13.7	13.6	13.5
5	16.3	13.3	12.1	11.4	11.0	10.7	10.5	10.3	10.2	10.1	9.89	9.72	9.55	9.47	9.38	9.29	9.20	9.11	9.02
6	13.7	10.9	9.78	9.15	8.75	8.47	8.26	8.10	7.98	7.87	7.72	7.56	7.40	7.31	7.23	7.14	7.06	6.97	6.88
7	12.2	9.55	8.45	7.85	7.46	7.19	6.99	6.84	6.72	6.62	6.47	6.31	6.16	6.07	5.99	5.91	5.82	5.74	5.65
8	11.3	8.65	7.59	7.01	6.63	6.37	6.18	6.03	5.91	5.81	5.67	5.52	5.36	5.28	5.20	5.12	5.03	4.95	4.86
9	10.6	8.02	6.99	6.42	6.06	5.80	5.61	5.47	5.35	5.26	5.11	4.96	4.81	4.73	4.65	4.57	4.48	4.40	4.31
10	10.0	7.56	6.55	5.99	5.64	5.39	5.20	5.06	4.94	4.85	4.71	4.56	4.41	4.33	4.25	4.17	4.08	4.00	3.91
11	9.65	7.21	6.22	5.67	5.32	5.07	4.89	4.74	4.63	4.54	4.40	4.25	4.10	4.02	3.94	3.86	3.78	3.69	3.60
12	9.33	6.93	5.95	5.41	5.06	4.82	4.64	4.50	4.39	4.30	4.16	4.01	3.86	3.78	3.70	3.62	3.54	3.45	3.36
13	9.07	6.70	5.74	5.21	4.86	4.62	4.44	4.30	4.19	4.10	3.96	3.82	3.66	3.59	3.51	3.43	3.34	3.25	3.17
14	8.86	6.51	5.56	5.04	4.70	4.46	4.28	4.14	4.03	3.94	3.80	3.66	3.51	3.43	3.35	3.27	3.18	3.09	3.00
15	8.68	6.36	5.42	4.89	4.56	4.32	4.14	4.00	3.89	3.80	3.67	3.52	3.37	3.29	3.21	3.13	3.05	2.96	2.87
16	8.53	6.23	5.29	4.77	4.44	4.20	4.03	3.89	3.78	3.69	3.55	3.41	3.26	3.18	3.10	3.02	2.93	2.84	2.75
17	8.40	6.11	5.19	4.67	4.34	4.10	3.93	3.79	3.68	3.59	3.46	3.31	3.16	3.08	3.00	2.92	2.83	2.75	2.65
18	8.29	6.01	5.09	4.58	4.25	4.01	3.84	3.71	3.60	3.51	3.37	3.23	3.08	3.00	2.92	2.84	2.75	2.66	2.57
19	8.19	5.93	5.01	4.50	4.17	3.94	3.77	3.63	3.52	3.43	3.30	3.15	3.00	2.92	2.84	2.76	2.67	2.58	2.49
20	8.10	5.85	4.94	4.43	4.10	3.87	3.70	3.56	3.46	3.37	3.23	3.09	2.94	2.86	2.78	2.69	2.61	2.52	2.42
21	8.02	5.78	4.87	4.37	4.04	3.81	3.64	3.51	3.40	3.31	3.17	3.03	2.88	2.80	2.72	2.64	2.55	2.46	2.36
22	7.95	5.72	4.82	4.31	3.99	3.76	3.59	3.45	3.35	3.26	3.12	2.98	2.83	2.75	2.67	2.58	2.50	2.40	2.31
23	7.88	5.66	4.76	4.26	3.94	3.71	3.54	3.41	3.30	3.21	3.07	2.93	2.78	2.70	2.62	2.54	2.45	2.35	2.26
24	7.82	5.61	4.72	4.22	3.90	3.67	3.50	3.36	3.26	3.17	3.03	2.89	2.74	2.66	2.58	2.49	2.40	2.31	2.21
25	7.77	5.57	4.68	4.18	3.86	3.63	3.46	3.32	3.22	3.13	2.99	2.85	2.70	2.62	2.53	2.45	2.36	2.27	2.17
30	7.56	5.39	4.51	4.02	3.70	3.47	3.30	3.17	3.07	2.98	2.84	2.70	2.55	2.47	2.39	2.30	2.21	2.11	2.01
40	7.31	5.18	4.31	3.83	3.51	3.29	3.12	2.99	2.89	2.80	2.66	2.52	2.37	2.29	2.20	2.11	2.02	1.92	1.80
60	7.08	4.98	4.13	3.65	3.34	3.12	2.95	2.82	2.72	2.63	2.50	2.35	2.20	2.12	2.03	1.94	1.84	1.73	1.60
120	6.85	4.79	3.95	3.48	3.17	2.96	2.79	2.66	2.56	2.47	2.34	2.19	2.03	1.95	1.86	1.76	1.66	1.53	1.38
∞	6.63	4.61	3.78	3.32	3.02	2.80	2.64	2.51	2.41	2.32	2.18	2.04	1.88	1.79	1.70	1.59	1.47	1.32	1.00

Degrees of freedom for denominator

TABLE A.4 (continued)

5 percent level of significance, $\alpha = 0.05$

Degrees of freedom for denominator (rows) / Degrees of freedom for numerator (columns)

	1	2	3	4	5	6	7	8	9
1	161.4	199.5	215.7	224.6	230.2	234.0	236.8	238.9	240.5
2	18.51	19.00	19.16	19.25	19.30	19.33	19.35	19.37	19.38
3	10.13	9.55	9.28	9.12	9.01	8.94	8.89	8.85	8.81
4	7.71	6.94	6.59	6.39	6.26	6.16	6.09	6.04	6.00
5	6.61	5.79	5.41	5.19	5.05	4.95	4.88	4.82	4.77
6	5.99	5.14	4.76	4.53	4.39	4.28	4.21	4.15	4.10
7	5.59	4.74	4.35	4.12	3.97	3.87	3.79	3.73	3.68
8	5.32	4.46	4.07	3.84	3.69	3.58	3.50	3.44	3.39
9	5.12	4.26	3.86	3.63	3.48	3.37	3.29	3.23	3.18
10	4.96	4.10	3.71	3.48	3.33	3.22	3.14	3.07	3.02
11	4.84	3.98	3.59	3.36	3.20	3.09	3.01	2.95	2.90
12	4.75	3.89	3.49	3.26	3.11	3.00	2.91	2.85	2.80
13	4.67	3.81	3.41	3.18	3.03	2.92	2.83	2.77	2.71
14	4.60	3.74	3.34	3.11	2.96	2.85	2.76	2.70	2.65
15	4.54	3.68	3.29	3.06	2.90	2.79	2.71	2.64	2.59
16	4.49	3.63	3.24	3.01	2.85	2.74	2.66	2.59	2.54
17	4.45	3.59	3.20	2.96	2.81	2.70	2.61	2.55	2.49
18	4.41	3.55	3.16	2.93	2.77	2.66	2.58	2.51	2.46
19	4.38	3.52	3.13	2.90	2.74	2.63	2.54	2.48	2.42
20	4.35	3.49	3.10	2.87	2.71	2.60	2.51	2.45	2.39
21	4.32	3.47	3.07	2.84	2.68	2.57	2.49	2.42	2.37
22	4.30	3.44	3.05	2.82	2.66	2.55	2.46	2.40	2.34
23	4.28	3.42	3.03	2.80	2.64	2.53	2.44	2.37	2.32
24	4.26	3.40	3.01	2.78	2.62	2.51	2.42	2.36	2.30
25	4.24	3.39	2.99	2.76	2.60	2.49	2.40	2.34	2.28
26	4.23	3.37	2.98	2.74	2.59	2.47	2.39	2.32	2.27
27	4.21	3.35	2.96	2.73	2.57	2.46	2.37	2.31	2.25
28	4.20	3.34	2.95	2.71	2.56	2.45	2.36	2.29	2.24
29	4.18	3.33	2.93	2.70	2.55	2.43	2.35	2.28	2.22
30	4.17	3.32	2.92	2.69	2.53	2.42	2.33	2.27	2.21
40	4.08	3.23	2.84	2.61	2.45	2.34	2.25	2.18	2.12
60	4.00	3.15	2.76	2.53	2.37	2.25	2.17	2.10	2.04
120	3.92	3.07	2.68	2.45	2.29	2.17	2.09	2.02	1.96
∞	3.84	3.00	2.60	2.37	2.21	2.10	2.01	1.94	1.88

TABLE A.4 (concluded)

			Degrees of freedom for numerator							
10	12	15	20	24	30	40	60	120	∞
241.9	243.9	245.9	248.0	249.1	250.1	251.1	252.2	253.3	254.3
19.40	19.41	19.43	19.45	19.45	19.46	19.47	19.48	19.49	19.50
8.79	8.74	8.70	8.66	8.64	8.62	8.59	8.57	8.55	8.53
5.96	5.91	5.86	5.80	5.77	5.75	5.72	5.69	5.66	5.63
4.74	4.68	4.62	4.56	4.53	4.50	4.46	4.43	4.40	4.36
4.06	4.00	3.94	3.87	3.84	3.81	3.77	3.74	3.70	3.67
3.64	3.57	3.51	3.41	3.41	3.38	3.34	3.30	3.27	3.23
3.35	3.28	3.22	3.15	3.12	3.08	3.04	3.01	2.97	2.93
3.14	3.07	3.01	2.94	2.90	2.86	2.83	2.79	2.75	2.71
2.98	2.91	2.85	2.77	2.74	2.70	2.66	2.62	2.58	2.54
2.85	2.79	2.72	2.65	2.61	2.57	2.53	2.49	2.45	2.40
2.75	2.69	2.62	2.54	2.51	2.47	2.43	2.38	2.34	2.30
2.67	2.60	2.53	2.46	2.42	2.38	2.34	2.30	2.25	2.21
2.60	2.53	2.46	2.39	2.35	2.31	2.27	2.22	2.18	2.13
2.54	2.48	2.40	2.33	2.29	2.25	2.20	2.16	2.11	2.07
2.49	2.42	2.35	2.28	2.24	2.19	2.15	2.11	2.06	2.01
2.45	2.38	2.31	2.23	2.19	2.15	2.10	2.06	2.01	1.96
2.41	2.34	2.27	2.19	2.15	2.11	2.06	2.02	1.97	1.92
2.38	2.31	2.23	2.16	2.11	2.07	2.03	1.98	1.93	1.88
2.35	2.28	2.20	2.12	2.08	2.04	1.99	1.95	1.90	1.84
2.32	2.25	2.18	2.10	2.05	2.01	1.96	1.92	1.87	1.81
2.30	2.23	2.15	2.07	2.03	1.98	1.94	1.89	1.84	1.78
2.27	2.20	2.13	2.05	2.01	1.96	1.91	1.86	1.81	1.76
2.25	2.18	2.11	2.03	1.98	1.94	1.89	1.84	1.79	1.73
2.24	2.16	2.09	2.01	1.96	1.92	1.87	1.82	1.77	1.71
2.22	2.15	2.07	1.99	1.95	1.90	1.85	1.80	1.75	1.69
2.20	2.13	2.06	1.97	1.93	1.88	1.84	1.79	1.73	1.67
2.19	2.12	2.04	1.96	1.91	1.87	1.82	1.77	1.71	1.65
2.18	2.10	2.03	1.94	1.90	1.85	1.81	1.75	1.70	1.64
2.16	2.09	2.01	1.93	1.89	1.84	1.79	1.74	1.68	1.62
2.08	2.00	1.92	1.84	1.79	1.74	1.69	1.64	1.58	1.51
1.99	1.92	1.84	1.75	1.70	1.65	1.59	1.53	1.47	1.39
1.91	1.83	1.75	1.66	1.61	1.55	1.50	1.43	1.35	1.25
1.83	1.75	1.67	1.57	1.52	1.46	1.39	1.32	1.22	1.00

TABLE A.5 Binomial Probability Distribution

n= 1
PROBABILITY

r	.05	.1	.2	.3	.4	.5	.6	.7	.8	.9	.95
0	.950	.900	.800	.700	.600	.500	.400	.300	.200	.100	.050
1	.050	.100	.200	.300	.400	.500	.600	.700	.800	.900	.950

n= 2
PROBABILITY

r	.05	.1	.2	.3	.4	.5	.6	.7	.8	.9	.95
0	.903	.810	.640	.490	.360	.250	.160	.090	.040	.010	.003
1	.095	.180	.320	.420	.480	.500	.480	.420	.320	.180	.095
2	.003	.010	.040	.090	.160	.250	.360	.490	.640	.810	.903

n= 3
PROBABILITY

r	.05	.1	.2	.3	.4	.5	.6	.7	.8	.9	.95
0	.857	.729	.512	.343	.216	.125	.064	.027	.008	.001	.000
1	.135	.243	.384	.441	.432	.375	.288	.189	.096	.027	.007
2	.007	.027	.096	.189	.288	.375	.432	.441	.384	.243	.135
3	.000	.001	.008	.027	.064	.125	.216	.343	.512	.729	.857

n= 4
PROBABILITY

r	.05	.1	.2	.3	.4	.5	.6	.7	.8	.9	.95
0	.815	.656	.410	.240	.130	.063	.026	.008	.002	.000	.000
1	.171	.292	.410	.412	.346	.250	.154	.076	.026	.004	.000
2	.014	.049	.154	.265	.346	.375	.346	.265	.154	.049	.014
3	.000	.004	.026	.076	.154	.250	.346	.412	.410	.292	.171
4	.000	.000	.002	.008	.026	.063	.130	.240	.410	.656	.815

n= 5
PROBABILITY

r	.05	.1	.2	.3	.4	.5	.6	.7	.8	.9	.95
0	.774	.590	.328	.168	.078	.031	.010	.002	.000	.000	.000
1	.204	.328	.410	.360	.259	.156	.077	.028	.006	.000	.000
2	.021	.073	.205	.309	.346	.313	.230	.132	.051	.008	.001
3	.001	.008	.051	.132	.230	.313	.346	.309	.205	.073	.021
4	.000	.000	.006	.028	.077	.156	.259	.360	.410	.328	.204
5	.000	.000	.000	.002	.010	.031	.078	.168	.328	.590	.774

TABLE A.5 *(continued)*

n= 6
PROBABILITY

r	.05	.1	.2	.3	.4	.5	.6	.7	.8	.9	.95
0	.735	.531	.262	.118	.047	.016	.004	.001	.000	.000	.000
1	.232	.354	.393	.303	.187	.094	.037	.010	.002	.000	.000
2	.031	.098	.246	.324	.311	.234	.138	.060	.015	.001	.000
3	.002	.015	.082	.185	.276	.313	.276	.185	.082	.015	.002
4	.000	.001	.015	.060	.138	.234	.311	.324	.246	.098	.031
5	.000	.000	.002	.010	.037	.094	.187	.303	.393	.354	.232
6	.000	.000	.000	.001	.004	.016	.047	.118	.262	.531	.735

n= 7
PROBABILITY

r	.05	.1	.2	.3	.4	.5	.6	.7	.8	.9	.95
0	.698	.478	.210	.082	.028	.008	.002	.000	.000	.000	.000
1	.257	.372	.367	.247	.131	.055	.017	.004	.000	.000	.000
2	.041	.124	.275	.318	.261	.164	.077	.025	.004	.000	.000
3	.004	.023	.115	.227	.290	.273	.194	.097	.029	.003	.000
4	.000	.003	.029	.097	.194	.273	.290	.227	.115	.023	.004
5	.000	.000	.004	.025	.077	.164	.261	.318	.275	.124	.041
6	.000	.000	.000	.004	.017	.055	.131	.247	.367	.372	.257
7	.000	.000	.000	.000	.002	.008	.028	.082	.210	.478	.698

n= 8
PROBABILITY

r	.05	.1	.2	.3	.4	.5	.6	.7	.8	.9	.95
0	.663	.430	.168	.058	.017	.004	.001	.000	.000	.000	.000
1	.279	.383	.336	.198	.090	.031	.008	.001	.000	.000	.000
2	.051	.149	.294	.296	.209	.109	.041	.010	.001	.000	.000
3	.005	.033	.147	.254	.279	.219	.124	.047	.009	.000	.000
4	.000	.005	.046	.136	.232	.273	.232	.136	.046	.005	.000
5	.000	.000	.009	.047	.124	.219	.279	.254	.147	.033	.005
6	.000	.000	.001	.010	.041	.109	.209	.296	.294	.149	.051
7	.000	.000	.000	.001	.008	.031	.090	.198	.336	.383	.279
8	.000	.000	.000	.000	.001	.004	.017	.058	.168	.430	.663

TABLE A.5 (continued)

n= 9
PROBABILITY

r	.05	.1	.2	.3	.4	.5	.6	.7	.8	.9	.95
0	.630	.387	.134	.040	.010	.002	.000	.000	.000	.000	.000
1	.299	.387	.302	.156	.060	.018	.004	.000	.000	.000	.000
2	.063	.172	.302	.267	.161	.070	.021	.004	.000	.000	.000
3	.008	.045	.176	.267	.251	.164	.074	.021	.003	.000	.000
4	.001	.007	.066	.172	.251	.246	.167	.074	.017	.001	.000
5	.000	.001	.017	.074	.167	.246	.251	.172	.066	.007	.001
6	.000	.000	.003	.021	.074	.164	.251	.267	.176	.045	.008
7	.000	.000	.000	.004	.021	.070	.161	.267	.302	.172	.063
8	.000	.000	.000	.000	.004	.018	.060	.156	.302	.387	.299
9	.000	.000	.000	.000	.000	.002	.010	.040	.134	.387	.630

n= 10
PROBABILITY

r	.05	.1	.2	.3	.4	.5	.6	.7	.8	.9	.95
0	.599	.349	.107	.028	.006	.001	.000	.000	.000	.000	.000
1	.315	.387	.268	.121	.040	.010	.002	.000	.000	.000	.000
2	.075	.194	.302	.233	.121	.044	.011	.001	.000	.000	.000
3	.010	.057	.201	.267	.215	.117	.042	.009	.001	.000	.000
4	.001	.011	.088	.200	.251	.205	.111	.037	.006	.000	.000
5	.000	.001	.026	.103	.201	.246	.201	.103	.026	.001	.000
6	.000	.000	.006	.037	.111	.205	.251	.200	.088	.011	.001
7	.000	.000	.001	.009	.042	.117	.215	.267	.201	.057	.010
8	.000	.000	.000	.001	.011	.044	.121	.233	.302	.194	.075
9	.000	.000	.000	.000	.002	.010	.040	.121	.268	.387	.315
10	.000	.000	.000	.000	.000	.001	.006	.028	.107	.349	.599

n= 11
PROBABILITY

r	.05	.1	.2	.3	.4	.5	.6	.7	.8	.9	.95
0	.569	.314	.086	.020	.004	.000	.000	.000	.000	.000	.000
1	.329	.384	.236	.093	.027	.005	.001	.000	.000	.000	.000
2	.087	.213	.295	.200	.089	.027	.005	.001	.000	.000	.000
3	.014	.071	.221	.257	.177	.081	.023	.004	.000	.000	.000
4	.001	.016	.111	.220	.236	.161	.070	.017	.002	.000	.000
5	.000	.002	.039	.132	.221	.226	.147	.057	.010	.000	.000
6	.000	.000	.010	.057	.147	.226	.221	.132	.039	.002	.000
7	.000	.000	.002	.017	.070	.161	.236	.220	.111	.016	.001
8	.000	.000	.000	.004	.023	.081	.177	.257	.221	.071	.014
9	.000	.000	.000	.001	.005	.027	.089	.200	.295	.213	.087
10	.000	.000	.000	.000	.001	.005	.027	.093	.236	.384	.329
11	.000	.000	.000	.000	.000	.000	.004	.020	.086	.314	.569

TABLE A.5 (continued)

n= 12
PROBABILITY

r	.05	.1	.2	.3	.4	.5	.6	.7	.8	.9	.95
0	.540	.282	.069	.014	.002	.000	.000	.000	.000	.000	.000
1	.341	.377	.206	.071	.017	.003	.000	.000	.000	.000	.000
2	.099	.230	.283	.168	.064	.016	.002	.000	.000	.000	.000
3	.017	.085	.236	.240	.142	.054	.012	.001	.000	.000	.000
4	.002	.021	.133	.231	.213	.121	.042	.008	.001	.000	.000
5	.000	.004	.053	.158	.227	.193	.101	.029	.003	.000	.000
6	.000	.000	.016	.079	.177	.226	.177	.079	.016	.000	.000
7	.000	.000	.003	.029	.101	.193	.227	.158	.053	.004	.000
8	.000	.000	.001	.008	.042	.121	.213	.231	.133	.021	.002
9	.000	.000	.000	.001	.012	.054	.142	.240	.236	.085	.017
10	.000	.000	.000	.000	.002	.016	.064	.168	.283	.230	.099
11	.000	.000	.000	.000	.000	.003	.017	.071	.206	.377	.341
12	.000	.000	.000	.000	.000	.000	.002	.014	.069	.282	.540

n= 13
PROBABILITY

	.05	.1	.2	.3	.4	.5	.6	.7	.8	.9	.95
0	.513	.254	.055	.010	.001	.000	.000	.000	.000	.000	.000
1	.351	.367	.179	.054	.011	.002	.000	.000	.000	.000	.000
2	.111	.245	.268	.139	.045	.010	.001	.000	.000	.000	.000
3	.021	.100	.246	.218	.111	.035	.006	.001	.000	.000	.000
4	.003	.028	.154	.234	.184	.087	.024	.003	.000	.000	.000
5	.000	.006	.069	.180	.221	.157	.066	.014	.001	.000	.000
6	.000	.001	.023	.103	.197	.209	.131	.044	.006	.000	.000
7	.000	.000	.006	.044	.131	.209	.197	.103	.023	.001	.000
8	.000	.000	.001	.014	.066	.157	.221	.180	.069	.006	.000
9	.000	.000	.000	.003	.024	.087	.184	.234	.154	.028	.003
10	.000	.000	.000	.001	.006	.035	.111	.218	.246	.100	.021
11	.000	.000	.000	.000	.001	.010	.045	.139	.268	.245	.111
12	.000	.000	.000	.000	.000	.002	.011	.054	.179	.367	.351
13	.000	.000	.000	.000	.000	.000	.001	.010	.055	.254	.513

TABLE A.5 *(continued)*

n= 14
PROBABILITY

r	.05	.1	.2	.3	.4	.5	.6	.7	.8	.9	.95
0	.488	.229	.044	.007	.001	.000	.000	.000	.000	.000	.000
1	.359	.356	.154	.041	.007	.001	.000	.000	.000	.000	.000
2	.123	.257	.250	.113	.032	.006	.001	.000	.000	.000	.000
3	.026	.114	.250	.194	.085	.022	.003	.000	.000	.000	.000
4	.004	.035	.172	.229	.155	.061	.014	.001	.000	.000	.000
5	.000	.008	.086	.196	.207	.122	.041	.007	.000	.000	.000
6	.000	.001	.032	.126	.207	.183	.092	.023	.002	.000	.000
7	.000	.000	.009	.062	.157	.209	.157	.062	.009	.000	.000
8	.000	.000	.002	.023	.092	.183	.207	.126	.032	.001	.000
9	.000	.000	.000	.007	.041	.122	.207	.196	.086	.008	.000
10	.000	.000	.000	.001	.014	.061	.155	.229	.172	.035	.004
11	.000	.000	.000	.000	.003	.022	.085	.194	.250	.114	.026
12	.000	.000	.000	.000	.001	.006	.032	.113	.250	.257	.123
13	.000	.000	.000	.000	.000	.001	.007	.041	.154	.356	.359
14	.000	.000	.000	.000	.000	.000	.001	.007	.044	.229	.488

n= 15
PROBABILITY

r	.05	.1	.2	.3	.4	.5	.6	.7	.8	.9	.95
0	.463	.206	.035	.005	.000	.000	.000	.000	.000	.000	.000
1	.366	.343	.132	.031	.005	.000	.000	.000	.000	.000	.000
2	.135	.267	.231	.092	.022	.003	.000	.000	.000	.000	.000
3	.031	.129	.250	.170	.063	.014	.002	.000	.000	.000	.000
4	.005	.043	.188	.219	.127	.042	.007	.001	.000	.000	.000
5	.001	.010	.103	.206	.186	.092	.024	.003	.000	.000	.000
6	.000	.002	.043	.147	.207	.153	.061	.012	.001	.000	.000
7	.000	.000	.014	.081	.177	.196	.118	.035	.003	.000	.000
8	.000	.000	.003	.035	.118	.196	.177	.081	.014	.000	.000
9	.000	.000	.001	.012	.061	.153	.207	.147	.043	.002	.000
10	.000	.000	.000	.003	.024	.092	.186	.206	.103	.010	.001
11	.000	.000	.000	.001	.007	.042	.127	.219	.188	.043	.005
12	.000	.000	.000	.000	.002	.014	.063	.170	.250	.129	.031
13	.000	.000	.000	.000	.000	.003	.022	.092	.231	.267	.135
14	.000	.000	.000	.000	.000	.000	.005	.031	.132	.343	.366
15	.000	.000	.000	.000	.000	.000	.000	.005	.035	.206	.463

TABLE A.5 *(continued)*

n=16
PROBABILITY

r	.05	.1	.2	.3	.4	.5	.6	.7	.8	.9	.95
0	.440	.185	.028	.003	.000	.000	.000	.000	.000	.000	.000
1	.371	.329	.113	.023	.003	.000	.000	.000	.000	.000	.000
2	.146	.275	.211	.073	.015	.002	.000	.000	.000	.000	.000
3	.036	.142	.246	.146	.047	.009	.001	.000	.000	.000	.000
4	.006	.051	.200	.204	.101	.028	.004	.000	.000	.000	.000
5	.001	.014	.120	.210	.162	.067	.014	.001	.000	.000	.000
6	.000	.003	.055	.165	.198	.122	.039	.006	.000	.000	.000
7	.000	.000	.020	.101	.189	.175	.084	.019	.001	.000	.000
8	.000	.000	.006	.049	.142	.196	.142	.049	.006	.000	.000
9	.000	.000	.001	.019	.084	.175	.189	.101	.020	.000	.000
10	.000	.000	.000	.006	.039	.122	.198	.165	.055	.003	.000
11	.000	.000	.000	.001	.014	.067	.162	.210	.120	.014	.001
12	.000	.000	.000	.000	.004	.028	.101	.204	.200	.051	.006
13	.000	.000	.000	.000	.001	.009	.047	.146	.246	.142	.036
14	.000	.000	.000	.000	.000	.002	.015	.073	.211	.275	.146
15	.000	.000	.000	.000	.000	.000	.003	.023	.113	.329	.371
16	.000	.000	.000	.000	.000	.000	.000	.003	.028	.185	.440

n= 17
PROBABILITY

r	.05	.1	.2	.3	.4	.5	.6	.7	.8	.9	.95
0	.418	.167	.023	.002	.000	.000	.000	.000	.000	.000	.000
1	.374	.315	.096	.017	.002	.000	.000	.000	.000	.000	.000
2	.158	.280	.191	.058	.010	.001	.000	.000	.000	.000	.000
3	.041	.156	.239	.125	.034	.005	.000	.000	.000	.000	.000
4	.008	.060	.209	.187	.080	.018	.002	.000	.000	.000	.000
5	.001	.017	.136	.208	.138	.047	.008	.001	.000	.000	.000
6	.000	.004	.068	.178	.184	.094	.024	.003	.000	.000	.000
7	.000	.001	.027	.120	.193	.148	.057	.009	.000	.000	.000
8	.000	.000	.008	.064	.161	.185	.107	.028	.002	.000	.000
9	.000	.000	.002	.028	.107	.185	.161	.064	.008	.000	.000
10	.000	.000	.000	.009	.057	.148	.193	.120	.027	.001	.000
11	.000	.000	.000	.003	.024	.094	.184	.178	.068	.004	.000
12	.000	.000	.000	.001	.008	.047	.138	.208	.136	.017	.001
13	.000	.000	.000	.000	.002	.018	.080	.187	.209	.060	.008
14	.000	.000	.000	.000	.000	.005	.034	.125	.239	.156	.041
15	.000	.000	.000	.000	.000	.001	.010	.058	.191	.280	.158
16	.000	.000	.000	.000	.000	.000	.002	.017	.096	.315	.374
17	.000	.000	.000	.000	.000	.000	.000	.002	.023	.167	.418

TABLE A.5 (continued)

n= 18
PROBABILITY

r	.05	.1	.2	.3	.4	.5	.6	.7	.8	.9	.95
0	.397	.150	.018	.002	.000	.000	.000	.000	.000	.000	.000
1	.376	.300	.081	.013	.001	.000	.000	.000	.000	.000	.000
2	.168	.284	.172	.046	.007	.001	.000	.000	.000	.000	.000
3	.047	.168	.230	.105	.025	.003	.000	.000	.000	.000	.000
4	.009	.070	.215	.168	.061	.012	.001	.000	.000	.000	.000
5	.001	.022	.151	.202	.115	.033	.004	.000	.000	.000	.000
6	.000	.005	.082	.187	.166	.071	.015	.001	.000	.000	.000
7	.000	.001	.035	.138	.189	.121	.037	.005	.000	.000	.000
8	.000	.000	.012	.081	.173	.167	.077	.015	.001	.000	.000
9	.000	.000	.003	.039	.128	.185	.128	.039	.003	.000	.000
10	.000	.000	.001	.015	.077	.167	.173	.081	.012	.000	.000
11	.000	.000	.000	.005	.037	.121	.189	.138	.035	.001	.000
12	.000	.000	.000	.001	.015	.071	.166	.187	.082	.005	.000
13	.000	.000	.000	.000	.004	.033	.115	.202	.151	.022	.001
14	.000	.000	.000	.000	.001	.012	.061	.168	.215	.070	.009
15	.000	.000	.000	.000	.000	.003	.025	.105	.230	.168	.047
16	.000	.000	.000	.000	.000	.001	.007	.046	.172	.284	.168
17	.000	.000	.000	.000	.000	.000	.001	.013	.081	.300	.376
18	.000	.000	.000	.000	.000	.000	.000	.002	.018	.150	.397

n= 19
PROBABILITY

r	.05	.1	.2	.3	.4	.5	.6	.7	.8	.9	.95
0	.377	.135	.014	.001	.000	.000	.000	.000	.000	.000	.000
1	.377	.285	.068	.009	.001	.000	.000	.000	.000	.000	.000
2	.179	.285	.154	.036	.005	.000	.000	.000	.000	.000	.000
3	.053	.180	.218	.087	.017	.002	.000	.000	.000	.000	.000
4	.011	.080	.218	.149	.047	.007	.001	.000	.000	.000	.000
5	.002	.027	.164	.192	.093	.022	.002	.000	.000	.000	.000
6	.000	.007	.095	.192	.145	.052	.008	.001	.000	.000	.000
7	.000	.001	.044	.153	.180	.096	.024	.002	.000	.000	.000
8	.000	.000	.017	.098	.180	.144	.053	.008	.000	.000	.000
9	.000	.000	.005	.051	.146	.176	.098	.022	.001	.000	.000
10	.000	.000	.001	.022	.098	.176	.146	.051	.005	.000	.000
11	.000	.000	.000	.008	.053	.144	.180	.098	.017	.000	.000
12	.000	.000	.000	.002	.024	.096	.180	.153	.044	.001	.000
13	.000	.000	.000	.001	.008	.052	.145	.192	.095	.007	.000
14	.000	.000	.000	.000	.002	.022	.093	.192	.164	.027	.002
15	.000	.000	.000	.000	.001	.007	.047	.149	.218	.080	.011
16	.000	.000	.000	.000	.000	.002	.017	.087	.218	.180	.053
17	.000	.000	.000	.000	.000	.000	.005	.036	.154	.285	.179
18	.000	.000	.000	.000	.000	.000	.001	.009	.068	.285	.377
19	.000	.000	.000	.000	.000	.000	.000	.001	.014	.135	.377

TABLE A.5 *(concluded)*

n= 20
PROBABILITY

r	.05	.1	.2	.3	.4	.5	.6	.7	.8	.9	.95
0	.358	.122	.012	.001	.000	.000	.000	.000	.000	.000	.000
1	.377	.270	.058	.007	.000	.000	.000	.000	.000	.000	.000
2	.189	.285	.137	.028	.003	.000	.000	.000	.000	.000	.000
3	.060	.190	.205	.072	.012	.001	.000	.000	.000	.000	.000
4	.013	.090	.218	.130	.035	.005	.000	.000	.000	.000	.000
5	.002	.032	.175	.179	.075	.015	.001	.000	.000	.000	.000
6	.000	.009	.109	.192	.124	.037	.005	.000	.000	.000	.000
7	.000	.002	.055	.164	.166	.074	.015	.001	.000	.000	.000
8	.000	.000	.022	.114	.180	.120	.035	.004	.000	.000	.000
9	.000	.000	.007	.065	.160	.160	.071	.012	.000	.000	.000
10	.000	.000	.002	.031	.117	.176	.117	.031	.002	.000	.000
11	.000	.000	.000	.012	.071	.160	.160	.065	.007	.000	.000
12	.000	.000	.000	.004	.035	.120	.180	.114	.022	.000	.000
13	.000	.000	.000	.001	.015	.074	.166	.164	.055	.002	.000
14	.000	.000	.000	.000	.005	.037	.124	.192	.109	.009	.000
15	.000	.000	.000	.000	.001	.015	.075	.179	.175	.032	.002
16	.000	.000	.000	.000	.000	.005	.035	.130	.218	.090	.013
17	.000	.000	.000	.000	.000	.001	.012	.072	.205	.190	.060
18	.000	.000	.000	.000	.000	.000	.003	.028	.137	.285	.189
19	.000	.000	.000	.000	.000	.000	.000	.007	.058	.270	.377
20	.000	.000	.000	.000	.000	.000	.000	.001	.012	.122	.358

TABLE A.6 Poisson Distribution: Probability of Exactly x Occurrences

					μ				
x	0.1	0.2	0.3	0.4	0.5	0.6	0.7	0.8	0.9
0	0.9048	0.8187	0.7408	0.6703	0.6065	0.5488	0.4966	0.4493	0.4066
1	0.0905	0.1637	0.2222	0.2681	0.3033	0.3293	0.3476	0.3595	0.3659
2	0.0045	0.0164	0.0333	0.0536	0.0758	0.0988	0.1217	0.1438	0.1647
3	0.0002	0.0011	0.0033	0.0072	0.0126	0.0198	0.0284	0.0383	0.0494
4		0.0001	0.0003	0.0007	0.0016	0.0030	0.0050	0.0077	0.0111
5					0.0002	0.0004	0.0007	0.0012	0.0020
6							0.0001	0.0002	0.0003

TABLE A.6 (continued)

x	2.0	3.0	4.0	5.0	6.0	7.0	8.0	9.0
0	0.1353	0.0498	0.0183	0.0067	0.0025	0.0009	0.0003	0.0001
1	0.2707	0.1494	0.0733	0.0337	0.0149	0.0064	0.0027	0.0011
2	0.2707	0.2240	0.1465	0.0842	0.0446	0.0223	0.0107	0.0050
3	0.1804	0.2240	0.1954	0.1404	0.0892	0.0521	0.0286	0.0150
4	0.0902	0.1680	0.1954	0.1755	0.1339	0.0912	0.0573	0.0337
5	0.0361	0.1008	0.1563	0.1755	0.1606	0.1277	0.0916	0.0607
6	0.0120	0.0504	0.1042	0.1462	0.1606	0.1490	0.1221	0.0911
7	0.0034	0.0216	0.0595	0.1044	0.1377	0.1490	0.1396	0.1171
8	0.0009	0.0081	0.0298	0.0653	0.1033	0.1304	0.1396	0.1318
9	0.0002	0.0027	0.0132	0.0363	0.0688	0.1014	0.1241	0.1318
10		0.0008	0.0053	0.0181	0.0413	0.0710	0.0993	0.1186
11		0.0002	0.0019	0.0082	0.0225	0.0452	0.0722	0.0970
12		0.0001	0.0006	0.0034	0.0113	0.0264	0.0481	0.0728
13			0.0002	0.0013	0.0052	0.0142	0.0296	0.0504
14			0.0001	0.0005	0.0022	0.0071	0.0169	0.0324
15				0.0002	0.0009	0.0033	0.0090	0.0194
16					0.0003	0.0014	0.0045	0.0109
17					0.0001	0.0006	0.0021	0.0058
18						0.0002	0.0009	0.0029
19						0.0001	0.0004	0.0014
20							0.0002	0.0006
21							0.0001	0.0003
22								0.0001

TABLE A.7 Critical Values of the Durbin-Watson Test Statistic

$\alpha = 0.05$

n	k = 1		k = 2		k = 3		k = 4			
	d_L	d_U	d_L	d_U	d_L	d_U	d_L	d_U	d_L	
15	1.08	1.36	0.95	1.54	0.82	1.75	0.69	1.97	0.56	
16	1.10	1.37	0.98	1.54	0.86	1.73	0.74	1.93	0.62	
17	1.13	1.38	1.02	1.54	0.90	1.71	0.78	1.90	0.67	
18	1.16	1.39	1.05	1.53	0.93	1.69	0.82	1.87	0.71	
19	1.18	1.40	1.08	1.53	0.97	1.68	0.86	1.85	0.75	
20	1.20	1.41	1.10	1.54	1.00	1.68	0.90	1.83	0.79	
21	1.22	1.42	1.13	1.54	1.03	1.67	0.93	1.81	0.83	1.96
22	1.24	1.43	1.15	1.54	1.05	1.66	0.96	1.80	0.86	1.94
23	1.26	1.44	1.17	1.54	1.08	1.66	0.99	1.79	0.90	1.92
24	1.27	1.45	1.19	1.55	1.10	1.66	1.01	1.78	0.93	1.90
25	1.29	1.45	1.21	1.55	1.12	1.66	1.04	1.77	0.95	1.89
26	1.30	1.46	1.22	1.55	1.14	1.65	1.06	1.76	0.98	1.88
27	1.32	1.47	1.24	1.56	1.16	1.65	1.08	1.76	1.01	1.86
28	1.33	1.48	1.26	1.56	1.18	1.65	1.10	1.75	1.03	1.85
29	1.34	1.48	1.27	1.56	1.20	1.65	1.12	1.74	1.05	1.84
30	1.35	1.49	1.28	1.57	1.21	1.65	1.14	1.74	1.07	1.83
31	1.36	1.50	1.30	1.57	1.23	1.65	1.16	1.74	1.09	1.83
32	1.37	1.50	1.31	1.57	1.24	1.65	1.18	1.73	1.11	1.82
33	1.38	1.51	1.32	1.58	1.26	1.65	1.19	1.73	1.13	1.81
34	1.39	1.51	1.33	1.58	1.27	1.65	1.21	1.73	1.15	1.81
35	1.40	1.52	1.34	1.58	1.28	1.65	1.22	1.73	1.16	1.80
36	1.41	1.52	1.35	1.59	1.29	1.65	1.24	1.73	1.18	1.80
37	1.42	1.53	1.36	1.59	1.31	1.66	1.25	1.72	1.19	1.80
38	1.43	1.54	1.37	1.59	1.32	1.66	1.26	1.72	1.21	1.79
39	1.43	1.54	1.38	1.60	1.33	1.66	1.27	1.72	1.22	1.79
40	1.44	1.54	1.39	1.60	1.34	1.66	1.29	1.72	1.23	1.79
45	1.48	1.57	1.43	1.62	1.38	1.67	1.34	1.72	1.29	1.78
50	1.50	1.59	1.46	1.63	1.42	1.67	1.38	1.72	1.34	1.77
55	1.53	1.60	1.49	1.64	1.45	1.68	1.41	1.72	1.38	1.77
60	1.55	1.62	1.51	1.65	1.48	1.69	1.44	1.73	1.41	1.77
65	1.57	1.63	1.54	1.66	1.50	1.70	1.47	1.73	1.44	1.77
70	1.58	1.64	1.55	1.67	1.52	1.70	1.49	1.74	1.46	1.77
75	1.60	1.65	1.57	1.68	1.54	1.71	1.51	1.74	1.49	1.77
80	1.61	1.66	1.59	1.69	1.56	1.72	1.53	1.74	1.51	1.77
85	1.62	1.67	1.60	1.70	1.57	1.72	1.55	1.75	1.52	1.77
90	1.63	1.68	1.61	1.70	1.59	1.73	1.57	1.75	1.54	1.78
95	1.64	1.69	1.62	1.71	1.60	1.73	1.58	1.75	1.56	1.78
100	1.65	1.69	1.63	1.72	1.61	1.74	1.59	1.76	1.57	1.78

TABLE A.6 (concluded)

x	1.0	2.0	3.0	4.0	5.0	6.0	7.0	8.0	9.0
0	0.3679	0.1353	0.0498	0.0183	0.0067	0.0025	0.0009	0.0003	0.0001
1	0.3679	0.2707	0.1494	0.0733	0.0337	0.0149	0.0064	0.0027	0.0011
2	0.1839	0.2707	0.2240	0.1465	0.0842	0.0446	0.0223	0.0107	0.0050
3	0.0613	0.1804	0.2240	0.1954	0.1404	0.0892	0.0521	0.0286	0.0150
4	0.0153	0.0902	0.1680	0.1954	0.1755	0.1339	0.0912	0.0573	0.0337
5	0.0031	0.0361	0.1008	0.1563	0.1755	0.1606	0.1277	0.0916	0.0607
6	0.0005	0.0120	0.0504	0.1042	0.1462	0.1606	0.1490	0.1221	0.0911
7	0.0001	0.0034	0.0216	0.0595	0.1044	0.1377	0.1490	0.1396	0.1171
8		0.0009	0.0081	0.0298	0.0653	0.1033	0.1304	0.1396	0.1318
9		0.0002	0.0027	0.0132	0.0363	0.0688	0.1014	0.1241	0.1318
10			0.0008	0.0053	0.0181	0.0413	0.0710	0.0993	0.1186
11			0.0002	0.0019	0.0082	0.0225	0.0452	0.0722	0.0970
12			0.0001	0.0006	0.0034	0.0113	0.0264	0.0481	0.0728
13				0.0002	0.0013	0.0052	0.0142	0.0296	0.0504
14				0.0001	0.0005	0.0022	0.0071	0.0169	0.0324
15					0.0002	0.0009	0.0033	0.0090	0.0194
16						0.0003	0.0014	0.0045	0.0109
17						0.0001	0.0006	0.0021	0.0058
18							0.0002	0.0009	0.0029
19							0.0001	0.0004	0.0014
20								0.0002	0.0006
21								0.0001	0.0003
22									0.0001

TABLE A.7 Critical Values of the Durbin-Watson Test Statistic

$\alpha = 0.05$

	k = 1		k = 2		k = 3		k = 4		k = 5	
n	d_L	d_U	d_L	d_U	d_L	d_U	d_L	d_U	d_L	d_U
15	1.08	1.36	0.95	1.54	0.82	1.75	0.69	1.97	0.56	2.21
16	1.10	1.37	0.98	1.54	0.86	1.73	0.74	1.93	0.62	2.15
17	1.13	1.38	1.02	1.54	0.90	1.71	0.78	1.90	0.67	2.10
18	1.16	1.39	1.05	1.53	0.93	1.69	0.82	1.87	0.71	2.06
19	1.18	1.40	1.08	1.53	0.97	1.68	0.86	1.85	0.75	2.02
20	1.20	1.41	1.10	1.54	1.00	1.68	0.90	1.83	0.79	1.99
21	1.22	1.42	1.13	1.54	1.03	1.67	0.93	1.81	0.83	1.96
22	1.24	1.43	1.15	1.54	1.05	1.66	0.96	1.80	0.86	1.94
23	1.26	1.44	1.17	1.54	1.08	1.66	0.99	1.79	0.90	1.92
24	1.27	1.45	1.19	1.55	1.10	1.66	1.01	1.78	0.93	1.90
25	1.29	1.45	1.21	1.55	1.12	1.66	1.04	1.77	0.95	1.89
26	1.30	1.46	1.22	1.55	1.14	1.65	1.06	1.76	0.98	1.88
27	1.32	1.47	1.24	1.56	1.16	1.65	1.08	1.76	1.01	1.86
28	1.33	1.48	1.26	1.56	1.18	1.65	1.10	1.75	1.03	1.85
29	1.34	1.48	1.27	1.56	1.20	1.65	1.12	1.74	1.05	1.84
30	1.35	1.49	1.28	1.57	1.21	1.65	1.14	1.74	1.07	1.83
31	1.36	1.50	1.30	1.57	1.23	1.65	1.16	1.74	1.09	1.83
32	1.37	1.50	1.31	1.57	1.24	1.65	1.18	1.73	1.11	1.82
33	1.38	1.51	1.32	1.58	1.26	1.65	1.19	1.73	1.13	1.81
34	1.39	1.51	1.33	1.58	1.27	1.65	1.21	1.73	1.15	1.81
35	1.40	1.52	1.34	1.58	1.28	1.65	1.22	1.73	1.16	1.80
36	1.41	1.52	1.35	1.59	1.29	1.65	1.24	1.73	1.18	1.80
37	1.42	1.53	1.36	1.59	1.31	1.66	1.25	1.72	1.19	1.80
38	1.43	1.54	1.37	1.59	1.32	1.66	1.26	1.72	1.21	1.79
39	1.43	1.54	1.38	1.60	1.33	1.66	1.27	1.72	1.22	1.79
40	1.44	1.54	1.39	1.60	1.34	1.66	1.29	1.72	1.23	1.79
45	1.48	1.57	1.43	1.62	1.38	1.67	1.34	1.72	1.29	1.78
50	1.50	1.59	1.46	1.63	1.42	1.67	1.38	1.72	1.34	1.77
55	1.53	1.60	1.49	1.64	1.45	1.68	1.41	1.72	1.38	1.77
60	1.55	1.62	1.51	1.65	1.48	1.69	1.44	1.73	1.41	1.77
65	1.57	1.63	1.54	1.66	1.50	1.70	1.47	1.73	1.44	1.77
70	1.58	1.64	1.55	1.67	1.52	1.70	1.49	1.74	1.46	1.77
75	1.60	1.65	1.57	1.68	1.54	1.71	1.51	1.74	1.49	1.77
80	1.61	1.66	1.59	1.69	1.56	1.72	1.53	1.74	1.51	1.77
85	1.62	1.67	1.60	1.70	1.57	1.72	1.55	1.75	1.52	1.77
90	1.63	1.68	1.61	1.70	1.59	1.73	1.57	1.75	1.54	1.78
95	1.64	1.69	1.62	1.71	1.60	1.73	1.58	1.75	1.56	1.78
100	1.65	1.69	1.63	1.72	1.61	1.74	1.59	1.76	1.57	1.78

TABLE A.7 *(concluded)*

$\alpha = 0.01$

	$k=1$		$k=2$		$k=3$		$k=4$		$k=5$	
n	d_L	d_U	d_L	d_U	d_L	d_U	d_L	d_U	d_L	d_U
15	0.81	1.07	0.70	1.25	0.59	1.46	0.49	1.70	0.39	1.96
16	0.84	1.09	0.74	1.25	0.63	1.44	0.53	1.66	0.44	1.90
17	0.87	1.10	0.77	1.25	0.67	1.43	0.57	1.63	0.48	1.85
18	0.90	1.12	0.80	1.26	0.71	1.42	0.61	1.60	0.52	1.80
19	0.93	1.13	0.83	1.26	0.74	1.41	0.65	1.58	0.56	1.77
20	0.95	1.15	0.86	1.27	0.77	1.41	0.68	1.57	0.60	1.74
21	0.97	1.16	0.89	1.27	0.80	1.41	0.72	1.55	0.63	1.71
22	1.00	1.17	0.91	1.28	0.83	1.40	0.75	1.54	0.66	1.69
23	1.02	1.19	0.94	1.29	0.86	1.40	0.77	1.53	0.70	1.67
24	1.05	1.20	0.96	1.30	0.88	1.41	0.80	1.53	0.72	1.66
25	1.05	1.21	0.98	1.30	0.90	1.41	0.83	1.52	0.75	1.65
26	1.07	1.22	1.00	1.31	0.93	1.41	0.85	1.52	0.78	1.64
27	1.09	1.23	1.02	1.32	0.95	1.41	0.88	1.51	0.81	1.63
28	1.10	1.24	1.04	1.32	0.97	1.41	0.90	1.51	0.83	1.62
29	1.12	1.25	1.05	1.33	0.99	1.42	0.92	1.51	0.85	1.61
30	1.13	1.26	1.07	1.34	1.01	1.42	0.94	1.51	0.88	1.61
31	1.15	1.27	1.08	1.34	1.02	1.42	0.96	1.51	0.90	1.60
32	1.16	1.28	1.10	1.35	1.04	1.43	0.98	1.51	0.92	1.60
33	1.17	1.29	1.11	1.36	1.05	1.43	1.00	1.51	0.94	1.59
34	1.18	1.30	1.13	1.36	1.07	1.43	1.01	1.51	0.95	1.59
35	1.19	1.31	1.14	1.37	1.08	1.44	1.03	1.51	0.97	1.59
36	1.21	1.32	1.15	1.38	1.10	1.44	1.04	1.51	0.99	1.59
37	1.22	1.32	1.16	1.38	1.11	1.45	1.06	1.51	1.00	1.59
38	1.23	1.33	1.18	1.39	1.12	1.45	1.07	1.52	1.02	1.58
39	1.24	1.34	1.19	1.39	1.14	1.45	1.09	1.52	1.03	1.58
40	1.25	1.34	1.20	1.40	1.15	1.46	1.10	1.52	1.05	1.58
45	1.29	1.38	1.24	1.42	1.20	1.48	1.16	1.53	1.11	1.58
50	1.32	1.40	1.28	1.45	1.24	1.49	1.20	1.54	1.16	1.59
55	1.36	1.43	1.32	1.47	1.28	1.51	1.25	1.55	1.21	1.59
60	1.38	1.45	1.35	1.48	1.32	1.52	1.28	1.56	1.25	1.60
65	1.41	1.47	1.38	1.50	1.35	1.53	1.31	1.57	1.28	1.61
70	1.43	1.49	1.40	1.52	1.37	1.55	1.34	1.58	1.31	1.61
75	1.45	1.50	1.42	1.53	1.39	1.56	1.37	1.59	1.34	1.62
80	1.47	1.52	1.44	1.54	1.42	1.57	1.39	1.60	1.36	1.62
85	1.48	1.53	1.46	1.55	1.43	1.58	1.41	1.60	1.39	1.63
90	1.50	1.54	1.47	1.56	1.45	1.59	1.43	1.61	1.41	1.64
95	1.51	1.55	1.49	1.57	1.47	1.60	1.45	1.62	1.42	1.64
100	1.52	1.56	1.50	1.58	1.48	1.60	1.46	1.63	1.44	1.65

Source: Reproduced by permission from J. Durbin and G. S. Watson, "Testing for Serial Correlation in Least Squares Regression, II," *Biometrika* 38 (1951), pp. 159–78.

TABLE A.8 Critical Values of the Studentized Range Distribution (Tukey test)

$\alpha = 0.05$

n−r	2	3	4	5	6	7	8	9	10	11	12	13	14	15	16	17	18	19	20
1	18.0	27.0	32.8	37.1	40.4	43.1	45.4	47.4	49.1	50.6	52.0	53.2	54.3	55.4	56.3	57.2	58.0	58.8	59.6
2	6.08	8.33	9.80	10.9	11.7	12.4	13.0	13.5	14.0	14.4	14.7	15.1	15.4	15.7	15.9	16.1	16.4	16.6	16.8
3	4.50	5.91	6.82	7.50	8.04	8.48	8.85	9.18	9.46	9.72	9.95	10.2	10.3	10.5	10.7	10.8	11.0	11.1	11.2
4	3.93	5.04	5.76	6.29	6.71	7.05	7.35	7.60	7.83	8.03	8.21	8.37	8.52	8.66	8.79	8.91	9.03	9.13	9.23
5	3.64	4.60	5.22	5.67	6.03	6.33	6.58	6.80	6.99	7.17	7.32	7.47	7.60	7.72	7.83	7.93	8.03	8.12	8.21
6	3.46	4.34	4.90	5.30	5.63	5.90	6.12	6.32	6.49	6.65	6.79	6.92	7.03	7.14	7.24	7.34	7.43	7.51	7.59
7	3.34	4.16	4.68	5.06	5.36	5.61	5.82	6.00	6.16	6.30	6.43	6.55	6.66	6.76	6.85	6.94	7.02	7.10	7.17
8	3.26	4.04	4.53	4.89	5.17	5.40	5.60	5.77	5.92	6.05	6.18	6.29	6.39	6.48	6.57	6.65	6.73	6.80	6.87
9	3.20	3.95	4.41	4.76	5.02	5.24	5.43	5.59	5.74	5.87	5.98	6.09	6.19	6.28	6.36	6.44	6.51	6.58	6.64
10	3.15	3.88	4.33	4.65	4.91	5.12	5.30	5.46	5.60	5.72	5.83	5.93	6.03	6.11	6.19	6.27	6.34	6.40	6.47
11	3.11	3.82	4.26	4.57	4.82	5.03	5.20	5.35	5.49	5.61	5.71	5.81	5.90	5.98	6.06	6.13	6.20	6.27	6.33
12	3.08	3.77	4.20	4.51	4.75	4.95	5.12	5.27	5.39	5.51	5.61	5.71	5.80	5.88	5.95	6.02	6.09	6.15	6.21
13	3.06	3.73	4.15	4.45	4.69	4.88	5.05	5.19	5.32	5.43	5.53	5.63	5.71	5.79	5.86	5.93	5.99	6.05	6.11
14	3.03	3.70	4.11	4.41	4.64	4.83	4.99	5.13	5.25	5.36	5.46	5.55	5.64	5.71	5.79	5.85	5.91	5.97	6.03
15	3.01	3.67	4.08	4.37	4.59	4.78	4.94	5.08	5.20	5.31	5.40	5.49	5.57	5.65	5.72	5.78	5.85	5.90	5.96
16	3.00	3.65	4.05	4.33	4.56	4.74	4.90	5.03	5.15	5.26	5.35	5.44	5.52	5.59	5.66	5.73	5.79	5.84	5.90
17	2.98	3.63	4.02	4.30	4.52	4.70	4.86	4.99	5.11	5.21	5.31	5.39	5.47	5.54	5.61	5.67	5.73	5.79	5.84
18	2.97	3.61	4.00	4.28	4.49	4.67	4.82	4.96	5.07	5.17	5.27	5.35	5.43	5.50	5.57	5.63	5.69	5.74	5.79
19	2.96	3.59	3.98	4.25	4.47	4.65	4.79	4.92	5.04	5.14	5.23	5.31	5.39	5.46	5.53	5.59	5.65	5.70	5.75
20	2.95	3.58	3.96	4.23	4.45	4.62	4.77	4.90	5.01	5.11	5.20	5.28	5.36	5.43	5.49	5.55	5.61	5.66	5.71
24	2.92	3.53	3.90	4.17	4.37	4.54	4.68	4.81	4.92	5.01	5.10	5.18	5.25	5.32	5.38	5.44	5.49	5.55	5.59
30	2.89	3.49	3.85	4.10	4.30	4.46	4.60	4.72	4.82	4.92	5.00	5.08	5.15	5.21	5.27	5.33	5.38	5.43	5.47
40	2.86	3.44	3.79	4.04	4.23	4.39	4.52	4.63	4.73	4.82	4.90	4.98	5.04	5.11	5.16	5.22	5.27	5.31	5.36
60	2.83	3.40	3.74	3.98	4.16	4.31	4.44	4.55	4.65	4.73	4.81	4.88	4.94	5.00	5.06	5.11	5.15	5.20	5.24
120	2.80	3.36	3.68	3.92	4.10	4.24	4.36	4.47	4.56	4.64	4.71	4.78	4.84	4.90	4.95	5.00	5.04	5.09	5.13
∞	2.77	3.31	3.63	3.86	4.03	4.17	4.29	4.39	4.47	4.55	4.62	4.68	4.74	4.80	4.85	4.89	4.93	4.97	5.01

TABLE A.8 *(concluded)*

$\alpha = 0.01$

n − r	2	3	4	5	6	7	8	9	10	11	12	13	14	15	16	17	18	19	20
1	90.0	135	164	186	202	216	227	237	246	253	260	266	272	277	282	286	290	294	298
2	14.0	19.0	22.3	24.7	26.6	28.2	29.5	30.7	31.7	32.6	33.4	34.1	34.8	35.4	36.0	36.5	37.0	37.5	37.9
3	8.26	10.6	12.2	13.3	14.2	15.0	15.6	16.2	16.7	17.1	17.5	17.9	18.2	18.5	18.8	19.1	19.3	19.5	19.8
4	6.51	8.12	9.17	9.96	10.6	11.1	11.5	11.9	12.3	12.6	12.8	13.1	13.3	13.5	13.7	13.9	14.1	14.2	14.4
5	5.70	6.97	7.80	8.42	8.91	9.32	9.67	9.97	10.2	10.5	10.7	10.9	11.1	11.2	11.4	11.6	11.7	11.8	11.9
6	5.24	6.33	7.03	7.56	7.97	8.32	8.61	8.87	9.10	9.30	9.49	9.65	9.81	9.95	10.1	10.2	10.3	10.4	10.5
7	4.95	5.92	6.54	7.01	7.37	7.68	7.94	8.17	8.37	8.55	8.71	8.86	9.00	9.12	9.24	9.35	9.46	9.55	9.65
8	4.74	5.63	6.20	6.63	6.96	7.24	7.47	7.68	7.87	8.03	8.18	8.31	8.44	8.55	8.66	8.76	8.85	8.94	9.03
9	4.60	5.43	5.96	6.35	6.66	6.91	7.13	7.32	7.49	7.65	7.78	7.91	8.03	8.13	8.23	8.32	8.41	8.49	8.57
10	4.48	5.27	5.77	6.14	6.43	6.67	6.87	7.05	7.21	7.36	7.48	7.60	7.71	7.81	7.91	7.99	8.07	8.15	8.22
11	4.39	5.14	5.62	5.97	6.25	6.48	6.67	6.84	6.99	7.13	7.25	7.36	7.46	7.56	7.65	7.73	7.81	7.88	7.95
12	4.32	5.04	5.50	5.84	6.10	6.32	6.51	6.67	6.81	6.94	7.06	7.17	7.26	7.36	7.44	7.52	7.59	7.66	7.73
13	4.26	4.96	5.40	5.73	5.98	6.19	6.37	6.53	6.67	6.79	6.90	7.01	7.10	7.19	7.27	7.34	7.42	7.48	7.55
14	4.21	4.89	5.32	5.63	5.88	6.08	6.26	6.41	6.54	6.66	6.77	6.87	6.96	7.05	7.12	7.20	7.27	7.33	7.39
15	4.17	4.83	5.25	5.56	5.80	5.99	6.16	6.31	6.44	6.55	6.66	6.76	6.84	6.93	7.00	7.07	7.14	7.20	7.26
16	4.13	4.78	5.19	5.49	5.72	5.92	6.08	6.22	6.35	6.46	6.56	6.66	6.74	6.82	6.90	6.97	7.03	7.09	7.15
17	4.10	4.74	5.14	5.43	5.66	5.85	6.01	6.15	6.27	6.38	6.48	6.57	6.66	6.73	6.80	6.87	6.94	7.00	7.05
18	4.07	4.70	5.09	5.38	5.60	5.79	5.94	6.08	6.20	6.31	6.41	6.50	6.58	6.65	6.72	6.79	6.85	6.91	6.96
19	4.05	4.67	5.05	5.33	5.55	5.73	5.89	6.02	6.14	6.25	6.34	6.43	6.51	6.58	6.65	6.72	6.78	6.84	6.89
20	4.02	4.64	5.02	5.29	5.51	5.69	5.84	5.97	6.09	6.19	6.29	6.37	6.45	6.52	6.59	6.65	6.71	6.76	6.82
24	3.96	4.54	4.91	5.17	5.37	5.54	5.69	5.81	5.92	6.02	6.11	6.19	6.26	6.33	6.39	6.45	6.51	6.56	6.61
30	3.89	4.45	4.80	5.05	5.24	5.40	5.54	5.65	5.76	5.85	5.93	6.01	6.08	6.14	6.20	6.26	6.31	6.36	6.41
40	3.82	4.37	4.70	4.93	5.11	5.27	5.39	5.50	5.60	5.69	5.77	5.84	5.90	5.96	6.02	6.07	6.12	6.17	6.21
60	3.76	4.28	4.60	4.82	4.99	5.13	5.25	5.36	5.45	5.53	5.60	5.67	5.73	5.79	5.84	5.89	5.93	5.98	6.02
120	3.70	4.20	4.50	4.71	4.87	5.01	5.12	5.21	5.30	5.38	5.44	5.51	5.56	5.61	5.66	5.71	5.75	5.79	5.83
8	3.64	4.12	4.40	4.60	4.76	4.88	4.99	5.08	5.16	5.23	5.29	5.35	5.40	5.45	5.49	5.54	5.57	5.61	5.65

Reprinted by permission of the *Biometrika* Trustees from E. S. Pearson and H. O. Hartley, eds., *Biometrika Tables for Statisticians*, vol. 1, 3rd ed. (Cambridge University Press, 1966).

TABLE A.9 Cumulative Distribution Function (Runs test)

$F(r)$ for the total number of runs (R) in samples of sizes n_1 and n_2

(n_1, n_2)	2	3	4	5	6	7	8	9	10
(2,3)	0.200	0.500	0.900	1.000					
(2,4)	0.133	0.400	0.800	1.000					
(2,5)	0.095	0.333	0.714	1.000					
(2,6)	0.071	0.286	0.643	1.000					
(2,7)	0.056	0.250	0.583	1.000					
(2,8)	0.044	0.222	0.533	1.000					
(2,9)	0.036	0.200	0.491	1.000					
(2,10)	0.030	0.182	0.455	1.000					
(3,3)	0.100	0.300	0.700	0.900	1.000				
(3,4)	0.057	0.200	0.543	0.800	0.971	1.000			
(3,5)	0.036	0.143	0.429	0.714	0.929	1.000			
(3,6)	0.024	0.107	0.345	0.643	0.881	1.000			
(3,7)	0.017	0.083	0.283	0.583	0.833	1.000			
(3,8)	0.012	0.067	0.236	0.533	0.788	1.000			
(3,9)	0.009	0.055	0.200	0.491	0.745	1.000			
(3,10)	0.007	0.045	0.171	0.455	0.706	1.000			
(4,4)	0.029	0.114	0.371	0.629	0.886	0.971	1.000		
(4,5)	0.016	0.071	0.262	0.500	0.786	0.929	0.992	1.000	
(4,6)	0.010	0.048	0.190	0.405	0.690	0.881	0.976	1.000	
(4,7)	0.006	0.033	0.142	0.333	0.606	0.833	0.954	1.000	
(4,8)	0.004	0.024	0.109	0.279	0.533	0.788	0.929	1.000	
(4,9)	0.003	0.018	0.085	0.236	0.471	0.745	0.902	1.000	
(4,10)	0.002	0.014	0.068	0.203	0.419	0.706	0.874	1.000	
(5,5)	0.008	0.040	0.167	0.357	0.643	0.833	0.960	0.992	1.000
(5,6)	0.004	0.024	0.110	0.262	0.522	0.738	0.911	0.976	0.998
(5,7)	0.003	0.015	0.076	0.197	0.424	0.652	0.854	0.955	0.992
(5,8)	0.002	0.010	0.054	0.152	0.347	0.576	0.793	0.929	0.984
(5,9)	0.001	0.007	0.039	0.119	0.287	0.510	0.734	0.902	0.972
(5,10)	0.001	0.005	0.029	0.095	0.239	0.455	0.678	0.874	0.958
(6,6)	0.002	0.013	0.067	0.175	0.392	0.608	0.825	0.933	0.987
(6,7)	0.001	0.008	0.043	0.121	0.296	0.500	0.733	0.879	0.966
(6,8)	0.001	0.005	0.028	0.086	0.226	0.413	0.646	0.821	0.937
(6,9)	0.000	0.003	0.019	0.063	0.175	0.343	0.566	0.762	0.902
(6,10)	0.000	0.002	0.013	0.047	0.137	0.288	0.497	0.706	0.864
(7,7)	0.001	0.004	0.025	0.078	0.209	0.383	0.617	0.791	0.922
(7,8)	0.000	0.002	0.015	0.051	0.149	0.296	0.514	0.704	0.867
(7,9)	0.000	0.001	0.010	0.035	0.108	0.231	0.427	0.622	0.806
(7,10)	0.000	0.001	0.006	0.024	0.080	0.182	0.355	0.549	0.743
(8,8)	0.000	0.001	0.009	0.032	0.100	0.214	0.405	0.595	0.786
(8,9)	0.000	0.001	0.005	0.020	0.069	0.157	0.319	0.500	0.702
(8,10)	0.000	0.000	0.003	0.013	0.048	0.117	0.251	0.419	0.621
(9,9)	0.000	0.000	0.003	0.012	0.044	0.109	0.238	0.399	0.601
(9,10)	0.000	0.000	0.002	0.008	0.029	0.077	0.179	0.319	0.510
(10,10)	0.000	0.000	0.001	0.004	0.019	0.051	0.128	0.242	0.414

TABLE A.9 (concluded)

	Number of Runs, r									
(n_1, n_2)	11	12	13	14	15	16	17	18	19	20
(2,3)										
(2,4)										
(2,5)										
(2,6)										
(2,7)										
(2,8)										
(2,9)										
(2,10)										
(3,3)										
(3,4)										
(3,5)										
(3,6)										
(3,7)										
(3,8)										
(3,9)										
(3,10)										
(4,4)										
(4,5)										
(4,6)										
(4,7)										
(4,8)										
(4,9)										
(4,10)										
(5,5)										
(5,6)	1.000									
(5,7)	1.000									
(5,8)	1.000									
(5,9)	1.000									
(5,10)	1.000									
(6,6)	0.998	1.000								
(6,7)	0.992	0.999	1.000							
(6,8)	0.984	0.998	1.000							
(6,9)	0.972	0.994	1.000							
(6,10)	0.958	0.990	1.000							
(7,7)	0.975	0.996	0.999	1.000						
(7,8)	0.949	0.988	0.998	1.000	1.000					
(7,9)	0.916	0.975	0.994	0.999	1.000					
(7,10)	0.879	0.957	0.990	0.998	1.000					
(8,8)	0.900	0.968	0.991	0.999	1.000	1.000				
(8,9)	0.843	0.939	0.980	0.996	0.999	1.000	1.000			
(8,10)	0.782	0.903	0.964	0.990	0.998	1.000	1.000			
(9,9)	0.762	0.891	0.956	0.988	0.997	1.000	1.000	1.000		
(9,10)	0.681	0.834	0.923	0.974	0.992	0.999	1.000	1.000	1.000	
(10,10)	0.586	0.758	0.872	0.949	0.981	0.996	0.999	1.000	1.000	1.000

Reproduced from F. Swed and C. Eisenhart, "Tables for Testing Randomness of Grouping in a Sequence of Alternatives," *Annals of Mathematical Statistics* 14 (1943) by permission of the authors and of the Editor, *Annals of Mathematical Statistics*.

Appendix B

Computerized Business Statistics

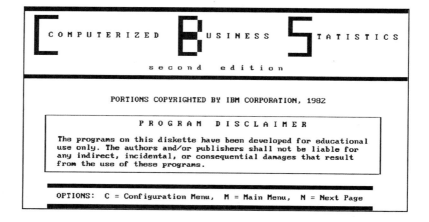

Overview

The growing demand for increased productivity throughout both business and government has brought about a growing interest in statistical analysis. These developments, in turn, have prompted an increased focus on statistics in many business and engineering schools. The primary purpose of this courseware package is to provide the student with the capability to solve a wide range of statistically based problems quickly and accurately. The authors believe that substituting computer modeling for traditional hand calculation methods not only increases the amount of time available for problem formulation and for a more indepth analysis of the results.

The material in this appendix is reprinted from O. P. Hall, Jr., and H. E. Adelman, *Computerized Business Statistics,* 2nd ed. (Homewood, Ill.: Richard D. Irwin, 1990), pp. iii and 1–16. Used with permission.

1.0 INTRODUCTION

COMPUTERIZED BUSINESS STATISTICS (CBS) is a collection of thirteen mathematical models which have been designed to facilitate a wide range of statistical analyses. The CBS package also contains a data management module for creating and editing data files. CBS is a user-friendly, menu driven package that requires very little computer knowhow to operate.

Presented below is a list of the models found on the CBS diskette:

- Data Base Management
- Descriptive Statistics
- Probability Theory
- Probability Functions
- Survey Design
- Random Sampling and Estimation
- Hypothesis Testing
- Simple Correlation and Regression
- Multiple Regression Analysis
- Time Series and Forecasting
- Chi-Square Analysis
- Analysis of Variance
- Nonparametric Methods
- Decision Analysis

1.1 EQUIPMENT

The minimum system configuration requires an IBM PC (or true compatible) with 128k memory and one double-sided disk drive, plus a copy of MS-DOS 2.0 +. A second disk drive, a printer, and a color monitor are optional. Graphic displays use the special text characters so a graphics card is not needed.

A separate diskette is required to store files generated by the data disk operations and MUST be PRE-FORMATTED or PRE-INITIALIZED. Note that the MS-DOS will store a maximum of approximately 120 files on a directory. IBM users who wish to store more files on a diskette will have to use tree-structured directories.

You should also note that when your data disk is full, the screen listing of files will sometimes show <u>only 84 file names</u>. (This happens because the program overwrites those at the bottom.) You will then need to either exit to DOS and use the run a "directory", or use the Data Base program (#2) and the "catalogue" program as described in chapter 2.

1.2 KEYBOARD

The keyboard is your method of communicating with CBS. The CBS package is easy to use. The programs will display messages informing you of the input that is expected whenever interaction with the computer is required.

To operate CBS: simply type in numbers or letters appropriate to your current activity and then press the <enter> or <return> key. CBS will expect you to press the <enter> or <return> key after you type any complete response (i.e. a number from a menu, a file name, a number for a data value). Forgetting to press <enter> is a common source of frustration for many new users. If you do not press <enter> after you are done typing in a response, the computer will sit and wait, and wait...

On the "standard" IBM keyboard, the <enter> key has an arrow pointing to the left and the tail of the arrow makes a right angle with a vertical line going upwards. Some IBM compatible keyboards have a <return> key to the right of the alphanumeric keys and an <enter> key to the right of the numeric keypad; both of which act as <enter> On the Apple keyboard, the <return> key says RETURN. IBM terminology is used throughout this text to describe the keyboard functions.

In general, any typing mistakes can usually be easily corrected by pressing the left arrow key and retyping the offending character(s). Do not enter a comma as part of a number. Commas are special characters to the computer. If you do enter a comma, the value entered will be equal to the digits to the left of the comma (i.e. entry of 12,500 would equal the value 12).

The keyboard is designed to be sensitive to a light touch and has auto-repeat, which means that it will repeatedly generate a character if you press it long enough (or too hard). The editing keys are the left, right, up, and down arrows, [home], [end], [pg up], [pg dn], [insert], and [delete] keys on the numeric keypad, and the [backspace] key. The "standard" IBM keyboard uses the same keys for a cursor pad and a 10 key pad. The [num lock] key controls these keys. **You may wish to set the [num lock] key to enable the cursor pad, and use the row of numbers on top of the letter keys.**

1.3 PRINTER

CBS supports a standard parallel printer, and uses the device label LPT1. Other printer configurations can be used with CBS if the output normally sent to LPT1 is redirected for the specific configuration. To use a printer other than the one connected to LPT1 you will need to execute several commands after booting MS-DOS, before starting CBS. These commands may be entered from the keyboard or executed from an autoexec.bat file. Please see your printer manual and DOS manual for specific commands and options to redirect the output for your particular printer and DOS version.

For any printer, if printing is attempted and the printer is not hooked up, you will get a message asking you if you want to proceed. If the printer is hooked up and turned on, but is not 'on-line', the computer will stop and wait (and wait...) for you to push the printer's on-line button.

1.4 DISKETTE HANDLING

The CBS program diskette is a circular disk, 5 & 1/4 inches in diameter, which is made of a flexible (floppy) plastic material. This material is coated so that information can be stored on and erased from its surface. It is permanently sealed in a square cover with several openings. This cover should never be tampered with. The sealed diskette is also kept in a paper pocket.

When using the diskette, be careful of the openings in the cover which expose the diskette itself. NEVER let anything touch those exposed areas. When the diskette is not in the disk drive, protect it by keeping it in its paper pocket.

The technology used to store information on the plastic material involves magnetics so you must take care to keep the diskette away from magnetic fields. Do not place diskettes near anything with an electric motor or magnet. Exposure can erase the information from the diskette.

To place the program diskette in the disk drive:

(1) Open the drive door (pulling outward on bottom edge of drive door).
(2) Pick up diskette and grasp it by its label only, positioning it so that it is horizontal to the floor, with the LABEL UP and the OVAL CUTOUT (from cover) OPENING is facing AWAY from you (towards disk drive).
(3) Slide diskette into opening in disk drive, keeping the LABEL UP, making sure that the OVAL CUTOUT goes in FIRST and that the LABEL goes in LAST.
(4) Push diskette gently until it is entirely in the drive. DO NOT BEND THE DISKETTE.
(5) Close the drive door by pushing it down gently. The drive door is operated by a spring mechanism which will loosen if mistreated.

The disk drive has a red light which is on while the drive is in use. NEVER, NEVER, NEVER remove or insert a diskette while this 'in use' light is on. This will surely destroy the information on the diskette and may ruin the drive.

1.5 GETTING STARTED

1) Boot Computer (insert MS-DOS 2.0 + System Diskette in 'A' drive, close drive door, turn power on).
2) Re-direct printer output, if necessary (see section 1.3 Printer).
3) Insert the CBS program diskette in 'A' drive.
4) Insert data diskette in 'B' drive, if 2 drive system.
5) Set default directory for data disk, if 2 drive system and tree-structured directory option used.
6) Type the 3 characters: CBS and press the enter key.

   IBM users with two disk drives may find it convenient to use a system disk for their data disk. This is accomplished by simply formatting the data disk with the /S option (see your DOS manual), and then using this disk to boot the system as well as to store data upon.

   If your system has two disk drives, You may set up a turnkey system to start CBS automatically. This is accomplished by using the EDLIN editor to create an autoexec.bat file. An autoexec.bat file is particularly convenient if you need to configure a printer before you start CBS.

   The following section outlines the steps for creating an autoexec.bat file, assuming the two drives are labeled A and B. To minimize swapping, the file assumes the CBS program disk is in drive B and the data disk is in drive A. The file prompts for the date and the time, then starts up CBS. Additional commands can be added to redirect printer output and/or set the default directory on the data disk to a subdirectory. Commands can also be added to read the date and time from a clock card, set up a print buffer and/or a ram disk, load memory resident programs such as SideKick or NicePrint.

   Step 1: Boot computer with MS-DOS system diskette in drive A. The EDLIN file must be available to create an autoexec.bat file, so if your normal boot diskette is not a complete copy of the original MS-DOS system diskette, put a copy of the original MS-DOS system diskette in drive A.

   Step 2: Prepare a diskette to serve as the new boot diskette.   1) Insert a blank diskette in drive B and close drive door 2) Type FORMAT B:/S/V and press the enter key.
   Follow the system prompts until the formatting process is finished. The /V parameter specifies a volume label, so enter something like CBS BOOT when prompted for label.

   Step 3: Set a path to EDLIN and change drives so DOS can find EDLIN, and so the new autoexec.bat file doesn't overwrite an existing file.
   1) Type PATH=A:\ and press the enter key.
   2) Type B: and press the enter key.

   Step 4: Start EDLIN: type EDLIN AUTOEXEC.BAT and press the enter key. The computer should respond with two words: new file, and change the prompt to an asterisk (*). If you get the * prompt but it doesn't say new file, type a Q <enter>, then type a Y, and restart this process.

Step 5:	Now that EDLIN has been started, enter the contents of the new autoexec.bat file by typing the following. See the sections in the DOS book on EDLIN and Batch files for further information.
1) I <enter>
2) PROMPT PG <enter>
3) CLS <enter>
4) DATE <enter>
5) TIME <enter>
6) PAUSE INSERT CBS PROGRAM DISK IN DRIVE B <enter>
6) B: <enter>
7) CBS <enter>
8) Press Control-C (hold down the control key and type the letter C)
9) E <enter>

Commands for redirecting printer output or for initializing goodies should be inserted between lines 5 & 6 above.

Step 6:	If redirecting printer output etc., copy any files needed (i.e. MODE.COM) to the new boot diskette. Remove new boot diskette from drive B and label it.

Your turnkey disk is now ready for use. If you wish to start CBS now, put the new diskette in drive A, and either reboot computer or type autoexec <enter>.

1.6 MODEL SELECTION

A specific statistical model can be selected by simply moving the highlight bar to the model of interest listed on the main menu and pressing return. After selecting a model, the red 'in-use' light will appear on the drive containing the program diskette while the selected program is loaded into memory. At that point, the Program Options Menu will appear on the screen.

1.7 OPTIONS MENU

Each of the thirteen statistical modules contain a Main Program Options Menu. This menu is nearly identical for every module. There are a few exceptions where certain selections are not applicable and these are noted at the end of this section. Each model option is numbered. To execute a selection, either move the highlight bar to the desired selection or type the specific number and press the enter key. Note that any option requiring data (such as view, edit or run current problem) will respond with the message 'ENTER DATA ' if you have not entered data. Figure 1 presents an overview of the Main Program Options Menu.

Figure 1 Main Program Options Menu Overview

The full Program Options Menu is as follows:

1. Enter Problem from Keyboard
2. Enter Problem from Data Diskette
3. Enter Example Problem
4. View Current Problem
5. Edit Current Problem
6. Quick Reviews
7. Run Problem
8. Exit to Main Menu
9. Exit to Operating System

Enter Problem from Keyboard: CBS prompts you through the steps for entering the data values for a new problem. Data currently stored in memory will be erased. The entered data can be saved on a data diskette.

Enter Problem from Data Diskette: CBS guides you through the steps for loading a data file from your data diskette. Data currently stored in memory will be erased. Note: **output files can not be used to enter problems.**

Enter Example Problem: An example problem will be loaded into memory. This problem is generally the first demonstration exercise in the chapter for the specific model. Data currently stored in memory will be erased.

View Current Problem: This option displays the definition, input statistics, and data values for the problem. An output options menu provides selections for screen viewing or printing.

Edit Current Problem: An editing options menu provides selections for changing the individual data values, the column labels, and the model sub-options. The data structure cannot be modified. Editing is either fully prompted and/or follows the structure of the Data Base Management Program (see chapter 2). An edited file can be saved.

Quick Reviews: This option summarizes the important aspects of the statistical model.

Run Current Problem: This option solves the problem currently residing in memory. An output menu contains options for viewing the solution on the screen, sending the solution to the printer, or storing the solution on an output text file. The output from the run includes the problem definition, the data values and the resultant calculations.

Exit to Main Menu: The program returns to the main menu. Data currently stored in memory will be erased.

Exit to Operating System: The program exits to your operating system. Data currently stored in memory will be erased. (Note : Exit to operating system effectively means Exit to DOS.)

Y/N/Q: This option will appear in many programs. It means Yes, No, or Quit. All you need to do is push in the first letter of each. Be sure to realize that Q means Quit to DOS.

1.8 OUTPUT OPTIONS

The user can choose from the following output options:

1) Screen - output is displayed only on the console (CRT).
2) Printer - output is sent to a standard line printer.
3) Disk - output is saved on data diskette (standard ASCII format).

A separate pre-formatted data disk is required to store the output disk file. An output file may be viewed via the VIEW option in the DBAS module or via a wordprocessor which accepts standard ASCII files.

764 *Appendix B*

2.0 DATA BASE MANAGEMENT

2.1 PROGRAM DESCRIPTION

CBS has the capability to create, view, transform, edit and save data files for use with one or more of the statistical programs. A pre-formatted (or pre-initialized) data disk is needed to save data on a diskette. This program is a collection of tools for creating and manipulating data files.

DO NOT ATTEMPT TO STORE DATA ON THE PROGRAM DISKETTE

2.2 OVERVIEW

The CBS Database Management Model(DBAS) offers the following options:

0. CBS Configuration
1. Create Data File
2. View File
3. Data Tranformations
4. Edit Data File
   o Edit Data
   o Restructure File
   o Merge File
5. Erase File
6. Catalog Data Diskette
7. Quick Reviews
8. Exit to Main Menu
9. Exit to Operating System

To select an option from this menu type the number next to the specific option and press the enter key. Figure 2 presents a schematic overview of the DBAS operating menu.

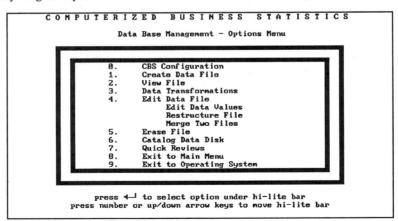

Figure 2 Overview of DBAS Operating Menu

2.3 DATA FILE OVERVIEW

Data files can be thought of as orderly groups of boxes. Each group of boxes is organized in a specific manner, and has a name to identify it and separate it from other groups of boxes. These groups are called files because you retrieve and store them in a manner logically similar to opening and closing a paper file in a filing cabinet.

Data files have four distinct characteristics:

1) NAME: You must give each file a unique name. It is recommended that file names be related to the specific problem. The file name can be 1-10 characters in length. A two-character prefix (I- or O-) will be added to the data file name. The purpose of this prefix is to distinguish between input (I-) data files, and output (O-) text files.

2) SIZE: Each data file is structured in a matrix format with columns representing the variables, and rows the data points. A file may have from 1 to 10 variables(columns) and 1 to 200 data points(rows) for any variable.

The basic format for a data file is as follows:

DATA FILE FORMAT

```
        Variable #1 .....   Variable #M
Row #1
......
......
Row #N
```

3) VARIABLE NAMES: You are given the option of labeling each variable(column) with any 0-5 character name.

4) DATA VALUES: The values of the data points (numbers inside the boxes) are the last (and most important) component of the data file. While the values of the data points can be of any magnitude, it is recommended that you avoid a data set where all the numbers are in the millions or billions. If this is the case, you may wish to go through and divide all numbers by 1000, or some appropriate number.

For a two disk drive system, the data diskette should be placed in the second disk drive during the start-up procedure. For a one disk drive system, the CBS program diskette and the separate data diskette will need to be switched before and after each data disk operation. CBS will prompt you for these diskette changes. Do not change diskettes unless you are prompted.

2.4 CBS CONFIGURATION

Select Option #0 to access the CONFIGURATION Menu. This is menu that sets the disk drive labels and the color selections.

2.5 CREATING A DATA FILE

Select Option #1 to CREATE a new data file. You will then be prompted for each of the four characteristics of the file as follows:

o   Number of Variables(Columns) in Data File

o   Number of Data Points:

Number of Data Points for Variable(Column) #1

.
.

Number of Data Points for Variable(Column) #M

o   Variable Names:

Name for Variable(Column) #1

.
.

Name for Variable(Column) #M

> Note : Once you are "in the spreadsheet," you can end the data entry process by pushing the "F" key <u>at any time</u>. If you have not entered numbers for any cells, CBS will save the zeros, or if you are editing a file, whatever number is currently in the remaining cells.

o   Variable Values:

Numerical Value of Data Point in Row #1, Column #1

.
.
.

Numerical Value of Data Point in Row #N, Column #1

.
.
.

Numerical Value of Data Point in Row #1, Column #M

.
.
.

Numerical Value of Data Point in Row #N, Column #M

o   Name of Data File

2.5.1 EXAMPLE PROBLEM #1 (Descriptive Statistics)

- NAME: DS-EX
- NUMBER OF VARIABLES: 1
- NUMBER OF DATA POINTS: 8
- VARIABLE NAME: X1
- DATA VALUES: Enter data values for X1

    29.7
    31.1
    34.7
    38.9
    44.8
    47.9
    50.2
    56.8

2.5.2 EXAMPLE PROBLEM #2 (Multiple Regression Analysis)

- FILE NAME: MR-EX
- NUMBER OF VARIABLES: 4
- NUMBER OF DATA POINTS: VAR 1. 7
                          VAR 2. 7
                          VAR 3. 7
                          VAR 4. 7
- VARIABLE NAMES: VAR 1. X1
                  VAR 2. X2
                  VAR 3. X3
                  VAR 4. X4
- DATA VALUES: Enter data values for each variable

X1	X2	X3	X4
2.1	0	102	47
2.5	0	107	35
2.9	1	100	51
3.1	0	99	42
2.8	1	108	38
2.9	0	103	44
3.0	1	104	46

2.5.3 EXAMPLE PROBLEM #3 (Nonparametric Methods)

- NAME: NP-EX
- NUMBER OF VARIABLES: 2
- NUMBER OF DATA POINTS: VAR 1. 10
  VAR 2. 12
- VARIABLE NAMES: VAR 1. POP A
  VAR 2. POP B
- DATA VALUES: Enter values for POP A and POP B.

POP A	POP B
20	12
25	11
15	15
19	19
30	8
28	10
27	12
21	9
22	16
18	9
	5
	8

2.6 VIEWING A DATA FILE

Select Option #2 to VIEW an existing data file. The following prompts will guide you through the process of viewing a file:

1) Specify whether you wish to view an input file (data values only) or an output file (problem data plus results).

2) Enter the name of the file.

3) Select an option from the output menu. You may either view the data on the display screen or you may send a copy to the printer. The output menu has the following three selections:

      S = Screen
      P = Printer
      R = Return to Program Menu

Enter the appropriate letter and press the enter key. If you are using a printer, be sure it is plugged in, turned on, and 'on-line'. Output for data files containing more than five variables is presented in two sections. Section #1 will contain all values for variables 1-5, and Section #2 will contain all values for variables 6-10.

2.7 TRANSFORMING A DATA FILE

Select Option #3 to TRANSFORM an existing data file. The following prompts will guide you through the processes of transforming one or more variables:

1) Specify whether you wish to transform a current file in memory or to load a stored file.

2) Enter the name of the file.

3) Specify the column number of the variable to be transformed.

4) Select from one of the following transformation options:

    1. Move (lag or lead)   2. Nth Root
    3. Nth Power   4. Logarithm

(Note : When you transform a data file, the most recent version of the data is held in memory, whether you have saved the data to the disk or not. This is similar to the way the "current problem" is held in memory with the other programs.)

The first option allows the user to move the data values for a selected column up or down by a specified amount. This option is useful when analyzing the impact of lag or lead variables. The resultant zero values can be removed using the edit option. The second option takes the nth root of the current variable values where n is specified by the user. For example, if you wish to take the square root of the current values for a given variable you would select option #1 and then input a two when prompted. In the case of negative values, the transformation takes the nth root of the absolute value which is then multiplied by negative one. The third option raises the current variable values to the nth power where n is specified by the user. For example, if you wish to cube the current values for a given variable you would select option #2 and then input a three when prompted. In the event the resultant transformed numbers exceed 1.703E+38 (the largest available number), you will be given the option of either scaling the original values or terminating the process. The fourth option takes the logarithm (base 10) of the current variable values. The logarithm of a variable value of zero will be assigned the value -1.703E-38 (the smallest available number). In the case of negative values, you will be given the option of either scaling the original values or terminating the process.

Each variable in the data file can be transformed using the above procedure. The transformed variables can then be saved in a file for subsequent analysis. This option also provides <u>scatter plots</u> for up to three variables. The user must specify which variable is to be plotted on the X-axis and which variable(s) are to be plotted on the Y-axis. This option can be terminated by pressing the F key. An example is shown below.

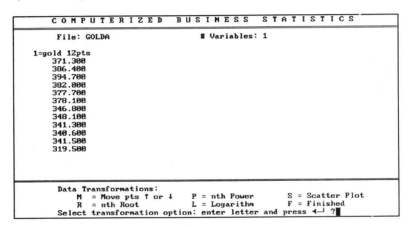

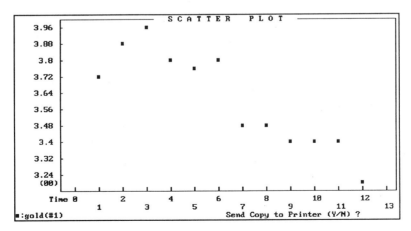

2.8 EDITING A DATA FILE

Select Option #4 to EDIT an existing data file. DBAS's Edit Menu will be displayed with the following four options:

1) Edit Data - Edits the numerical data values and/or the alphanumeric variable(column) labels.

2) Restructure - Adds or deletes variables(columns), or changes the number of data points in a variable(column).

3) Merge Files - Merges two data files side-by-side or above-below each other.

4) Quit Edit - Returns to the DBAS Main Menu.

The next step in the editing process is to select the desired data file. If a file already resides in memory, DBAS will ask you if you wish to use the current data. If you wish to use a different file, or if there is no data in memory, DBAS will prompt you for the name of the file to edit. When DBAS asks which file to edit, enter the name you gave the data file when you created it (then press the enter key).

Select option #1 from the DBAS Edit Menu to edit data values and variable names. The selections on the editing menu are:

1) Change Data Values - Edits numerical data using a spreadsheet like format.

2) Change Variable Names - lets you edit the variable/column labels.

3) Return to DBAS Menu.

To input, change, or view any portion of the data that is not currently on the screen simply press the arrow key that points in the direction you want to move the display until the information you desire moves onto the screen. Moving the display back and forth or up and down is called scrolling. For convenience, the enter key behaves like a down arrow key when you are simply moving the cursor around the table. To change any numerical value in the matrix simply position the cursor to the beginning of the space allocated for that data point and type the appropriate numeric characters. The cursor can be moved around the input/editing table using the quick movement commands. If you try to use one of the editing commands (i.e. page up or end) and the computer does not respond, you probably have one of two problems: the num lock key is set wrong, or you never pressed enter to finish the last number entered.

Select option #2 from the DBAS Editing Menu to restructure a data file. The restructuring commands are 'A' (to add a column), 'D' (to delete a column), 'L' (to enter a new length for a column), and 'F' (to signal that you are finished). Attempting to create a data base with less than one or more than ten variables (columns) will result in a reprompt for another restructure command.

Select option #3 from the DBAS Editing Menu to merge two data files. You will be prompted for the name of the 2nd file once the 1st file is read into memory. You will be prompted. If you are organizing the files side-by-side and the total number of variable(columns) in both files exceeds 10, columns will be dropped from the 2nd file to make a maximum of 10 variable(columns). If you are organizing the files below each other and the total number of data points/rows in any column exceeds 125, rows will be dropped from the 2nd file to make a maximum of 125 data points(rows).

2.9 ERASING A FILE

Select Option #5 to ERASE a file from the data diskette. DBAS will guide you through the following procedure:

1) Specify whether you wish to erase an input file (data values only) or an output file (problem data plus results.)

2) Enter the name of the file to erase. Type the name of the file and then press the enter key.

3) The computer will first verified whether the file exists. Next a prompt will ask you to verify that you do wish to delete the file. A erased file is lost forever (unless you're one of those sophisticated users with a DOS tool kit).

2.10 CATALOG DATA DISK

Select Option #6 to CATALOG the files on the data diskette. CBS will provide a list of the files in "screen pages." You will have to press the enter key after viewing each screen page.

2.11 QUICK REVIEWS

Select Option #7 to view the QUICK REVIEW screens for each of the statistical analysis modules. Use the right and left arrow keys to "page" through the quick review screens.

2.12 EXITING DBAS

The last two selections on the DBAS Menu are used to exit the data base module. Select option #8 if you wish to EXIT to the Main Program Menu. Select option #9 if you wish to EXIT to the operating system.

Appendix C

Solutions to Selected Odd-Numbered Problems

Chapter 2

1. *a.* Banning tobacco smoke in public facilities will reduce the incidence of lung cancer in the general population.
   b. Lowering the maximum automobile speed limit will improve fuel economy.
   c. Reducing the capital gains tax will promote economic growth.

11. *a.* Qualitative
    b. Qualitative
    c. Quantitative
    d. Quantitative
    e. Qualitative

13. *a.* Ordinal
    b. Interval
    c. Ratio
    d. Nominal
    e. Nominal

15. Survey data from the various districts around the country

17. *a.* Nominal
    b. Nominal
    c. Nominal
    d. Ratio
    e. Interval

19. Neither has a real zero point.

21. Statistical analysis provides the information needed to support the TQM philosophy.

25. Process control focuses on monitoring and maintaining quality during the manufacturing and service operations. Acceptance sampling is used to determine whether to accept or reject an incoming or outgoing lot based on specific standards.

27. By indicating to each employee how his or her efforts contribute to the overall success of the organization through strict attention to product or service quality.

Chapter 3

1. *a.* The effectiveness of various headache drugs (histogram)
   b. The preference for various soft drink products (pie chart)

3. The sample standard deviation has a correction ($N - 1$ in the denominator) to make it more reflective of the population standard deviation (N in the denominator).

5. *a.* Kurtosis: platykurtic. Skewness: skewed right
   b. Kurtosis: mesokurtic. Skewness: symmetrical
   c. Kurtosis: leptokurtic. Skewness: skewed left

7. *b.* Skewed right

9.
Group	Frequency
18–28	3
28–38	9
38–48	10
48–58	5
58–68	3

11. *b.*
Scores	Cumulative Frequency
50–59	3
50–69	10
50–79	20
50–89	41
50–100	55

13. *b.*
Therms	Cumulative Frequency
0–30	2
0–60	5
0–90	10
0–120	14
0–150	15

c. 67%

15. It appears that businessmen average more trips per year than businesswomen. This should not be surprising since there are currently more men than women in positions that require travel. Women might be inclined to select a hotel nearer their business meeting, which in many cases could be more expensive. Additionally, they tend to bring family members, which would further increase the cost.

17.
	Raw	Group
Mean	41	41.9
Median	41.5	42.0
Mode	35, 45, 43	42–50 (modal class)
Range	49	Interval – 8
Variance (s)	138	143.47
Standard dev. (s)	11.75	11.98

19. *a.* Not all respondents answered in the top five categories.

21. *c.*
	Sales	Market Share
Mean	660,128.94	7.05
Median	358,403	3.85
Mode	225,000	2.40

23. *a.*
	Total Plays	Total Yards	Total Yards/Game
Mean	604.4	3,978.1	496.5
Median	601.5	3,915.5	493.5
Mode	—	—	—
Range	192	1,496	73
Variance	3,856.36	183,766.89	429.65
SD	62.1	428.68	20.73

b. Mean

c. See table at bottom of page.

23. *c.*
Total Plays		Total Yards		Total Yards/Game	
Wyoming	698	Wyoming	4856	Wyoming	540
Nebraska	693	Nebraska	4489	Utah	520
Utah	652	Utah	4160	Wash. St.	510
BYU	628	Wash. St.	4081	Nebraska	499
UCLA	601	W. Va.	3963	W. Va.	495
Wash. St.	596	Air Force	3868	Oklahoma St.	492
W. Va.	589	UCLA	3822	Air Force	484
Air Force	547	BYU	3739	Houston	480
Houston	528	Oklahoma St.	3443	UCLA	478
Oklahoma St.	506	Houston	3360	BYU	467

25. a.

	CD Yield		FMR
Rep. Fed.	8.58	Cal. Fed.	11.00
West. Fed.	8.46	Coast	10.90
Cal. Fed.	8.27	GW	10.87
Imperial	8.19	Home	10.87
Far West	8.17	Glen. Fed.	10.77
Home	8.15	Columbia	10.76
Val. Fed.	8.14	Val. Fed.	10.71
World	8.12	Imperial	10.63
Glen. Fed.	8.11	World	10.63
Coast	8.11	Far West	10.53
GW	8.09	West. Fed.	10.53
Columbia	8.00	Rep. Fed.	10.52

b. Higher CD yield, lower FMR

d.

	CD	FMR
Mean	8.1992	10.7267
Median	8.145	10.7350
Mode	8.11	10.53, 10.63, 10.87
Range	0.58	0.48
Variance	0.0272	0.262
SD	0.1648	0.1618

27. $\overline{X} = 51$, $s = 22.39$

29. $\overline{X} = 9.15$, $s = 1.04$

31. Largest average demand: soyburgers
Smaller standard deviation: hamburgers

35. a.

	Production		Price
Kansas	433.2	California	3.44
N. Dakota	323.2	Washington	3.40
Texas	187.2	N. Dakota	3.32
Oklahoma	165.0	Idaho	3.26
Minnesota	142.4	Minnesota	3.23
Colorado	139.3	Oklahoma	3.20
Washington	128.3	Texas	3.10
Nebraska	89.7	Kansas	3.05
Idaho	72.0	Nebraska	3.00
California	68.9	Colorado	2.91

b. Higher production, lower price
c. Production
d. Production: SD = 116.6997
Price: SD = 0.1739

37. a. 1989, $675.73; 1990, $741.64
b. See table at bottom of page.

39. a. 89.8%
b. 89.4%

41. 215 mph.

45. a. 2.29
b. 0.91

51. a. p-chart
c. In control

53. a. p-chart
c. Unstable (out of control)

Chapter 4

1. a. Subjective
b. Objective
c. Subjective

37. b.

	1988		1989		1990
North	975	North	1,048.13	North	1,116.27
East	825	East	866.25	East	905.23
West	515	West	571.65	West	648.82
South	425	South	463.25	Overseas	525.99
Overseas	375	Overseas	429.38	South	511.89

d. Objective
e. Objective

3. a. Dependent
   b. Independent
   c. Mutually exclusive
   d. Independent
   e. Mutually exclusive

5. Bayes' Theorem is used to revise prior probability data with new information. It is used extensively in updating marketing estimates with the results of survey studies.

7. a. Totally exhaustive
   b. Neither
   c. Mutually exclusive
   d. Totally exhaustive

9. $5! = 120$

11. a. Yes, $8!/3! = 6,720$
    b. Yes, $8!/3!5! = 56$

13. a. $S = (2, 3, 4, 5, 6, 7, 8, 9, 10, 11, 12)$
    b. 7

15. a. ~2.6 million $(52!/5!47!)$
    b. $4/2.6$ million
    c. $48/2.6$ million
    d. Bluffing

17. a. .375
    b. .0015
    c. .72

19. a. Mutually exclusive
    b. Dependent
    c. .227
    d. .544

21. b. 0.1225
    c. .31

23. a. .42, .19
    b. .675
    c. Maybe

25. a. .48
    b. .82

27. a. 1/24
    b. 0
    c. 4/6
    d. 8/24
    e. 11/24

29. a. 775
    b. 100
    c. 325

31. Coke = 11, Pepsi = 12, 7up = 2

33. Five classifications = 31, six classifications = 63

35. a. .0150
    b. .5129
    c. .0156

37. a. Objective
    b. .00005
    c. Small

39. $P = 1 - \dfrac{365 \times 364 \times \cdots \times (365 - (n-1))}{365^N}$

where N is the number of individuals in the group.

41. a. $11 \times (1.05)^N > 17.5$, $N = 10$ years
    b. Reverse numbers and letters

43. a. .57
    b. .06

45. a. .51, .33, .16
    b. .941
    c. .125
    d. .875

47. a. .931
    b. .273
    c. .725

49. a. .162
    b. .998
    c. .117

51. a. Model 300 (.93)
    b. 0

53. a. .0004
    b. .9924

Chapter 5

3. a. Typically, when the number of trials exceeds 20

b. Typically, when the population size is very large and the sample size exceeds 20
c. Typically, when the mean exceeds 20

5. *a.* 5
   b. 4
   c. 2
   d. 1
   e. 3

7. *a.* .42
   b. .028
   c. .192

9. *a.* .0654
   b. .1586
   c. .3413

11. *a.* Approximately 23% of the cases cost more than $350.
    b. Net on 77% of the cases

13. *a.* $640
    b. $825

15. *a.* 0.97 gal.
    b. .3821

17. *a.* .8
    b. .4
    c. .2

19. *a.* .092
    b. .059
    c. .997

21. *a.* 99%
    b. 18%
    c. 1.66%

23. *a.* Number of cigarettes smoked
    b. Continuous
    c. Binomial
    d. .5833
    e. .0008

27. *a.* 0
    b. .1509
    c. .6964
    d. .1964

29. *a.* .9999
    b. .5398

c. 0
d. 0

31. $X = 45, s = 4.97$
    a. .9987
    b. .1558
    c. .0122

33. *a.* .3457
    b. .0115
    c. .9770

35. *a.* .0143
    b. .7571
    c. .5143

37. .72 (summary) vs. .719 (computed)

39. *a.* .3940
    b. .9098
    c. .3000

41. *a.* 24
    b. .135
    c. .632

43. *a.* .1248
    b. .0000
    c. App. 22

45. .053, .271

47. *a.* .0981
    b. .0000
    c. App. 21

49. 0, .9874

51. *a.* .6967
    b. .0331
    c. 0

53. *a.*

Yield	f	Relative Frequency	Expected Value
80–89	2	.167	14.11
90–99	3	.250	23.72
100–109	4	.333	34.80
110–119	2	.167	19.12
120–129	1	.083	10.33

b. Expected value = 102.084
c. Probability of 100 or more = .583

55. a.

Sales	f	Relative Frequency	Expected Value
0–19	22	.172	1.634
20–39	47	.367	10.826
40–59	33	.258	12.771
60–79	19	.148	10.286
80–99	7	.055	4.922

    b. Expected value = 40.439, SD = 22.15
    c. $X = 40.4375$, .1885
    d. Under 40 = .457

57. a. False, $p = .2374$
    b. False, $p = .7769$
    c. True, $p = .0013$
    d. False, $p = .0202$
    e. False, $p = .1866$

59. a. .1250
    b. .9657
    c. 0

61. 60 missiles

Chapter 6

1. a. Mail generally is too slow a method for the immediate needs for trend recognition.
    Advantages:
        1. Least costly
        2. No interviewer effect
        3. As objective as the items

3. The survey is more accurate when certain populations are difficult to identify in household determinations.

5. a. Survey
    b. Population census
    c. Survey
    d. Census

9. Instrument design entails the development of a questionnaire to collect the data that will fulfill the objectives. The instrument may contain scales for the measurement of variables.

11. a. *Regularly* is not defined, nor is the term *exercise*.
    b. The term *large* is indefinite—dollars, period of time, and other aspects are not dealt with.
    c. The term *think* is too vague.

13. Demographic data allow for:
    a. Representative sampling.
    b. The source of independent variables for analysis.

17. A hierarchical data system organizes data into distinct levels through the use of an inverted tree structure. A network data system organizes the data such that multiple relationships are permitted between data records. A relational data system organizes data into a table where each row represents a specific record or file.

19.

5	4	3	2	1
Strongly Agree	Agree	Neither	Disagree	Strongly Disagree

    a. Citizens should be prohibited from carrying assault weapons.
    b. Assault weapons can be used on ranges.
    c. No assault weapons should be allowed to be manufactured.
    d. Owning assault weapons should be a felony.
    e. The government should not interfere with the citizen's right to bear arms.

23. a. Do you need to use the computer center?
    b. How many hours per week would you use it?
    c. What hours should the center remain open?
    d. During what weekend hours do you need the center?
    e. What kind of equipment will you require?
    f. What software will you require?
    g. What manuals should be available?
    h. What staff support do you require?
    i. What training do you require?
    j. What kind of reservation system will serve your purpose?

27. Pretesting is important:
    a. To test wording and understandability.
    b. To determine sampling error.
    c. To check measurement capacity.
    d. To check reliability and validity.

29. *a.* High Ridge Customer Questionnaire

Dear Customer: Please take a few minutes to fill out this questionnaire while your car is being washed. Your response will help us improve the overall effectiveness of our service. [Some sample questions are shown here.]

1. Evaluate the courtesy and friendliness of the staff:

    |⎯⎯⎯⎯⎯|⎯⎯⎯⎯⎯|⎯⎯⎯⎯⎯|⎯⎯⎯⎯⎯|
    Very   Good   Average   Poor   Very
    Good   Poor

2. Evaluate the helpfulness of the staff:

    |⎯⎯⎯⎯⎯|⎯⎯⎯⎯⎯|⎯⎯⎯⎯⎯|⎯⎯⎯⎯⎯|
    Very   Good   Average   Poor   Very
    Good   Poor

7. Evaluate the extent of the current services:

    |⎯⎯⎯⎯⎯|⎯⎯⎯⎯⎯|⎯⎯⎯⎯⎯|⎯⎯⎯⎯⎯|
    Very   Good   Average   Poor   Very
    Good   Poor

8. Identify any new services you would like to see.

    ⎯⎯⎯⎯⎯⎯⎯⎯⎯⎯⎯⎯⎯⎯⎯⎯⎯⎯⎯⎯
    ⎯⎯⎯⎯⎯⎯⎯⎯⎯⎯⎯⎯⎯⎯⎯⎯⎯⎯⎯⎯
    ⎯⎯⎯⎯⎯⎯⎯⎯⎯⎯⎯⎯⎯⎯⎯⎯⎯⎯⎯⎯

Chapter 7

1. *a.* The proposition that the sampling distribution approaches a normal curve as the sample size increases
   b. A probability distribution of sample means for a given population
   c. A single numerical value used for estimating an unknown population parameter
   d. A confidence interval used for estimating an unknown population parameter

5. The basic trade-off involves accuracy versus cost.

7. *a.* Systematic
   b. Neither
   c. Stratified
   d. Cluster

11. *a.* Mean response time estimate = 14.118 minutes
    b. Standard deviation estimate = 4.98 minutes

13. *a.* .45 Democratic, .55 Republican
    b. The shopping center does not provide a representative sample of the electorate.

15. *a.* 50% probability
    b. 34.13% of the receivables will fall between $90,000 and $100,000; 19.15% fall between $100,000 and $105,000. Thus, 53.28% should fall between $90,000 and $105,000.

17. *a.* The sample is from a normal population.
    b. The measurement is linear and reliable.
    c. Measures are independent.

19. *a.* 491.77 to 508.225
    b. 490.20 to 509.00
    c. 487.10 to 512.90
    d. $-\infty$ to $+\infty$

21. *a.* .0934 to .6566
    b. .0395 to .7105
    c. 0 to .8159
    d. 0 to 1

23. *a.* 192.90 to 198.96
    b. 192.24 to 199.62
    c. 190.80 to 201.06
    d. $-\infty$ to $+\infty$

25. *a.* 46,133.84 to 53,866.16
    b. 47,673.62 to 52,326.38

27. *a.* Let $P = .5$ (worst case); then $n = 2,401$.
    b. With $P = .15$, $n = 1,225$.

29. From 12.063 to 12.937 years

31. $P = .675$; .654 to .696 for proportion supported

33. $SD = .020$, $\bar{X} = .30$
    a. .4

35. 363.22 to 436.78

37. 14.22 to 15.78 seconds

39. Mean difference = 6.4, SE = 1.76(2.262) = 3.98
    a. $6.4 \pm 3.98 = 2.42$ to 10.38
    b. The null hypothesis is rejected. The diet worked.

41. $P_1 = .4, P_2 = .55$
    $.15 + .05(1.6579) = .15 + .0824$
    90% limits $= .068$ to $.232$

43. $N = 246$

45. a. $N = 25$
    b. $N = 35$

47. a. $P_s = .42, s_{\bar{x}} = .0044$
    b. $.42 \pm .0087$ at .95

49. a. $\bar{X}_c = 176.8, s_{\bar{x}} = 4.43$
    b. 176.8 ± 8.68 at .95

51. a. $\bar{X}_s = \$37,650, s_{\bar{x}} = \$430$
    b. $\$37,650 \pm 843$ at .95

Chapter 8

1. a. Reject H_0.
   b. Reject H_0.
   c. Do not reject H_0.

3. a. Null: The program does not reduce absenteeism below 2%.
     Alt.: The program reduces absenteeism below 2%.
   b. Null: The radial tires last for 30,000 miles or more.
     Alt.: The radial tires last for less than 30,000 miles.
   c. Null: Shipment of chips will have a defective rate of up to 3%.
     Alt.: Shipment will have a rate of defectiveness in excess of 3%.
   d. Null: Membership will be 12% or higher.
     Alt.: Membership will be below 12%.

5. a. H_0: GNP growth $\leq 4\%$
     H_1: GNP growth $> 4\%$
   b. H_1: Increasing speed to 65 will increase deaths.
     H_0: Increasing speed to 65 will not increase deaths.
   c. H_0: No cure will be found by 1995.
     H_1: A cure will be found.
   d. H_0: Smog levels reduced.
     H_1: Smog levels not reduced.

9. a. Two-sided
   b. One-sided
   c. Two-sided
   d. One-sided

11. Reject H_0: $t_c = 10.96$
13. Do not reject H_0: $t_c = 0.55$
15. Do not reject H_0: $Z_c = 0.79$
17. Do not reject H_0: $Z_c = -0.73$
21. a. .0002
    b. 0.61
23. H_0: $\mu = 500$; reject H_0: $Z_c = 6$
25. H_0: $\mu = 25.5$; reject H_0: $Z_c = 11.2$
27. a. $.09 \pm .0272$ (5% both sides)
    b. No, $Z_c = 3.03$
29. 0
31. a. Too close to call; $Z_c = 0.56$
    b. No, due to local characteristics
33. a. H_0: $\mu_1 = \mu_2$
    c. Do not reject H_0: $Z_c = 1.84$
35. a. Some internal variables are controlled.
    b. Only one estimate of population variance is needed.
    c. One of the assumptions of the normal curve is met.
37. a. The courseware does not improve performance ($t = 0.556$).
    b. Do not reject the null hypothesis ($t = 1.595$).
41. a. The null hypothesis is rejected. The West is significantly larger in salary ($t = -3.83$). Variances are equal.
    b. The null hypothesis is rejected, with same results as above.
43. The results show a significant preference for Pepsi.
45. Fuel economy is significantly less in the 65 mph time period ($t = 5.19$).
47. $\bar{X} = 214.08, s = 40.21, N = 12$
    Reject H_0: $Z_c = -4.38$
49. a. H_0: $\mu_1 = \mu_2$
    b. Do not reject H_0: $Z_c = -0.96$
51. a. H_0: $\mu(1989) = \mu(1990)$: do not reject H_0

53. a. H_0: μ(home) = μ(away)
   b. Do not reject H_0: $t_c = -0.444$

55. a. H_0: μ(U.S.) = μ(local)
   b. Matched samples
   c. Reject H_0: $t_c = 3.887$

57. a. H_0: μ(before) = μ(after)
   b. Reject H_0: $t_c = 2.975$

65. $\bar{X}$: L = 123.63, U = 127.22; R: L = 0, U = 3.12
   Process out of control

67. $\bar{X}$: L = 0, U = 0.060; R: L = 0, U = 0.065
   Process in control

Chapter 9

3. Yes, two variables can be causally related yet yield a low correlation coefficient.

5. A linear relationship between two variables can be described using a straight line. A curvilinear relationship between two variables requires the use of a nonlinear function.

7. a. *Linear relationship:* A linear relationship exists between X and Y in the population.
   b. *Normally distributed error terms:* The values of Y are normally distributed around the regression line.
   c. *Independent error terms:* One error term cannot be predicted from another.
   d. *Constant error term variance:* The dispersion of data points around the regression line remains constant.

9. a. No (more likely nonlinear)
   b. Theoretically, compensation should be reduced. However, this does not always happen in practice.
   c. The intercept indicates the average compensation when earnings are equal to zero. The slope indicates the change in compensation ($000) for every ($000) change in earnings.

11. a. $Y = 49{,}063.67 + 4.69X$
    b. $R^2 = .78$

13. a. The null hypothesis has been rejected, which tells us that the nature of the relationship between personnel and gross sales is linear.
    b. Number of items on sale, length and timing of open business hours, price of merchandise, disposition of sales staff, layout of merchandise, the economy, the season, and other factors.
    c. The estimated monthly sales if 12 full time equivalent employees are used is $48{,}829.87 \pm 5{,}381.224$

15. a. The correlation coefficient is given in the results of running the problem on CBS as .9592.
    b. A high degree of linear association between the two variables
    d. Statistically significant at the .05 level.
    e. $655.69/month
    f. The 90% confidence interval for a 4-year-old fire engine is 655.67 ± 35.78/month.

17. a. Yes
    b. 12.456 ± 1.865
    c. 8.535 ± 2.251

19. Because it includes both the variance of Y around the regression line and the variance of individual Y values.

21. $Y = 258.64 \pm 9.38$ (mean), $Y = 258.64 \pm 32.96$ (individual)

23. a. There is a linear relationship between years of service and invoices generated.
    b. The correlation coefficient was equal to .5831, which shows that there is a positive and linear relationship between the two variables.
    c. The equation is $Y = 578.9249 \pm 40.2637X$, and for an employee with seven years service, the invoice forecast is 860.776 ± 197.607.

25. a. No
    b. The correlation coefficient of $-.2504$ indicates a negative relationship between GNP and passenger car retail sales.
    c. Developing a forecast from a nonsignificant model cannot be justified.
    e. Important ones include: Interest rates, inflation, unemployment, future expectations about the economy, and oil (gas) prices.

27. a. Yes
    b. 67% of the variation in sunshine is explained by the temperature.
    c. 65.66% ± .059
    d. The model would not produce good results if it were used for June or December.
    e. Proximity to the equator and amount of rain/snowfall in the region, for example.

29. a. Yes, there is a positive, linear relationship between employment level and value added.
    b. The correlation coefficient was .9953, which shows that the two variables in this study are virtually perfectly correlated.
    c. Overtime, variable costs, fixed costs, overhead associated with production, etc.

31. There appears to be no relationship between the two variables at the 0.05 level.

33. a. The relationship is curvilinear, with marginal costs declining, until production reaches 20,000; costs then begin to rise in a fairly linear fashion.
    c. The linear regression model does not seem to match the apparent curvilinearity of the relationship.
    f, g. Neither model appears to explain the actual relationship between production and marginal costs. Any transformation applied to the entire data range would not reflect the two-tiered nature of the function.

35. $X = 20{,}000$, $Y = 12.64 \pm 3.10$, $X = 30{,}000$, $Y = 16.11 \pm 3.61$

37. a. Positively sloped scatter graph, with a slope of nearly 45°.
    c. The model matches the data very well (correlation coefficient of .9366).
    d. Not needed due to the high degree of linear correlation.

41. The development model is statistically significant since the computed t-value (3.13) is greater than the critical value (2.09).

$$Y(\text{app}) = 1{,}266.3 \pm 807.5$$

43. The development model is statistically significant since the computed t-value (16.48) is greater than the critical value (2.07).

$$Y(\text{app}) = 672.6 \pm 12.2$$

47. 179.79 ± 17.81 ($X = 507.93$)

49. a. 0.5292
    b. Positive slope
    c. $Y = -2.195 + 5.092X$
    d. Yes
    e. 1.476 ± 0.023

51. b. No
    d. Neither

53. a. $Y(\text{profits/assets}) = 0.0195 + 0.0005 \times (\text{sales})$
    b. This model is statistically significant at the .05 level.

55. a. $r = .185$
    b. This model is not statistically significant at the .05 level.

Chapter 10

1. a. Stepwise regression analysis.
   b. Full regression analysis.
   c. Collinearity.

3. The correlation matrix shows the amount of correlation between each of the variables in the model.

5. a. Newspaper
   b. Newspaper (X_1) and trade journal (X_3)
   d. Y (sales) $= 25{,}376.81 + 9.77X_1 + 14.71X_3$, $R^2 = .47$ $Y = 61{,}161.84 \pm 2{,}546.83$
   e. General economic conditions and competition

7. None of the predictor variables is statistically significant at the .05 level.

9. a. Site
   b. Price $= 67{,}500 + 13{,}333.334(\text{site}) + 3.750(\text{size}) + 4{,}166.667(\text{bedrooms})$
   c. Yes, the two approaches yield the same model.
   e. In regard to this model, a change of 1 standard deviation in site will result in a 0.667 standard deviation increase in price.

11. a. None
    b. No
    c. Large R^2 is the result of small data base with respect to the number of predictor variables.
    d. Increase the size of the data base and run stepwise analysis.

13. a. Hospital, rate, employees
    b. Yes
    c. For hospitals, 1 standard deviation produces a 0.769 standard deviation increase in costs. Average daily census, when increased by 1 standard deviation, produces a large increase in costs (3.867 standard deviations).

15. a. The most visible change is that ERA now has a higher correlation (with ranking as opposed to percent).
    b. ERA

17. a. Data base is too small for the number of candidate predictor variables.
    b. Yes, $r = .98$.
    c. Value of shipments.

19. a. Annual compensation
    b. Dummy
    d. Both sex and experience

21. a. National data, if the coefficients have been determined.
    b. Positive for all
    c. There will be a high degree of collinearity.

23. $Y = 23.22 \pm 4.05$ (mean)

25. a. Assets (both), profit growth (veterans)
    b. Compensation increases as the log of assets increases.

27. The data base contains a relatively large number of variables compared to the number of data points. This condition, which is called "overfitting," often leads to large r^2 even though the predictor variables are not significant.

29. a. $Y(E/S) = 1.79 + 0.60X_2(P/A)$, $R^2 = .40$
    b. $Y(E/S) = 5.50 \pm 1.09$ (mean)

31. Typically, the t-test is used to measure the significance of each predictor variable individually.

33. One possible classification scheme is as follows:

    Education (X_1):   Race (X_2):
    1. Grade School   1. Caucasian
    2. High School   2. Hispanic
    3. College   3. Black
    4. Undergraduate   4. Asian
    5. Graduate

    Four dummy variables will be required for each factor, for a total of eight.

35. $Y(E/S) = 0.63 + 2.89 \times \log(P/A)$, $R^2 = .29$
    The linear model in problem 29 yields a larger R^2.

37. a. Model 1: None of the variables is significant.
    b. Model 2: $Y = -27.69 + 0.90X_1X_2$, $R^2 = .63$

39. a. $Y = 60.99 + 5.52X_2$, $R^2 = .88$
    b. $Y = 51.31 + 22.77X_2 + 0.21X_1X_3$, $R^2 = .89$

41. a. $R^2 = .83$
    b. Families with children tend to use public hospitals more than private hospitals.
    c. The computed F of 9.8 is greater than the critical F of 5.32. Therefore, the model is statistically significant.

43. $Y(\text{comp}) = 19.51 + 3.18X_3$; $R^2 = .86$. The interactive effect between experience and sex (X_3) is statistically significant.

45. The general model for this problem is:

    $$Y(\text{Prod}) = b_0 + b_1X_1 + b_2X_2 + b_3X_3 + b_4X_1X_2 + b_5X_1X_3 + b_6X_1X_2X_3,$$

    where $X_1 = $ Apt., $X_2 = $ Exp., and $X_3 = $ Sex.
    Variables X_2 and X_6 are significant. This would suggest that an interactive effect exists among all three factors.

47. a. $R^2 = .64$ (linear) vs. $R^2 = .71$ (nonlinear)
    b. $Y = 0.44 \pm 0.0826$ (mean)

49. a. $Y(\text{Trade}) = -986.69 - 27.66\text{FI}$, $R^2 = .49$
    b. Yes, foreign investment (FI)
    c. $Y = -3{,}794.36 \pm 1{,}702.36$

51. $Y(\text{GNP}) = -918.74 + 11.75\text{POP} - 0.02\text{CPI}$, $R^2 = .96$

53. $Y(\text{GNP}) = -156.16 + 0.78\text{POP} + 1.54\text{GDI}$, $R^2 = .96$

55. a. $Y(E/S) = -0.900 + 0.026\text{Profits}$
    b. Profits
    d. $R^2 = .24$
    e. $Y = 4.27 \pm 2.49$ (mean)

57. a. $Y(E/S) = -0.457 + 0.340\text{Sales} + 0.372P/A - 0.069P/E$
    b. All three
    d. $R^2 = .74$
    e. $Y = 2.10 \pm 0.619$ (mean)

59. a. Sale and PPE
    b. $R^2 = .70$
    c. Sufficiently close
    d. The overall model is statistically significant, as indicated by the F values.
    e. Significant collinearity exists between PPA and PPE.
    f. $Y(E/S) = 4.55 \pm 0.86$ (mean)

Chapter 11

3. Methods for evaluating the effectiveness of a given forecast include MAD, MSE, and MAPE.

5. a. April
   b. November

7. The basic strength of both moving averages and exponential smoothing is simplicity. The basic weakness of both methods is that they are technically limited to forecasting only the next period. Experience indicates that exponential smoothing tends to yield somewhat better forecasts.

9. b. Exponential smoothing (MAD = 58.5)
   c. $\alpha = 1$
   d. MA = 603.75, ES = 595.37

11. b. $\alpha = .6$ yields the lowest MAD value.
    c. $\alpha = 1$

13. c. These data show a cyclic trend every seven years.

15. a. The following results are based on the smoothing with trend factoring model.
    b. The industry sales forecast of $54,111 is slightly lower than the actual figure of $54,993.
    g. The company sales for 1984 are estimated at $3,554.

17. a. MAD = 664.3, forecast = 2,945.8
    b. MAD = 449.1, forecast = 2,313.7

19. $Q_1 = 15,041$, $Q_2 = 17,155$, $Q_3 = 18,490$, $Q_4 = 19,234$

21. a. MAD = 8,562.5, forecast = 15,875
    b. $S_1 = 1.27$, $S_2 = 0.69$, $S_3 = 0.29$, $S_4 = 1.76$

23. Quarterly seasonal effects are minimal ($Q_1 = 0.93$, $Q_2 = 0.99$, $Q_3 = 1.01$, $Q_4 = 1.08$).

25. a. No, $R^2 = .10$
    b. Yes, $R^2 = .98$
    c. $F(85) = 10.1$

27. AWI = $15{,}282.08 - 23.01$Prod, $R^2 = .55$, $t_c = 2.72$

29. A nonstationary time series involves a trend.

31. a. Negative slope
    b. EPS = $5.677 - 0.313$Time, $R^2 = .61$

33. Dummy variables (0/1) are used to isolate a specific seasonal period. A total of three would be required.

35. a. The relationship seems to be getting stronger over time.
    b. Unclear; there is a need to consider the general business climate.

37. b. Positive slope
    c. 8.744

39. b. Negative slope
    c. $F(13) = 33.97$, $F(14) = 32.23$
    d. $R^2 = .4896$
    e. 1991 is considerably outside the data range.

41. e. MAD(a) = 22,616, MAD(b) = 16,672, MAD(c) = 13,413, MAD(d) = 14,072

43. e. MAD(a) = 15,648, MAD(b) = 28,175, MAD(c) = 7,958, MAD(d) = 6,452

45. e. MAD(a) = 175.4, MAD(b) = 166.7, MAD(c) = 83.7, MAD(d) = 153.4

47. a. $n = 3$
    e. MAD(a) = 54.3, MAD(b) = 116.3, MAD(c) = 50.2, MAD(d) = 23.3

49. a. $Q_1 = 2.03$, $Q_2 = 0.50$, $Q_3 = 0.13$, $Q_4 = 1.35$

51. MAD(.4, .2) = 37.6, MAD(1, .01) = 22.8

53. a. $r = -.05$
    b. No, $t_c = 0.16$

55. b. 10.15
    c. MAD = 0.14
    d. MAD = 0.15

57. a. MAD = 3.2
    b. MAD = 2.5
    d. No, $t_c = 0.8$

59. 3

61. a. MAD = 5.2
    b. MAD = 2.6
    c. MAD = 5.2
63. None

Chapter 12

5. A manufacturing firm would like to determine if production yield is related to the quality of the raw material used, the average seniority of employees in the firm, and production shift.

7. a. *Null hypothesis:* There is no relationship between sex and mathematics scores.
   Research hypothesis: Males score higher on mathematics tests.
   c. Do not reject H_0: $\chi^2 = 0.824$, $p = .28$

9. a. *Research hypothesis:* Employee turnover and job classification are related.
   b. *Null hypothesis:* There is no relationship between job classification and employee turnover.
   c. Reject H_0: $\chi^2 = 154$

11. a. *Research hypothesis:* A physical fitness program reduces absenteeism.
    Null hypothesis: There is no relationship between a physical fitness program and employee absenteeism.
    b. 7.79
    c. Reject H_0

13. Reject H_0: $\chi^2 = 12$, $p = .004$

15. Reject H_0: $\chi^2 = 28.9$, $p = .0001$

17. Observed frequency appears binominal. Do not reject H_0: $\chi^2 = 2.89$.

19. H_0: Population test scores are normally distributed with $\bar{X} = 53.6$, $s = 19$, and $n = 25$. Used six groups. Do not reject H_0: $\chi^2 = .5929$.

21. Observed frequency appears Poisson: $\chi^2 = 10.47$.

23. Observed frequency appears Poisson: $\chi^2 = 2.5$.

25. The test results show that music preference and media type are related. Informally, there appears to be a difference between those who preferred modern music on TV (less) versus radio (more). This trend is reversed with respect to those that preferred no music.

27. Reject H_0: $\chi^2 = 35.38$, $p = .0001$

29. a. $0.047 < \sigma < 0.089$
    b. Sample standard deviation falls within range.

31. $8.40 \leq \sigma \leq 12.12$

33. a. Test of independence
    c. Do not reject H_0

35. a. Test of independence
    c. Reject H_0

37. a. H_0: Same repair time
    c. Reject H_0: $\chi^2 = 20.68$

39. a. Yes, $\chi^2 = 49.92$
    b. Yes, $\chi^2 = 13.07$

41. a. H_0: No relationship between level of education and interest in a bank loan.
    b. Because the row totals were fixed in advance
    c. Reject H_0: $\chi^2 = 42.1$, $p = .0001$

Chapter 13

1. a. Random
   b. Fixed
   c. Fixed
   d. Fixed

5. a. The effects of level of on-the-job training (low vs. high) and the leadership style of the management (autocratic vs. participative) on product quality
   b. The effects of site location (poor, good, excellent) and extent of amenities (below average, average, above average) on residential real estate prices.

7. Yes, computed $F = 22.5$, critical $F = 3.29$

9. a. Higher R&D expenditures appear to generate higher earnings since computed F (16.7) > critical F (5.72).
   b. A lagged relationship seems to make sense in light of the time required to bring new products into the marketplace.

11. a. H_0 (model type): No impact
    H_0 (salesperson): No impact
    H_0 (model type and salesperson): No impact
  c. Reject H_0 (model) since computed F (9.09) > critical F (6.01). Do not reject H_0 (salesperson) since computed F (1.21) < critical F (6.01). Reject H_0 (model & salesperson) since computed F (20.3) > critical F (4.58).

13.

Sources	SS	df	MS	F
Rows	276,125	1	276,125	131.7
Columns	385,583	2	192,792	91.9
Inter	7,552	2	3,776	1.8
Error	25,166	12	2,097	
Total	696,426	17		

Critical F (rows) = 9.33
Critical F (columns) = 6.93
Critical F (inter) = 6.93
Conclusion: Reject H_0.
Conclusion: Reject H_0.
Conclusion: Do not reject H_0.

17. a. Yes, there is a difference in prices since the computed F (24.9) > critical F (6.01).
  b. Use the Tukey test to identify which stores have different prices.

19. a. H_0: Sales(LA) = Sales(Chi) = Sales(Atl)
  b. No, since computed F (0.11) < critical F (3.68).

21. There appears to be an interactive effect between the type of radio transmission and the region, since computed F (7.78) > critical F (5.14).

23. a. No, computed F (3.27) < critical F (3.49).
  b. Yes, computed F (415.9) > critical F (3.89).
  c. No interactive effects appear to be present.

25. a. Let $X_1 = 1$ if California, $X_2 = 1$ if New York, and $X_1 = 0$, $X_2 = 0$ if Florida. Let $X_3 = 1$ if corporation, $X_4 = 1$ if partnership, and $X_3 = 0$, $X_4 = 0$ if proprietorship.
  b. Both the corporation and New York dummy variables appear to be statistically significant. This suggests that the fees are different. The resultant r^2 for the model is .69. This model does not include interactive effects.

27. a. Let $X_1 = 1$ if urban and $X_1 = 0$ if rural. Let $X_2 = 1$ if view and $X_2 = 0$ if no view. Let $X_3 = 1$ if modern and $X_3 = 0$ if Spanish.
  b. All three dummy variables appear statistically significant at the .05 level using stepwise analysis. The resultant r^2 is .95. This model does not include interactive effects.

31. a. Let A = Geographic location, B = Site location, C = Enterprise type, and D = Equipment. The major factors and interactions are: A, B, C, D, AB, AC, AD, BC, BD, CD, ABC, ABD, BCD, $ABCD$.

35. a. Let $X_1 = 1$ for 1–3 years, $X_2 = 1$ for > 3 years, and $X_1 = 0$ and $X_2 = 0$ for < 1 year. Let $X_3 = 1$ for Apple, $X_4 = 1$ for DEC, and $X_3 = 0$, $X_4 = 0$ for IBM.
  b. The variables X_1, X_2, and X_4 were found to be statistically significant at the .05 level using the stepwise approach. The resultant r^2 is .73. This model does not include interactive effects.

39. a. Let $X_1 = 0$ if male and $X_1 = 1$ if female. Let $X_2 = 1$ if medium, $X_3 = 1$ if high, and $X_2 = 0$, $X_3 = 0$ if low.
  b. The variables X_2 and X_3 are statistically significant at the .05 level using the stepwise approach. The resultant r^2 is .79. This model does not include interactive effects.

43. Applications consisting of distinct groups such as education level, race, marital status, and location.

47. a. See table at bottom of page.

47. a.

Source of Variance	Sum of Squares	Degrees of Freedom	Mean Squared	Computed F Value
Columns	110	2	55.5	16.8
Error	40	12	3.3	
Total	150	14		

b. The variable effect is statistically significant since the computed F (16.8) > critical F (3.88).

49. a. See table at bottom of page.
   b. All three effects are statistically significant since the computed F > critical F in each case.

51. a. MSB = 404.6, MSE = 17.6
   b. $F = 22.9$
   c. $H_0: \mu_1 = \mu_2 = \mu_3$ $H_1: \mu_1 \Leftrightarrow \mu_2 \Leftrightarrow \mu_3$
   d. The null hypothesis of no difference can be rejected, which suggests that the training programs yielded different results.

53. a. MSR = 601.1, MSC = 333.5, MSI = 3.0, MSE = 36.4
   b. F(rows) = 16.5, F(cols) = 9.16, F(int) = 0.08
   d. Management level does have an impact.
   e. Sex does have an impact.
   f. No effect.

55. a. One-factor
   b. $\mu_1 = \mu_2 + \mu_3 + \mu_4$
   c. Yes, F(computed) = 6.14 > F(critical) = 2.81
   d. 20th century

57. a. Time and industry
   b. Two-factor
   c. Industry factor since F(computed) = 8.2 > F(critical) = 3.89

59. a. Planning level and planning horizon
   b. Two-factor
   c. Both planning level and planning horizon are statistically significant. There is no interactive effect.

Chapter 14

5. a. H_0: Fuel economy is the same for both types of cars.
   b. Reject null hypothesis of no difference since sum of group 1 (77.5) < lower critical limit (85.3).

7. a. Spearman rank correlation ($r = .42$): Relationship is not statistically significant at the .05 level.
   b. Pearson correlation coefficient ($r = .40$): relationship is not statistically significant at the .05 level.
   c. The results obtained in parts a and b agree.

11. The results of the test indicate that there was a significant increase in weight as a result of the smoking withdrawal program (i.e., reject the null hypothesis).

15. a. The sign test indicates that there is insufficient evidence to reject the null hypothesis of no change in the number of new plants and expansions. The sum of the signed ranks of 22 is less than the critical upper limit of 23.19.
   b. The key assumption, based on the small sample size, is that the samples came from a normally distributed population.
   c. The results from a standard matched-pairs t-test indicate that the null hypothesis can be rejected. Notice that this conclusion differs from the one developed in part a.

17. a. Reject H_0: The sum of the signed ranks of 25 is greater than the upper limit of 23.19.
   b. The key assumption, based on the small sample size, is that the samples came from a normally distributed population.
   c. The results from a standard matched-pairs t-test also indicate that the null hypothesis can be rejected.

49. a.

Source of Variance	Sum of Squares	Degrees of Freedom	Mean Squared	Computed F Value
Columns	900	2	450	23.4
Rows	1,100	3	366.6	19.1
Inter	750	6	125	6.5
Error	250	13	19.2	
Total	3,000	24		

19. a. Do not reject H_0: The Spearman coefficient of $-.17$ is within the critical limits.
    b. The computed correlation coefficient of $-.43$ is not statistically significant. This result supports the conclusions drawn in part a.

21. a. Reject H_0: The Spearman coefficient of .89 is greater than the upper critical limit of .59.

23. a. Traffic level and service are not related.
    b. Reject H_0: r_s of .87 is greater than the upper critical limit of .41.
    c. Because the data involve ordinal ranking

25. Do not reject H_0: $r_s = -.34$

27. Reject H_0: KW $(13.56) > \chi^2 (5.99)$

29. Reject H_0: KW $(13.15) > \chi^2 (9.49)$

31. Do not reject H_0: $r_s = .43$

33. Reject H_0: $r_s = .89$

35. a. Kruskal-Wallis
    b. Spearman
    c. Wilcoxon signed rank
    d. Wilcoxon ranked sum

37. The KW statistic (9.38) is greater than the critical chi-square (5.99). Therefore, reject H_0.

39. There is no significant difference among the depreciation schedules.

41. The sum of the ranks (50) is less than the lower critical limit of 68.7. Therefore, reject the null hypothesis.

43. Reject H_0 of no relationship.

45. Process is occurring randomly.

47. a. $\bar{X} = 5.253$.
    b. Seq = (B, B, A, A, A, B, B, A, B, A, B, B, A, B, A, B, B, A, A, B)
    Process is occurring randomly.

49. a. Spearman rank correlation
    b. H_0: No relationship between rankings
    c. Reject H_0: $r_s = .86$
    d. The relative rankings between 1988 and 1989 have not changed.

51. a. Wilcoxon rank sum test
    b. H_0: Men and women spend the same.
    c. Reject H_0: A difference exists.

53. a. Spearman rank correlation
    b. H_0: No relationship between price and view.
    c. Reject H_0.

Chapter 15

3. a. Uncertainty
   b. Certainty
   c. Risk
   d. Risk

5. This is true because the EPPI for each alternative is the same. Consequently, since EMV + EOL equals EPPI, the option that maximizes EMV must minimize EOL.

7. a. A_6 (125)
   b. A_5 (50)
   c. A_5 (68.75)
   d. A_5 (D_1)

9. a. A_5 (550)
   b. A_1 (275)
   c. A_1 (275)
   d. A_4 (250)

11.
Weighting Coefficient	Weighted Payoff	Optimal Decision
0	50	A_5
.2	60	A_5
.4	70	A_5
.6	80	A_5
.8	100	A_6
1	125	A_6

13.
Weighting Coefficient	Weighted Payoff	Optimal Decision
0	275	A_1
.2	275	A_1
.4	275	A_1
.6	300	A_5
.8	440	A_5
1	550	A_5

15.

Decision	Lower Limit	Current Value	Upper Limit
A_1	No limit	100	366.7
	No limit	150	264.3
A_2	No limit	200	483.3
	No limit	100	221.4
A_3	No limit	300	541.6
	No limit	75	178.6
A_4	No limit	150	366.7
	No limit	150	242.9
A_5	33.3	250	No limit
	107.1	200	No limit

19.

Decision	Lower Limit	Current Value	Upper Limit
A_1	35	35	35
	75	75	75
	100	100	100
	50	50	50
	25	25	25
A_2	50	50	50
	50	50	50
	50	50	50
	50	50	50
	50	50	50
A_3	25	25	25
	35	35	35
	100	100	100
	125	125	125
	0	0	0

21.

Decision	Lower Limit	Current Value	Upper Limit
A_1	No limit	100	450
	No limit	150	500
	No limit	200	550
	No limit	175	350
A_2	No limit	200	1000
	No limit	100	900
	No limit	50	850
	No limit	0	400
A_3	No limit	300	575
	No limit	75	350
	No limit	100	375
	No limit	200	337.5
A_4	No limit	150	550
	No limit	150	550
	No limit	150	550
	No limit	150	350
A_5	-25	250	No limit
	-75	200	No limit
	-175	100	No limit
	162.5	300	No limit

23. Optimal decision A_5 (EMV = 230)

25.

Decision	Lower Limit	Current Value	Upper Limit
A_1	No limit	-20	135
	No limit	15	170
A_2	No limit	-15	115
	No limit	35	165
A_3	-180	-50	No limit
	70	200	No limit

27. Banzai should operate six cafes, with a net expected value of 20.7 million yen.

29. a. The 3,000-square-foot store
    b. p (demand) = .2
    c. 160

35.

Decision	Limit	Value	Limit
A_1	No limit	200	900
	No limit	420	1200
	No limit	640	2200
A_2	220	1,000	No limit
	-380	400	No limit
	-920	600	No limit
A_3	-560	1,800	No limit
	$-1,160$	1,200	No limit
	$-4,120$	600	No limit

37. a. 0
    b. 11.50

39. Cost (100%) = \$200,000, Cost (0%) = \$220,000
    Conclusion: Go with 100% inspection.

41. a. Optimal decision: A_5 (EUV 0.83)
    b. Same optimal decision as in problem 14.

43. a. Optimal decision: D5 (EUV = 10.3)
    b. Same optimal decision as in problem 14.

45. a. Optimal decision: A_2 (EMV = 48)
    b. Optimal decision: A_2 (EMV = 0.64)

47. Use a decision tree. Optimal decision is to continue with current development program. The expected value for the optimal decision is \$6,125,000.

49. Let 1 = Win and 0 = Lose for the payoffs. The optimal decision is to allocate the resources to the new design. The payoff is 0.58.

51.
Weighting Coefficient	Optimal Decision
0	A_1
.5	A_3
1	A_3

53. a. 8,000 units
    b. $1,250,000
    c. 280,000 units

55. a. $10.8 million
    b. 0.0228

57. a. .6915
    b. $70,000
    c. $44,000
    d. $26,000
    e. 207

59. a. Drilling for oil where the chances of a dry hole are high, which could result in bankruptcy.
    b. Investing in a new firm where the market potential is very large, but the chances for success are quite small.

Index

A

Acceptance sampling, 21
  practical application of, 175–76
Achenwall, Gottfried, 12
Addition law, 100–102
Adjusted coefficient of
    determination, 405–6
Albright Lighting Company, 555–57
Alpha, 294
  correlation analysis and, 354
  exponential smoothing and, 481
  optimizing sample size and, 310
Alternative hypothesis, 292
  correlation analysis and, 354
  optimizing sample size and, 310
  sign test and, 640
  two population tests and, 309
Alternatives, evaluation of, 19
Amex Machine Shop, 357
Ampex Manufacturing Company, 152
Analysis of variance (ANOVA), 408, 603
  block designs and, 590–91
  introduction to, 580–81
  model formulation, 583–84
  one-factor ANOVA and, 584–87
  practical applications for, 598–600
  recognizing problems, 582–83
  regression analysis and, 592–93
  two-factor ANOVA, 587–90
ANOVA; *see* Analysis of variance (ANOVA)
Applied statistics, 11
Arithmetic mean, 41
Asymmetrical scales, 204
Attribute control chart, 313

B

Audit Supply, Inc, 576–77
Autocorrelation, 406–7, 485
Automobile fuel economy standards
    case study, 114–17
Automobile product reliability case,
    130–32
Autoregression, 485, 487
Autoregressive analysis, 476
  practical analysis, 500–502
Auto-stepwise solution, 414–15
Avoidance of power, 312

B

Backwards elimination solution, 415
Bacon, Francis, 579
Bank customer survey, 216–18
Bar charts, 32
Basic risk preference model, 699
Bayes, Thomas, 678
Bayesian analysis, 106–10
  decision making and, 694–96
  practical application, 112–13
Bernoulli process, 153–55
Beta, 294
  optimizing sample size and, 310
Beta coefficient, 404
Bidirectional process, 296
Binomial probability distribution,
    153–59
  acceptance sampling and, 175–76
  approximating with normal curve,
    157–59
  practical application, 173–74
Biorhythms theory, 10

Bivariate correlation coefficient, 350
Bivariate regression model, 358–60
  forecasting and, 476
  limitations of, 365–66
Blackman Tool and Die, 558–59
Bland Foods, 548
  chi-square distribution for, 549
  hypothesis test for, 550
Block designs, 590–91
  computer analysis, 597–98
Bowen, Freddie R., Jr., 287
Bozart Investments Corporation case
    study, 270–71
Branches (decision trees), 689
Break-even analysis, 706–7
Brookline Consulting Group
  dummy variables and, 424–25
  forecast of absenteeism for, 428
  multiple regression and, 408–9
  stepwise regression analysis of,
    426–27
Brookline Robotics Corporation,
    284–85
Budget Computer, Inc., 356–57
Business failures, 398–99
Business statistics
  defined, 2–3
  new developments in, 4
  teaching trends in, 4–5
Butler, B. J., 133

C

California case study, 533–34
California lottery, 90, 110–11, 187–88

791

792 *Index*

Causal relationships, 353
CDBMS; *see* Computer data base management systems (CDBMS)
Census, 193
Central limit theorem (CLT), 247–49
Central tendency, 38–44
  group measures of, 50–51
Certainty, decision making under, 679
Chi-square analysis, 311, 561
  coin flip experiment and, 543
  goodness of fit, 543–45
  introduction to, 538
  model formulation, 540–43
  population variance, 548–50
  practical applications, 553–59
  recognizing problem, 540
  test of independence, 545–48
Chi-square distribution, 542–43
Chi-square statistic, 541
Citrus Association Exchange, 498
Civil Rights Act of 1964, 467–68
Cleanall Corporation case study, 709–11
Cluster sampling, 240
Coefficient of determination, 360–62, 366, 405
  regression model and, 361
Coefficient of nondetermination, 361
Coin flip experiment 543–44
  goodness of fit and, 551
Collectively exhaustive event, 92
Combinations, 97, 99–100
Compound events, 92
Computer analysis, 169–71, 411
  ANOVA problems and, 593–98
  chi-square analysis and, 551–53
  decision analysis problems and, 703–5
  descriptive statistics and, 56–58
  hypothesis testing and, 314–18
  multivariable regression and, 416–22
  nonparametric statistics and, 644–47
  point and interval estimates, 264–67
  probability and, 110–11
  regression/correlation problems and, 366–74
  time series problems and, 496–97
Computer-based planning system, 1
Computer Business Statistics (CBS), 5
Computer data base management system (CDBMS), 211–14, 223–24
Computer Interrogation, 210–11
Conclusions from hypotheses, 298–99
Conditional probabilities, 102
Confidence intervals, 251

Consistent estimators, 251
Consumer's Buyers Guide, 260
  interval estimates for, 265–66
Content analysis, 203
Contingency table, 545, 554
Continuous probability distribution, 135
  practical application of, 706–8
Correlation analysis, 346, 349–53, 376
  correlation analysis, 353–55
  practical application, 422–24
Correlation coefficient, 346, 349, 363, 366
Counting rules, 97–100
Craig Investments International
  block design for, 591
  computer application, 597–98
Critical value, 295
Cross-sectional data, 468
Cumulative probability distribution (CPD), 138–39, 152
Curvilinear relationships, 352
Customer preference surveys, 637–38
Cyclical analysis, 487–89
  practical application, 498–500
Cyclic component, 473

D

Danderberg Auditing International, 163–64
  hypergeometric probability distribution and, 171–73
Darwin, Charles, 9
Data, 192, 211
  collection of, 18
  methods of collection, 209
  presentation of, 34–35
Data base management; *see* Computer data base management systems (CDBMS)
Data collection/analysis, 18, 209
  hypothesis testing and, 298–99
  methods of, 155–56
dBase IV, 213
Decision making, managerial, 677–78
  additional information and, 691–94
  decision trees, 689–91
  introduction to, 678–80
  practical applications, 705–8
  recognizing problems, 682
  sensitivity analysis and, 700–703
  under risk, 686–91
  under uncertainty, 682–86
  utility and, 696–700

Decision-making process, managerial, 18–20, 232
  data collecting/processing, 192–196
Decision making under certainty, 679
Decision rules, 295–98
Decision trees, 689–91
Decomposition model, classical, 471–72
  cyclic component, 473
  irregular component, 475
  seasonal component, 473–75
  trend component, 472–73
Degrees of freedom, 255, 297
  chi-square distribution and, 542
Department of Housing and Urban Development (HUD), 27–28
Department of Labor, U.S., 263
  sample size and, 266–67
Dependence, 545
Dependent events, 102
Dependent sample design (matched pairs), 306–9
Dependent variable, 14, 347, 349
  multiple regression analysis and, 400–401
Descriptive analysis, 31
  data presentation, 34–35
  summary measures, 38–50
Descriptive statistics, 3, 28–31
  practical applications of, 58–61
  quality control and, 52–56
Design principles, 15–16
Dialnet Telephone Exchange, 469–70
  centered moving average model and, 490
  deseasonalized sales for, 492
  exponential smoothing and, 481, 497
  installation data for, 488–89
  MAD/MSE error measurements for, 495
  moving averages for, 478, 496
  regression analysis, 498
  seasonal index computations, 491, 499
  trend/cyclic components for, 493
  weighted moving averages for, 479
Dichotomous scale, 202–3
Dietfast Company, 317–18
Discrete distributions, 705–6
Discrete probability distribution, 135
Discriminant analysis, 413–14
  practical application of, 429
Dow Jones average, 401
Drack Industries case study, 176–78
Dummy variables, 407–9
  practical application, 424–25
Durbin-Watson (DW) test, 486

E

Earthquake Real Estate, 355
Economic forecasters, evaluating, 343–44
Efficient estimator, 251
Employee absenteeism, 428, 579–80
Employee termination interview, 214–16
EMV; *see* Expected monetary value (EMV)
Engineering studies, 555–57
EOL; *see* Expected opportunity loss (EOL)
EPPI; *see* Expected payoff with perfect information (EPPI)
Equal-likelihood criteria, 684–85
Error variance, 409
Estimate, 249
Estimation, 235–36
Estimation of proportions, 269
Estimator, 232, 249, 692
Estimators and estimates, 249–50
  selection of, 250–51
Events, 89
EVPI; *see* Expected value of perfect information (EVPI)
Expected monetary value (EMV), 686–87
  computer analysis and, 703
Expected opportunity loss (EOL), 687–88
  computer analysis and, 703–4
Expected payoff of additional information (EPAI), 696
Expected payoff with perfect information (EPPI), 688–89
Expected utility value (EUV), 700
Expected value, 140
Expected value analysis, 686–87
Expected value of additional information (EVAI), 696
Expected value of perfect information (EVPI), 688
Experiment, 193
Exponential smoothing, 476, 480–84

F

Face-to-face procedures, 210
Factors, 581
Far Filtration Company, 401–2
  multiple regression computer analysis, 417–22
Fechner, G. T., 29
Fedin, Konstantin, 537
Fill-in-the-blanks, 205–6
Financial analysis, 84–86
Finite-population multiplier, 244
First-order correlation coefficient, 350
Fisher, R. A., 192, 581
Fixed-effects model, 584
Fluoride test program, 129–30
Forecasting, 3, 468
  basic categories of, 469
  methods overview, 476
Forecasting and interval estimation, 368–71
  practical application for, 372–74
Forecasting models, 476
Forecast validation, 494–95
Formal planning system, 1
Frequency curves, 35–36
Frequency tables, 33
Fujiti Motors case study, 651–53
Full regression solution, 414

G

Gallop, George, 192
Galton, Francis, 231, 346
Gauss, Carl Fredrich, 134
General law of addition, 101
Geometric mean, 42
Goodfaith Emergency Clinic case study, 62–64
Goodness of fit, 538
  example of, 551
Goodness-of-fit test, 543–45, 561
Gossett, William S., 233, 255

H

Hard data, 196
Harmonic mean, 42–43
Hartwell Music Experiment case study, 321–23
Hawthorne study, 15
Health and Human Services (HHS), Department of, 303
  testing one population proportions and, 317
Hericlitus, 27
Heyel, C., 15
Hierarchical structure, 212
Histograms, 33–34, 221
Hollywood Park (wagering), 557–58
Holt two-parameter exponential smoothing model, 482
Home Real Estate, 422–24
  forecast of housing price for, 425
Horse racing at San Yenz, 577–78
Hot stuff, Inc., 553–55
Hurwicz rule, 708
Hypergeometric distribution, 159–64
Hypothesis, 14, 292
  drawing conclusions from 298–99
Hypothesis testing, 288–89, 362
  chi-square methodology and, 550
  limitations of, 311–12
  operating characteristics and power curves, 300–302
  practical applications for, 319–21
  process of, 292–300
  recognizing a problem, 290–91
  several populations and, 580–81
  statistical inference and, 323–324
  tests of proportions, 303–5
  two-population tests, 305–9

I

IDMV; *see* Iowa Department of Motor Vehicles (IDMV)
Implementation, 19–20
Independence, 545
Independence of variables, 561
Independent events, 101
Independent sample design (unmatched pairs), 305–7
Independent variable, 14, 346, 349
  multiple regression analysis and, 400–401
Indicators, 692
Inductive reasoning, 13
Inferential statistics, 3
Information, 193
Insignificant results, 311–12
Instrument pretest, 207
Interactive effects, 588
Interactive terms, 411–13
  practical application, 427–28
Internal Revenue Service, 320–21
Interval estimate, 232
  chi-square analysis and, 548–49
Interval estimates, 251–55
Interval scale, 17
Iowa Department of Motor Vehicles (IDMV), 289–90
  hypothesis testing and, 292–94, 302–3
  sample size and, 310–11
  testing one population mean and, 315–17
Irregular component, 475

J

Japanese-U.S. exchange rates/trade, 535–36

Jensen Musical Instrument
  Company, 188–189
Johnson, Samuel, 629
Johnson Carbide, 587
  data base for, 588
  two-factor ANOVA and, 596–97
Johnson Pharmaceutical, 157
  binomial distribution and, 171
Joint probabilities, 102
  practical application, 111–12

K

Kettering, C. F., 191
Kurskal-Wallis test, 643
  computer analysis for, 646–647
Kurtosis, 50
Kuznets, Simon, 469
Kwan Bottling Company, 539–40
  accounts receivable analysis and, 552
  chi-square analysis for, 544, 552
  contingency table for, 546

L

Lagged effects, 355
Lagged variables, 372
Landon, Alfred M., 191–92
Law of large numbers, 96
Leaky Pen Company case study, 559–61
Learning curve, 353
Least squares method, 358
Leptokurtic, 50
Level of acceptable error, 293
Level of significance, 294
Life-cycle approach, 629
Likert symmetrical 5-point scale, 204
Linear association, 349
Linear correlation analysis, 346–47
Linear regression lines, 363
Literary Digest, 191–92
Longfellow, Henry Wadsworth, 677
Lotus 1-2-3, 213
Lower confidence limit, 253

M

Mcquine Electric Limited, 175–76
MAD; *see* Mean absolute deviation (MAD)
Mail surveys, 209–10
Management composition, 288
Management information system (MIS), 235, 629–30

Mann-Whitney rank-sum test, 633–34
  computer analysis and, 644–45
Marginal analysis, 705–8
Marginal probabilities, 102
Market survey, 553–55
Mastermind, 99
Matched pairs, 260–62, 307–9
  test for, 308
Matched sample design, 307
Maximax decision criteria, 684
Maximum decision criteria, 684
Maximum-likelihood criteria, 684, 686
Mean absolute deviation (MAD), 45, 494
Mean percentage error (MPE), 494
Means, 38–39
  alternative decision rules using, 297
  estimation of, 268–69
  matched/unmatched testing, 317
  sampling distribution for, 241–44
Mean square error (MSE), 494
Mean square for between-group variance (MSB), 585–86
Mean square of the error variance (MSE), 586
Means square of the error, 415
Means square of the regression, 415
Measurement and numbers, 16–17
Measuring R&D performance, 396–97
Median, 39–40
Mehta Corporation, 625–26
Method of centered moving averages (CMA), 490
Metrics, 31–32
Microchips Electronics, Inc. case study, 600–602
Minimax decision criterion, 685
Missing data, 356
Mitterand Cable Company, 581–82
  computer analysis, 593–96
  one-factor ANOVA and, 585–86
Mode, 40
Model formulation, 18, 540–43
  ANOVA model, 583–84
  autoregressive models, 485
  basic nonparametric models, 633
  bivariate regression, 358–360
  centered moving averages model, 490
  chi-square model, 540–43
  correlation analysis, 349–53
  decision analysis and, 682
  decomposition model, 471–72
  forecasting models, 476
  Holt two-parameter exponential smoothing model, 482
  Kruskal-Wallis test, 643

Model formulation—*Cont.*
  multiple regression, 402–5
  rank-sum test model, 633–35
  risk preference model, 699
  Spearman rank correlation, 641–42
  subjective decision analysis model, 683
  Wilcoxen signed-rank test, 635–37
  Winter's three-parameter model, 484
Moivre, Abraham de, 88
Moving average, 477–80
MSE; *see* Mean square error (MSE)
Multicollinearity, 406
Multiple bar charts, 32–33
Multiple choice questions, 204–5
Multiple correlation coefficient, 406
Multiple regression analysis, 431
  ANOVA model and, 592–93
  credit evaluation and, 399–400
  introduction to, 400–402
  model formulation and, 402–5
  practical applications, 422–30, 509
  recognizing problem type, 402
Multiple regression model, 402
  formulation of, 402–5
  measurements of, 405–7
  model extensions, 407–14
Multiplication law, 100, 102–5
Mutually exclusive events, 92

N

Naive forecast, 476
National Baseball League case study, 430–31
National Cancer Institute, 11–12
National census, 231–32
National Football Conference, 647–48
National Weather Service, 107, 110
Nautilus Health Spa, Inc., 196–97, 207
  data schema for, 212
  prototype questionnaire for, 208
Negative correlation, perfect, 351
Network structure, 212–13
New York Stock Exchange, 91–92
Neyman, Jerzy, 289
Nodes (decision trees), 689
Nominal scale, 17
Nondescriptive scales, 205
Nonlinear effects, 363–65
  multiple regressions and, 409–11
  practical application of, 426–27
Nonlinear relationships, 352
Nonparametric statistics
  advantages/disadvantages of, 632
  introduction to, 630–31

Nonparametric statistics—*Cont.*
  Kruskal-Wallis test, 643
  practical applications, 647–51
  rank-sum test models, 633–35
  recognizing problems, 632–33
  sign test and, 637–41
  Spearman rank correlation, 641–42
  Wilcoxen signed-rank test and, 635–37
Nonparametric tests, 631
Nonresponse bias, 195
Nonstationary time series, 476
Normal curve, 144
  Poisson distribution and, 168–69
Normal distribution, 140–49
Null hypothesis, 292–93
  avoidance of power and, 312
  block designs and, 591
  chi-square methodology and, 538
  correlation analysis and, 354
  Kruskal-Wallis test and, 643
  multiple regression and, 416
  optimizing sample size and, 310
  sign tests and, 638
  two population tests and, 309

O

Objective analysis, 679
Objective probability, 94–96
One-factor ANOVA, 584–87
  computer analysis and, 594–96
  structure for, 586
One-sided hypothesis, 293
Open-ended questions, 203
Operating characteristics, 300–303
Oppenheimer, J. R., 467
Ordinal scale, 17
  rank-sum tests and, 633
Outcomes, 89

P

Pacific Construction Company, 347, 354, 400
  residual analysis for, 368
  scatter diagram/regression line for, 359
Palmer Video, 112–13
Parametric tests, 630
Payoff table, 682
Peakedness (Pk), degree of, 48–50
Pearson, E. S., 289
Pearson, Karl, 192, 539, 581
Pearson correlation coefficient, 350
  Spearman rank correlation and, 641
Permutations, 97–99

Perpetual Investments Corporation, 680–81
  computer analysis and, 703–4
  decision table for, 684
  loss table for, 685
  net payoff table for, 683
  probability sensitivity analysis for, 702
  probability tables for, 692–93
  probability tree for, 695
  utility function of, 701
Perpetual Savings and Loan, 631
  computer analysis of, 644–45
  data ranking for, 634
  Wilcoxen/Mann-Whitney analyses for, 635
Personal computer, 7–8
Personality differences, analyzing, 674–76
p-charts, 313–14
Pictographs, 37
Pie charts, 33
  quality control and, 54
p-values, 299–300
Plato, 345
Platykurtic, 50
Plumbing West Inc., 463–64
Point estimate, 249
Poisson probability distribution, 164–69
  practical application, 558–59
Pooled estimator, 259
Population parameters, 232, 249, 310–11
Population regression equation, 358
Populations, 13
  differences between, 258–62
  hypothesis testing and, 291
Population variance, inference about
  hypothesis testing and, 550
  interval estimates, 548–49
Positive autocorrelation, 485–87
Positive correlation, perfect, 350
Posterior probabilities, 694–96
Posterior probability, 108–9
Power, 294
Power curves, 300–303
  hypothesis tests and, 301
Predictor, 692
Price indexes, 61–62
Principle of indifference, 702
Prior probability, 106–7
Pritikin, Nathan, 374
Pritikin Diet case study, 374–76
Probability, 87–89
  basic rules of, 90–91
  Bayesian analysis and, 106–10
  computer analysis and, 110–11
  counting rules and, 97–100
  definition of, 94–97

Probability—*Cont.*
  laws of, 100–106
  practical applications, 111–13
  random events, 91–93
Probability density function (PDF), 138
  normal distribution and, 146–47
  uniform distribution and, 150–51
Probability distributions, 133–37
  basic concepts of, 137–40
  binomial distribution, 153–59
  computer analysis, 169–73
  hypergeometric distribution, 159–64
  introduction to, 135
  normal distribution, 140–49
  Poisson distribution, 164–69
  practical applications, 173–76
  uniform distributions, 149–52
Probability mass function (PMF), 138
  binomial probability function and, 155–56
  hypergeometric distribution and, 160–62
Probability tree, 105–6, 109, 161
Problem definition, 18
Process control, 21
Product placement (two-way nova), 598
Pronet Tennis Company, 464–66
Proportions
  estimation of, 269
  one population testing and, 315–17
  sampling error for, 244–47
  tests for, 303–5
  two population proportions, 309

Q

Qualitative variable, 21; *see also* Dummy variable
Quantitative variables, 16–17
Quartiles, 45
Questionnaires
  design of, 197–98
  instrument pretest, 207
  item construction, 201–2
  objectives/data categories of, 199–200
  practical applications of, 214–18
  primary purpose of, 198–99
  sample plan, 200–201
  scaling for response, 202–7

R

Random-effects model, 584
Random errors, 194

Random events, 91–93
Random experiments, 91, 97
Random numbers, 236–37
Random sampling, 236–38
Random variable, 138
   summary measures of, 140
Range, 45
Rank correlation; *see* Spearman rank correlation
Rank/order scale, 206–7
Rank-sum tests, 633–35
   computer analysis and, 644
Ratio scale, 17
Reasonable doubt, 88
Reddin Test of Management, 14–15
Regression analysis, 346, 357–58, 376
   extensions of, 407–14
   forecasting and, 484–487, 508
   model for, 358–60
   model measurements, 405–7
   model significance, 415–16
   practical applications for, 371–72
   solution methods, 414–15
Regression coefficient, 363
Regression problems, recognizing, 347–49
Relational structure, 213
Relative frequency, 95
Replications, 589
Research hypothesis, 292–93
Research statistics, 11
Residual analysis, 367–68
Residual value, 359
   multiple regressions and, 407
Response bias, 195
Results, interpretation of, 19
R charts, 314
Risk, 679
   decision making under, 686–89
Rocket Chemical Company, 626–28
Roosevelt, Franklin Delano, 191–92
Ross International case study, 83–84
Rules of evidence, 15

S

Sample, 13
Sample plan, 200–201
Sample size, 266–67
   determination of, 262–64
   optimizing, 310
Sample space, 91
Sample survey, 193
Sampling, 232–33
   determining size of sample, 262–64
   distributions and, 240–47
   methods for, 238–40
   process of, 235–36
   random sampling, 236–38

Sampling distributions, 240–47
Sampling error, 252
Sampling error for proportions, 244–47
Scaling for response, 202–7
Scatter diagrams, 37, 349, 411
Schema, 212
Scientific method, 10–11
   design principles, 15–16
   formulation of questions and, 13–15
   measurement and numbers, 16–17
   overview of, 12
Seasonal analysis, 489–94
Seasonal component, 473–75
Seasonal variations, 487–88
Second Federal Bank, 429
Securities option market, 537–38
Selection bias, 195
Self-stepwise solution, 414
   practical application, 426–27
Seven-Day Tire Company, 136
   computer analysis and, 169–71
   probability distribution and, 147–49
Significance, testing for, 362–63
Sign test, 637–41
Simple correlation problems, recognizing, 347–49
Simple moving average forecast, 476–77
Simple random sample, 201
Simple random sampling, 236–38
Skewness (Sk), degree of, 48
Sky High Aircraft, 317
   hypothesis testing and, 318
Smoothing coefficient (alpha), 481
Soft data, 196
Software package, 211
South Eastern Charter Airlines, 174
Southern California Edison, 282–84
Spearman, C. E., 641
Spearman rank correlation, 641–42
   computer analysis, 645–46
   practical applications of, 647–51
Special law of addition, 101
Sports Shoe, Inc., 227–28
SQC; *see* Statistical quality control (SQC)
Standard deviation, 47
Standard error of estimate, 360, 406
Standard error of the difference, 306
Standard error of the mean, 243
Standard normal distribution, 143–46
States of nature, 679
Stationary time series, 476
Statistical Abstract of The United States, 468
Statistical inference, 232
   hypothesis testing and, 323–24

Statistical quality control (SQC), 312–13
Statistics, genesis of, 12
Stock market trends analysis, 395–96
Stratified proportionate sampling, 200
Stratified sampling, 239
Struxs Corporation, 579–80
Student t-distribution, 255–57
   application of, 257–58
   two-populations tests and, 306–7
Subjective analysis, 682
   using weighted averages, 708
Subjective decision analysis model, 683
Subjective probability, 94, 96–97
Suboptimal decisions, 696–98
Sufficient estimator, 251
Summary measures
   binomial distribution and, 156–57
   hypergeometric distribution and, 162–63
   measures of central tendency, 38–44
   measures of dispersion, 44–47
   measures of grouped data, 50–51
   measures of shape, 48–50
   Poisson distribution and, 167–68
   uniform distribution and, 151–52
Suppression variables, 409
Survey bias, summary of, 195
Surveys, 193
   administration and analysis of, 207–9
   mail surveys, 209–10
Systematic errors, 194
Systematic sampling, 238–39

T

Telephone procedures, 210
Test for independence, 538
Test of homogeneity, 547
Test of independence, 545–58
   computer application of, 552
Thermhouse Insulation Corporation case study, 502–3
Three-point symmetrical scale, 204
Ticket forecasting, 174
t-distribution; *see* Student t-distribution
Time series, 468–69, 510–11
   decomposition model and, 471–75
   exponential smoothing, 480–84
   forecasting models, 476
   moving average and, 477–80
   practical applications, 497–502
   problem recognition, 471
Time series charts, 37
Tippet, L. H. C., 192

Total quality management (TQM), 4, 11, 20–22
Transpacific Airlines case study, 218–23
Treasury bonds, U.S., 500–502
  autoregressive analysis, 503
Treatment, specific, 581
Trend charts, 37
Trend component, 472–73
Two-factor ANOVA, 587–90
  block designs and, 590–91
  computer analysis and, 596–97
  practical applications, 598–600
Two population proportions, difference between, 309
Two-population tests, 305–9, 320–21
  examples of, 305
  testing means, 317, 319–20
Two-sided hypothesis, 293
Type I error, 294, 300, 314
Type II error, 294, 314

U

Unbiased estimators, 251
Uncertainty, 679
  decision making under, 682–85
Unidirectional process, 296
Uniform distribution, 149–52

United Airlines, 193
United Electronics, 410–11
  nonlinear effects and, 426–27
Unmatched pairs, 256–60, 306–7
Upper confidence limit, 253
Utility
  decision making and, 696–98
  utility function, 698–700

V

Variability, group measures of, 52
Variance, 45–47
Venn, John, 93
Venn diagram, 93–94
Vernadsky, V. I., 287
Violation of assumptions, 312
Von's Supermarket, Inc., 268–69

W

Wainwright Forge Company, 412
  interactive effects and, 427–28
War Games International, 228–229
Weber Health Spa, 266
Weighted average, 41–42
  subjective decision analysis and, 708

Weighted moving average forecast, 476
Weighted moving average (WMA), 478–79
Wells, H. G., 1
Western Telephone Exchange Company, 173
Wilcox Accounting Service, 233–35
  random sampling and, 240–41
  relative-frequency histogram of means, 242
  single population interval estimate, 265
  t-distribution and, 257–58
Wilcoxen, Frank, 630
Wilcoxen rank-sum test, 633–35
Wilcoxen signed-rank test
  computer analysis and, 645
  process for, 635–37
  sign tests and, 637
Win/no win concept, 105–6
Winter's three-parameter model, 484
Women's International Golf Association, 674
Worldwide Insurance, 319–20

X–Z

$\overline{X}$-charts, 54–56, 314
Z-test, 295